AF655982

JOSHI's
Process Equipment Design

JOSHI'S
Process Equipment Design

Fifth Edition

V V Mahajani
S B Umarji

TRINITY
PRESS
(An Imprint of Laxmi Publications Pvt. Ltd.)
An ISO 9001:2015 Company
BENGALURU • CHENNAI • GUWAHATI • HYDERABAD • JALANDHAR
KOCHI • KOLKATA • LUCKNOW • MUMBAI • RANCHI
NEW DELHI

PROCESS EQUIPMENT DESIGN

Copyright © by the Author

All rights reserved including those of translation into other languages. In accordance with the Copyright (Amendment) Act, 2012, no part of this publication may be reproduced, stored in a retrieval system, translated into any other language or transmitted in any form or by any means, electronic, mechanical, photocopying, recording or otherwise. Any such act or scanning, uploading, and or electronic sharing of any part of this book without the permission of the publisher constitutes unlawful piracy and theft of the copyright holder's intellectual property. If you would like to use material from the book (other than for review purposes), prior written permission must be obtained from the publishers.

Printed and bound in India

Fifth Edition

First Reprint : 2014, *Second Reprint* : 2016, 2017, 2019, *Edition* : 2022

ISBN : 978-93-5138-019-1

Limits of Liability/Disclaimer of Warranty: The publisher and the author make no representation or warranties with respect to the accuracy or completeness of the contents of this work and specifically disclaim all warranties. The advice, strategies, and activities contained herein may not be suitable for every situation. In performing activities adult supervision must be sought. Likewise, common sense and care are essential to the conduct of any and all activities, whether described in this book or otherwise. Neither the publisher nor the author shall be liable or assumes any responsibility for any injuries or damages arising here from. The fact that an organization or Website if referred to in this work as a citation and/or a potential source of further information does not mean that the author or the publisher endorses the information the organization or Website may provide or recommendations it may make. Further, readers must be aware that the Internet Websites listed in this work may have changed or disappeared between when this work was written and when it is read.

All trademarks, logos or any other mark such as Vibgyor, USP, Amanda, Golden Bells, Firewall Media, Mercury, Trinity, Laxmi appearing in this work are trademarks and intellectual property owned by or licensed to Laxmi Publications, its subsidiaries or affiliates. Notwithstanding this disclaimer, all other names and marks mentioned in this work are the trade names, trademarks or service marks of their respective owners.

Branches

©	Bengaluru	080-26 75 69 30
©	Chennai	044-24 34 47 26
©	Guwahati	0361-254 36 69
©	Hyderabad	040-27 55 53 83
©	Jalandhar	0181-222 12 72
©	Kochi	0484-405 13 03
©	Kolkata	033-40 04 77 79
©	Lucknow	0522-430 36 13
©	Ranchi	0651-224 24 64

PUBLISHED IN INDIA BY

TRINITY PRESS

(An Imprint of Laxmi Publications Pvt. Ltd.)
An ISO 9001:2015 Company
113, GOLDEN HOUSE, GURUDWARA ROAD,
DARYAGANJ, NEW DELHI-110002, INDIA
Telephone : 91-11-4353 2500, 4353 2501
www.laxmipublications.com
info@laxmipublications.com

C—00000/022/03
Printed at: Ajit Printing Press, Delhi.

Dedicated to

Late Shri M V JOSHI

गुरुर्ब्रह्मा गुरुर्विष्णुः गुरुर्देवो महेश्वरः।
गुरुः साक्षात्परब्रह्म तस्मै श्रीगुरुवे नमः।।

Preface to the Fifth Edition

This edition of the book, like the previous edition, is dedicated to late Professor MV Joshi, who was the architect and the author of this book.

This edition brings you few updates and additions, like chapter 12 gives you a new write up on hydrocyclone, in Auxiliary Process Vessels; Chapter 14 gives a write up on Jet Mixer and Eductor, in Mixers; and Appendix J, gives a calculation of volume of a horizontal tank, with formulae, and so on.

We are very much thankful to Dr VV Chavan, Managing Director, Mamko Design and Engineering Private Ltd Mumbai 400 063 for pictures of Jet Mixers and hydrocyclone. His suggestions on brief write up of these items are thankfully acknowledged. Suggestions by Professor (Dr) S P Deshmukh of ICT (UDCT) Mumbai 400 019 were very much valuable.

We acknowledge with thanks the support of our friends and student community and now look forward to their comments and suggestions for the next edition.

V V Mahajani

S B Umarji

Preface to the Fourth Edition

This revised edition is dedicated to late Professor M V Joshi who was the main architect and the author of this book.

The emphasis in this fourth revised edition is on mechanical design, as in the earlier editions, and not on process design. In this edition, an attempt has been made to eliminate all types of errors missed out in earlier editions. The chapters have been revised wherever necessary. The reference list at the end of each chapter has been updated with some very old references being deleted. The chapter on 'Agitators' (Chapter 14) has been renamed as 'Mixers' taking into account static mixers also.

All numerical problems have been reworked on to make them more comprehensive without altering the main definition. Illustrations have been presented to appreciate equipment, particularly the mechanically agitated reactor.

We take this opportunity to thank all those who gave feedback to bring forth this revision.

Our thanks are due to Dr P D Vaidya and research scholars, Mr T L Gunale for going through illustrations and the revision, and Mr M S Talekar for updating literature search. Their constructive suggestions from the students' point of view have been invaluable.

We thank Mr S M Mokashi, former Managing Director of Xytel India Ltd and Director (then) Davy Powergas India Ltd, for making the opening Chapter 1 more appealing. Our sincere thanks are due to Dr V V Chavan, Director, Mamko Design and Engineering Pvt Ltd, Mumbai for their inputs on static mixers. We thank with equal sincerity Mr Kekatpure of Evergreen Technologies Pvt Ltd Mumbai for permitting to use photographs of structured packings and mist eliminators.

We also thank Professor (Dr) J B Joshi, Director, Institute of Chemical Technology, Matunga, Mumbai, where this revision was designed and executed.

We also wish to express our sincere gratitude to the entire team of Macmillan Publisher India Ltd, in particular Mr Suresh Gopal, Senior Editorial Consultant and Ms Garima Rao, the Development Editor, for their enormous help and cooperation.

Finally, we would like to acknowledge the support of the student community to further improve this book.

Mumbai

V V Mahajani

S B Umarji

Preface to the Third Edition

During the last ten years, several engineering institutions have started offering courses in chemical engineering. During this period, a large number of chemical industries have been set up, leading to the application of a variety of chemical equipment. This has caused an increased demand for this book both by students and design engineers. To meet this growing demand, the second edition of the book was reprinted three times.

However, keeping in view the developments in this field, it was necessary to incorporate some significant and essential features of process equipment in the original text. Hence, the second edition of the book has been thoroughly revised and updated by certain additions and alterations wherever necessary. In the present edition, certain additions relating to items of equipment such as filters, dryers, evaporators, etc., have been made. A chapter on auxiliary process equipment has been incorporated. These items of equipment are functionally essential as a complement to the main items of equipment, data sheets have also been included. A chapter on process flow sheets has been included, since it is desirable for an equipment designer to understand not only the function of the equipment but also its location, necessary piping and instrumentation in overall process.

Lately, computer aided design has become a specialized topic of study, and many good books are available in this area. Hence, in the present edition the chapter on computer aided design has been deleted. SI units have been used in all numerical problems and tables.

While preparing this edition, information has been obtained from various sources such as textbooks, handbooks, journals, etc. Indebtedness to these sources is freely acknowledged. Special mention must be made of the help received from Mr S M Mokashi and Mr V R Thakur of Davy Powergas India Pvt Ltd for which the authors are grateful to them. Thanks are due to Mr Anand Kulkarni for preparing some of the sketches, and to Prof P K Jadhav for checking the numerical problems.

M V Joshi
V V Mahajani

Preface to the Second Edition

The response to the first edition of the book was most encouraging, both from the students and design engineers from the field of chemical engineering. To fulfil the demand, the book was reprinted in 1979. The purpose of this edition is to update the previous text and to revise certain topics, based upon the comments received from experts in the field.

Certain changes will be found in most chapters. In particular, greater emphasis has been laid on topics pertaining to corrosion and its prevention and pressure vessel codes. Pressure vessels form the most important component of chemical equipment. Codes on pressure vessels have been evolved by different countries. The design engineer has, therefore, a wide choice. The information given in this volume will help him to understand the basic difference in the design approach adopted by four important codes. With these additions it is expected that this edition will provide a sound basis for the study of the fundamentals of the subject and will be of some value to practicing engineers.

In the preparation of this edition, information has been obtained from various sources such as textbooks, handbooks, journals, etc. Indebtedness to these sources is freely acknowledged. Special mention must be made of the help received from Shri J P Roy, for which the author is grateful to him. Thanks are due to Shri V R Laud, who prepared the sketches and Shri S R Tuvekar who typed the manuscript.

M V Joshi
V V Mahajani

Preface to the Second Edition

Preface to the First Edition

There are at present only a few books available on the subject of process equipment design. The scope of these books is essentially limited to design of pressure vessel components and some machine elements. However, if the subject is to be made sufficiently comprehensive, much more emphasis is needed on design and construction features of different items of equipment. The present book is an attempt to fill this gap. It is expected to satisfy the long-felt need for a more comprehensive book, which sets out to embrace a wider field, viz., application of fundamental principles of machine design to specific items of equipment and elaborate details of construction illustrated through sketches and drawings. The book is intended both for the undergraduate students of chemical engineering and for practicing engineers in the chemical industry. It is the outcome of experience gained by the author in classrooms and industry.

Chemical industries involve problems in process design, unit operations, equipment design and overall plant design. In a design of a chemical plant these problems cannot be segregated. However, in spite of their interdependence, these problems may be advantageously segregated for study and development because of different principles involved in each. Process problems are primarily physicochemical in nature. Unit operation problems are for the most part physical, while equipment design problems are to a large extent mechanical. The fundamentals and theory of chemical engineering process design and unit operations are well covered in a number of books and handbooks. Overall sizing of equipment and its components is no doubt determined by the above considerations. Equipment design is, therefore, essentially limited to mechanical aspects of design and construction features of process equipment.

The first few chapters are devoted to a review of materials of construction, corrosion and protective coatings, stresses arising out of different loading conditions and factors which influence design. An outline of the design features of some machine elements, which form a part of chemical equipment, is presented in Chapter 5. Chemical equipment may be classified on the basis of certain common features with somewhat similar design procedures. Such classification leads to three groups, namely, pressure vessel group, structural group and rotational motion group. Chapters 6 to 16 deal with design and construction features of equipment from these three groups. Hazards and methods of protection relevant to equipment design are briefly discussed in Chapter 17. Use of computers in a design organisation is becoming an economical advantage. Fundamentals of the application of computers in design problems are presented in the last chapter. Numerical problems have been incorporated to illustrate the application of equations. At the end of each chapter, references have been cited to enable the reader to locate the sources of information. The method of presentation is particularly suited to direct solutions of design problems. The theoretical development of equations is excluded. Details of constructional features are described and illustrated by a large number of sketches and drawings.

In view of the variety of equipment used in the chemical industry and its continuous development, it is difficult to state just what topics should be included in a book of this scope. The field is vast and ever-expanding. It is virtually impossible for one individual to be familiar with all the facets. It was, therefore, necessary to omit certain categories of equipment. The decision to include or omit is somewhat arbitrary, based on practical considerations of time, space and personal experience. However, an attempt has been made to include the important items of equipment.

Certain omissions or errors may have inevitably occurred in the present volume. Readers are requested to bring these to the notice of the author and make suggestions for rendering the book more useful.

M V Joshi

Acknowledgements

The author wishes to acknowledge with sincere appreciation the encouragement received from Prof. N R Kamath and assistance given by friends and colleagues, in particular Mr C S Raje, Mr R S Barve, Dr A S Dighe, Mr D K Korgonkar, Mr L V Dixit and Mr V V Mahajani. Thanks are due to Mr D V Tamhankar who helped the author in checking the manuscript and press proofs and also to Messrs V R Laud, A N Joshi and S R Kadam who prepared the sketches and drawings. Special thanks are due to Mr K B Patil who typed the manuscript.

The author is particularly grateful to the National Book Trust for granting financial subsidy, which helped a great deal in the publication of this book.

In the preparation of this book, information has been obtained from various sources such as text-books, handbooks, journals and catalogues. Indebtedness to these sources is freely acknowledged. Particular mention may be made of the following books whose publishers have permitted incorporation of figures, tables, etc.

1. *Unit Operations* by G G Brown, John Wiley & Sons, Figs 264, 268 and 270.
2. *Process Equipment Design* by L E Brownell and E H Young; John Wiley & Sons, Figs 10.5, 11.19 and 11.20.
3. *Pressure Vessel Engineering Technology* by R W Nichols; Applied Science Publishers Ltd, Tables 5.2a, 5.2b, 5.5a and 5.5b.
4. *Pressure Vessel Design and Analysis* by M B Bickell and C Ruiz; Macmillan and Co. Ltd, (London), Eqs 15.3 and 15.4; Figs 15.1 and 15.2; Table 1.1.
5. *Chemical Engineering Progress;* December 1954, American Institute of Chemical Engineers, Fig. 13.

6. *Indian Standard Specifications;* Indian Standards Institution, New Delhi. IS-803, 1962—Figs 6, 9, 10, 11 and 12 (Page nos. 15, 18, 19 and 23); IS-4072, 1967—Tables 2 and 3; IS–54.03, 1967 — Tables 1, 4 and 7; IS–2825, 1969—Tables 3.1, 3.2, 3.3, 3.4, 4.1, 4.3, 4.5, 6.3 and Appendix A-1; Figs 3.22, 4.2, G.9, G.51, G.54, G.55, G.60 and G.61.

7. Davy Powergas India Pvt Ltd, Mumbai

   For process data sheets and process flow sheets

8. Sulzer India Ltd, Pune

   For sketch of packed column

(These standards are available from Indian Standards Institution, New Delhi and its branch offices at Ahmedabad, Bangalore, Bombay, Calcutta, Chandigarh, Hyderabad, Kanpur, Madras and Patna.)

M V Joshi

Contents

1 Chapter

Basic Considerations in Process Equipment Design

1.1 INTRODUCTION

Chemicals and hence chemical producing plants (Chemical Projects) have become very essential for giving us a quality of life leading to sustainable development. In the modern competitive chemical industry, new plants are continuously being set up, existing ones modified and modernized to meet the market demands for improvement in performance or to accommodate the latest technology. A chemical plant consists of many different types of equipment in order to manufacture the desired chemical products. These different equipment consist of static and fabricated machineries (reactors, heat exchangers, absorption, distillation and extraction columns, storage tanks, etc.) and rotary machines (pumps, compressors, blowers, turbines, and so on) and systems (dryers, filters, ejectors). In a chemical plant, all equipment is made functional with the help of the following:

Piping interlinks the various equipments to transfer mass and energy and to regulate/control flows. By piping we mean the interconnecting pipes, flanges, valves and bends.

Electricals include power generation devices/systems (turbines, diesel generators), power receiving systems (transformers) from the state grid and distributes it in the plant for the rotating equipment like motors and associated electrical devices such as motor control cabinets, cables, etc.

Instrumentation to measure, indicate and control the process parameters such as flow, pressure, temperature and also includes the master control of the Programmable Logic Control (PLC) and Distributed Control Systems (DCS).

In a chemical project, the cost of the item – 'Plant and Machinery', which includes all the above hardware, is the highest cost component. Further in 'Plant and Machinery' the equipment costs contribution is the highest. Hence a cost effective optimum design is very vital to the chemical project economics and this involves both technical and economic evaluation (Mahajani and Mokashi, 2005).

The knowledge of various technical subjects such as thermodynamics, reaction kinetics, unit operations, process design, equipment design, safety, health and environmental aspects of the plant, etc., is a prerequisite for the establishment or the development of any chemical plant. Broadly speaking, the design activity starts with the technology development or selection and involves:

- Process design
- Detailed mechanical design. (For the sake of simplicity we shall call it as equipment design. Now onwards equipment design means detailed mechanical design).

The study of the unit processes and unit operations leads us to process design. The mechanical design is required for the fabrication, erection, installation and commissioning of the equipment. Thus, mechanical design of the equipment constitutes a major activity of detailed engineering of a chemical project under construction. While process designing has been kept out of the purview of this book, attention is focused on mechanical designing. The purpose of this book is to present the methods and the procedures adopted in the mechanical design of process equipment. Though there are standards for equipment design, the illustrations provided here will be helpful in designing non-standard equipment and also enable appreciation of standards in a more comprehensive way. The emphasis here is not so much on the study of the actual process, but on specifying the functioning and operation of the equipment as also on the choices of material of construction and strength considerations. Please note that the materials of construction, from the process compatibility point of view, are decided during the process design by the process chemical engineer.

Projects in Chemical Engineering which require designing can be divided into three types:

(a) Modifications and additions to the existing plant.

(b) Increasing the capacity of an existing plant.

(c) Development of a new process based on laboratory research and pilot plant studies.

Each piece of equipment is expected to serve a specific function, although in some cases it can be suitably modified for some different function. Conditions such as temperature, pressure, etc., under which the equipment is expected to perform are stipulated by the process requirements. Although the maximum capacity or the size of the equipment is specified, it is necessary to ensure a satisfactory performance even under certain amount of overload. The overall satisfactory performance and reliability of the equipments are dependent on the following factors:

- Optimum processing conditions.
- Appropriate materials of construction.
- Strength and rigidity of components.
- Satisfactory performance of mechanisms and an adequate operating range.
- Reliable methods of fabrication.
- Ease of maintenance and repairs.
- Ease of operation and control.
- Safety requirements.

1.2 THE GENERAL DESIGN PROCEDURE

It is difficult to suggest a standard design procedure. Such a procedure, however, would involve the following steps:

(a) Specifying the process or operational requirements.

(b) Analysing the probable methods and procedures to achieve them.

(c) Application of chemical process principles and theories of mechanics for designing a device or a system for optimum performance.

(d) Selecting the proper materials for the construction to suit the process conditions.

(e) Evaluation and optimization of the design.

(f) Preparing the specifications and or drawings for vendors for preparing their quotations and subsequent fabrication.

The first step is the recognition and understanding of the purpose or objective. There is a need to specify the problem as precisely as possible. The solution of the problem may be based on a new method, scheme or idea, or an old idea applied in a new way. This requires a great deal of imagination, ingenuity and innovativeness. The solution may involve a system, a process and/or a special equipment to accomplish the specified task, subject to certain constraints. The method, scheme or idea must be analysed quantitatively to ensure that it can be made to work satisfactorily and optimally. After analysing the problem, appropriate theories pertaining to the process principles and mechanics must be applied. Generally, basic principles are used, and complex techniques or equations are applied where necessary. After the selection of suitable processing conditions, materials and stresses, the problem can be resolved. The calculation process may require simple arithmetic, algebra or differential or integral calculus. In many cases it may not be feasible to have exact solutions and, therefore, various approximation techniques, such as graphical and numerical methods are used.

Sometimes approximations are made due to the lack of accurate values or methods of calculation. Methods involving close approximations are also used because exact treatments would require very long and laborious calculations giving little gain in accuracy. In some equipments, only a portion of the total number of parts is designed on the basis of an analytical calculation. The form and the size of the remaining parts are then usually determined by practical considerations based on experience.

Every step needs to be checked both in respect of mathematical calculations and engineering feasibility. It is also necessary to ascertain whether the results are consistent with experience and are practical. In some problems the final result is obtained by optimization by taking into account the influences of all the controllable parameters. It may take several iterations before a satisfactory solution is obtained.

Experience in designing similar machines or equipment will be helpful in choosing suitable sizes. A designer should realize the practical limitations involved in the application of his/her work. To arrive at the final dimensions of any component, based on an optimum theoretical and economic consideration, the designer should ascertain the exact shape and the standard sizes in which the materials are available and adopt these as far as possible. In the case of a new equipment, fabrication and testing of a prototype for ensured satisfactory performance is some times essential. Before the design is finalized, sketches must accompany all design calculations. On the basis of these sketches, detailed assembly drawings can be made, properly specifying all dimensions, selected materials, tolerances, fits, limits, the required surface finish, etc.

Provisions must be made during the different stages in the development of the design to facilitate assembly and installation of the machine or equipment. Similarly, ease of problems associated with dismantling and maintenance activities should be considered in the design and layout.

1.3 COMPUTER AIDED DESIGN

Computer aided designing system is being used as an efficient and extremely advanced technique for assessing several parameters and the final results of many design problems.

The system consists of:

- Data files of properties of materials and products.
- Database covering standard specifications of equipment and materials, cost indices, etc.
- Subsystems for modeling and simulation for the execution of the data for the evaluation of optimum design parameters.

The system is helpful in creating a more logical design procedure. The designing and analysis can be conducted simultaneously. Optimum designs can be worked out with the help of a computer program speedily. The speed and accuracy of computations obtainable by computers are impossible by manual efforts.

Databases can be prepared, which can be made accessible for the specific data needed during the designing procedure.

For details of Computer Aided Design of process equipment, please refer to the book by Bhattacharyya and Narayanan (1992).

1.4 FABRICATION TECHNIQUES

An economical and reliable fabrication technique must be adopted to manufacture the various components of the machine or equipment. In fact, the selection of materials and an appropriate choice of the method of fabrication do form an important aspect influencing the design considerations. The fabrication techniques may be classified into two groups:

(a) Techniques adopted to give an approximate shape to the material. Casting, forging, rolling, extrusion, drawing and stamping are some examples of such techniques.

(b) Techniques adopted to have precise dimensions, ensure the desired surface finish. Such methods are shaping, turning, milling, drilling, boring, broaching, grinding, honing, polishing, electroplating, coating, etc.

1.5 EQUIPMENT CLASSIFICATION

Classification of the chemical equipment is generally based on the particular type of the unit operation or unit processes. Each equipment is designed to carry out a specific unit operation such as distillation, evaporation, filtration, etc., as also processes such as hydrogenation, chlorination and so on. Equipment may also be classified to emphasize certain common features, which require similar design procedures. Such classification leads to three groups.

- ***Pressure vessel group*** This group of equipment has a cylindrical or spherical vessel as the main component, which has to withstand variations of pressure and temperature, in addition to other loading conditions. The design procedure, therefore, involves satisfying a number of criteria involving the different loading conditions. Such equipment, which covers a major part of this book is included in Chapters 5 to 12.
- ***Structural group*** This group consists of equipment or components, which are stationary and have to sustain not only the dead load but also natural phenomena like earthquakes. They are generally made up of structural sections and must satisfy conditions of elastic and structural stability. Such items are covered in Chapter 13 and a few others in Chapters 15 and 16.
- ***Group involving rotational motion*** This group covers equipment or components where rotational motion is necessary to satisfy process requirements. A drive system and power supply are the essential features. Considerations of torque, dynamic stresses, apart from other loading conditions form the basis of design. Such items are covered in Chapters 14 to 16.

Types of equipment, under design consideration, may not strictly belong to only one group, but may have features involving combination of the above groups. In such cases, it may be possible to design components belonging to each group independently and then combine them to form complete equipments.

1.6 POWER FOR ROTATIONAL MOTION

Rotational motion often involves the assessment of power (horsepower, hp) requirement. To arrive at the maximum horsepower, an analysis should be made of the mechanical operation to the equipment in terms of torque at the motor shaft, and also the speed requirements.

- ***Torque*** Torque is defined as the rotational force the driver (motor) develops at the shaft. It is expressed as Nm.

  The value of the torque varies according to the stipulated conditions of working. For example,

  (a) *Starting torque:* torque required to overcome the static friction to produce motion.

  (b) *Accelerating torque:* torque required to accelerate the driven equipment to full speed.

  (c) *Running torque:* torque required to drive the equipment or machine under normal conditions at a specified speed.

On the basis of the above torques, it is possible to classify different types of mechanical loads:

- ***Friction loads*** In this type of load, static or dynamic friction is the main component responsible for torque requirements. Types of equipment subject to such loads are conveyers, grinders, rotating filters, rotary dryers, etc.
- ***Fluid viscosity loads*** In this type of load, viscous friction between layers of fluid is responsible for torque. Types of equipment with fluid viscosity loads are agitators, fans, pumps, etc.
- ***Accelerating loads*** In certain loads, accelerating motion forms a major portion of the torque apart from frictional torque. Types of such equipment are cranes, elevators, etc.

1.7 DRIVES FOR PROCESS EQUIPMENT

We shall simply overview two types of drive commonly used in chemical plants namely, electric motors and turbines. While electric motors are the commonly used drivers, turbines are used in big petrochemical complexes and refineries to drive compressors, cooling water pumps or seawater / creek water pumps. We shall broadly review them and introduce some basics here to make subsequent reading more appealing. Detailed discussion on this subject is out of the scope of this book.

1.7.1 Electric motors

Electric motor driven equipment are the most common in any plant. Motors are divided into two types based on electric power supply namely,

- AC motors (alternate current)
- DC motors (direct current)

DC motors were popular in services where there was a demand for a variation of speed. However, with the introduction of a device known as Variable Frequency Drive (VFD), DC motors are becoming obsolete in chemical plants and are being replaced by AC motors with VFD. Patzler (1992) has presented basics of the VFD device to have energy savings. Lucchesi (2000) has presented selection criteria and operational aspects of VFD drives. It should be noted that VFD driven motors are slightly derated.

Motors are further classified based on type of the power input, i.e., single phase or three-phase. While domestic appliances generally work single phase, industrial motors are mostly of the three-phase type at 415 V and 50 Hz (frequency). As a thumb rule motors of up to say, 160 kW ratings are on three-phase supply are known as LT (low tension, 415 V) motors and above this are HT (high tension) motors (3.3 kV or 6.6 kV).

Three phase motors are further classified as:

- Synchronous motor
- Induction (squirrel cage) motor

In a synchronous motor there is no variation in speed as the load changes. On the other hand, in induction motors there is a variation in speed with load. This variation is known as slip and is expressed as a percentage of the full load speed. Induction motors are commonly used in chemical plants. As a thumb rule, synchronous motors are used at low speed and power greater than say, 35 kW.

Motors are also rated for continuous applications (continuous plant) or for intermittent (short time) application.

It is obvious that some of the electrical energy put into motors will get converted into heat and, therefore, motors need to be cooled. Air is the cheapest cooling medium available. The motor has either an open enclosure or a closed enclosure. The open enclosure allows free air from surroundings to cool motor windings. In chemical plants for safety reasons we need closed enclosures. Therefore, Totally Enclosed Fan Cooled motors known as TEFC are used. The motor windings are isolated from the surrounding hostile environment. In TEFC motors there are two types namely, increased safety and explosion proof motors. The explosion proof motor has an enclosure to confine any internal explosion. Such motors are required to handle flammable materials (flash point < 40°C). For very dangerous materials such as hydrogen we have special motors having enclosures under an inert gas atmosphere like nitrogen circulating at a few mm of water gauge pressure. Thus, all explosion proof motors are TEFC type but all TEFC motors are not explosion proof motors. Basso (1992) has given a comprehensive account of motors used in Chemical Process Industries (CPI). The discussion presented is valid for Indian conditions by and large except for the fact that our electric power supply voltage and frequency are different from USA.

In the selection of a motor, the starting torque and current must match with the driven equipment. Motors are classified on the basis of high or normal starting torque and a high or low starting current.

Further, for the final choice of motors, the following points need to be considered:

- ❖ Torque response of the driven unit.
- ❖ Basic speed of the driven unit.
- ❖ Required shaft output.
- ❖ Type of enclosures.
- ❖ Input voltage and frequency in case motor is an integral part of an imported equipment.
- ❖ Motor size/power.
- ❖ Required dynamic response of the driven system.
- ❖ Need for electric braking.
- ❖ Speed adjustment range in the case of a variable speed AC motor with VFD.

1.7.2 Turbines

The steam, gas and hydraulic turbines are the second most widely used driving systems. They are more often used in the refinery and petroleum industry than in the chemical industry. Turbines are available from, say, 10 kW onwards with speeds ranging from 1000 to 7000 RPM.

In many chemical industries due to the cost and the non-availability of power, cogeneration is done. By cogeneration we mean the simultaneous generation of power and energy to heat process fluids. By the expansion of steam through a turbine to the desired pressure, power is generated and then this exhaust steam is used for heating purposes. By doing this, we use the latent heat of steam resulting in higher energy efficiency. When high pressure superheated steam is expanded through the turbine, instead of generating power via an alternator, the turbine shaft is directly coupled to the compressor or blower or pump via a speed governor. Thus, without generating electric power, one can drive the machinery, say a compressor. This method is commonly used in methanol and ammonia plants.

If a liquid stream is available under pressure, one can use a hydraulic turbine to drive a pump.

Turbines are classified as either condensing or non-condensing. A governor is required to control the speed of turbines. Handbooks of power plant engineering present more detailed information on this subject.

Selection of motors for some items of Process Equipment is indicated in Table 1.1.

Table 1.1

Equipment	*Type of Motors*
Agitator	1. Squirrel cage AC induction motor:
	(a) normal torque, normal starting current.
	(b) normal torque, low starting current.
	(c) multispeed constant torque.
	2. Direct current, constant speed.
Compressor	1. Squirrel cage AC induction motor:
	(a) normal torque, normal starting current.
	(b) normal torque, low starting current.
	(c) high torque, low starting current.
	2. Wound rotor AC induction motor.
	3. Synchronous motor.
	4. Direct current, constant speed compound wound.
Pump, centrifuge	Same as compressor.
Rotary filter	1. Squirrel cage AC induction motor multispeed constant torque.
Rotary dryer	2. Direct current, adjustable speed.
Centrifuge	1. Squirrel cage AC induction motor multispeed variable torque.
	2. Direct current, variable speed.

References and Recommended Readings

Basso, D., 'Electric Motors Power Up The CPI', *Chemical Engineering*, **99** (7) 65 (1992).

Bhattacharyya, B.C., and Narayanan, C.M., *Computer Aided Design of Chemical Process Equipment*, New Central Book Agency (1992).

Couper, J.R., Roy Penny, W., Fair, J.R., and Walas, S.M., *Chemical Process Equipment Selection & Design*, Second Edition, Elsevier, Amsterdam (2005).

Cremar, H.W., *Chemical Engineering Practice*, Vol. IX, Butterworth, London (1965).

Dixon, J.R., *Design Engineering*, McGraw-Hill Book Co., New York (1966).

Elliott, T.C., *Standard Handbook of Power Plant Engineering*, McGraw-Hill Publishing Company, NY (1989).

Lucchesi, L.J., 'Match the Variable Frequency Drive to the Industrial Fan', *Chemical Engineering*, **107** (10) 127 (2000).

Mahajani, V.V. and Mokashi, S.M., *Chemical Project Economics*, Macmillan India Ltd., New Delhi (2005).

Mallev, V.L., and Hartman, I.B., *Machine Design*, Scranton International (1946).

Patzler, K. 'Electronics Enter the Electric Motor', *Chemical Engineering*, **99** (7) 71 (1992).

2

Chapter

Materials of Construction

2.1 MECHANICAL PROPERTIES

A designer of machines or equipment must be well acquainted with a large number of available materials. In choosing such materials, the most important criteria are obviously the various properties of the materials. The final choice will, however, depend on the ease of fabrication and the overall cost. Certain properties of materials can be well defined, others can be stated in relative terms, while a few may be rather vague. The basic properties are composition, structure, specific weight, thermal conductivity, coefficient of expansion and resistance to corrosion. Properties such as strength, elastic limit, moduli of elasticity, endurance limit, resilience, toughness, ductility, brittleness and hardness are termed as mechanical properties. From the point of view of fabrication, machinability, weldability and malleability might be considered as relevant properties.

Properties of all materials arise essentially from their structure, which varies from simple to complex crystalline bodies according to the type of bonds present, the relative sizes of atoms and their geometrical arrangement in the crystal. Furthermore, many substances may exist in more than one stable crystalline form or in an amorphous state, such as various glasses and vitreous bodies. Most of the properties are cooperative properties of the aggregate rather than of the constituent atoms and molecules. Amongst the various properties, mechanical properties are of major interest to the designer of equipment. These properties determine the behaviour of materials under applied forces and loads. The response of materials to applied forces depends on the type of bonding and the structural arrangement of atoms or molecules in the solid. These arrangements may greatly be modified by manufacturing processes and operations, thereby resulting in highly variable properties even in the materials of the same chemical composition. In addition, the type of stress and the mode of its application may change the behaviour of a material to a greater degree than the chemical composition, heat treatment of temperature. Some of the important mechanical properties are discussed in the following paragraphs.

2.1.1 Strength

Strength represents the capacity of the material to withstand external forces. Depending on the nature of the force strength can be classified as tensile, compressive, shear and impact. External forces are resisted by the material and, therefore, induce stresses and deformations. Stress may be simply defined as internal force of resistance per unit area, while deformation in the original size and/or shape of the material is defined as strain.

In linear elastic deformation, stress and strain are related by Hooke's law

$$\text{Strain} = \frac{\text{Stress}}{\text{Constant}} \quad \text{or } \varepsilon = \frac{f}{E} \tag{2.1}$$

where E is known as Young's modulus of elasticity. In non-linear elastic behaviour

$$\varepsilon = \left(\frac{f}{\lambda}\right)^n \tag{2.2}$$

where λ – pseudoplastic modulus
n – constant

The stress-strain relationship of different materials can be determined by uniaxial tension or compression tests. Some typical stress-strain diagrams are shown in Fig. 2.1.

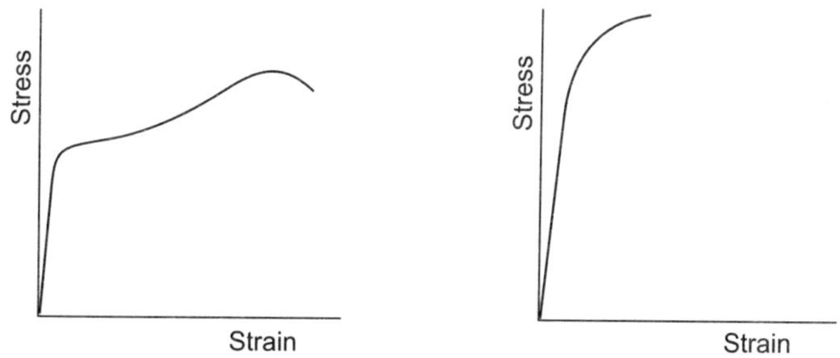

Fig. 2.1 (a) and (b)

Figure 2.1(a) indicates the deformation produced in a ductile material, while Fig. 2.1(b) represents deformation of a brittle material, showing very limited deformation.

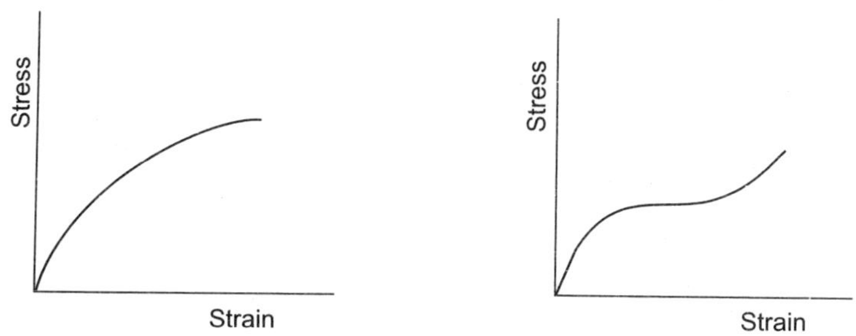

Fig. 2.1 (c) and (d)

Large non-linear deformations, representing highly elastic materials are shown in Fig. 2.1(c) and (d).

2.1.2 Stiffness and Rigidity

It is a measure of the ability of the material to resist deformation.

2.1.3 Elasticity

Elasticity is the ability of the material to regain its original shape as soon as the load is removed. In the design of most of the components, permanent deformations are generally avoided and the material is utilised with a view to retaining its elasticity.

2.1.4 Proportional Limit

Figure 2.2 represents a stress-strain graph of an elastic material. Point 'a' on the graph indicates the greatest stress up to which the Hooke's law of proportionality of stress of strain is observed. This point is therefore known as the proportional limit.

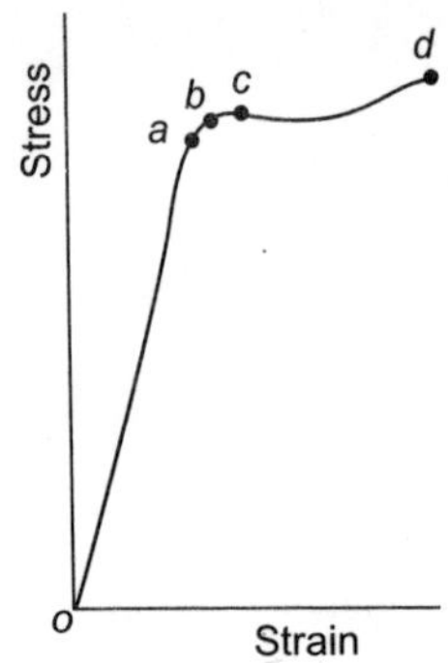

Fig. 2.2

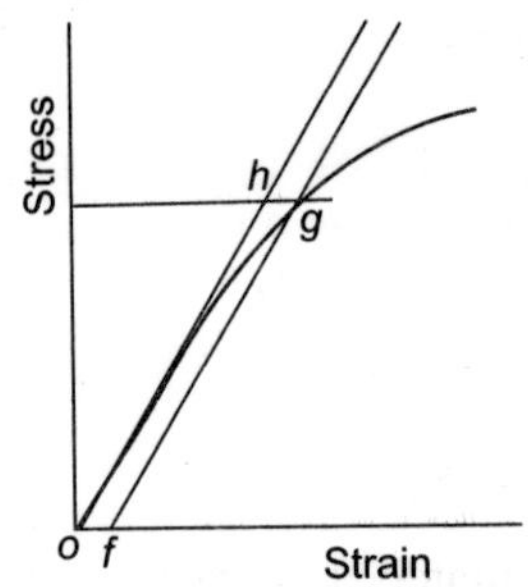

Fig. 2.3

2.1.5 Elastic Limit

It is the maximum stress, indicated by the point 'b' in Fig. 2.2, which the material can withstand without a permanent deformation. It is generally very close to the proportional limit.

2.1.6 Yield Stress

It is the stress at which the resistance of the molecules of the material begins to break down rapidly and a sudden large increase in strain occurs without an increase in stress. This is indicated by point 'c' in Fig. 2.2.

2.1.7 Proof Stress

For certain materials there is no well defined stress at which yielding occurs, the stress-strain diagram being curved almost from the origin (Fig. 2.3). If a tangent to the curve at the origin is drawn 'oh' and a line 'fg' cutting at curve at 'g', such that 'of' = 0.1% then the stress at 'g' is called 0.1% proof stress. It is the stress at which the strain has departed 0.1% of the gauge length from the line of proportionality 'oh'.

2.1.8 Ultimate Stress

It is the greatest stress at which the failure of the material takes place. The portion of the stress-strain curve between the yield stress 'c' and ultimate stress 'd' in Fig. 2.2 is termed as plastic range. Hooke's law is not applicable to this range and the deformation produced is permanent.

2.1.9 Ductility

It is a measure of the deformability of the material, determined by the percentage of elongation or reduction of area.

2.1.10 Resilience

It is the elastic energy released by the material, as a result of the stress removal. It is measured by finding out the area under the stress-strain curve within the elastic limit. The stress can be tensile, compressive or shear.

2.1.11 Toughness

It is the ability of the material to absorb energy in deformation in the plastic range and is measured by finding out the total area of the stress-stain curve.

2.1.12 Hardness

It represents the surface characteristics of a material and is assessable by the resistance it offers to indentation and scratching.

2.1.13 Fatigue

Fatigue represents the failure of the material when subjected to the application of a cyclic load, which causes a progressive enlargement of an initial small crack.

2.1.14 Creep

It is a slow and progressive deformation of a material with time under constant stress.

The mechanical properties stated so far are affected by the temperature of the conditions under which the material has to function. In general, an increase in temperature reduces the strength of the material. In addition, other conditions such as the state of stress, strain rate, size and shape of the component and method of fabrication are likely to have a significant effect on the mechanical properties.

2.2 MATERIALS

Materials of Construction (MOC) may be divided into two general categories, viz., metals and non-metals. Table 2.1 gives physical constants of some common materials. For properties of materials, see Appendices C, D and E.

Table 2.1 Physical Constants of Some Common Materials

Material	*Modulus of Elasticity (E) expressed in 10^3 N/mm^2*	*Modulus of Rigidity (G) expressed in 10^3 N/mm^2*	*Poisson's Ratio (μ)*	*Density kg/m^3*
Aluminium (alloys)	71.0	2.6	0.334	2,730
Beryllium	287.1	–	--	1,820
Beryllium copper	124.2	48.3	0.285	8,230
Brass	95.1	34.3	0.30–0.40	8,450
Bronze	109.0	–	–	8,730
Carbon steel	202.0	78.5	0.292	7,850
Cast iron-grey	100.0	41.4	0.211	7,200
Copper	120.6	38.3	0.260	8,960
Inconel	214.0	76.0	0.290	8,960
Lead	15.7	7.5	0.450	11,340
Magnesium	44.8	16.6	0.350	1,800
Molybdenum	332.0	117.2	0.307	10,200
Monel metal	179.3	65.6	0.320	8,830
Nickel silver	127.5	48.3	0.322	8,690
Nickel steel	196.1	75.6	0.291	7,750
Phospher bronze	111.0	41.4	0.349	8,160
Stainless steel (18:8)	190.0	73.1	0.305	8,020
Titanium	103.5	–	–	4,480
Tungsten	437.3	173.6	0.170	19,300
Zirconium	68.4	–	–	6,500

2.2.1 Metals

2.2.1.1 Ferrous Metals

These are the most important metals, which are known for their strength, wide availability and ease of forming and machining. A wide range of properties can be obtained by suitable alloying. According to the percentage of carbon present these are classified as wrought iron, cast iron and steel.

Wrought iron This is almost pure iron with a low carbon and manganese content. Wrought iron is not normally considered a corrosion resistance material. It has, however, excellent resistance to atmospheric corrosion. Similarly, alkalis and alkaline solutions, concentrated and cold sulphuric acid and organic compounds cause very little attack on the metal. It finds extensive use at elevated temperatures. Steam supply lines, condensate return lines, heating coils, exhaust lines, smoke stacks are some of the typical high temperature applications. The material can be easily formed to desired shape, forged, machined and welded.

Cast iron This has usually 2.5 to 4.5% carbon. In general, it is cheap and easy to cast even in complicated shapes because of its fluidity in the molten state. Cast iron can be classified as: (i) grey cast iron, (ii) white cast iron, depending on the form in which the carbon is distributed in iron. In grey cast iron the carbon is in the form of graphite, distributed as free crystals throughout the iron. In white cast iron there is a chemical combination of iron and carbon. The formation of the two types of cast irons depends on the rate at which it is cooled from the molten state.

Grey cast iron The compressive strength of this is about three to four times its tensile strength, and hardness number between 180 and 240 on Brinnel Scale (BHN). Generally cast iron has no outstanding resistance to acids or acid solutions. It is not corroded by alkali solutions of less than 30% concentration at any temperature. Its applications are in pumps and valves, special piping, acid coolers, acid reaction and concentration vessels. Many reaction kettles and vats were made of enamelled cast iron in the older days.

In general, cast iron is used in producing components whose complicated shapes introduce fabricating difficulties and in cases where resistance to abrasion and corrosion is desired. The components produced must have sufficient thickness for casting purposes.

White cast iron Due to its highly brittle nature, it is seldom used as such. Usually, grey cast iron is chilled to produce a hard superficial depth of white cast iron. It is used in such parts as furnace bars, teeth and jaws of crushing and grinding machines. Such components have a good wear resistance. White cast iron can be made malleable by heat treatment. The resultant castings are able to withstand shocks better, and are used for pipe fittings, gear wheels, housings, pulleys, etc.

A high duty cast iron can be produced by the use of small quantities of calcium silicide. The resultant structure after heat treatment has a better tensile and shear strength. It is used for the fabrication of pressure vessels and high pressure and temperature pipes and fittings. It has a higher rate of contraction than ordinary cast iron.

Alloy cast iron The basic properties of cast irons have been improved by alloying and by changes in foundry techniques, which have extended their range of applications. Since alloy castings possess good mechanical properties coupled with a good corrosion resistance, they are exceedingly useful to process industries.

High silicon iron is very hard and brittle, which makes it exceedingly difficult while machining as well as vulnerable to mechanical shock and excessive strain. The superior degree of chemical resistance make these cast irons extremely useful in severe applications. They are practically unaffected by sulphuric acid at all strengths and at all temperatures. Similarly, hot strong nitric acid does not affect this material. Addition of 3% molybdenum imparts a high degree of resistance to attack by hydrochloric acid and chlorine-containing chemicals. High silicon irons are not attacked by atmospheric conditions.

Although high-silicon cast irons have excellent chemical resistance they have poor mechanical properties. The addition of silicon and a small percentage of copper and nickel give an alloy of superior mechanical properties and good chemical resistance. It has resistance to most organic compounds and solvents, alkalis and phosphoric acid. It is also resistant to atmospheric oxidations, flue gases, hydrocarbons as well as wet and dry ammonia.

A series of alloy cast irons containing varying percentages of high nickel with copper and chromium are known as 'Ni-resist' which possess a unique combination of engineering properties such as corrosion resistance, wear resistance, strength, heat resistance, toughnees, good castability and machinability. Alkalis and neutral salt solutions are readily handled by these materials.

Steel Steel is iron alloyed with carbon between 0.05 and 2.0%. In addition, it contains smaller proportions of phosphorus, sulphur, silicon and manganese. These steels are known as plain carbon steels. In some cases certain other alloying elements are added; such steels are known as alloy steels.

Plain carbon steels According to the percentage of carbon these steels are classified as low or mild, medium and high carbon steels.

Low carbon steel or mild steel It has 0.05 to 0.3% carbon. It is undoubtedly one of the most versatile materials available. It has good strength and ductility. It can be easily rolled, forged and drawn. Fabrication by welding and machining is also easily carried out. For chemical equipment it is therefore used on a large scale. Some of the applications are in pressure vessels, pipes and fittings, machine components and structural sections. The corrosion resistance of mild steel is low. If the rate of corrosion is slow it is certainly economical to use such steel. In cases, where contamination of the product with products of corrosion cannot be accepted, the surface of steel may be protected by paint, metal spraying or by other protective linings. This type of protection is not satisfactory for severe corrosive conditions, because it is usually not possible to obtain a perfect coating. Please note that allowable stress remains fairly constant at 1230 kg/cm^2 over a wide temperature range 30 to 340 °C.

Medium carbon steel It has a carbon content between 0.3 and 0.5%, which helps to improve its hardness, strength and fatigue resistance. The material is used for fabrication of shafts, springs, gears, bolts and certain structural sections. This is difficult to weld.

High carbon steel It has a carbon content of more than 0.5%, up to a maximum of 2%. It is much harder and stronger than the two preceding grades, but much less ductile. The material is primarily used for cutting tools and dies.

Alloy steels A number of elements, such as nickel, chromium, silicon, manganese, molybdenum, tungsten, beryllium, vanadium, cobalt, and titanium are used as alloying elements. Generally, certain properties are associated with certain elements.

Manganese – abrasion resistance and toughness.

Nickel and chromium – corrosion and high temperature resistance.

Chromium, tungsten, vanadium, molybdenum, cobalt – cutting action.

Nickel, chromium, titanium – creep resistance.

Silicon, manganese – elasticity.

A few of the important alloys used in the chemical industry are mentioned here.

Low alloy steels These contain small percentages (below 10%) of nickel, chromium, molybdenum, manganese, etc., in different combinations. These are used primarily for machined components and to a limited extent for pressure vessels. In these steels the carbon content is usually of the order of 0.15%, manganese about 1% and silicon 0.3%. Steels with 0.5%, 1%, 2.25%, 3% and 5% chromium have a better corrosion resistance than ordinary mild steels and are used for high temperature service under mild corrosive conditions. For bolting, medium carbon steel with 1% Cr-Mo with or without vanadium is most suitable.

In particular, quenched and tempered low alloy steels are in extensive use, due to their special characteristics such as high strength, high toughness, weldability, high corrosion resistance, heat resistance, abrasion resistance and good workability. The yield stress values of these steels are about three times that of ordinary mild steel of structural or boiler quality. Hence, there is considerable saving in weight of equipment. Pressure vessels are, therefore, fabricated with an economy of material, and are also useful under high and low temperature service conditions.

High alloy steels These are of two types: straight chromium steels, with chromium content ranging from 13 to 17% and chromium-nickel steels or stainless steels, with a chromium content between 18 and 25%, and nickel content between 8 and 20%. The carbon content varies between 0.03 and 0.25%. According to the carbon content, straight chromium steels with 0.08% carbon are used for plates, 0.1% for forgings, 0.2% for castings and 0.15% for bolting. Only those varieties with a carbon content higher than 0.1% are hardenable by heat treatment.

Special mention must be made of stainless steels. These are essentially high chromium or high nickel-chromium alloys of iron containing small amounts of other essential elements. They have excellent corrosion resistance and heat resistance properties. The most common variety of stainless steels contains approximately 18% chromium and 8% nickel. Addition of molybdenum to the alloy increases the corrosion resistance and high temperature strength. Certain varieties of stainless steel are hardenable by cold working, others by heat treatment.

Stainless steels exhibits the best resistance to corrosion when the surface is oxidized to a passive state. This condition can be obtained at least temporarily by a passivation operation in which the surface is treated with dilute nitric acid and then rinsed with chloride free water. Localized corrosion can occur at places where foreign material collects, such as scratches, crevices and corners.

There are more than a hundred different types of stainless steels, which can be classified according to the alloy content, microstructure and major characteristics. According to microstructure, they can be classified as follows:

(a) Austenitic stainless steels These are the most important types, containing mainly iron, nickel 6 to 22% and chromium 16 to 26% (Tables 2.2 and 2.6).

Table 2.2 Standard AISI Type Steels

Type	*C (max)*	*Cr*	*Ni %*	*Mo*
304	0.08	18 to 20	8 to 10.5	-
316	0.08	16 to 18	10 to 14	2 to 3

The American Iron and Steel Institute (AISI) numbering system is widely accepted for the identification of the chemical composition. AISI 300 series are the predominant type, known for their combination of high strength to weight ratios, ductility and ease of fabrication; resistance to heat, cold, mechanical damage, fatigue and corrosion. They are widely used in the chemical industry and form 70% of the consumption of stainless steels. They have a much better corrosion resistance than the other types but have a certain susceptibility to stress corrosion cracking. The austenitic stainless steel with carbon content less than 0.03% are identified by a numerical number followed by letter 'L' say, SS 316 L. These materials exhibit a high corrosion resistance even under as welded condition (Without any heat treatment). Therefore, heat exchanger tubes are SS304L or SS316L. However, "L" grade steel is weaker at higher temperatures despite of good stress corrosion cracking resistance exhibited. While the ASME code (section II part D) 2007, recommends maximum temperature for SS304 plate 815 °C that for SS304 L is 650 °C.

(b) Martensitic stainless steels These have generally the lowest alloy content. Chromium content ranging 11% upwards, nickel 1.25 to 2.5% or a higher carbon content (0.1% to 0.6%). The higher carbon steels are generally poor in corrosion service.

(c) Ferrite stainless steels These are alloys containing 13 to 30% chromium as a major alloying element with low carbon. Due to its good oxidation resistance, it finds many uses at high temperatures. These are identified by AISI 400 series. They are suitable for very mildly corrosive service.

(d) Duplex steels As the name suggests it is a two-phase mixture (hence the name Duplex) of approximately 35–55% ferrite and 65–45% austenitic steel. The two-phase microstructure has a very fine grain size imparting high strength. The corrosion due to organic acid is substantially reduced with duplex steel.

(e) Nitrogen-added stainless steels When nitrogen of the order 0.15 to 0.20% is added to stainless steels the yield strength is raised substantially. These are called 'high proof stainless steels' and are most commonly used in moderately high temperature application. These are produced in 304, 316 variations.

Table 2.3 gives the composition of special stainless steels for high temperature service.

Table 2.3 Composition of Various Steels

AISI Type No.	*C max*	*Cr % w/w*	*Ni*	*Mo*	*Others*
302	0.15	17-19	8-10		
304	0.08	18-20	8-10.5		
304 H	0.1	18-20	9		M_n = 2, Si = 0.75
316	0.08	16-18	10-14	2.0-3.0	
316 H	0.1	16-18	12	2.0-3.0	Si = 0.75, M_n = 2
310	0.08	24-26	19-22		
310 H	0.1	24-26	20.5		Si = 0.75, M_n = 2
*H, stands for high temperature application such as in a refinery.					
High temperature alloy used for reformer tubes.					
HK 40	0.4	25	20		
HK 40 Nb	0.4	25	35		Nb = 1.5

The major application of stainless steels in the chemical industry are:

(a) Vessels such as storage tanks, reactors, absorption and distillation columns and heat exchangers.

(b) Machinery such as pumps, fans, compressors, centrifuges, dryers, coolers and filters.

(c) Material handling equipment such as pipes, conveyers and tankers.

The American Iron and Steel Institute (AISI) numbering system for standardization of stainless steels is widely accepted. These are indicated in Tables 2.3 and 2.4. A comparative study of the Indian and foreign standards are presented in Table 2.5. For fabrication purposes ASTM nomenclature based on strength is more appropriate.

Stainless steel is used as a cladding material for plain carbon or low alloy steel. The thickness of stainless steel may vary between 2 and 5 mm. The austenitic stainless steel has a peculiar property wherein allowable stress strongly depends upon temperature. For example, at 30°C, SS304 has allowable stress 129.5 N/mm^2 which is reduced to 109.5 N/mm^2 at 340°C. Therefore, specifying unduly high design temperature might result in costlier equipment. Economically, it is important to choose a suitable thickness of cladding. Generally, it will be about 2 to 4 mm in thickness.

Table 2.4 AISI Numbering System

Group	*Structure*	*Hardenability*	*AISI Type*	*Analysis Built up from Basic Types*
Chromium Nickel	Austenitic	Hardenable by Cold Work	302	Cr 18% + Ni 8% basic type
			301	Cr and Ni lower for more work hardening
			304	C lower to avoid carbide precipitation
			305	Ni higher for less work hardening
			302B	Si higher for more scaling resistance
			303	S or Se added for easier machining
			316	Mo added for more corrosion resistance
			317	Mo higher for more corrosion resistance and strength at heat
			318	Cb added to avoid carbide precipitation
			321	Ti added to avoid carbide precipitation
			347	Cb added to avoid carbide precipitation
			347F	Se added to improve machinability
Chromium	Martensitic	Hardenable by Heat Treatment	308	Cr and Ni higher with C for more corrosion and scaling resistance
			309	Cr and Ni still higher for more corrosion and scaling resistance
			309S	C lower to avoid carbide precipitation
			309C	Cd added to avoid carbide precipitation
			310	Cr and Ni highest to increase scaling resistance
			410	Cr 12% basic type
			403	Cr 12% adjusted for special physicals
			414	Ni added to increase corrosion resistance and physicals
			416	S or Se added for easier machining
			418	W added to improve high temperature properties
			420	C higher for cutting purposes
			420F	S or Se added for easier machining
			431	Cr higher and Ni added for better resistance and property
			440A	C higher for cutting applications
			440C	C still higher for wear resistance
			440F	S or Se added for easier machining
	Ferritic	Non-hardenable	405	Al added to Cr 12% to prevent hardening
			430	Cr 17% basic type
			430F	S or Se added for easier machining
			442	Cr higher to increase scaling, resistance
			446	Cr much higher for improved scaling resistance

2.2.1.2 Non-Ferrous Metals

Aluminium and its alloys These materials are light and easy to fabricate. Commercial pure grade aluminium is a relatively weak material. Alloys of aluminium have better mechanical properties. The most common alloying elements are copper, silicon, zinc, magnesium, nickel and manganese. Aluminium resists attack by acids because a surface film of inert hydrated aluminium oxide is formed. This film adheres to the surface and offers a good protection. Aluminium alloys are satisfactory for low temperature service up to –150°C. Some of the most important alloys are duralumin with 4% copper, 0.6% magnesium, 0.5% manganese; aluminium-silicon alloys; and aluminium-zinc-magnesium alloys. The last ones are the strongest alloys of aluminium. These alloys are used for fabrication of storage vessels, reaction vessels, heat exchangers and absorption columns.

Table 2.5 Comparative Study of Standards

IS	*BS*	*AISI*	*DIN*
		405	
1570–07Cr13		410	(1.4001) × 7 Cr 14
1570–15Cr13		420	(1.4024) × 15Cr 13
1570–07Cr17	970En60	430	(1.4016) × 8 Cr 18
1570–20 Cr18 Ni 2	970En57	431	(1.4057) × 22 Cr Ni 17
1570–04 Cr19 Ni 9	970En58E	304	(1.4302) × 5 Cr Ni199
	3014 gr2	302	(l.4324) × 12 Cr Ni 177
1570–04Cr19Ni9Ti20	3014 gr3	321	(1.4544) × 10 Cr Ni Ti 189
1570–04Cr19Ni9Nb40	1501–821 Nb	347	(1.4544) × 10 Cr Ni Nb 189
1570–05Cr18Ni 11 Mo3	1501–845B	316	(1.4436) × 5 Cr Ni Mo18 12
1570–10Cr25 Ni 18	2901–A 11	310	(1.4843) Cr Ni 25 20

ASTM is also commonly used in India.

Table 2.6 Stainless for the Chemical Industry

Types 304, 304L, 321, 347	general corrosion resistance
Types 308, 309, 309S, 310, and 310S	largely for high-temperature work, such as preheaters and vessel liners, as well as for welding rods.
Types 316, 316L	general corrosion resistance especially when Types 304 and 304L are subject to pitting or excessive corrosion.
Type 317	for application where a somewhat better corrosion resistance is required; than that can be obtained from Types 316 and 316L

Note: Type 304 is the general purpose grade for the Process Industries.

Copper and its alloys Pure copper has good ductility, malleability, high electrical and thermal conductivity. It has fair mechanical strength and can be fabricated easily into a wide variety of shapes. It can be easily joined by brazing, welding and soldering. Copper exhibits good corrosion resistance to strong alkalis, and organic solvents. It is resistant to atmospheric moisture or oxygen because a protective coating of copper oxide is formed on the surface.

Copper is alloyed with zinc, tin, nickel, aluminium and lead. In special cases beryllium and manganese are also used. These alloys have greater strength and provide greater case in machining and coating. Copper-zinc alloys, known as brasses, have up to 45% zinc and are suitable for tubes, wires and sheets, which can be cold worked. Copper-tin alloys, known as bronzes, contain tin up to 5% for wrought work and about 10% for castings. These alloys are used widely for pumps, valves, pipe fittings, etc. Copper-aluminium alloys contain up to 14% aluminium, and are known as aluminium bronzes. They are highly resistant to oxidation and scaling and for that reason, are amongst the best copper alloys for services at moderately elevated temperatures. They include some of the strongest copper alloys. These alloys are used for pump castings, condenser tubes, valve seats, etc. Copper-nickel alloys contain 10 to 30% nickel. They have good strength, ductility and corrosion resistance. These alloys are extensively used for heat exchangers where sea water is used as a cooling medium. Copper and its alloys have been used in the chemical industry over a long period. Copper itself has been always used in the wrought form as a sheet or tube, while fittings such as valves, pumps and other cast parts were fabricated from one of the copper alloys. They are used for evaporators, stills, condensers and heat exchangers.

Nickel and its alloys Nickel exhibits high corrosion resistance to most alkalis. The strength and hardness of nickel is almost as great as that of carbon steel and the metal can be fabricated easily. For economical use of nickel, steel clad with nickel is used extensively in the production of caustic soda and alkalis. An alloy of nickel and copper, known as monel contains 67% nickel and 30% copper. The resistance of monel of molten salts and alkali metals is inferior to that of pure nickel, but the copper content imparts superior resistance to non-oxidizing chloride solutions and to hot dilute solutions of non-oxidizing acids. The alloy is used in food industries. In chemical industries it is used in evaporators, heat exchangers, etc. An alloy of nickel-chromium-iron, known as inconel (Table 2.7), contains 76% nickel, 15% chromium and the balance mainly iron. The presence of chromium in this alloy greatly increases its resistance to oxidizing conditions. It also has a good heat resistance. Applications of inconel include heat exchangers, digesters, etc. Nickel-molybdenum alloys, known as hastelloys, have high resistance to corrosion over a wide range of materials with higher concentrations and temperatures. Hastelloy C has 54% nickel, 17% molybdenum, 16% chromium, 5% iron and 4% tungsten (Table 2.9). It has excellent corrosion resistance even at high temperatures. It is used for valves, piping, heat exchangers and vessels. To meet very severely corrosive conditions nickel-molybdenum alloys are often the only alternatives to precious metals like platinum and silver (Table 2.8).

Table 2.7 Nominal Composition of Some Nickel Alloys

	Carbon	*Copper*	*Iron*	*Silicon*	*Others*	
Wrought						
Nickel	0.10	0.15	0.15	0.15	Manganese	0.25
'Monel'	0.15	30.00	1.40	0.10	Manganese	1.00
'Inconel'	0.15	0.50	8.00	0.50	Chromium	15.50
					Manganese	1.00
'Incoloy'	0.10	0.50	44.40	1.00	Chromium	20.50
					Manganese	1.50
Cast						
Nickel	0.80	0.50	0.50	1.50	Manganese	0.80
'Monel'	0.15	30.00	1.50	1.60	Manganese	0.75
'Inconel'	0.20	0.50	8.00	2.00	Chromium	16.00
					Manganese	1.00

Balance: Nickel plus small amounts of sulphur.

Table 2.8 Nominal Composition (Expressed as %) of Some Medium Alloys

Alloy	*Carbon*	*Nickel*	*Chromium*	*Molybdenum*	*Copper*	*Others*	
'Durimet'	0.07 max	29.00	20.00	2.00	3.00 min	Manganese	1.50
						Silicon	1.50
'Worthite'	0.07 max	24.00	20.00	3.00	1.75	Manganese	0.60
						Silicon	3.25
'Carpenter 20'	0.07 max	29.00	20.00	2.00	3.00 min	Manganese	0.75
						Silicon	1.00
'Ni-o-nel'	0.05 max	42.00	21.50	3.00	2.25	Titanium	0.90
						Manganese	1.00
						Silicon	0.50
'Hastelloy F'	0.05 max	44.50	22.00	6.50	—	Cobalt	2.50
						Tungsten	1.00
						Manganese	1.50
						Silicon	1.00
						Columbium, Tantalum	2.15
'Illium G'	0.20	56.00	22.50	6.40	6.50	Manganese	1.25
						Silicon	0.65
'Illium R'	0.08 max	64.00	22.00	5.00	2.50	Manganese	0.30
						Silicon	0.15

Balance: Iron with normal amount of sulphur and phosphorus.

Table 2.9 Nominal Composition of Some High Alloys

	Carbon	Nickel	Molybdenum	Iron	Chromium	Others	
'Hastelloy B'	0.05 max	61.00	28.00	5.50	1.00	Manganese	1.00
						Silicon	1.00
						Cobalt	2.50
'Chlorimet 2'	0.10	63.00	32.00	3.00	—	Manganese	1.00
						Silicon	1.00
'Hastelloy C'	0.10 max	54.00	17.00	5.00	15.50	Manganese	1.00
						Silicon	1.00
						Cobalt	2.50
						Tungsten	3.90
'Chlorimet 3'	0.07	60.00	18.00	2.00	18.00	Manganese	1.00
						Silicon	1.00

Chromium and its alloys Chromium is used either for plating of iron and steel, or as an alloying element in low and high alloy steels. It helps to improve corrosion resistance, creep resistance, wear resistance and the cutting action of steels. The above combination of properties makes chromium stubbornly resistant to abrasion, enabling it to preserve surfaces at exact dimension? Electroplated chromium can produce a surface which is either mirror-bright, dulled grain or an intermediate texture, is maintainable over a long time.

Lead It is one of the oldest metals, known for its resistance to corrosion. It has a low creep and fatigue resistance. It is soft and malleable. Due to these properties, lead is used as a lining or cladding for metal equipment. The corrosion resistance of lead depends on the fact that a protective surface film is formed. If the surface film is one of the highly insoluble lead salts, such as the sulphate, carbonate or phosphate, good corrosion resistance is obtained. Lead sheets and pipes are used for fabrication of equipment. The use of lead as a material of construction is nowadays discouraged due to its adverse effect on health particularly during fabrication.

Titanium This material is available over a wide range of tensile strengths, ranging from that obtainable from carbon steel up to those of highly alloyed steels. However, its elastic modulus is only about half that of steel. It resists corrosion caused by sea-water, brine, chlorides, and is unaffected by oxidants, such as nitric acid and many organic acids. The corrosion resistance of titanium depends on the formation and maintenance of a thin oxide film. The material can be used for pumps, valves and heat exchangers. It can also be used for lining of mild steel vessels. In all such applications it is necessary to choose the correct grade of the metal to suit the particular component and conditions.

Beryllium This has low ductility, but its modulus of elasticity is usually high. It is stronger than steel and retains its strength at high temperatures. Beryllium is used primarily as an alloying element due to its corrosion resistance, and excellent wear and fatigue properties. Due to its high strength to weight ratio combined with corrosion resistance, it is used in special applications, such as reactor components.

Zirconium This has good mechanical properties at room temperature, but retains strength at elevated temperature only if it is highly alloyed. It is completely resistant to hydrochloric acid and oxidizing media such as solutions of nitric and chromic acid. The material is used in special applications such as thermometer pockets, sheaths, etc.

Tantalum This has good mechanical properties at room and elevated temperatures. The chemical behaviour of tantalum which is scarcely attacked by most of the acids, but is corroded by hot strong alkalis, is often compared with that of glass. It is resistant to both oxidizing and non-oxidizing media. The metal is used for the construction of special heat exchangers.

2.2.2 Non-Metals

2.2.2.1 Carbon and Graphite

These comprise a family of materials. Each member is essentially pure carbon, but differs from others in such things as basic structure, orientation of the crystallites, size and number of pore spacings, degree of graphitization, etc. Consequently it is possible to manufacture many different grades of carbon and graphite, ranging from coarse-grained relatively weak materials to the fine-grained strong and purest materials. Impervious graphite and carbon are made by impregnating them with a suitable resin.

These materials are isotropic and, therefore, their properties vary with grain orientation. The corrosion resistance of both carbon and graphite is almost unsurpassed. Graphite is generally inert to all chemicals except in strongly oxidizing applications. Unlike metals, these materials do not depend for corrosion resistance on the formation of an inherent oxide or other film. This is particularly important in chemical process equipment in which erosion and corrosion can combine to destroy the metal that is protected by a film. Carbon and graphite have an unusual range of thermal conductivity. Carbon is a fairly good insulator, while graphite is a good conductor of heat. The low thermal conductivity of carbon is advantageous when carbon bricks are used to line vessels for high temperature reactions. On the other hand, the high coefficient of thermal conductivity of graphite makes it possible to manufacture efficient, compact heat exchangers.

2.2.2.2 Glass

Commonly available glass has a low tensile strength, but a high compressive strength. It is brittle and is damaged by thermal shock. Glass has excellent resistance to corrosion and is subject to attack only by hydrofluoric acid and hot alkaline solutions. Coating of steel by glass gives a material which has the corrosion resistance of glass and the strength of steel.

2.2.2.3 Rubber

Natural and synthetic rubbers are primarily used as linings for chemical equipment. Some other components such as gaskets and bushes are also made of rubber. The hardness of rubber can be improved by addition of various ingredients. Natural rubber is resistant to dilute acids, alkalis and salts, but is affected by oxidizing media, oils, benzene, etc. Some of the important synthetic rubbers are polychloroprene, or neoprene, nitrile, butyl, silicone, hypalon fluro-elastomer like viton and polyurethane. They are resistant to oils and solvents. Silicon and viton have excellent high temperature resistance, while urethane rubber has good abrasion resistance, hardness and resilience. Due to its resistance to acids and alkalis, rubber lining is used for stirrers, fans, centrifuge baskets, pumps, valves, storage vessels, pickling vats, etc.

2.2.2.4 Plastics

These materials are classified in two main categories: thermoplastics, which soften under the application of heat and thermosetting plastics, which under pressure and heat are cured and become hard and will not deform when subjected to heat again. The important properties of these materials are their low weight, good thermal and electrical insulation, ease of fabrication and pleasing appearance. However, they can be used only at moderate pressures and temperatures. Their strength is low and they have only a fair resistance to solvents. They have excellent resistance to weak mineral acids and salt solutions and are not corroded by atmospheric conditions.

Thermoplastic The most common thermoplastics are: cellulose acetate, polyethylene, polystyrene, polyvinyl chloride, polypropylene, nylon, methacrylate, etc. These materials are generally available in the form of films, sheets, rods and pipes. They can be moulded and extruded to various shapes and sizes. The main applications are in reaction vessels, storage tanks, gas washing towers, ducting, ventilating systems, fans, pipes, pipe fittings, gaskets, etc.

Thermosetting plastics These comprise, phenol formaldehyde, urea formaldehyde, melamine formaldehyde, polyester, epoxy and silicone resins. In general, these materials are comparatively hard and brittle and are therefore added with fillers such as wood flour, cotton flock, asbestos, etc. Polyester and epoxy resins with glass fibre reinforcements are used for fabrication of pipes and fittings, ducts, vessels and tanks. They are also used as lining materials. Thermosetting plastics are moulded to give them specific shapes. Laminates can also be formed.

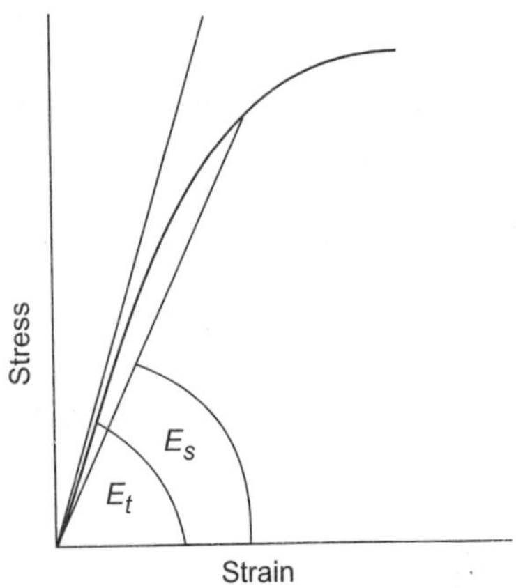

Fig. 2.4 Tangent and Secant Modulii E_t – Tangent Modulus, E_s – Secant Modulus

Mechanical properties of plastics In respect of the mechanical properties, plastics differ considerably from metals. For instance, when a stress-strain diagram is made for a thermoplastic, it is found that the thermoplastic does not follow Hooke's law of elasticity. The modulus is not constant but varies with stress. The best measure of the stiffness in this case is the average stiffness at a given stress, which is given by the secant modulus (Fig. 2.4). Also the results of the tensile strength test are significantly affected by the rate of loading and the temperature and the elongation at rupture often varies considerably. Furthermore, at all practical temperatures a thermoplastic will continue to deform when a constant load is applied for some time, in other words thermoplastics are invariably subject to creep (see Appendix C).

The basic creep date are obtained by determining the elongation as a function of time for a series of stress levels, each for a series of temperatures. These date are usually presented in the form of graphs (Fig. 2.5).

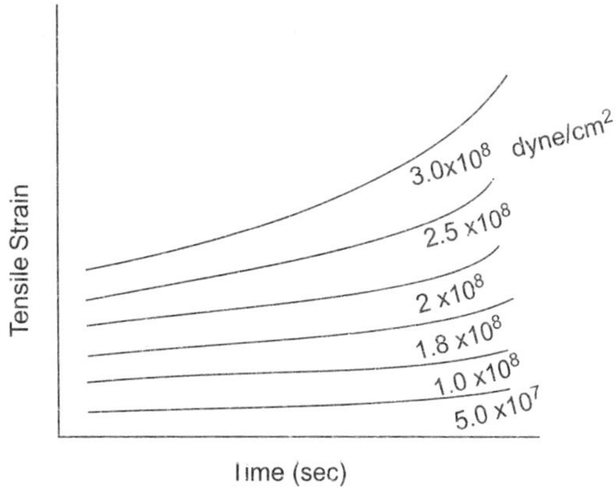

Fig. 2.5

Reinforced plastics Both thermoplastics and thermosetting plastics are reinforced to improve their mechanical strength. Reinforcements are usually in the form of fibres, mat, shreddings, etc., which are made from glass, asbestos, jute, sisal, and carbon. The percentage of reinforcement varies from 5 to 30%. Glass fibres of different diameter and lengths are most commonly used. More than 90% of reinforced plastics are thermosets and amongst these 90% are polyesters. In particular for large size vessels and tanks, it is convenient to use polyester which cures at almost atmospheric temperature and pressure. The main limitations are low service temperature, and mechanical damage due to impact.

2.3 CORROSION

Apart from the mechanical properties of materials stated so far, another important property is that of corrosion resistance. All metals are thermodynamically unstable and tend to react with their environment to produce compounds such as oxides or carbonates. This reaction involves movement of electrons and is called electrochemical reaction. The corrosion products are usually insoluble and may form a protective skin on the metal which gives the metal a high degree of corrosion resistance as in the case of aluminium and stainless steel. On the other hand if the corrosion product is loose and non-adherent as in the case of steel, the corrosion resistance is poor. Corrosion of metals is influenced by their metallurgical composition and micro-structure. Metals can be arranged in order of their electrode potentials. This is known as galvanic or electrochemical series (Table 2.10). The extent of corrosion between two dissimilar metals depends on their electrode potential difference and the distance apart and relative areas of the metals. Non-metals in general, have good corrosion resistance under atmospheric conditions. However, they are affected by other environmental conditions.

To understand the phenomenon of corrosion and the extent to which materials can resist corrosion, it is desirable to get requisite knowledge of the common forms of corrosion.

These are as follows:

(a) *Uniform corrosion*

This is the most common form of corrosion, which is characterised by chemical or electrochemical reaction which proceeds evenly and uniformly over the entire exposed area. It is indicated by general wasting of surface. The corrosive product may either form a protective layer on the metal or as in the case of direct chemical attack, the corroded material will dissolve in the corrosive environment. This type of corrosion can be prevented or reduced by (i) Choice of proper materials, (ii) Coatings, (iii) Inhibitors or (iv) Cathodic protection.

Table 2.10 Galvanic Series of Metals (Anode end to Cathode end)

Anode	(1) Magnesium	(9) Nickel (active)	(17) Stainless steel 18-8 type 304 (passive)
↓	(2) Zinc	(10) Inconel 600 (active)	(18) Stainless steel 18-8-3 type 316 (passive)
↓	(3) Aluminium alloys	(11) Brasses	(19) Titanium
↓	(4) Aluminium	(12) Copper	(20) Hastealloy C 276
↓	(5) Mild steel	(13) Bronzes	(21) Silver
↓	(6) Cast iron	(14) Nickel (passive)	(22) Graphite
↓	(7) Stainless steel 18-8 type 304 (active)	(15) Inconel 600 (passive)	(23) Gold
Cathode	(8) Stainless steel 18-8-3 type 316 (active)	(16) Monel 400	(24) Platinum

(b) *Galvanic corrosion*

This occurs when two dissimilar metals are placed in contact. The less resistant metal becomes anodic and the more resistant metal cathodic. Usually the corrosion of cathodic metal is nil or very little in this type of couple (Table 2.10). For example in heat exchangers with copper tubes and cast iron or steel, tube sheets, if galvanic corrosion occurs, it accelerates attack on heavy tube sheet instead of thin copper tubes. The following measures are useful in minimizing galvanic corrosion:

(i) Select the combination of metals as close together as possible in the galvanic series.
(ii) Avoid the unfavourable area effect of a small anode and a large cathode.
(iii) Insulate dissimilar metals whenever possible.
(iv) Apply coatings with caution.
(v) Add inhibitors, if possible, to decrease the aggressiveness of the environment.
(vi) Avoid threaded joints for materials far apart in the galvanic series.
(vii) Design a system for ease of replacement of anodic material. Also, one can have a thicker anode (sacrificial anode).
(viii) Install a third metal which is anodic to both metals in the galvanic contact.
(ix) Provide for cathodic protection.

(c) *Crevice corrosion*

This is an intense, localised corrosion, which occurs within crevices and other shielded areas on metal surfaces exposed to corrosives. This type of attack is usually associated with small volumes of stagnant solution caused by holes, gasket surfaces, lap joints, surface deposits and crevices under bolt and rivet heads. Stainless steels are particularly susceptible to crevice attack.

Crevice corrosion can be minimized by using welded joints instead of threaded or bolted joints. Similarly, in equipment, sharp corners and stagnant areas should be avoided.

(d) *Pitting corrosion*

This occurs between surfaces which are in close contact. It is extremely localised and occurs due to presence of impurities, rough spots and scratches. Pitting seldom occurs with extreme suddenness. It occurs due to stagnant conditions, which implies that in equipments where a continuous flow of liquid is present, often exhibits reduced pitting. Adding inhibitors to metals is helpful in reducing pitting.

(e) *Intergranular corrosion*

This consists of a localised attack or an intercrystalline cracking along the grain boundaries of the metal. It is caused by impurities at the grain boundaries, enrichment of one of the alloying elements, or a depletion of one of these elements in the grain-boundary areas.

(f) *Selective leaching*

In this type of corrosion, one element of a metal or alloy is singled out for attack. The common types are: dezincification, dealuminification and graphite erosion. For example, if copper-zinc alloys (brasses) containing less than 85 per cent copper are exposed to wet conditions, this type of corrosion occurs, for redeposited copper has little mechanical strength. Similarly, in cast iron, under certain conditions, iron corrodes and leaves a porous graphite residue. Breakage occurs suddenly without any indication.

(g) *Stress corrosion*

This corrosion is the result of internal or external stresses and corrosive environment. It manifests itself in the form of cracks and is, therefore, often known as stress corrosion cracking. Many metals and some plastics suffer from this effect. With plastics it is called environmental cracking. The important variables affecting stress-corrosion cracking are temperature, solution composition, metal composition and structure and the stress.

(h) *Fatigue corrosion*

As in the case of stress corrosion, where static stresses are linked up with corrosion, in this type of corrosion, cyclic loads combine with corrosion to cause corrosion fatigue. Usually fatigue failures occur at stress levels below yield point and after many cyclic applications of this stress.

(i) *Erosion corrosion*

This is caused due to combination of corrosive fluid and mechanical wear resulting from the impingment of liquid or abrasion of solid particles. This effect is mainly dependent on the liquid velocity and on account of the contained air, solids or any factors which affect the rate of formation of protective films. This process removes a layer from the surface of metal as dissolved ions, as it forms from the metal surface.

All types of equipment which are exposed to moving fluids, are subject to erosion corrosion. Some of these are piping systems, valves, pumps, blowers, centrifuges, agitators, heat exchanger tubings, equipment for spraying fluids, etc.

(j) *Cavitation corrosion*

This is a special form of corrosion, which occurs due to the repeated collapse of vapour bubbles on a metal surface. This causes mild physical damage to the protective films, severe deformation and fracture of surface. Cavitation occurs in equipments where a high velocity liquid flows and pressure changes are encountered, such as pump impellers, resulting in vapour bubbles.

(k) *Fretting corrosion*

This occurs at the contact areas between materials under load subjected to vibration and slip. It appears as pits or grooves in the metal surrounded by corrosion products. Fretting is also called friction oxidation or wear oxidation.

This type of corrosion can be minimized by lubricating contacts areas or by increasing the hardness of one or both of contacting materials.

(l) *Hydrogen damage*

This refers to the mechanical damage of metal caused by the presence of or by the interaction with hydrogen. This can be in the form of hydrogen blistering, hydrogen embrittlement, decarburization or hydrogen attack.

Table 2.11 Environments Causing Stress Corrosion

Material	*Environment*
Aluminium	Water and steam, NaCl, including sea atmospheres and waters, air water vapour.
Copper alloy (brass)	Ammonical solutions, amines, mercury-salt solutions.
Austenitic stainless steels	Chlorides, including $FeCl_2$, $FeCl_3$, NaCl, NH_4Cl, Sea environments, H_2SO_4, fluorides, condensing steam from chloride waters.
Ferritic stainless steels	Chlorides, including NaCl, fluorides, bromides, iodides, caustics, nitrates, water steam.
Carbon and low alloy steels	HCl, caustics, nitrates, HNO_3, HCN, molten zinc and Na-Pb alloys, H_2S, H_2SO_4-HNO_3, sea water.
High strength alloy	Sea and Industrial Environments.
Magnesium alloys	Atmosphere, tap water, sodium chloride solutions.
Lead	Lead acetate solutions.
Nickel	Bromides, caustics, H_2SO_4.
Monel	Fused caustics soda, hydrochloric and hydrofluoric acids.
Titanium	Sea environments, NaCl in environments, mercury, molten cadmium, silver and AgCl, fuming HNO_3, Chlorinated or fluorinated hydrocarbons.

(m) *Thermogalvanic corrosion*

When a metal is subjected to a thermal gradient by uneven heating or dissipation of heat, it produces a similar effect on metal as that of galvanic corrosion. The metal is then differentially polarised and anodic and cathodic areas are formed, thus causing an effect similar to thermogalvanic attack to develop.

The commonly used unit for measuring the rate of corrosion is 'mils per year' wherein mils $= 25.4 \times 10^{-3}$ mm or $= 1 \times 10^{-3}$ inch.

2.4 CORROSION PREVENTION

Corrosion is the biggest single cause of plant and equipment breakdowns in the chemical and allied industries. However, it is usually not necessary to design equipment which is completely free from corrosion. There are a few product areas in which this may be required, for example, in the manufacture of pharmaceutical and food products, equipment must be completely free from metallic contamination. Such absolute purity of products, however, is not usually required and some corrosion of equipment is acceptable provided it does not prejudice the integrity and safety of the plant during its planned life. The problem then is to choose the cheapest material which will be adequate for the conditions.

This approach is followed in the pressure vessel codes for equipment made from ferrous and non-ferrous metals, where provisions are made for corrosion allowances to be added to the design stress thickness to compensate for the expected corrosion rate over the planned life of the vessel. As these codes point out, however, there are difficulties in deciding the correct allowance.

If the corrosion is uniformly distributed over the surface of the process equipment, there should be no difficulty in making a suitable allowance. For most metals it is usually accepted that a corrosion penetration rate of less than 0.12 mm per year is satisfactory and corrosion rates between 0.12 and 1.2 mm per year may be economic, depending on the cost of alternate materials and the degree of contamination acceptable in the process.

However, corrosion is often not uniform and it is estimated that only 35% of failure of chemical plant are due to general corrosion. The most common causes of corrosion failure are as shown earlier, namely, galvanic corrosion, concentrated cell corrosion, pitting corrosion, intergranular corrosion, stress corrosion, erosion corrosion, etc. Most of these can be either controlled or eliminated if proper measures are adopted.

These measures are:

(a) Material selection

(b) Anodic or cathodic protection

(c) Engineering design of components

(d) Control of process environment

(e) Protection methods.

(a) *Material selection*

Since the materials available for fabrication of equipment are large, the factors such as resistance to corrosion in a given environment, peculiar tendency of materials to specific types of corrosion, and special treatments should be coordinated with the functional parameters. Strength, fabricability, ease of production, appearance, availability and cost should also be considered. It may be necessary to compromise and sacrifice some mechanical advantageous properties to satisfy corrosion requirements and vice-versa. Using fully corrosion resistant material is not always the correct choice; a balance between the first cost and the cost of subsequent maintenance should be found over the full estimated life of the desired utility.

Certain combinations of metal and compatibility are a natural choice:

Aluminium	– non-staining atmospheric exposure
Chromium containing alloys	– oxidizing solutions
Copper and alloys	– reducing and non-oxidizing environments
Hastealloys	– hot hydrochloric acid
Lead	– dilute sulphuric acid
Monel	– hydrofluoric acid
Nickel and alloys	– caustic, reducing and non-oxidizing environments
Steel	– concentrated sulphuric acid
Tin	– distilled water
Titanium	– hot strong oxidizing solutions
Tantalum	– ultimate resistance

Non-metals are not as widely used as metals. The use of plastic, for instance, is limited to applications such as gaskets and coatings mainly due to lack of strength, particularly at higher temperature.

(b) *Anodic and cathodic protection*

Dissimilar metals in intimate contact or connected by a conductive path should be used only when the functional design or other important consideration render them unavoidable. Wherever possible, select materials in contact, close together in the emf series. Exposed area of less noble metals should be kept large, relative to more noble metals. If joining of non-compatible metals is unavoidable, dielectric separation is imperative. This can be provided in many ways as for instance insulating gaskets and protective coatings. Figures 2.6 and 2.7 show ways in which corrosion is prevented in such cases. In Fig. 2.6 apart from the use of dielectric insulation separating the aluminium footing and steel pad, a zinc chromate primer may be used to cover surfaces in contact. In Fig. 2.7 instead of using a copper coil, a nickel plated copper coil is preferable.

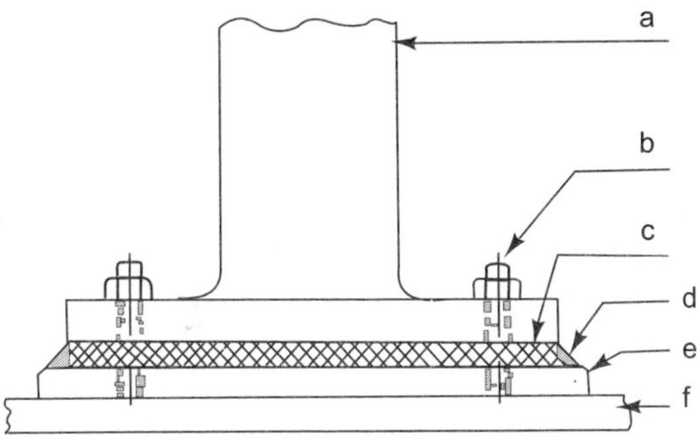

Fig. 2.6 **Dielectric Separation of Metals: (a) Aluminium Footing, (b) Stainless Steel Stud, (c) Dielectric Insulation, (d) Sealant, (e) Steel Pad, (f) Steel Plate**

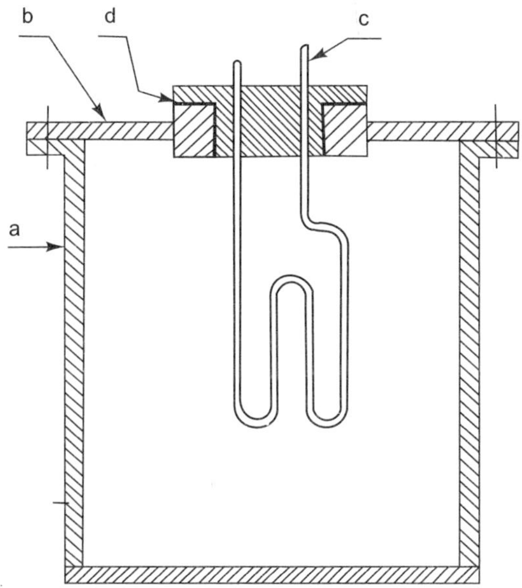

Fig. 2.7 **Dielectric Separation of Metals: (a) Galvanised Steel Tank, (b) Cover, (c) Nickel Plated Copper Coil, (d) Insulation**

(c) *Engineering design of components*

The main considerations which will help to prevent corrosion are essentially based on an appropriate shape of the component, concentration of stresses due to loading, the surface conditions and precautions in fabrication. Each component will have to be designed to avoid concentration cells and crevice corrosion. Crevices can be avoided by using welded joints in preference to bolted or riveted joints. Unavoidable crevices as shown in Fig. 2.8 may be caulked or sealed. Whenever possible, horizontal surfaces exposed to the atmosphere should be avoided, since these tend to hold moisture and dirt. Dish shaped heads for vessels are preferable to flat heads. Opportunities for the collection of concentrated solutions and corrosive materials at inaccessible positions in liquid-carrying equipment must be minimised. For example, provision is required to be made for the complete drainage in tanks and vessels (Fig. 2.9). It is also necessary to avoid pockets in which stagnant liquids may accumulate such as behind baffles or thermowells in reactors and heat exchangers. Inlet or outlet nozzles must not project into the vessels and tanks. Crevices, sharp corners and projections are causes for corrosion-erosion attack. Suitable radius of fillets and corners give a streamlined flow and help eliminate pockets. Corrosion rate increases as the ratio of exposed surface to metal volume is increased; for example, a wire corrodes faster than a sheet of the same thickness. Therefore, in a vessel, if the shell has a 2 mm corrosion allowance, we need to provide double the corrosion allowance for impellers, baffles, and shafts. Uneven stress distribution should be avoided in corrosive conditions. Localized stresses in any metal component intensify corrosion, viz., drilling of bolt holes sets up unequal stress distribution. The area surrounding the hole becomes anodic and is thus susceptible to attack in a corrosive environment. To relieve stresses in fabrication such as welding, machining, forging, etc., stress relief through heat treatment should be specified.

Rough surfaces are more susceptible to corrosion than smooth surfaces. Likewise, projections and sharp corners are prone to corrosion-erosion failures. All fabrication marks, rough joints, weld heads and pipe burrs should be removed.

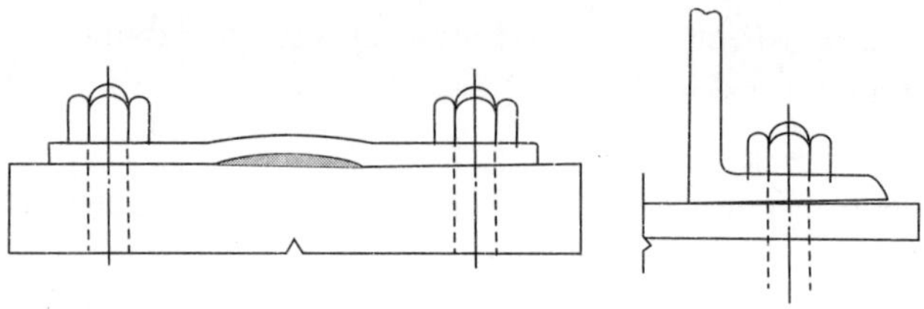

Fig. 2.8 Crevices

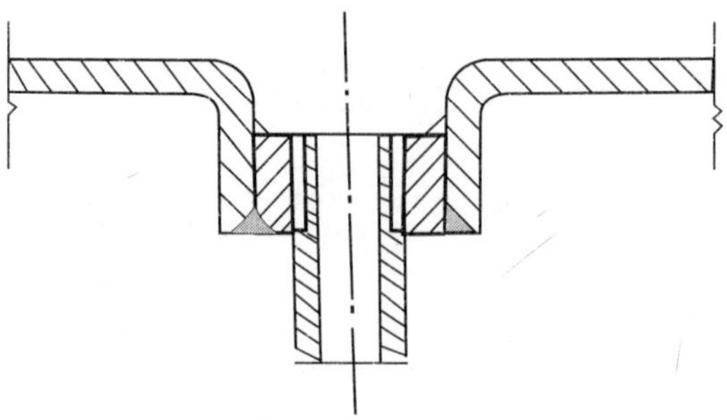

Fig. 2.9 Drainage from A Tank

(d) *Control of process environment*

It is certainly not possible to change the process composition. Even so there is always scope for control in certain auxiliaries such as corrosive cooling waters, steam condensate lines, etc. Many of these problems occur during specific periods of operation, e.g., local surges, blowdown on steamlines during cleaning or shutdown and include temperature fluctuations. In general, it is not the corrosion from the major process constituents or operating conditions but the unexpected minor impurities and excursions due to malfunctioning that initiate the problems of corrosion. In some cases corrosion originates with the moisture in some form. The effect of dissolved gases, specially oxygen, is detrimental to carbon steel, copper, and nickel but is beneficial to stainless steel and aluminium. These conditions should be controlled by removal of impurities, adverse process conditions and environmental conditions.

(e) *Protection methods*

Protective methods comprise of providing separation of surfaces from environment, cathodic protection or anodic polarization. These methods can be used individually or in various combinations. Separation of materials from environment, provided by application of metallic coatings, by jointing, coating with plastics or ceramics, by lining, sealing, enveloping, insulation and by application of temporary protectives (oils, greases, removable plastics, etc.) involves primarily a change in the surface composition and this change is brought about by additions of different materials in the form of an outer skin. Most of these processes involve a dimensional change and a weight change.

Cathodic protection is the application of a direct current from an auxiliary anode to prevent or retard the corrosion of metal which has been made the cathode in the corroding media. For example, by installing an anode near a metal component of the structure, the areas of the metal (anodic or cathodic) within the range of the electrolyte all become cathodic. This current, when properly applied, opposes the corrosion current from the surface of the protected metal and will minimize corrosion. Electrolytic anodes are used with external power source, and may be either sacrificial or non-sacrificial. Sacrificial or expendable anodes may be iron, steel or aluminium. Aluminium is preferred due to its light weight, uniform corrosion rate and white or colourless corrosion products. Non-sacrificial anodes may be platinum, high-silicon iron, graphite and carbon. These are more expensive but last longer. Cathodic protection is used to protect pipelines, tanks, heat exchangers, etc.

Inhibitors are also used for control of corrosion. This is in the form of a chemical, which, when added in small amounts to the corrosive medium, markedly reduces the rate of chemical attack. They are believed to form and maintain upon the metallic surface a protective film, which decreases the corrosion current and renders the environment non-reactive. Inhibitors can control either the anodic or cathodic corrosion reaction and are, therefore, known as either anodic or cathodic inhibitors. Anodic inhibitors include chromate, silicate, and nitric salt of sodium and potassium. Glassy phosphates and many organic compounds function as cathodic inhibitors. Cathodic and anodic inhibitors acting together are more effective than either of them acting separately. In addition to above, protective coatings and linings are used for vessels, piping (pipes, bends and valves) and storage tanks, as protective means. We shall discuss in brief below:

2.4.1 Protective Coatings

Quite often limitations are imposed by cost and fabrication requirements, on the choice of the most desirable material for resisting corrosion. Therefore, a protective coating on a less resistant metal or alloy represents a practical compromise. There are three general categories of coatings: metallic, inorganic and organic. A coating should be applied only on a clean surface. The presence of grease, oil, dirt or scale adversely affects the adherence, continuity and durability of coatings. The surface preparation is one of the most important acts for the durability of coatings.

An excellent compilation of papers published between 1970 and 1980 in Chemical Engineering magazine on coatings and linings is presented in a book *Materials Engineering II of Chemical Engineering*, edited by McNaughton (1980).

Metallic Coatings

Many different techniques are practised commercially for applying metallic coatings. These can be classified as anodic or cathodic. The anodic processes will protect the coated metal through their preferential corrosion, whereas the noble metal coatings are mainly used for their superior corrosion resistance properties. The most common methods are hot-dipping, cementation, mechanical cladding, electroplating, metal spraying and condensation of metal vapours. Table 2.12 indicates the principal methods, their characteristics and coating metals used. Carbon steels are some of the most commonly coated metals.

An important method of metallic coating is cladding, which is of interest in the pressure vessel industry. The cladding metals in the form of thin plates are bonded to the backing steel plates either by hot rolling or by explosion cladding. The backing plate is ordinarily of plain carbon steel, but for special applications low alloy steels can also be used. The cladding metals are stainless steels, mainly austenitic types, nickel, inconel, monel, copper and cupro-nickel. Cladded plates are economical, since the corrosion resistance of special cladding metals is obtained, by the use of only a thin layer of the material. Cladded plates can be fabricated by cutting, shearing, welding, etc.

Inorganic Coatings

Commercial inorganic coatings for protection of metals fall into two general categories, namely, chemical or electrochemical surface conversion treatments and vitreous enamels. Those in the first group often do not constitute the final protective layer but serve as an undercoating for paint or other organic materials. Chemical dip methods are employed to create protective oxide films on iron, steel, stainless steel, aluminium, copper and some of their alloys. Such films are usually very thin and are frequently coloured. Electrolytic coatings may be made, as in the case of anodising treatments for aluminium which produce a relatively thick, abrasion resistant coating.

Table 2.12 Protective Metallic Coating

Method	*Coating-metal Requirements*	*Coating Characteristics*	*Metal Used*
Hot Dipping			
Short-time immersion into bath of coating metal.	Low melting point alloying capability	Alloy layers Adherent, moderately uniform thickness difficult to control	Al, Zn, Sn, Pb, alloys
Cementation			
Powdered coating metal alloyed with base metal at temperature below melting point.	Alloying capability	Alloy throughout; hard, brittle, porous; poor corrosion resistance	Al, Zn, Cr, W, Si, Mo
Cladding			
Veneering and alloying of two or more metals under pressure; rolling of duplex ingots.	Rolling characteristics liks base metal, alloying capability.	Pure metal surface; impervious; controlled thickness	Cu, Ni, stainless steel, noble metals
Electroplating			
Electrodeposition of coating metal at cathode from solution or fused salt bath.	Waterable deposition potential; adhesion to base metal	Pure, uniform, impervious; no alloying; unstressed; controlled thickness, hardness	Ag, Au, Cu, Cd, Co, Cr, Ni, PB, Sn, Zn, Cu alloys
Metal Spraying			
Atomization of molten metal in hot gas stream. Wire of powder may be melted in gun by flame arc.	Non-refractory Not easily oxidized	Porous laminate structure; harder; no alloying with surface	Al, Cu, Fe, Pb, Sn, Zn alloys
Vapour Plating			
Surface condensation of metal vapour by pyrolysis of metal compounds cathode sputtering, vaccum condensation	Vapourisability	Costly; crystalline, fine grain; very thin; readily applicable to non-metals	Mo, W, Cr, Al, Au, Ag, Ni

Vitreous coatings although brittle, possess surface hardness and complete inertness to many corrosive environments. Enamelled or glass lined vessels and other equipments are available in a variety of shapes and sizes. These enamels are made from fused silicates of various compositions, containing colloidal suspensions of colouring materials. For metallic sheets, the enamel is applied as slurry. Subsequent heating conditions are dependent on enamel composition and intended service.

Organic Coatings

These represent a large variety of materials and are the most widely used methods of protecting metals against corrosion. More than a thousand different synthetic resins, as well as a wide variety of pigments, modifying oils and solvents are used in coating formulations. These coatings protect metals by interposing a continuous, adherent, inert film between the metal and its environment. They also markedly change the appearance of the metal. These coatings can be divided into three classes. Paint is a dispersion of pigments in a vehicle which consists of drying oils modified with a solvent or thinner to aid in the application. Enamel is a dispersion of pigments in varnish or resin vehicle which polymerises either by oxidation at room temperature or by application of heat. Lacquer is a pigmented natural or synthetic resin dissolved or suspended in solvents. Powder coating is effectively carried out by applying a polymer by a spray or an electrostatic spray followed by baking at high temperature. This results in a uniform continuous coating.

2.4.2 Linings for Chemical Plants and Equipment

Storage tanks, reaction vessels, pipes, ducting, etc., are covered with linings in order to (a) give the underlined structure protection against chemical attack, (b) prevent contamination of the materials being processed, and (c) minimize the effect of abrasion. The various materials commonly used for linings are as follows.

Rubber

This material is applied in the form of latex, sheets of partially or fully vulcanised soft, semihard and hard rubber. The surface to be lined is suitably prepared and treated, and adhesives or bonding agents are applied to the cleaned surface. Vulcanisation is carried out, using steam, hot water or hot air.

Latex-cement based rubber linings do not require the application of a specific bonding cement and are normally applied directly to the prepared surfaces. Lightly vulcanised rubber sheets have been applied with great success as a lining material for general industrial purposes, and for chemical plants. However, the most effective and the most commonly employed method of protecting corrodible surfaces with soft or hard rubber is to apply it over a bonding cement before vulcanisation.

After being rubber-lined, the equipment is usually allowed to remain for a day or two before being vulcanised. Any air entrapped between rubber and metal will then be diffused. Vulcanisation changes rubber from a plastic state to an elastic state. Tensile strength, resistance to abrasion, tearing and flexing increase due to vulcanisation.

Lead

Lead linings are principally used in two forms. In loose lead sheet lining, the lead sheet is applied to surfaces in thickness varying from 3 mm to 10 mm or more and the joints are lead-burnt in position. Proper anchoring of lead sheets is essential. Where process conditions are more rigorous a better form of homogeneous lead lining is prescribed. In this form lead is bonded to the cleaned surfaces with the aid of a flux and the thickness of lead lining required is thereafter deposited in drop-by-drop lead burning and peened after each run, which finally forms a perfectly fused homogeneously bonded mass. Homogeneous lead lining should have a minimum thickness of 4.5 mm up to a maximum of 8 mm. This type of lining is expensive and is, therefore, used only in those cases where temperatures up to 100°C are obtained, where good heat transfer properties are requied, or where the operation is under vacuum. Lead linings are discouraged nowadays due to health concern during fabrication.

Glass

The production of glass lined equipment requires the suitable basic materials. In addition to chemical composition, surface quality and absence of internal flaws contribute a lot to the process of lining. There are three methods for lining glass on iron and steel, viz., wet spray process, hot dust process and spray dust process.

(a) *Wet spray process*

After cleaning, the metal surface the ground coat enamel is applied in slip form by spraying. The coat is dried and the equipment is then fired in the enamelling furnace, after which it is allowed to cool. The process is repeated with subsequent coats till the desired thickness of lining is obtained.

(b) *Hot dust process*

This process is generally applicable to cast iron components. After cleaning the metal surface the ground coat is applied in a manner similar to that in the wet spray process. When firing of the compound is completed, the equipment is removed from the furnace and a dry powder cover coat enamel is dusted over the surface while it is still red hot. Several separate dustings of the cover coat may be applied.

(c) *Spray dust process*

This process is a combination of the two processes described above. A wet coat of enamel slip is sprayed on while the equipment is cold and powered enamel fit is dusted on this coat while it is still wet. In baking operations the lined equipment is air-dried and baked in a furnace at a temperature of 800-900°C. The firing time and temperature are suitably controlled depending on the size of equipment and the thickness of lining.

Plastics

The process involves cleaning of the metal surface, application of bonding agent, and lining with requisite sheet. The joints between sheets may be either welded of filled up with resin. Apart from the lining of sheets, there are other methods of depositing films of thermoplastic materials on metal surfaces. Three principal methods have been developed. These are described below: (a) the object to be coated is heated above the melting point of the thermoplastic and is immersed in a fluidized bed of finely divided polymer; (b) thermoplastic powder is sprayed under pressure, through a hot air or flame gun where the powder particles are softened and on impact with the surface to be coated, fuse together; (c) a suspension of the polymer in water or in a volatile organic medium is first deposited on the metal object which for evaporation of the liquid medium is fired, thus, fusing together the deposited particles. In all cases, the preparation of the surface plays an important part in the subsequent adhesion of the film to the metal.

Bricks

Acid proof brick lining of a vessel or storage tank is popular in the chemical process industry to combat corrosion problems, inspite of new corrosion resistance materials introduced in the market.

The acid proof bricks themselves are not liquid proof as far as permeance of liquid is concerned. Therefore, one has to have an impervious lining between the bricklayer and metal wall (carbon steel). This impervious lining is known as membrane. The membrane could be made up of bitumen or asphalt. Natural rubber, neoprene and butyl rubber are also used. The membrane could be applied as roller or laminate or the material could be applied by brush or spray in liquid form, which on drying becomes the membrane.

2.5 CHOICE OF MATERIAL

Various materials of constructions, their mechanical properties, corrosive properties and measures which help prevention of corrosion are indicated so far. For choice of such materials for chemical process equipment, it is essential to assess operating conditions such as temperature, pressure, flow rate, etc., under which the equipment is to work. For instance, glass lined vessels do not withstand thermal shock. Therefore, a good engineering practice is to have the glass lined reactor content and jacket heating medium at a temperature difference of say 25°C. The vendor may be consulted for this.

The task of selecting the best material cannot be made merely by choosing a suitable material having the requisite mechanical properties and anti-corrosive properties or availability of protective coating, but must be based on a sound economic analysis of competing materials. Many factors enter into economic considerations. A high materials cost may be accompanied by easier fabrication, with significant saving in labour, or it may be accompanied by longer life owing to a property such as corrosion resistance which would reduce the annual cost of replacement. Sometimes a material must be selected, even though it has inferior properties, because the right type of material is too expensive or simply is not available.

In the absence of data on corrosion or experience, one can conduct the corrosion coupon test. A piece of the possible material of construction of known width (w), breadth (b), and thickness (t) all expressed in meter, is taken. It is thoroughly cleaned to be free of grease, rust, etc. The cleaned and dried piece is then weighed in an electronic balance. This piece is then exposed to the environment for some specific period say, seven days. Afterwards, it is again cleaned and dried (to remove the moisture) and then weighed on the same electronic balance. We thus get the loss in weight (L) expressed in gms,

The corrosion rate is then computed as:

$$= \frac{L}{\{2(w)(b) + 2(w)(t) + 2(b)(t)\}\,(\text{no. of days})}$$

$$= X,\ \text{gms/(m}^2\ \text{day)}$$

It is desirable to have 3 coupons. One totally dipped in liquid, one partially dipped in liquid and the last should have a weld joint dipped in liquid. One may repeat the test to get statistically acceptable results. Apart from the above, there are other techniques which are followed. For instance, in the refinery, the dependency of corrosion on flow velocity, acid content (naphthenic acids), acid gas content H_2S and temperature can be quantified for the test specimen by the Jet impingement flow cell as illustrated by Kane and Cayard (1998). In sour gas treating facility and in the fertilizer industry (ammonia manufacture) corrosion by various alkanol amines and their blends in a formulated solvent used for H_2S and CO_2 removal can be assessed by electrochemical technique exemplified by Veawab et al., (1999).

Kane (2007) has presented an account of dynamic on line approach to corrosion monitoring in chemical plants. This new approach is expected to reduce unscheduled shutdown thereby increasing productivity of the plant.

References and Recommended Readings

Ambegonkar, M.S., 'Corrosion in Austenitic Stainless Steels under Chemical Conditionts', *Chemical Age of India*, **30** (3) 227 (1979).

Amos, R.S., and Townsend D.W., 'Stainless Steel for Corrosion Prevention in Process Plant', *The Chemical Engineer*, **331**, 270 (1978).

Batra, B. Borchert, C.A., Lewis, K.R., and Smith, A.R., 'Design Process Equipment for Corrosion Control', *Chemical Engineering Progress*, **89** (5) 68 (1993).

Bhattacharya, B.C., 'Design Considerations in Corrosion Control', *Chemical Age of India*, **30** (3) 247 (1979).

Brown, R.S., 'Select Alloys for Severely Corrosive Environment', *Chemical Engineering Progress*, **93** (3) 74 (1997).

Bukovinsky, S., '253 MA-A Special Stainless Steel', *Chemical Age of India*, **30** (3) 261 (1979).

Burns, R.H., Kumar P., 'Conquer Corrosion in Harsh Environments With Tantalum', *Chemical Engineering Progress*, **92** (3) 32 (1996).

Chari, K.S., Nickel Alloys in Fertilizer Industry', *Chemical Age of India*, **29** (3) 207 (1978).

Chemical Resistant Linings for Pipes and Vessels, Handbook No. 6, The Engineering Equipment Users Association Handbook.

Choudhary, S.G., 'Monitor and Control Cooling Water Corrosion', *Hydrocarbon Processing*, **83** (5) 55 (2004).

Dawson, J.L., 'Methods of Corrosion Control', *The Chemical Engineer*, **331** 266 (1978).

Deily, J.E., 'Use Titanium to Stand Up to Corrosives', *Chemical Engineering Progress*, **93** (6) 50 (1997).

Dorsey, J.S., 'Using Reinforced Plastics for Process Equipment', *Chemical Engineering*, **82** (19) 104 (1975).

Elliott, P., 'Choose Materials for High-Temperature Environments', *Chemical Engineering Progress*, **97** (2) 75 (2001).

Evans, L.S., *Selecting Engineering Materials for Chemical and Process Plant*, Business Books Ltd., London, (1974).

Gaugh, R.R., 'New Stainless for High Temperature', *Hydrocarbon Processing*, **55** (6) 87 (1976).

Habiby, F., Imtiaz, R.R., and Al-Mutairi, A.H., 'Reduce Underside Corrosion in Aboveground Storage Tanks, Follow these guidelines to avoid leaks', *Hydrocarbon Processing*, **82** (1) 59 (2003).

Henthrone, M., 'Cathodic and Anodic Protection for Corrosion Control', *Chemical Engineering*, **78** (20) 73 (1971).

Hepner, I.L., *Materials of Construction for Chemical Plants*, Leonard-Hill, London (1962).

Holmquist, M., 'Consider Duplex Stainless Steels for Corrosive Exchanger Service', *Hydrocarbon Processing*, **80** (10) 65 (2001).

Jastrzebski, Z.D., *Nature and Properties of Engineering Materials*, John Wiley and Sons, New York (1959).

Jenkins, L., Libert, S., and Lusvardi, V., 'Consider Using Fluoropolymers in Biological Applications', *Chemical Engineering Progress*, **100** (12) 39 (2004).

Kane R.D., A New Approach to Corrosion Monitoring, *Chemical Engineering*, **114** (6) 34 (2007).

Kane, R.D. and Cayard, M.S., 'Select Materials for High Temperatures', *Chemical Engineering Progress*, **91** (3) 83 (1995).

———, 'Assess Crude Oil Corrosivity', *Hydrocarbon Processing*, **77** (10) 97 (1998).

Kirby, G., 'Avoid Common Corrosion Mistakes for Better Performance', *Chemical Engineering Progress*, **93** (4) 76 (1997).

Krupavaram, 'Materials of Construction for Low Temperature and Cryogenic Processes', *Chemical Engineering*, **113** (7) 44 (2006).

Landils, H.H., and Stout, E., 'Glassed Steel Equipment - A Guide to Current Technology', *British Chemical Engineering*, **15** (10) 1289 (1970).

Mantell, E.L., *Engineering Materials Handbook*, McGraw-Hill, New York (1958).

McDonald, J.P., 'Corrosion Resistant Materials and Their Applications', *British Chemical Engineering* **16** (9) 801 (1971).

McNaughton, K.J., ed., *Materials Engineering I: Selecting Materials for Process Equipment, Chemical Engineering*, McGraw-Hill Publications Co., New York, N.Y. (1980); *Materials Engineering II: Controlling Corrosion in Process Equipment, Chemical Engineering*, McGraw-Hill Publications Co., New York, N.Y. (1980).

Morton, T.R., 'Fibre Glass Reinforced Plastics for Corrosion Control, Part I, *Chemical Engineering*, **80** (29) 70 (1973); Part II, Chemical Engineering, **81** (2) 140 (1974).

Naitoh, K., and Kihara T., Application of New Materials for Chemical Process Industries', *Chemical Age of India*, **30** (3) 249 (1979).

Nicolio C., 'Curb organic-acid Corrosion With Duplex Stainless Steel', *Chemical Engineering*, **108** (11) 94 (2001).

O'Connor, M., 'Plastics in Chemical Plant Fabrication', *The Chemical Engineer*, **306** 95 (1976).

Perry, R.H., Green, D.W., and Maloney, J.O., *Perry's Chemical Engineers' Handbook*, 6th Edition, International Edition McGraw-Hill Book Co., Singapore (1984).

Pitman, J.S., 'Rubber Lining', *The Chemical Engineer*, **306** 97 (1976).

Quik, J.M.A. and Geudeke, M., 'Switch to Duplex Stainless Steels', *Chemical Engineering Progress*, **90** (11) 49 (1994).

Rabald E.W., Corrosion Guide, Elsevier, Amsterdam, (1968).

Schillmoller, C.M., Control Organic Acid Corrosion with These Metals and Alloys', *Chemical Engineering Progress*, **93** (2) 66 (1997).

———, 'Select the Right Alloys for Caustic Soda Service', *Chemical Engineering Progress*, **92** (5) 48 (1996).

Tillack, D., and Bangoli, D., 'Understanding High-Temperature Failure Modes of Metals', *Chemical Engineering*, **107** (11) 62 (2000).

Tuthill, A.H., and Lamb, S., 'Combat Corrosion in Wastewater Treatment Plants With Stainless Steels', *Chemical Engineering Progress*, **91**(7) 59 (1999).

Veawab, A., Tontiwachwuthikul, P. and Chakma A., 'Influence of Process Parameters on Corrosion Behaviour in a Sterically Hindered Amine — CO_2 System', *Industrial Engineering Chemistry Research*, **38** 310 (1999).

3

Chapter

Design Considerations

3.1 INTRODUCTION

Several factors which influence the design of components, need consideration in order to understand the effect and the influence of each factor individually and also in the combination with other factors, so that the final design can be based on the proper application of these factors. In Chapter 1, general principles involved in designing of machine parts and equipment have been discussed and certain guidelines have been indicated. For a satisfactory design of any component, various factors which must be considered are:

(1) material selection
(2) corrosion prevention
(3) stresses created due to static and dynamic loads
(4) elasticity, instability
(5) combined stresses and theories of failure
(6) fatigue
(7) brittle fracture
(8) creep
(9) temperature effects
(10) radiation effects
(11) effects of fabrication methods
(12) economic considerations.

Amongst the various factors mentioned above, the first two, viz., material selection and corrosion prevention and protective coatings have been discussed in Chapter 2. The remaining factors are considered in this chapter.

3.2 STRESSES CREATED DUE TO STATIC AND DYNAMIC LOADS

3.2.1 Stresses Due to Static Loads

3.2.1.1 Direct Stresses

Tension or compression $f = \frac{W}{A}$ (3.1)

Shear in single plane $f_s = \frac{W}{A}$ (3.2)

Shear in two planes $f_s = \frac{W}{2A}$ (3.3)

Bearing or crushing stress $f_b = \frac{\text{force of crushing}}{\text{area of contact}}$ (3.4)

where, W — applied load

A — area of cross-section

f — tensile or compression stress

f_5 — shear stress

f_b — stress in bearing.

3.2.1.2 Strains

Tensile strain $\rho_t = \frac{f_t}{E}$ (3.5)

Shear strain $\rho_s = \frac{f_j}{G}$ (3.6)

where, $\rho_t\ \rho_s$ — strains

E — modulus of direct elasticity

G — modulus of rigidity

3.2.1.3 Thermal Stresses

$$E = \frac{\text{Stress}}{\text{Strain}} = \text{modulus of elasticity} \tag{3.7a}$$

where, $\text{Strain} = \Delta l/L = \frac{\text{increase/decrease in length}}{\text{total length}}$ (3.7b)

$$\Delta l = L\ \alpha\ \Delta t \tag{3.7c}$$

Therefore, $f = E\ \alpha\ \Delta t$ = thermal stress (3.8)

where, Δl — increase in length due to increase in temperature

L — original length
α — coefficient of linear expansion
Δt— increase in temperature
f — stress created if the expansion due to temperature rise is prevented. Thermal stresses are the result of material being subjected to a higher or lower temperature than the ambient prevailing temperature, at which, components have been assembled.

3.2.1.4 Stresses Caused by Bending

Bending is caused by forces acting normal to the axis of the beam. In such cases the distribution of the stress on the cross-section of the beam is not uniform.

A relation between the moment of resistance and the stress can be written as

$$\frac{M}{I} = \frac{f}{y} \quad \text{or} \quad M = \frac{f\,I}{y} = fZ \tag{3.9}$$

where, M— movement of resistance of the cross-section of the beam being considered (bending moment) (see Appendix B)
I — second moment of area of cross-section about the neutral axis (see Appendix A)
f — tensile or compressive bending (flextural) stress at a distance V from the neutral axis
y — distance from the neutral axis to the point where the stress f is to be determined
Z — $\frac{I}{y}$ is known as section modulus (see Appendix A for Z of various sections).
(The shaft of a paddle dryer could be one example.)

3.2.1.5 Deflection

Forces acting normal by to the axis of beam cause deflection of beam in addition to bending. Some cases of beams with loads and deflections caused are indicated in Appendix B.

3.2.1.6 Stresses Caused by Torsion

Components subjected to torque, are exerted by torsional shear stresses, which act tangentially over the cross-section. The relation between torque and shear stress for circular section is as follows:

$$\frac{T}{I_p} = \frac{f_s}{r} = \frac{G\theta}{l} \tag{3.10}$$

$$T = \frac{f_s I_p}{r} = f_s Z_p \tag{3.11}$$

where, T — torque applied
I_p — polar second moment of area
r — radius of circular section at which shear stress is to be determined

G — modulus of rigidity
θ — angle of twist over the length L
l — length of section under torsion

Z_p — polar modulus of section = $\dfrac{\pi d^3}{16}$

f_s — shear stress
d — diameter of shaft.

The maximum shear stress is at the outer radius of the circular section.

(For example, a typical stress on shaft in an agitated vesse.l)

3.2.1.7 Stresses in Struts

Components which are under longitudinal (axial) compression are known as struts.

Short struts or columns (slenderness ratio, for steel, < 30)

$$P = f_c A \tag{3.12}$$

where, P — external load
f_c — compressive stress
A — area of cross-section normal to the load.

In these struts the slenderless ratio is about 10

where, slenderness ratio = $\dfrac{\text{(maximum unsupported length)}}{\text{radius of gyration}} = l_e/k$

where, l_e — effective length of strut
k — least radius of gyration of cross-section.

Long struts (slenderness ratio > 120)

$$P = a\frac{\pi^2 EI}{l_e^2} \tag{3.13}$$

where, a — a constant depending on the conditions of restrain of strut
P — external load
E — modulus of elasticity
I — second moment of area of cross-section
l_e — effective length of strut.

Intermediate Struts

Slenderness ratio varies between 30 — 120 for steel

Rankine-Gordon formula

$$P = \frac{f_c A}{l + a\left(\dfrac{l_e}{k}\right)^2} \tag{3.14}$$

Johnson's parabolic formula

$$P = f_c A\left[1 - b\left(\frac{l_e}{k}\right)^2\right] \tag{3.15}$$

a and b are constants depending on the conditions of the load.

3.2.1.8 Stresses in Flat Plates – Solid Circular Plate

(a) Uniformly loaded, edge freely supported. Two stresses are created in the plane of the plate. These are perpendicular to each other. At the surface of the centre of the plate, the stress is equal to

$$f = \frac{3pR^2}{8t^2}\left(\frac{3\mu + 1}{\mu}\right) \tag{3.16}$$

where, f — maximum stress at the centre surface

p — uniformly distributed load = force/area

R — radius of plate

t — thickness of plate

μ — Poisson's ratio.

(b) Uniformly loaded plate and fixed at the edges. The stress at the centre surface of the plate

$$f = \frac{3}{8}\frac{pR^2}{t^2}\left(\frac{\mu + 1}{\mu}\right) \tag{3.17}$$

stress at the circumference is

$$f = \frac{3}{4}\frac{pR^2}{t^2} \tag{3.18}$$

(c) *Perforated plate:* The maximum stress in a perforated plate is given by

$$f_{max} = \frac{\text{maximum stress in a solid plate}}{\text{ligament efficiency}} \tag{3.19}$$

The ligament efficiency depends on the arrangement of holes. For the pitch (p) and the hole (d) ligament

$$\text{efficiency} = \frac{p - d}{p} \tag{3.20}$$

(For example, tube sheet of a heat exchanger, plate on an agitated nutsche fitter.)

(d) *Rectangular plate* Uniformly loaded and supported at its perimeter maximum stress

$$f_{max} = \frac{pb^2}{2t^2}\left(\frac{1}{1+\frac{b^2}{a^2}}\right) \tag{3.21}$$

where, p — uniformly distributed load

t — thickness of plate

a, b — two sides of the plate.

(For example, enclosures for furnace, bottom plate of a rectangular tank.)

3.2.1.9 Thin Cylinder Under Internal Pressure

Stresses are produced in three directions due to internal pressure.

$$\text{Circumferential or hoop stress } f_p = \frac{pD}{2t} \tag{3.22}$$

$$\text{Longitudinal or axial stress } f_a = \frac{pD}{4t} \tag{3.23}$$

The radial stress is small and can be neglected

where, p — internal pressure

D — internal diameter

t — thickness.

Thin spheried shell under internal pressure: In this case both the stresses are equal.

$$f_p = f_a = \frac{pD}{4t} \tag{3.24}$$

Thick cylinder under internal pressure: The stresses in the radial and circumferential direction vary over the cross-section. The axial stress is constant

$$f_R = A - \frac{B}{R^2} \tag{3.25}$$

$$f_p = A + \frac{B}{R^2} \tag{3.26}$$

$$f_a = \frac{pR_1^2}{(R_2^2 - R_1^2)} \tag{3.27}$$

where, A, B — constants

R — any radius

p — internal pressure

R_1, R_2 — internal and external radii respectively.

(For example, high pressure reactors.)

3.2.1.10 Thick Spherical Shell

Radial stress is given by

$$f_R = A - \frac{R}{D^3} \tag{3.28}$$

Circumferential or hoop stress is given by

$$f_p = A + \frac{B}{2D^3} \tag{3.29}$$

where, A, B — constants

D — diameter of shell.

The radial stress at internal surface of thick cylinder or sphere is equal to the internal operating pressure.

3.2.2 Dynamic Stresses

Stresses produced by variable loads changing in magnitude and direction are termed as dynamic stresses. These may be divided into two groups: (a) those produced by external forces, (b) those resulting from the inertia of the mass of the component.

3.2.2.1 Stresses in Rotating Rims and Disks

(a) *Rotating Rim*

Circumferential or hoop stress

$$f = \rho\omega^2 r^2 \tag{3.30}$$

where, m — mass, kg/m^3

ω — angular velocity of rim

r — radius at which the stress is determined.

(b) *Rotating Disk*

Circumferential Stress

$$f_y = \frac{\rho\omega^2}{8\mu}\left\{(3\mu + 1) f_0^2 - (\mu + 3) r^2\right\} \tag{3.31}$$

Radial stress

$$f_x = \frac{\rho\omega^2}{8\mu}\left\{(3\mu + 1)\,(r_0^2 - r^2)\right\} \tag{3.32}$$

where μ — Poisson's Ratio

r_0 — maximum radius

r — radius at which the stress is determined

ρ — density of the disk material

ω — angular velocity of disk.

The maximum values of f_y and f_x will be at $r_0 = 0$, i.e. at the centre of the disk.

(For example, certrifuges high shear dynamic mixers.)

3.2.3 Impact Stresses

If a moving body strikes another body the second body is subjected to an impact, which is equal to the kinetic energy of the moving body.

In the case of a rod when a weight W falling from a height h strikes the rod, the stress in the rod is:

$$f = \frac{W}{A}\left(1 + \frac{2hEA}{WL}\right) \tag{3.33}$$

where, A — area of cross-section of rod

E — modulus of elasticity

L — length of rod.

(For example, hammer mill, pulverizers.)

3.2.4 Compound Stresses

The stress resulting from combined action of several stresses acting simultaneously is called the resultant stress. It is possible to resolve this stress acting at a point in the material in three mutually perpendicular planes, called principal planes across each of which only one single normal stress, either tensile or compressive acts.

If at any point in the material with two-dimensional stress pattern, two perpendicular normal stresses f_x and f_y and shear stress f_s are acting then the principal stresses are given by

$$f_{RN} = \frac{1}{2}\left(f_x + f_y\right) \pm \frac{1}{2}\left[\left(f_x - f_y\right)^2 + 4f_s^2\right]^{1/2} \tag{3.34}$$

The +ve sign gives one principal stress and the –ve sign gives the second principal stress. The maximum shear stress is given by

$$f_{RS} = \frac{1}{2}\left\{\left(f_x + f_y\right)^2 + 4f_s^2\right\}^{1/2} \tag{3.35}$$

3.3 DESIGN STRESS

So far, several cases of stress created due to loads are considered. These loads are either constant, impact or alternating. To ensure adequacy of design it is necessary to assess the correct stress distribution in the various components. These stresses result in deformations which may interfere with the functional operation of the component. Sufficient rigidity must be incorporated in the design of the part to restrict the amount of deformation to a permissible value. The most widely used criterion to restrict deformation is to maintain the induced stresses within the elastic region of the material of construction in order to avoid plastic deformation resulting from the stresses exceeding the yield point. However, in special cases where plastic deformation is limited to small regions in the neighbourhood of structural discontinuities or stress raisers, failure will not occur provided that the material remains sufficiently ductile to accommodate these deformations without rupture. In such cases it is, therefore, of utmost importance to ensure that the ductility of the original material is not lost due to fabrication process or during service duration. The stresses created must be limited to a permissible value that is accepted as safe for a particular material and its applications. This value of the stress is known as a design stress or permissible or allowable stress. It is controlled by a number of factors such as, the accuracy with which the loads can be estimated, the reliability of the stresses computed, the uniformity of the material, the hazard if failure occurs, and other considerations like local stress concentrations, fatigue, etc.

In the case of ductile materials, where failure can be expected to occur as a result of plastic deformation, the design stress is usually obtained by dividing the yield stress of the material by a factor of safety.

$$\text{Design stress} = \frac{\text{Yield stress}}{\text{Factor of safety}}$$

Brittle materials are characterized by remaining elastic up to fracture. Such materials fail by fracture with little or no deformation. The design stress, therefore, is to be based on the ultimate strength.

$$\text{Design stress} = \frac{\text{Ultimate stress}}{\text{Factor of safety}}$$

Some values of the factor of safety for ductile and brittle materials are given in Table 3.1.

Table 3.1 Factor of Safety

Kind of Load	*Factor of Safety*		
	Ductile Material F.S. Based on Ultimate Strength	*Ductile Material F.S. Based on Yield*	*Brittle Material F.S. Based on Ultimate Strength*
1. Dead or steady load	3 – 4	1.5 – 2	5 – 6
2. Repeated load applied gradually but not reversed	6	3	7 – 8
3. Repeated load applied gradually and reversed	8	4	10 – 12
4. Load applied with medium or heavy shock	10 – 15	5 – 7	15 – 20

For certain materials there is no well defined stress point at which yielding occurs. For such materials, instead of yield stress, a proof stress is defined (Section 2.1.7). Similarly, due to cyclic loading conditions, a fatigue failure is likely to occur (Section 3.6). The design stress under such conditions is obtained by:

$$\text{Design stress} = \frac{\text{Average stress for 1\% creep in 1,00,000 hours}}{\text{Factor of safety}}$$

In the case of thermoplastics, as has been pointed out under material properties, the behaviour of such materials is not elastic and the elongations are relatively large. The design stress in these cases is based on creep data, and the deformations are to be less than 3%. The design stress should be determined by taking the lesser of the following two criteria:

(1) Average stress required to produce rupture in 10^5 hours at design temperature, divided by a factor of safety (usually between 1.5 to 2).

(2) Average stress required to produce total creep strain of 2 to 3% in 10^5 hours at design temperature.

3.4 ELASTIC INSTABILITY

This is caused due to sufficient stiffness or rigidity in a component or structure which is subjected to compression, bending, torsion or a combination of such loading conditions. The applied loads cause the body to buckle, wrinkle or otherwise collapse. Design of a component in these cases is generally executed by limiting the stresses, as well as the geometrical shape, deflection or twist.

In case of columns, tubes, plates or shells, the design is based both or critical buckling load and the permissible or design stress. Buckling may be: (1) in the form of bending and deflection of the component as in the case of medium and long columns, (2) local buckling or wrinkling as in the case of a tube or cylindrical shell with axial load and (3) deformation of shape or collapse as in the case of a shell with external pressure.

(1) In case of long columns with axial load failure occurs by buckling. The critical buckling load is given by Eq. (3.13), and the critical stress is given by

$$f_{cr} = \frac{K_1 E}{2} \cdot \frac{A}{l^2} \tag{3.36}$$

(2) In case of tubes or thin cylindrical shells with axial load, buckling takes place with formation of axial or circumferential corrugations on the surface. The critical buckling load can be calculated by Eq. (3.13) as in the case of long columns. The critical stress is given by

$$f_{cr} = \frac{K_2 E}{\sqrt{3(1-\mu)^2}} \frac{t}{(D/2)} \tag{3.37}$$

(For example, reactor jacket having steam at say 15 bar.)

(3) In case of cylindrical shells, with external pressure, the stability of the shell depends on the dimensions of the shell, the manner in which the ends are fixed and the material properties. In the case of long cylinders ($L > 8D$), the stability of portions which are far from end is practically not affected by the manner in which ends are fixed. The critical pressure can be determined from the equation

$$p_{cr} = \frac{E}{2(1-\mu^2)}\left(\frac{l}{D/2}\right)^3 \tag{3.38}$$

and the critical stress is given by

$$f_{cr} = \frac{p_{cr} D}{2t} \tag{3.39}$$

Buckling pressure in case of long cylinders is given by Eqs 3.13 and 3.14.

In all the above cases the critical stress should not exceed the yield point under compression of the material. To avoid failure and assure safe design, a factor of safety must be used for p_{cr} and f_{cr}. In the above equations:

- f_{cr} — critical compressive stress
- K_1 — end condition factor
- A — area of cross-section
- I — length of column
- E — modulus of elasticity
- K_2 — factor to account for initial imperfections of the cylinder
- t — thickness of cylinder
- D — diameter of thin cylinder
- p_{cr} — critical pressure
- μ — Poisson's ratio.

3.5 COMBINED STRESSES AND THEORIES OF FAILURE

For components which are subjected to combined stresses, a procedure will have to be established which will relate the resultant stress or principal stress to the yield stress obtained from a test of the material. This procedure stipulates yielding of the material and consequent failure, if the maximum value of the stress or strain created by the combination of stresses, exceeds the conditions at the yield point. Various theories for the failure of the material have been suggested and are made applicable according to the nature of the stresses and the type of material (ductile or brittle). Some of these theories are described in the following:

3.5.1 The Maximum Normal Stress Theory or Rankine Theory

According to this theory, failure of the component will take place if any of the resultant normal stresses reaches the value of the stress at the yield point, in simple tension. For ductile materials, the tension and compression properties are the same and the direction of the stresses (i.e., tensile or compressive) will, therefore, not affect the above condition. The condition of failure is

$$f_{RN} = f_y \tag{3.40}$$

where, f_{RN} — maximum resultant normal stress (as per Eq. 3.34)

f_y — yield stress in tension test.

3.5.2 The Maximum Shear Stress Theory (Guest Theory)

According to this theory failure will take place, if the maximum resultant shear stress in a component, subjected to combined stresses, reaches the value of the shear stress in simple tension at the yield point. The condition of failure is

$$f_{RS} = \frac{1}{2} f_y \tag{3.41}$$

where, f_{RS} — resultant shear stress.

3.5.3 The Maximum Strain Theory (Saint-Venant's Theory)

According to this theory, failure will occur, if the maximum strain created due to combined stresses, reaches the strain created in simple tension at the yield point. The condition of failure for a three-dimensional stress pattern is

$$\frac{f_{RN_1}}{E} - \frac{\mu}{E}\left(f_{RN_2} + f_{RN_3}\right) = \frac{f_y}{E} \tag{3.42}$$

where, $f_{RN_1}, f_{RN_2}, f_{RN_3}$, – Resultant normal stresses in three directions.

where, μ — Poisson's ratio
E — modulus of elasticity.

3.5.4 Strain Energy Theory

This theory stipulates that failure will occur when the strain energy stored, under the action of the combined stresses, reaches the strain energy at yield point in simple tension or compression,

i.e.
$$\frac{1}{2E}\left(f_{RN_1}^2 + f_{RN_3}^2 + f_{RN_3}^2\right) - \frac{\mu}{E}\left(f_{RN_1} f_{RN_2} + f_{RN_2} f_{RN_3} + f_{RN_3} f_{RN_1}\right) = \frac{f_y}{E} \tag{3.43}$$

3.5.5 Distortion Energy Theory or Shear Energy Theory (Hencky–Von Mises Theory)

According to this theory, failure of the material subjected to combined stresses, will take place when the elastic strain energy required for distortion of the material reaches the energy required to produce yielding under simple tension or compression. This gives for a three-dimensional stress pattern:

$$f_y = \frac{1}{\sqrt{2}}\left\{(f_{RN_1} - f_{RN_2})^2 + (f_{RN_2} - f_{RN_3})^2 + (f_{RN_1} - f_{RN_3})^2\right\}^{1/2} \tag{3.44}$$

For a two-dimensional stress pattern f_{RN_3} will be zero. The equation for two-dimensional stress pattern can also be written in terms of the actual stresses

$$f_y = \left\{(f_x)^2 + (f_y)^2 - f_x f_y + 3f_s^2\right\}^{1/2} \tag{3.45}$$

In making an appropriate choice from among the above theories it is necessary to consider the relative magnitude of the resultant normal and shear stresses in the components as also the properties of the material. When the resultant normal stresses are either all positive (tensile) or all negative (compressive) the application of maximum normal stress theory is satisfactory. In casts where some stresses are positive and some negative the maximum shear stress theory or shear energy theory should be used.

In all the above theories, the deformation and stress have to be restricted to elastic region, and hence the value of yield stress f_y, will have to be divided by a factor of safety.

3.6 FATIGUE

In arriving at a safe value of the design stress, one of the factors that need consideration is the nature of any cyclic load, which induces alternating stresses or fluctuating stresses. Failure due to fatigue is due to a slow but progressive enlargement of an initial crack subjected to cyclic load. It is, therefore, necessary to ascertain the greatest stress or range of stress which can be applied to a material for an unlimited number of times without causing failure. This is known as fatigue limit or endurance limit. The ratio of the endurance limit for reversed stresses to ultimate static stress is known as endurance ratio. The usual values of the endurance ratio are:

Steel with hardness less than 400 BHN (at 10^6 cycles)	0.5
Cast iron and cast steel (at 10^6 cycles)	0.4
Aluminium and magnesium (at 5×10^8 cycles)	0.3

Various components or machine elements have usually grooves, holes, fillets and notches. Such changes in shape interrupt the lines of stress and create concentrations of stress which cause reductions in the fatigue strengths of these components. These effects have to be taken into account by using appropriate stress concentration factors. Fatigue failures can, therefore, be avoided by providing smooth and gradual changes in shape, and eliminating sharp corners in a groove. The fatigue can be experienced more in a batch plant than in a continuously operated plant. Equipment in batch plant are subjected to cyclic temperature change.

Some factors which influence fatigue are: (a) material and material factors, (b) type of loading, (c) size of the number, (d) surface finish, (e) stress raisers, (f) surface stressing, (g) corrosion, (h) temperature.

3.7 BRITTLE FRACTURE

Ductile materials are likely to fail due to brittle fracture under the following conditions:

(a) Presence of a defect of sufficient size or notch

(b) High localised stresses in the vicinity of the notch

(c) Operation at sufficiently low temperature

(d) Wrong selection or treatment of materials.

Due to a notch in a ductile material the material surrounding the notch becomes brittle. The phenomenon is known as 'notch brittleness' for example, carbon steels are brittle below –30°C. Similarly, due to low temperatures the material becomes brittle. In all such cases, these materials should be treated as brittle materials and the design stress should be based on a higher factor of safety.

Brittle fracture can occur at levels of nominal stress below yield point, or even at ordinary permissible levels of stress. It can occur suddenly without prior indication of any deformation.

3.8 CREEP

Certain materials like thermoplastics and lead at room temperature and metals at high temperature, exhibit a slow but continuous plastic deformation which eventually results in failure. This process is termed as creep. It is a function of stress, temperature and time. With metals, creep is generally associated with high temperatures. It is essential to know, whether the rate of creep is compatible with the working life of the component. A maximum plastic deformation at the end of a certain period is selected to satisfy the working conditions and the corresponding stress may be considered as safe. Assuming a safety factor, the design stress may be found. In carbon steels up to 350°C, creep is not significant. Above 400°C temperature the design stress is reduced. In the creep range even a small increase in temperature often results in a large reduction in creep strength and consequently in the design stress.

3.9 TEMPERATURE EFFECTS

The effect of temperature on continuous deformation in the form of creep, has already been referred to. In general, with increase in temperature, there is a reduction in ultimate strength, modulus of elasticity, and hardness. Materials expand with increase in temperature. These changes must be taken into account in designing components at higher temperatures. Stresses are also created, if the expansion due to temperature is prevented, or in case the temperatures at different points in a component are not the same.

3.10 RADIATION EFFECTS

Nuclear reactor vessels are subject to material irradiation, due to neutron bombardment from the core. Neutrons are classified as fast neutrons and thermal neutrons. Fast neutrons cause damage by dislocation or displacement of the atomic structure of the metal, whereas the effect of thermal neutrons is one of transmutation of trace impurities that can substantially change the properties of the material. Typical changes in such cases are a marked increase in yield point, a smaller increase in tensile strength, and decrease in ductility. These factors are taken into account by a suitable factor of safety.

3.11 EFFECTS OF FABRICATION METHODS

Fabrication process such as casting, hot and cold rolling, sheet forming and welding give rise to stresses or changes, in properties of material. For instance, in cold forming operations, due to the stretching of the material beyond its yield point the material hardens and suffers a loss of ductility and toughness. This phenomenon is called *strain* or *work hardening*. The work of streching is not entirely transformed into heat, but part of it is retained in the form of strain energy. In joints made by welding certain amount of stress concentrations takes place, due to uneven temperature rise, structure of weld metal, and welding defects such as porosity, slag inclusions and shrinkage cracks. Work hardening and residual stresses may be relieved by heat treatments like annealing and normalising.

3.12 ECONOMIC CONSIDERATIONS

Mahajani and Mokashi (2005) in their book *Chemical Project Economics* have presented a very comprehensive treatment of cost estimation and economic evaluation. They have also presented in Chapter 9 on *Interest*, how the concept of capitalized cost can be used to select appropriate material of construction from various options available.

References and Recommended Readings

Barve, R.S., *Strength of Materials*, B.I. Bombay, (1970).

Brooks, W.H., *Strength and Elasticity of Materials*, Vols. I-IV, Asia Publishing House, Bombay (1963).

Faculty of Mechanical Engineering, PSG College, Coimbatore, *Design Data*, Revised Edition, DPV Printers and Publishers, Coimbatore (1978).

Faupel, J.H., *Engineering Design*, John Wiley and Sons, New York (1964).

Huckert, J., *Engineering Tables* (ASME Handbook), McGraw-Hill, New York (1956).

Jenson, A., *Applied Strength of Materials*, Tata McGraw-Hill, New Delhi (1976).

Mahadevan, K., Reddy, K.B., and Hunashikatti, H.G., *Design Data Handbook for Mechanical Engineers Ranges*, Bombay (1972).

Mahajani, V.V., and Mokashi, S.M., *Chemical Project Economics*, Macmillan India Ltd., New Delhi (2005).

Redford, G.D., *Mechanical Engineering Design*, Macmillan, London (1966).

Ryder, G.H., *Strength of Materials*, Macmillan, London (1973).

Timoshenko, S., *Strength of Materials*, Vols. I and II, McGraw-Hill, New York (1951).

4

Chapter

Design of Machine Elements

4.1 INTRODUCTION

A variety of machine elements usually form a part of the equipment used in the chemical industry. Such elements are shafts, keys, couplings, bearings, bolts, etc. A detailed account of the types, capacity and design of these elements are available in standard textbooks on the subject. It is proposed to give here an outline of the design procedure adopted for each element. In the overall design of chemical equipment, discussed in subsequent chapters, the design procedure for machine elements suggested here may be referred to wherever such machine elements are provided.

4.2 SHAFTS

These are used for transmission of mechanical power with the help of rotary motion. They form either a direct component of an equipment or are merely used for transmitting rotary motion from a prime mover to an equipment. The first type of shaft is known as machine shaft, while the second type is known as counter shaft or line-shaft. The majority of rotating shafts carry a fairly steady torque and bending moment, the loads remaining fixed in space in both direction and magnitude.

4.2.1 Strength under Steady Loads

The average torque, T_{avg}, transmitted by the shaft is assessed from the power

$$T_{avg} = \frac{p}{2\pi N} \tag{4.1}$$

where, p — power, W

N — rps, revolution per second (speed)

T_{avg} — N.m

The maximum torque (T_{max}) is taken as 20–25% more than the average torque for uniform load condition. However, for solid handling machines such as ball mills, ribbon blenders, design torque is 50 to 100% more than the maximum starting torque. Please note that devices like the VFD (variable frequency drive), fluid coupling can significantly reduce the starting torque by gradually increasing speed.

The maximum value of the bending moment is determined from the loads normal to the axis of the shaft, by drawing a bending moment diagram. The resultant normal and shear stresses are determined by combining the maximum torque and bending moment, on the basis of the maximum normal stress and maximum shear stress (see Eqs 3.34 and 3.35).

$$f_N = \frac{1}{2Z}\left\{M + (M^2 + T^2)^{1/2}\right\} \tag{4.2}$$

$$f_s = \frac{1}{Z_p}(M^2 + T^2)^{1/2} \tag{4.3}$$

where, f_N — resultant normal stress

f_s — resultant shear stress

Z — modulus of section of the shaft

Z_p — polar modulus of section of the shaft

M and T — maximum bending moment and torque respectively.

Shafts are usually circular in cross-section. In some cases hollow shafts are used. The diameters of shafts are determined by choosing permissible values for shear stresses, for materials which are ductile, in Eq. 4.3 and permissible values of maximum normal stresses, for brittle materials, in Eq. 4.2.

4.2.2 Strength under Suddenly Applied Loads and Fluctuating Loads

When the shafts have to withstand suddenly applied loads, or fluctuating loads the values of the maximum torque and bending moment are increased by applying combined shock and fatigue factors; namely, K_m and K_t. Values of these factors vary between 1.5 and 3 (Table 4.1).

4.2.3 Deflection

It is necessary to control deflection of the shaft due to loads acting normal to the shaft. Determination of deflection in some standard cases of beams is indicated in Appendix B. The deflection in shafts should be limited to a maximum value of 0.02 mm at important positions such as bearings and 0.1 mm at mechanical seals along the shaft.

Table 4.1 Shock and Fatigue Factors

Nature of Loading	Value for K_m	Value for K_t
Stationary Shafts		
Gradually applied load	1.0	1.0
Rotating Shafts		
Suddenly applied load	1.5 to 2.0	1.5 to 2.0
Steadily or gradually applied loads	1.5	1.0
Suddenly applied loads, minor shocks only	1.5 to 2.0	1.0 to 1.5
Suddenly applied loads, heavy shocks	2.0 to 3.0	1.5 to 3.0

4.2.4 Slope and Twisting of Shaft

The deflection of the shaft creates a slope, while the torque creats an angular twist. The slope of the shaft at the supporting bearing should be limited to 4 minutes and the angular twist should be limited to 1° in a shaft of length of 20 times the diameter (1 minute = 1/60 of a degree).

4.2.5 Shaft Speed

There are certain limitations on the speed range within which the shaft should rotate. At certain speeds, the shaft tends to whirl or vibrate violently. Such vibrations may be in resonance with the natural vibration frequency of the elastic system. At these speeds, known as critical speeds, the vibrations can cause complete and rapid failure of the shaft and equipment. The critical speed corresponds to the speed equal to natural frequency.

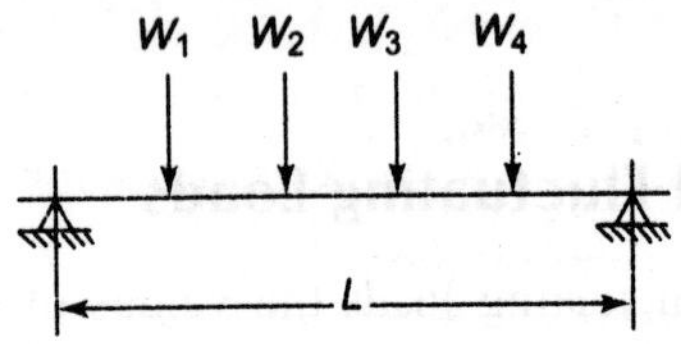

Fig. 4.1 Shaft Freely Supported

The determination of the critical speed is based on the static deflections of the shaft, created due to positions of impeller load (weight) and supporting bearings, as shown above. The critical speed N_c is given by the Rayleigh method (Shigley, 1961):

$$N_c = (60/(2 \times \pi))\{g\Sigma W_i\delta_i/\Sigma W_i\delta_i^2\}^{1/2}$$

$$= 946 \times \{\Sigma W_i\delta_i/\Sigma W_i\delta_i^2\}^{1/2} \quad (4.4)$$

where, I varies from 1 to n number of load points, i.e. W_1, W_2, W_3, etc. The other popular approach is by Dunkerly method wherein N_c is given by:

$$N_c = 946/\{\delta_1 + \delta_2 + \delta_3 + \delta_s/1.25\}^{1/2} \quad (4.5)$$

where, N_c — critical speed in rpm

W — load in kg

δ — deflection in mm due to load

δ_s — deflection of the shaft, mm due to its weight which is taken as uniformity distributed load, $w = W/L$

g — acceleration due to gravity = 9807 mm/s^2

The application of this is given in Chapter 14 on Mixers.

The operating speed of the shaft should not exceed 60% of the critical speed. The above situation (Fig. 4.1) arises in the case of ball mills, ribbon blender and rotary dryer wherein the shaft is supported at the both ends with bearing support. Even an agitator when supported at the bottom is under this class.

4.2.6 Effect of Keyway

A shaft is rigidly attached to hub or a machine part, by means of a key (Fig. 4.2). A keyway, in the form of an axial groove is made on the surface of the shaft, to accommodate the key. A keyway reduces both the strength and the rigidity of a shaft. This is taken into account by introducing a factor, similar to a stress concentration factor. The design of the shaft in such cases is based on a reduced value of the permissible stress, which is usually taken as 75 per cent of the normal permissible stress.

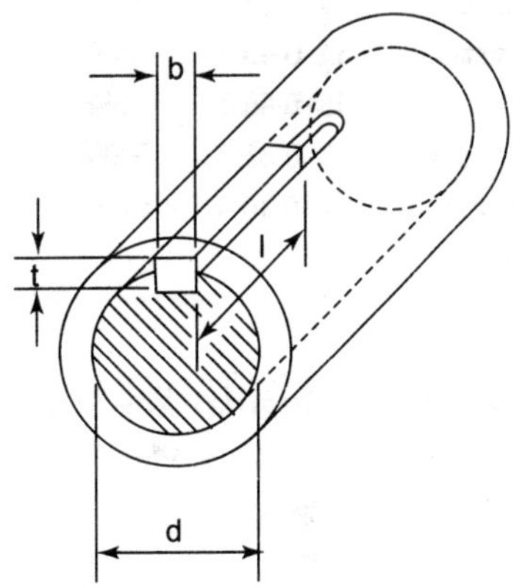

Fig. 4.2 Rectangular Key Fixed in Shaft and Hub

4.2.7 Materials

Ordinary shafts are fabricated from medium carbon steel having 0.15 to 0.4 per cent carbon. When greater strength is required shafts are made of alloy steels, the most common alloys being nickel, nickel-chromium and chrome vanadium steels. When resistance to corrosion is required shafts may be made of copper alloys or steel shafts may be lined with lead, glass or other materials.

Steel shafts are cold rolled or hot rolled. Most commercial shafting after being hot rolled is turned and polished. Alloy steel shafts are always heat treated. Properties of shafting materials are given in Table 4.2.

4.3 KEYS AND PINS

Keys are used to attach a machine part to a shaft and thereby transmit the torques preventing relative motion (both rotary and axial) between the two components. Some type of keys allow axial motion. According to the shape and function, keys are classified as square, rectangular, dovetailed, woodruff, round, flat, saddle, jib-headed, etc. Pins are cylindrical in shape and are either straight or tapered. They may be used either for locating the relative positions of two parts or for fixing two parts together.

Table 4.2 Properties of Shaft Material

Material (Steel)	*Percentage Carbon*	*Ultimate Strength kg/cm²*			*Elastic Limit kg/cm²*			*Percentage Elongation*
		Tension	*Compression*	*Shear*	*Tension*	*Compression*	*Shear*	
Commercial cold rolled	0.10–0.25	4920	4920	2460	2460	2460	1250	35
Commercial	0.10–0.25	4200	4200	2100	2100	2100	1050	35
turned	0.15–0.25	4600	4600	2300	2500	2500	1150	26
Hot rolled	0.25–0.35	4920	4920	2460	2800	2800	1230	24
or forged	0.34–0.45	5300	5300	2650	3200	3200	1325	22
	0.45–0.55	5640	5640	2820	3520	3520	1410	20
3½% Nickel	0.15–0.25	6000	6000	3000	3900	3900	1500	26
Chrome vanadium	0.25–0.35	6320	6320	3160	4220	4220	1580	25

Square or rectangular keys The cross-section of the key is square or rectangular, with a straight length. When the torque is transmitted by the key, the tangential force acts as a shearing and crushing force on the key. This can be expressed as in (Fig. 4.2):

$$\frac{T_{\max}}{\frac{d}{2}} = b.l.f_s = \frac{lt}{2} f_c \tag{4.6}$$

For the shaft $T_{\max} = \frac{\pi d^3}{16} f_s'$ (4.7)

where, $T_{\max}$ — maximum torque transmitted

d — diameter of shaft

l — length of key

b — width of key

t — thickness of key

f_s — shear stress in the key

f_c — stress in crushing of key

f_s' — shear stress in the shaft.

If the material for shaft and key is similar, then

$$f_s = f_s'$$

and

$$lb = \frac{\pi d^3}{8} \tag{4.8}$$

If l is taken as equal to 1.5 d, then

$$b \simeq d/4$$

For a rectangular key, the value of t may be taken as $(1/16)d$.

4.4 COUPLINGS

These are used to connect two shafts which are in line. Couplings are mainly of two types, namely, rigid and flexible. According to their construction they are known as sleeve, butt-muff, flange, clamp, oldham, or flexible disc couplings. Some of the more common types are considered here.

4.4.1 Rigid Flange Coupling

It consists to two flanges, one for each shaft, fixed by keys. Certain dimensions are based on practical and safety consideration. Figure 4.3 shows a rigid coupling, in which

$R = 1.5\ d$; $D = 2d$; $B = 1.5$ to $2d$

Number of bolts $(n) = 3 + 0.2d$ (d in cm)

(In normal practice, choose nearest whole even number.) However, in some exceptional cases odd number of bolts are seen as given in Fig. 4.3.

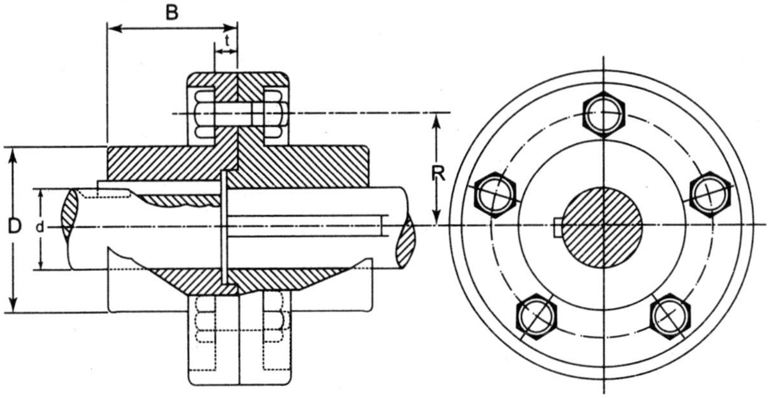

Fig. 4.3 Rigid Coupling

The bolts are under shearing and crushing stresses.

$$\text{Force on the bolt } (F) = \frac{T_{max}}{R} \tag{4.9}$$

$$F = \frac{\pi\delta^2}{4} f_s \times n = \delta\, t\, f_c \times n \tag{4.10}$$

where, δ — mean diameter of bolt

t — thickness of flange

$f_s f_c$ — shearing and crushing stresses.

n — number of bolts

Assuming permissible values of the stresses, the bolt diameter δ and the thickness of the flange t, can be calculated. It is usual to assume higher factor of safety for permissible stresses (f_s and f_c) to account for uneven loading, sudden application of load and bending moment.

The flange is usually made of cast iron, the stress in the hub and the portion of the flange can be checked by the following equations:

$$f_s\,(\text{hub}) = \frac{T_{max}}{\pi(D^4 - d^4)} \times 16D \tag{4.11}$$

$$f_s\,(\text{flange}) = \frac{T_{max}}{\pi D t \times D/2} \tag{4.12}$$

4.4.2 Split-Muff or Clamp Coupling

It consists of a sleeve generally made of cast iron split into two halves, which are clamped over the shaft by bolts (Fig. 4.4). This coupling is more suitable for vertical shafts, and is easy to dismantle. This transmission of torque is by pure friction which can be written in terms of shaft diameter and length, assuming uniform contact between shaft and sleeve.

$$T_{max} = \mu\pi d p \frac{l}{2} \times \frac{d}{2} \tag{4.13}$$

where, μ — coefficient of friction between shaft and sleeve

d — diameter of shaft

p — contact pressure

l — length of sleeve.

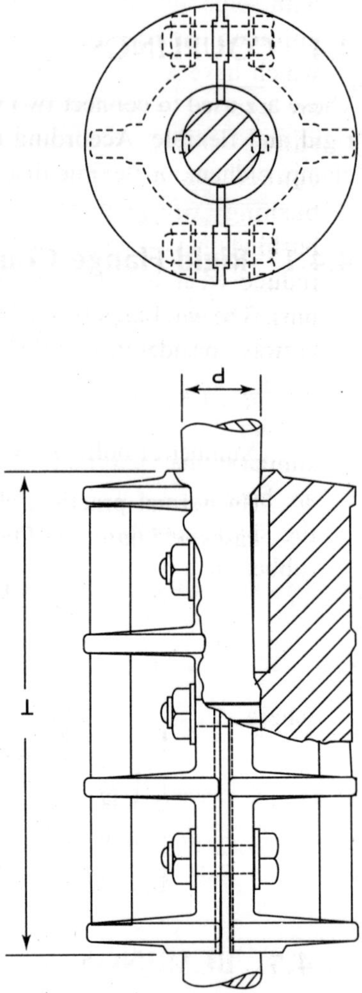

Fig. 4.4 Split-muff Coupling

$$\text{Force per bolt } (P) = \frac{p.d}{\left(\frac{n}{2}\right)}\frac{l}{2} = \frac{2T_{max}}{\pi\mu d\left(\frac{n}{2}\right)} \tag{4.14}$$

where, n — total number of bolts.

The number of bolts used are six for a shaft diameter of 50 mm and eight for larger diameters.

4.4.3 Flexible Coupling

This is essentially similar to a flange coupling with some modification to create flexibility. It is applicable particularly for connecting shafts which have a lack of alignment or shafts which are exerted with sudden changes of load. As shown in Fig. 4.5 the holes for the flange connecting pins are provided with rubber bushings fitting into the holes. These rubber bushings have a bronze bush cemented inside to reduce wear between the rubber bush and the pins. The thickness of the rubber bush is usually between 6 and 9 mm and that of the bronze bush is about 2 mm.

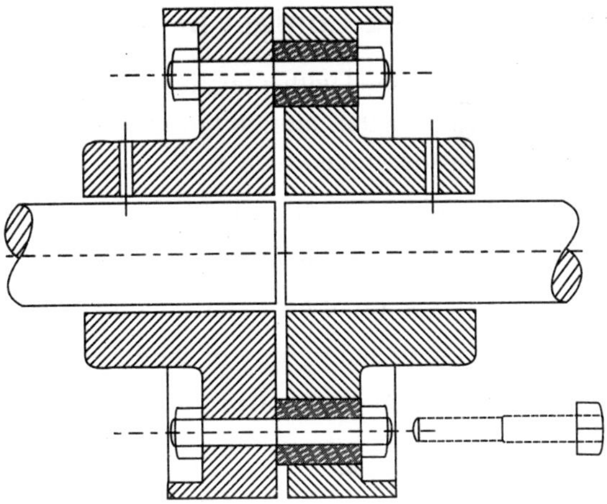

Fig. 4.5 Flexible Coupling

The design of this coupling is somewhat similar to that of a rigid flange coupling. The pins are however located at a radius of 2.5 to 3 times the shaft diameter. The bearing pressure on the rubber bush is given by

$$p_b = \frac{T_{max}}{R \times n} \times \frac{1}{d_B \times t_b} \tag{4.15}$$

where, T_{max} — maximum torque

R — radius at which the pins are located

n — number of pins

d_B — outside diameter of bush

t_b — length of bush

4.5 BEARINGS

A bearing is a machine component, which acts as a support for a moving part, having rotary, oscillating or sliding motion. Its function is mainly to support the load of the moving part, with minimum loss due to friction. A bearing which supports a load normal to the longitudinal axis is known as radial bearing or journal bearing. A portion of a rotating shaft which is supported by the bearing is known as a journal. If the load on the bearing acts in the axial direction, the bearing is known as thrust bearing.

If the contacting surfaces of the rotating member and the bearing cause a sliding action, the bearing is known as sliding contact bearing. If the action is of a rolling type the bearing is known as rolling contact bearing. In the first case, supporting contact is over a surface. In the latter case the supporting contact is on rollers or balls. In either case the friction between the rotating part and the bearing must be reduced by proper lubrication.

4.5.1 Sliding Contact Bearings

The bearing is lined with a relatively softer material than the shaft material. Such materials are brass, bronze, white metal, aluminium alloys, plastics, rubber, etc., (Table 4.3). The design is based on several considerations, such as

(a) Diameter, length and clearance
(b) Provision of sufficient lubrication
(c) Dissipation of heat generated by friction
(d) Strength and rigidity of shell, cap and bolts
(e) Minimum wear and provision for taking up wear.

Table 4.3 Bearing Materials

	Material	*Characteristics*	*Applications*
1.	Babbitt or white metal alloys (tin base)	Very expensive	Used for general purpose, resistant to corrosive effects of acids. I.C. engine bearings
2.	Lead base white metals	Low cost, lower strength and susceptibility, high coefficient of friction	General machinery purpose. Used at moderately high temperature. Load ranging from 10 kg/cm^2 to 22 kg/cm^2 at 60 m/min.
3.	Plastic bronze	Alloy of copper and lead, resistance to seizure, conformability and embedability. Less corrodible metal. Low coefficient of friction	Used in automobiles, locomotives and rolling mills where there are heavy loads, max. load carrying capacity is approx. 105 kg/cm^2 at 900 m/min
4.	Phosphor bronze	Good mechanical and anti-friction properties	All machines. The permissible load is 700 kg/cm^2. Max. rubbing speed 300 m/min.
5.	Gun metal	Chill casting gives a finer structure and is preferable to sand casting	Light loads

6.	Brass	Lower cost than babbitt	At high pressures
7.	Aluminium	Good corrosion, fatigue resistance and thermal conductivity, low embedability, high thermal expansion	Suitable for heavy load up to about 245 kg/cm^2 peak load and 140 kg/cm^2 continuous load at moderate speeds of 240 to 360 m/min
8.	Cast iron lubrication	Low friction, must have good lubrication	Cam shaft, light transmission. Max. speed 40 m per min., max. load 35 kg/cm^2
9.	Cadmium base bearing metals	Allow much higher operating temperatures than are safe with babbitt	Used for severe service in internal combustion engines
10.	Porous bearings	Oil impregnated bearings	Medium duty application in small size bearings. Where supply of lubricant is difficult, inadequate or infrequent. Rubbing speed from 22 to 450 m/min. with light load.
11.	Wood bearing (lignum vitae, rock maple or oak)	Low cost, self-lubricating	Conveyors
12.	Rubber	Low coefficient of rubbing friction	Hydraulic turbines, centrifugal and deep well pumps
13.	Nylon or metal filled Teflon	Extremely low coefficient of friction, resistance to corrosion	Light and medium load

4.5.1.1 Radial Bearing or Journal Bearing

(a) *The load* The load in radial bearing (Fig. 4.6) is supported on the projected bearing area given by $l \times d$

$$\text{Bearing pressure } p_b = \frac{P}{l \times d} \tag{4.16}$$

where, P — radial load

l — length of bearing

d — diameter of shaft.

The ration l/d varies between 1 and 3, and the permissible bearing pressure depends on the material and the type of bearing. Longer bearings have a lesser bearing pressure, but friction, torque and deflection increase. The clearance between the bearing and the journal (shaft) depends on the bearing pressure, rubbing velocity and materials of construction. In general, higher the rubbing velocity greater can be the clearance.

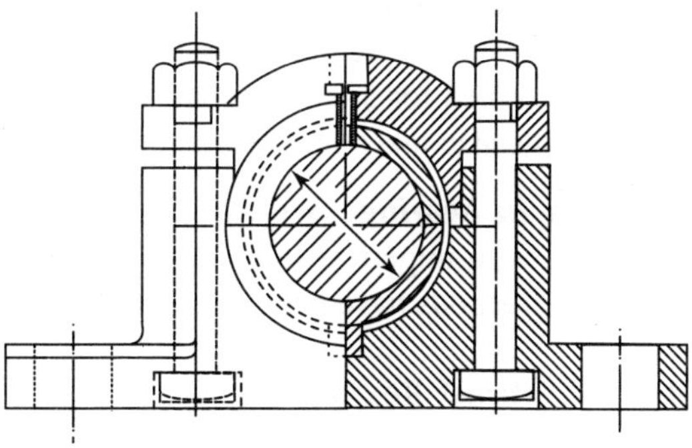

Fig. 4.6 Radial Pedestal Bearing

(b) *Lubrication* Oil admission to a bearing must be in the region of low pressure. Oil grooves cut on the surface of the lined material, help to distribute the oil evenly over the entire surface.

(c) *Dissipation of heat generation* The heat generated due to the friction between the journal (shaft) and the bearing depends on the coefficient of friction, load and the speed. The coefficient of friction is a function of several factors such as viscosity of the lubricant, bearing pressure, speed, clearance, etc. Heat generation can be written as

$$H/\text{minute} = \mu \times p_b \times d \times l \times \pi d N \tag{4.17}$$

where, μ — coefficient of friction

p_b — bearing pressure

l — diameter of shaft

N — rpm of shaft.

Heat is dissipated by flow of oil conduction to other parts of the machine. The temperature of the oil film should be between 28°C and 60°C. Lower temperature results in high operating viscosity and consequently high power loss. Higher temperatures may cause oil film breakdown, rapid vaporization and oxidation of the lubricant.

(d) *Strength and rigidity*

(i) *Shell:* The thickness of the brass or bronze shell is taken as $0.1d + 3$ mm (where d is the diameter of the shaft in mm). For large diameter bearings, the shells are made of cast iron with a layer of white metal or other soft material.

(ii) *Cap:* The covering cap of the bearing is not usually subjected to a heavy load. However, in some cases, the load acts on the cap. In such cases, the cap may be regarded as a simple beam loaded at the centre and supported by the holding down bolts. It is also desirable to check the deflection of the cap.

(iii) *Holding-down bolts:* The bolts used are for holding down the cap has to withstand a total load due to tightening of the cap in addition to the bearing load. Each bolt is, therefore, designed for a tension load which is 33 per cent greater than the normal load on each bolt. This also takes into account the uneven distribution of the load due to friction.

(e) *Wear in bearing* With perfect lubrication there should be no wear of the bearing surface. However, with imperfect lubrication, metal to metal contact causes a wear of the rubbing surfaces. Different methods have been suggested for take-up of wear.

4.5.1.2 Thrust Bearing

Such bearings (Fig. 4.7) operate with small clearance between the adjacent parts of the bearing and, therefore, adequate supply of oil between the rubbing surfaces is extremely important. The area of the bearing should be adequate to take the thrust load. Oil grooves reduce the actual area of contact.

$$\text{Load } W = \frac{\pi}{4}(D^2 - d^2)p_b$$

where, p_b — bearing pressure
D — diameter of thrust collar
d — diameter of shaft
W — thrust load.

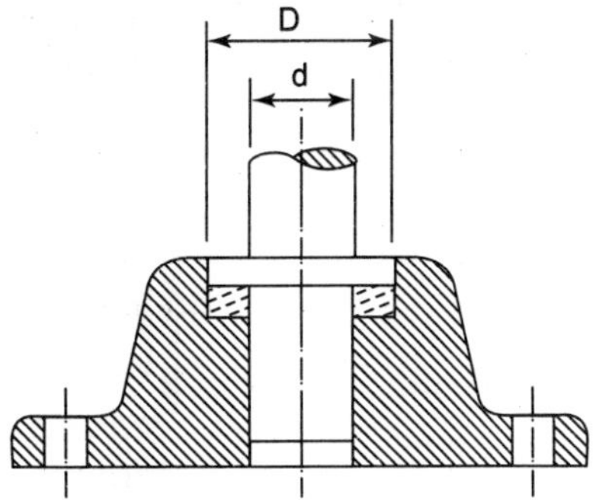

Fig. 4.7 Thrust Bearing

Heat generation in thrust bearing is given by

$$H/\text{minute} = \frac{2}{3}\mu W\left(\frac{D+d}{2}\right) \times 2\pi N \tag{4.18}$$

where, N — shaft speed in rpm
μ — coefficient of friction.

4.5.2 Bearings with Rolling Contact - Anti-friction Bearings

In sliding contact bearings there is considerable loss of friction, particularly when the oil film lubrication is not perfectly maintained. In rolling contact bearings friction is reduced. Ball bearings [Fig. 4.8(a)] have a lesser area of contact and have, therefore, limited load carrying capacity. For higher loads, roller bearings [Fig. 14.8(b)] with a roller length of 4 to 6 times the diameter are used. Such bearing may have cylindrical, conical or convex rollers.

Ball and roller bearings are standardized in respect of their capacity, dimensions, type, service conditions and life. In the selection of bearing it is usual to choose a bearing from standard tables available from the manufacturers. The equivalent load p_e, which can be sustained by the bearing is given by the following equation:

$$p_e = XVP + YW \tag{4.19}$$

where, X — radial factor
V — rotation factor
Y — thrust factor
P — radial load
W — axial load.

Values of factors X, V and Y are available in the ball and roller bearing catalogues (see also IS-3824 Part II).

4.6 BELTS AND PULLEYS

Belts and pulleys are used for transmitting power between two parallel shafts. In special cases, these elements can transmit power in shafts, which are perpendicular. Pulleys are fixed to the shafts and a flexible belt is made to envelope these pulleys. Transfer of power is due to the friction between the

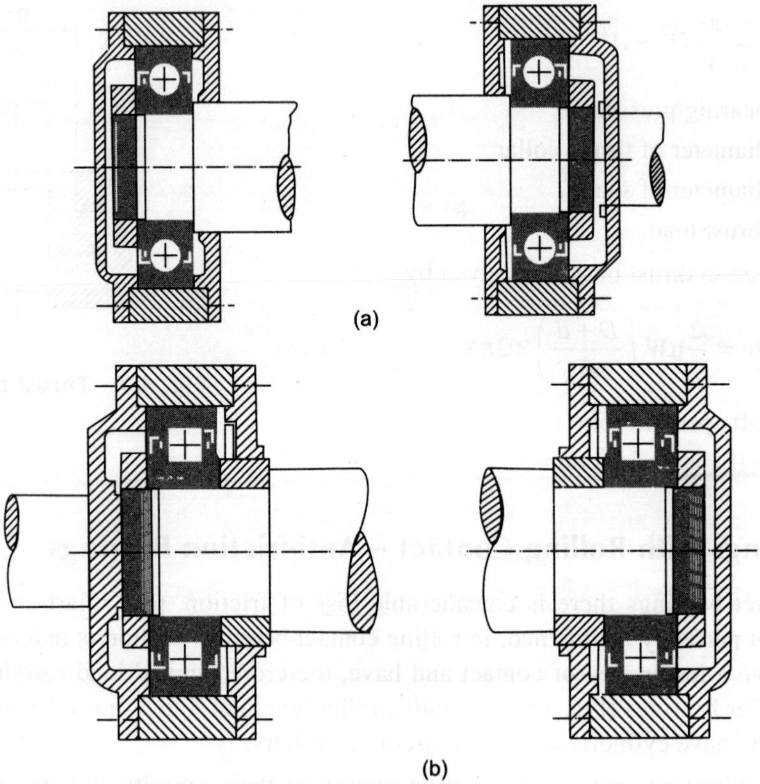

Fig. 4.8 Rolling Contact Bearings: (a) Ball Bearing, (b) Roller Bearing

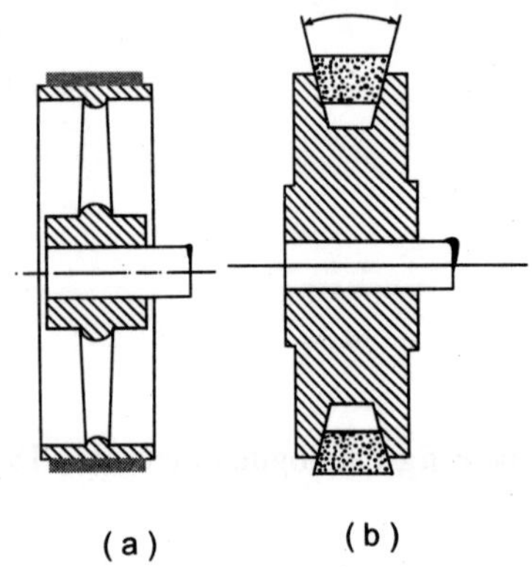

Fig. 4.9 Pulley with Belt: (a) Flat Belt, (b) V-Belt

pulley and the belt. Two types of pulleys and belts are generally used. Pulleys with flat rims and a belt of rectangular section are used for shafts with large centre distance. For short centre distance pulleys with V-grooves on their rim surfaces and with belts of trapezoidal section are used [Figs 4.9(a) and (b)].

The power transmitted between pulleys by use of a belt depends on the friction and the area of contact. Tensions on the two sides of the belt are related as follows (Fig. 4.10).

For flat belt

$$\frac{T_1}{T_2} = e^{\mu, \theta} \tag{4.20}$$

where, T_1 — tension on the tight side

T_2 — tension on the slack side

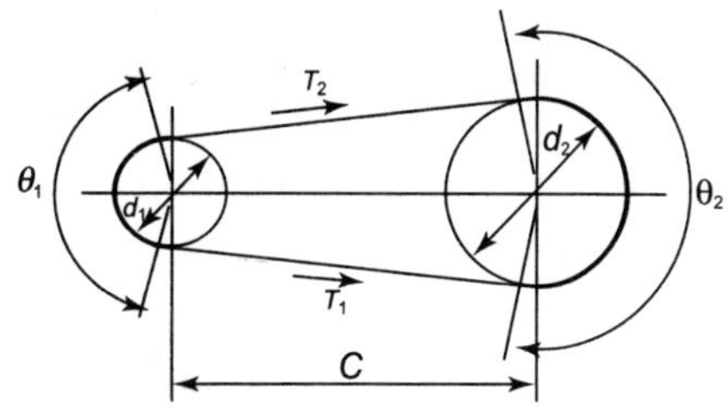

Fig. 4.10 Belt Tensions

θ — angle of wrap for the smaller pulley

μ — coefficient of friction between belt and pulley.

For V-belt

$$\frac{T_1}{T_2} = e^{\mu_1 \theta} \tag{4.21}$$

where, $\mu_1 = \mu$ cosec α and 2α is the angle of the V-groove [Fig. 4.9(b)]. The angle of lap between the belt and the pulley may be found from the following:

$$\theta = \pi \pm 2\sin^{-1}\left(\frac{D_2 - D_1}{2C}\right) \tag{4.22}$$

$$\simeq \pi \pm \frac{D_2 - D_1}{C}$$

where, C — centre distance between pulleys

θ — angle of contact in the open belt drive in radians.

The +ve sign gives the angle for larger pulley, while the –ve sign gives that for the smaller pulley.

Power transmitted is given by

$$hp = \frac{2\pi N_1 (T_1 - T_2)\frac{D_1}{2}}{60 \times 75} \tag{4.23}$$

where, D_1— diameter of the smaller pulley

N_1— rpm of the smaller pulley.

Speed ratio between the pulleys is

$$\frac{N_1}{N_2} = \frac{D_2}{D_1} \tag{4.24}$$

N_1, N_2, and D_1, D_2 are rpm and diameters of the two pulley respectively.

The cross-section of the belt can be determined by the capacity of the belt to withstand tension T_1 on the tight side

$$T_1 = f_t A \tag{4.25}$$

where, f_t — permissible tensile stress in the belt/factor of safety

A — area of cross-section of the belt.

Belts are made of leather, canvas or rubberised canvas. Flat belts of rectangular section are fabricated from plies of leather, with thickness of 4 to 6 mm in single ply and 7.5 to 10 mm in double ply. V-belts are fabricated from textile cord covered with rubber. The angle between the sides is 40° and the other dimensions are standardised.

According to the size, V-belts are classified as A, B, C, D and E.

Pulleys are made of cast iron, stamped sheet steel, aluminium, wood, and plastics. The hub of the pulley is attached to the shaft by a key. The rim of the pulley is made sufficiently wide to accommodate the belt on its surface. The hub and the rim are held in position by a series of arms, radiating out from the hub. The arms are circular or elliptical in cross-section. The belt tension creates a bending moment on the arms. The magnitude of the bending moment determines the size of the arm. V-pulleys are generally smaller in size and are fabricated from a solid piece of the material by providing a hole for inserting the shaft and grooves on the peripheral surface for accommodating the belts.

4.7 CHAIN DRIVE

Chains are used for power transmission, hoisting and conveying. The system consists of two sprocket wheels attached to the shafts and a chain meshing with the teeth on the sprocket wheels. The arrangement is essentially a belt built up of rigid links which are hinged together in order to provide the necessary flexibility for the wrapping action round the driving and the driven wheels. These wheels have projecting teeth, which fit into suitable recesses in the links of the chain and, thus, enable a positive drive to be obtained. Various types of chains are used. The roller chain consists of a series of rollers freely mounted on and connected by thin metallic links (Fig. 4.11).

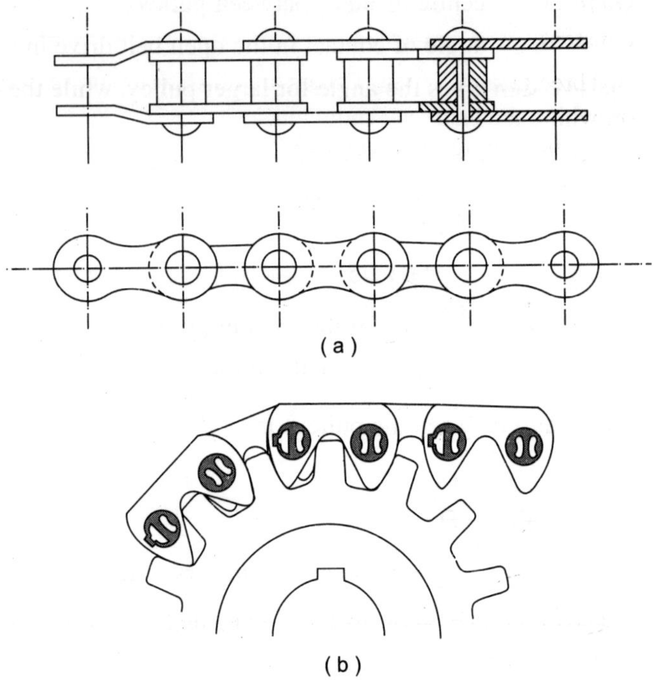

Fig. 4.11 Chain Drive: (a) Roller Chain, (b) Link Chain

The chains are made of medium carbon or alloy steel. The speed ratio of the chain drive depends on the number of teeth in the driving and the driven sprockets. Short centre drives with high speed ratio require a fine pitch chain, while a narrow large pitch chain is better adapted to low ratio long centre drives. When the speed ratio is less than 2.5, the minimum centre distance may be equal to one-half the sum of the sprocket diameters plus the pitch;

for speed ratio greater than 2.5, the minimum centre distance should be equal to the sum of the sprocket diameters. The maximum centre distance should not exceed eighty times the pitch.

Chains and sprocket wheels are standardised, according to the pitch, number of teeth, power transmission capacity, strength, etc. The maximum allowable pitch for a given speed may be found from

$$p_m < 10\left[\frac{3640}{N_1}\right]^{2/3} \tag{4.26}$$

where, p_m — pitch in mm

N_1 — speed of smaller sprocket in rpm.

4.8 GEAR DRIVES

Gear drives are used for transmission of power between parallel, intersecting or skew shafts. Two gear wheels are fixed to two corresponding shafts and remain in mesh throughout the rotations of the shafts. Gear drives are preferred to belt drives for transmitting medium or large amounts of power at constant speed ratios.

Three general types of gears are classified with respect to relative position of the axis of the shafts on which the gears are mounted.

(a) Spur gearing (Figs 4.12 and 4.13) is used for connecting shafts whose axes are parallel, which include external spur gearing, internal gearing, rack and pinion, helical gearing and herringbone gearing.

(b) Bevel gearing (Fig. 4.14) is used for connecting intersecting shafts which include straight bevel gears, miter gears and crown gears.

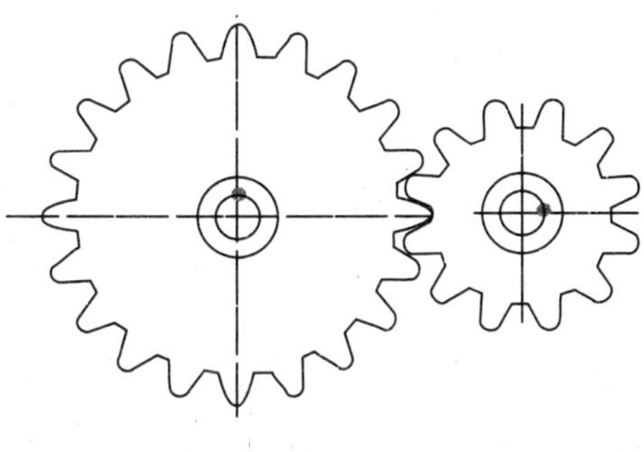

Fig. 4.12 Spur Gears

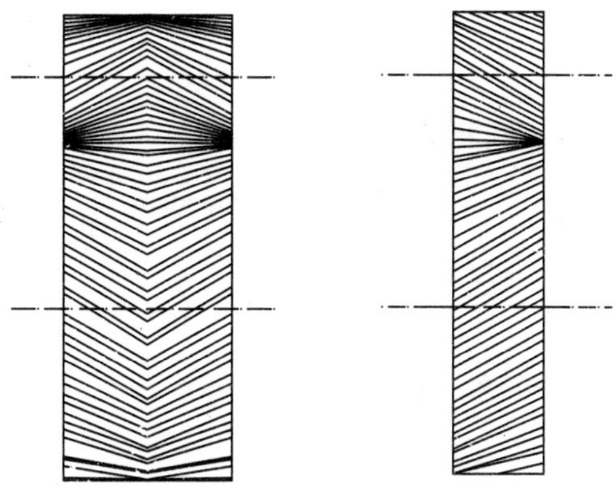

Fig. 4.13 Helical Gears

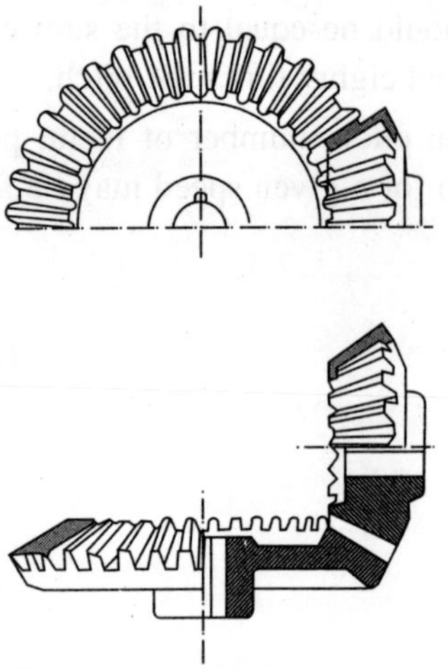

Fig. 4.14 Bevel Gears

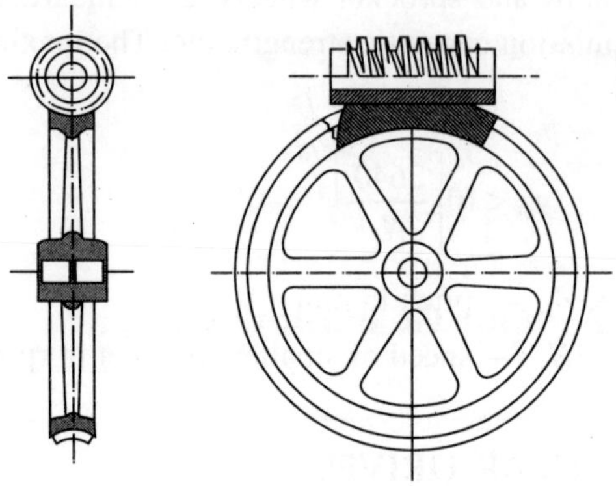

Fig. 4.15 Worm and Worm Gear

(c) The third type is used for connecting shafts whose axes are neither parallel nor intersecting, which include worm and worm wheels (Fig. 4.15), hypoid gearing and spiral gearing.

4.8.1 Spur Gear Terminology

Figure 4.16 shows details of spur gear teeth. The profile of the tooth is an involute curve. Some of the terms used are indicated in the figure.

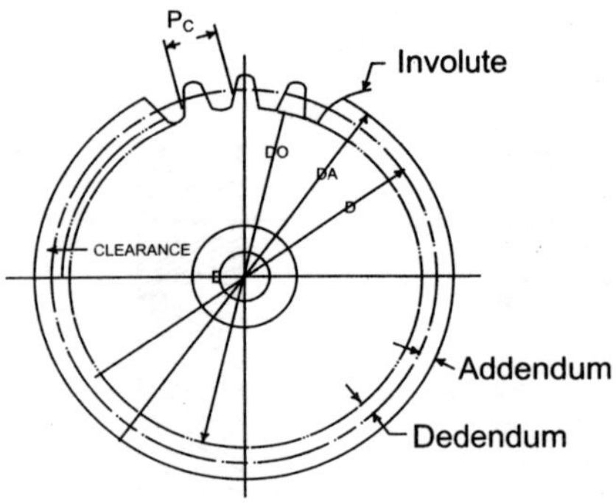

Fig. 4.16 Gear Terminology

(a) *Pitch circle*: It is the circle of a cylindrical body, which by pure rolling action transmits motion to another cylindrical body, same as the motion of two gears.

(b) *Pitch diameter* (*D*): It is the diameter of the pitch circle.

(c) *Circular pitch* (p_c): It is distance between similar points of adjacent teeth measured along the pitch circle.

(d) *Diametral pitch* (p_d): It is the number of teeth per unit diameter.

(e) *Module:* It is the length of diameter per tooth. Teeth are standardised according to module size in mm.

(f) *Pressure angle* (α): It is the angle which the line of action of the force of pressure makes with the common tangent to the pitch circle at the pitch point. Commonly used pressure angles are 14.5° and 20°.

The above terms gives the following relations:

$$p_c = \frac{\pi D}{T}$$

where, T — number of teeth.

$$p_d = \frac{D}{T}; \quad m = \frac{D}{T}$$

$$p_c \times p_d = \pi; \quad p_c = \pi m$$

4.8.2 Spur Gear Tooth Design

Force on the gear tooth $F = \dfrac{hp \times 4500}{2\pi N(D/2)}$ (4.27)

where, N — rpm

D — pitch circle diameter, in metres.

F — kgf

The stress under static loading conditions is given by

$$F = fbp_cY \qquad (4.28)$$

where, f — permissible stress due to bending

b — face width which is

$2.5 \times p_c$ — for cast teeth

$3.5 \times p_c$ — for finished steel teeth

Y — Lewis factor

$$Y = 0.124 - \frac{0.684}{T} \text{ for } 14.5° \text{ pressure angle}$$

$$Y = 0.154 - \frac{0.912}{T} \text{ for } 20° \text{ pressure angle}$$

Under dynamic loading, which is actually the type of loading expected, Eq. 4.30 has to be multiplied by a velocity factor (k_v) and a service factor (k_s) (Table 4.4).

where, $k_v = \dfrac{10}{10 + v}$

v — pitch line velocity in metres/sec.

Table 4.4 Service Factor (k_s)

Type of Load	*Intermittent*	*Continuous*
Steady load	1	0.8
Light and medium shock	0.8	0.65
Heavy shock	0.6	0.5

Out of the two wheels, which remain in mesh while transmitting motion, the smaller wheel is known as pinion and the larger is known as gear wheel. The minimum number of teeth for the pinion should be sufficiently large to reduce the specific loading per tooth to a minimum value, so that the wheels will work smoothly without excessive wear and heating. In general, the pinion should have a pitch circle diameter of 2.2 d for steel, 2.5 d for cast iron and 2.8 d for plastics, where d is the diameter of the shaft. A velocity ratio of about 4 to 6 is usually adopted. Standard modules vary from 0.3 mm to higher values. Gears are made of cast iron, steel, brass, bronze, plastics and other non-metallic materials.

4.9 RIVETED JOINTS

Rivets are used for connecting permanently structural sections, plates and machine components. The joints made between structural sections or between machine components must have sufficient strength and rigidity. Joints for storage tanks, drums and pressure vessels must have enough strength, rigidity and must also be leakproof. In recent years riveted joints are being replaced by welded joints in structural and pressure vessel work. However, riveted joints are still in use to a very limited extent, in pressure vessels fabricated from non-ferrous, ductile materials such as copper, brass, aluminium, etc., and for fabrication of certain machine components such as frames, bases, etc.

4.10 WELDED JOINTS (see Table 4.5)

This method of joining components of machines or structural sections is now being used extensively in place of riveting or bolting. In may cases it has replaced even castings, since parts such as machinery bases, frames, brackets, etc., are now being fabricated by welding from standard sections. This method produces parts which are much greater in strength and rigidity than parts made of cast iron.

In welding, no holes are required to be made as in the case of riveting. The entire cross-section of the component is, therefore, effective. The weight of the material is, therefore, less by as much as 25 per cent, thus, effecting a saving not only in cost of material but also in the cost of handling. In cases where the thickness of the plates to be joined is large, this method is the only solution. Pressure vessels, tanks and and other similar equipments are most conveniently and efficiently fabricated by welding.

4.10.1 Type of Joints

There are two types of joints made by the method of welding (Table 4.5):

(i) *Butt weld,* which is made by placing two pieces to be welded, close to each other and filling the gap by a suitable welding material.

(ii) *Fillet weld,* which is made by overlapping two pieces to be welded and filling the corners formed between them by a suitable welding material.

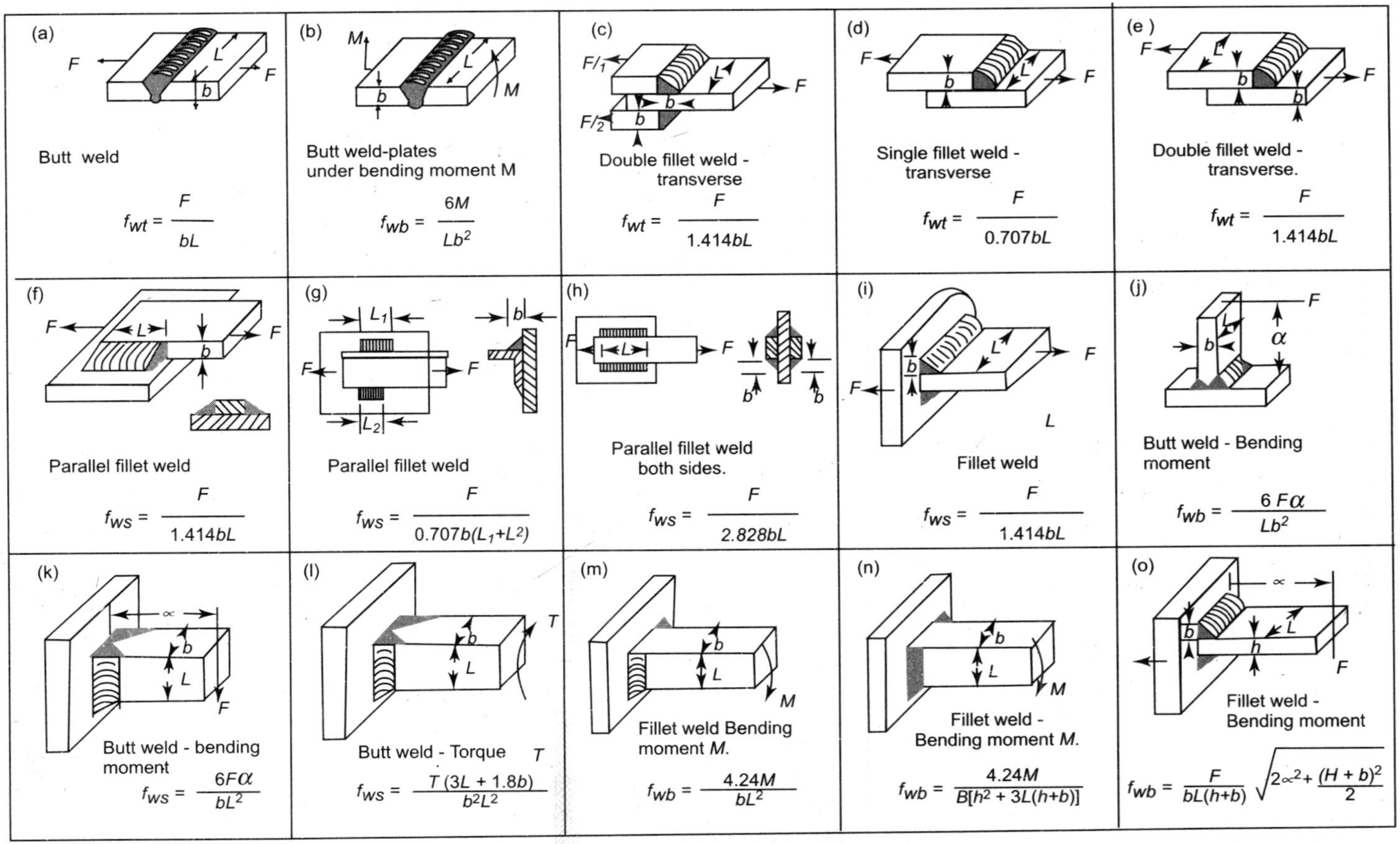

Fig. 4.17 Welded Joint

Table 4.5 Welded Joints

Joint	*Name*	*Joint Efficiency*	*Application*	
			Longitudinal	*Circumferential*
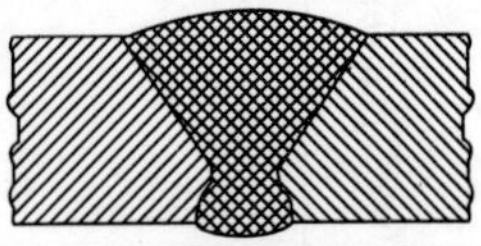	Double-welded butt joint with single 'V'	0.7 to 1.0	All thickness	All thickness
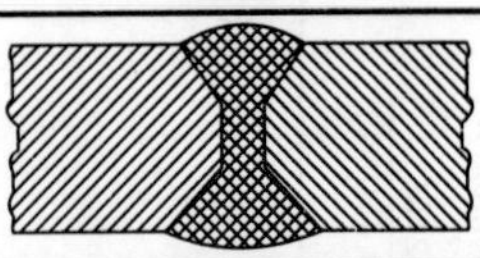	Double-welded butt joint with double 'V'	0.7 to 1.0	All thickness	All thickness
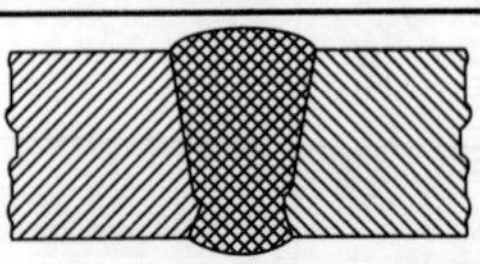	Double-welded butt joint with single 'U'	0.7 to 1.0	All thickness	All thickness
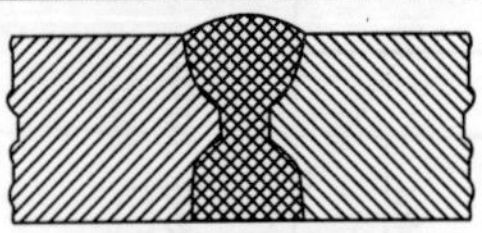	Double-welded butt joint with double 'U'	0.7 to 1.0	All thickness	All thickness
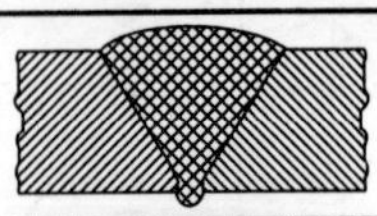	Single-welded butt joint with 'V' groove without backing strip	May be 0.5 to 0.85		Not exceeding 15 mm plate
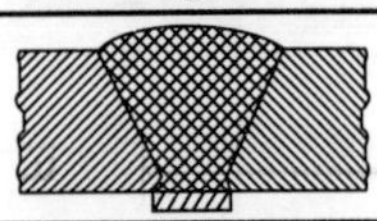	Single-welded butt joint with 'V' with backing strip	0.65-0.85	Not exceeding 30 mm plate	Not exceeding 30 mm plate
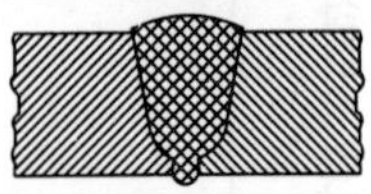	Single-welded butt joint with 'U' groove without backing strip	0.55 to 0.85		Not exceeding 15 mm plate
	Single-welded square butt joint without backing strip	0.55 to 0.85		Not exceeding 6 mm plate
	Single-welded square butt joint with backing strip	0.60-0.85	Not exceeding 6 mm plate	Not exceeding 6 mm plate
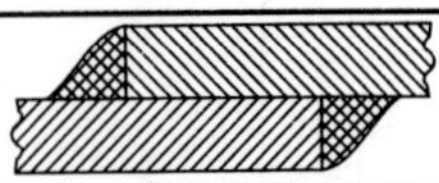	Double full-fillet lap joint	0.45-0.70	Not exceeding 8 mm plate	Not exceeding 15 mm plate
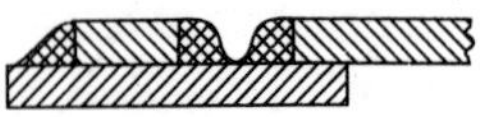	Single full-fillet lap joint with plug weld		Not allowed	Not exceeding 14 mm plate. Plugs shall be proportioned to take 20 per cent of total load
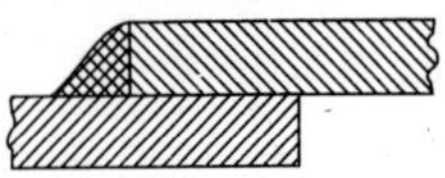	Single full-fillet lap joint		Not allowed	Not exceeding 12 mm plate

The joint efficiency depends upon the extent of Non Destructive Test (NDT), for instance, Double welded, butt joint,

NDT	*efficiency*
No radiography	0.7
Spot radiography	0.85
100% radiography	1.00

4.10.2 Design of Welded Joints

The design of a welded joint should satisfy the following basic requirements:

(1) Give good accessibility of welding and inspection.

(2) Require minimum amount of weld metal.

(3) Good penetration of the weld metal; from both sides of the joint, if practicable.

(4) Incorporate sufficient flexibility to avoid cracking due to differential thermal expansion.

(5) Create minimum amount of stress during cooling.

Types of welds which are designed are:

(a) Butt welds: When the load is applied to the plates or machine components by use of butt welds, it is necessary to ensure that the joint will withstand the load without failure. The nature of the stresses, namely tensile, compressive, shear, etc., created in the portion are similar to those in the original plate or component to which the load is applied. The magnitude of the stress depends on the size of the welded metal.

(b) Fillet welds: These welds are of two types: (i) transverse fillet welds which are under tension and (ii) parallel fillet welds which are under shearing action.

4.10.3 Joints for Pressure Vessels

For details of pressure vessel components and welded connections applicable to pressure vessels, see Chapter 5. Typical weld joints are exhibited in Fig. 4.17 and Table 4.5 shows the efficiency of the welded joints and their range of application.

4.11 THREADED FASTENERS

Bolts, studs and screws are used for fastening and making detachable joints between plates, sections, machine components, etc. The standard V-threads formed on such elements are either fine or coarse. The fine threads are used where greater strength is required or where the connected parts are subjected to vibrations or for fine adjustments. Coarse threads are used where threads are to be formed in weaker materials. The following threaded fastenings are generally made (Fig. 4.18).

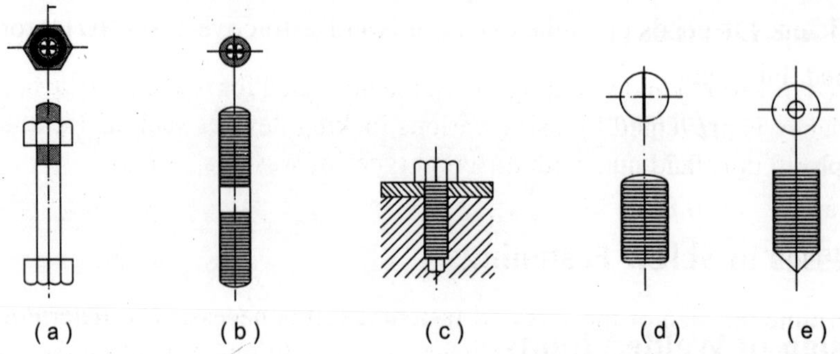

Fig. 4.18 Threaded Fasteners: (a) Bolt and Nut, (b) Stud, (c) Tap Bolt, (d) and (e) Set Screws

4.11.1 Bolts and Nuts

The size of the bolt is specified by the external diameter of the thread and the length is measured under the head. The height of the nut is generally 0.8 times diameter of the bolt for steel, 1.5 times diameter for bronze and 1.5 to 2 times diameter for cast iron. The pertinent Indian Standard is IS 1367.

4.11.2 Tap Bolts or Cap Screws

These devices do not require a nut, but are screwed directly into one of the components to be connected. These are generally used for joints which are not to be detached frequently and where through bolts are not possible because of lack of space. The threaded length is equal to d for steel, $1.5d$ for cast iron and $3d$ for aluminium.

4.11.3 Set Screws

These have a primary function to prevent relative motion.

4.11.4 Studs

These have threads at both the ends. They are preferred when the joint is to be detached or disassembled.

4.11.5 Special Bolts: Swing Bolts, Anchor Bolts, Eye Bolts and T-Head Bolts

These are used in special applications such as pressure vessel covers, foundations, lifting devices, etc.

4.11.6 Locking Devices

In parts, subjected to varying and vibrating loads, the nuts used for tightening the bolts or studs work loose. This tendency is prevented by using various locking devices such as lock nuts, slotted nuts, castle nuts, nuts with pins, and nuts with different types of washers.

4.11.7 Stresses in Screw Fastenings

In order to determine the size of the screwed fastenings, it is necessary to determine stresses under static loading as well as dynamic loading. Under static conditions of loading the following stresses should be considered:

(a) Initial stresses due to tightening
(b) Stresses due to external forces
(c) Stresses due to combination of (a) and (b).

4.11.7.1 Initial Stresses

When a bolt, stud or screw is tightened the stresses created are:

(a) Tensile stress due to stretching of the bolt
(b) Compression or bearing stresses on the thread
(c) Shearing stress across threads
(d) Torsional shear stresses caused by the frictional resistance of the threads
(e) Bending stresses if the surfaces under the head or nut are not perfectly normal to the bolt axis.

None of the above mentioned stresses can be accurately determined. So the bolts or screws are designed on the basis of direct tensile stress induced in the bolt or screw by a reasonable tightening action with a standard wrench. This initial tightening force is given by (3150 d) in kg, where d is the diameter of a high-tensile steel bolt in cm. This value should be used only for leakproof joints. It may be reduced for fastenings not made as tight as leakproof joints.

$$\text{Stress due to tightening } f = \frac{3150d}{A} \tag{4.29}$$

where, A – mean area of cross-section of bolt. This value should be less than the permissible stress, determined with a high factor of safety.

It is difficult to tighten a bolt by hand wrenches to develop full strength of the bolt, if it is larger than 5 cm diameter. Beyond this size, three other methods are used, namely, (1) bolt tensioners, (2) bolt heaters, (3) impact wrenches.

4.11.7.2 Stresses Due to External Forces

Bolts, studs, and screws are subjected to tensile, shear or bearing stresses due to external forces. These can be calculated as follows:

$$\text{axial tensile force} = f_t \times A$$

$$\text{shear force} = f_s \times A$$

$$\text{bearing or crushing force} = f_c \times d_c \times t$$

where, A — area of cross-section of the bolt

$$= \frac{\pi}{4}\left(\frac{d_o^2 + d_i^2}{2}\right)$$

t — length of bolt under bearing or crushing load

d_o— outside diameter

d_i — core diameter.

4.11.7.3 Stresses Due to Combination of (a) and (b)

In a number of applications, where the tightening of the bolt is limited, the stresses induced are mainly due to external load. In such cases, the practice is to design the bolt for 1.25 to 1.75 times the external load or use a lower value of the permissible stress. In some applications, design of the bolts based on initial stresses is desirable; in others, design for a combination of initial and load stresses is necessary. Such applications are cases where bolts and studs are used for tightening two flanges with gaskets, made to ensure leakproof joints. If the gasket is soft, the resultant load on the bolt will be approximately the sum of the external load and initial tightening load. In cases where the gasket is hard or when there is a metal to metal contact between the flanges the resultant load is either the initial tightening load or the external load, whichever is greater. Detailed design procedure for flanged joints is given in Chapter 5.

4.12 PACKING AND GASKETS

Packings are used for dynamic sealing of moving parts, which have reciprocating, rotary or helical motion. The packing is installed in a stuffing box, surrounding the moving part and is then compressed by an adjustable component called a gland. Gaskets are static seals, which are used for making a pressure-tight joint between two rigid elements, usually in the form of flanges. Various gasket types and their applications are considered in Chapter 5.

The general arrangement of the stuffing boxes with packing is shown in Fig. 4.19. These indicate the shape of the stuffing box, the type of packings and the gland used for compressing the packing close to the moving part, so as to prevent leakage. Packing throttles the leakage between a moving and stationary part. However, it does not stop leakage completely. If it runs dry, the packing will get heated and will wear out fast. It will also create considerable friction. A leakage rate of 1 to 2 drops per minute, which flows through the clearance between the moving shaft and packing, acts as lubricant. Packings are made of five basic materials, namely, fibres, metals, plastics, rubber and leather, which are combined with various impregnants and lubricants, depending on the application.

Asbestos is the most widely used material because of its strength, temperature resistance and versatility. However, asbestos is now banned in many countries. The hazard feared by asbestos application may be overcome by having a thin 'teflon' coating. Synthetic fibres such as rayon and nylon are good for water service at temperatures below 120°C. Nylon has a low coefficient of friction. Compression metal packings are usually in foil form, although sometimes in braided or shredded forms. Lead, copper and aluminium are common metals for packings that serve as end rings. The most common plastics for packings are teflon and kel-F, both of which are highly resistant to chemical attack. They can withstand temperatures up to 250°C. Natural rubber, neoprene, buna-N, butyle and silicone are some of the synthetic rubbers used for packings. Leather packings have a high tensile strength. Plastics, rubber and leather are mounted and machined to form shapes like, V-, U- and O-rings.

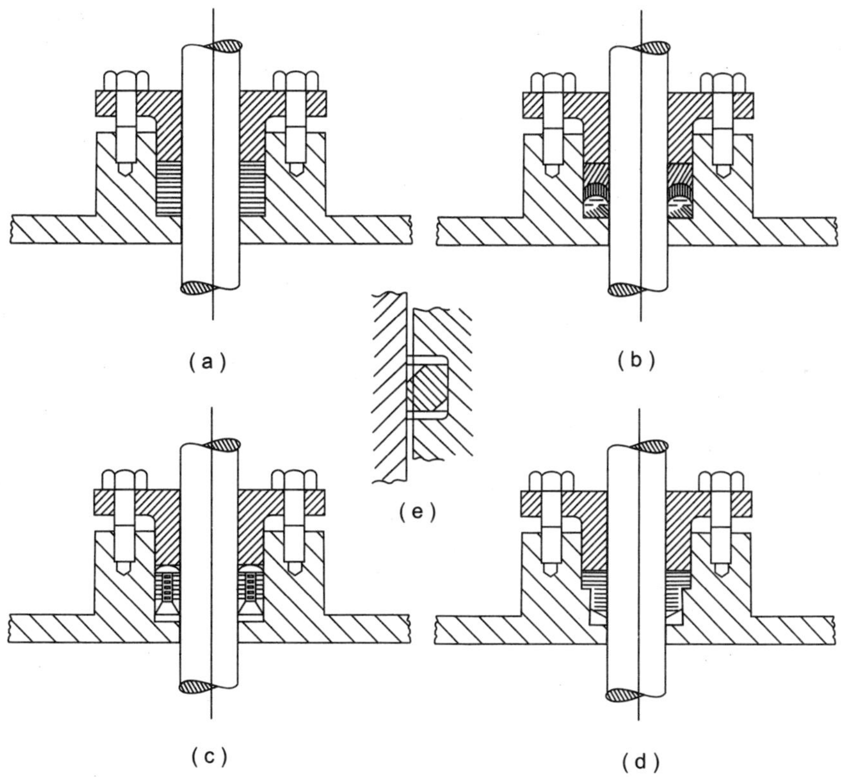

Fig. 4.19 Arrangement of Packing in a Stuffing Box: (a) Ring Packing, (b) V-ring or Chevron Packing, (c) U-cup Packing, (d) Hat Packing, (e) O-ring Packing

Impregnated packings cannot provide enough lubricant when high temperatures are involved and when conditions create high frictional heat. External lubrication must be supplied through a lantern ring, which is usually located near the middle of the stuffing box. The sealing liquid may be injected into the packing at a pressure higher than that of the fluid being contained. They are also required for packings to prevent leakage of gases, which leak easily and are poor lubricants (refer to Figs 14.14, 14.15 and 14.16).

4.13 DESIGN OF STUFFING BOX AND GLAND

Figure 4.20 shows the details of the arrangement with a shaft either rotating or reciprocating. Some of the important dimensions may be determined as follows:

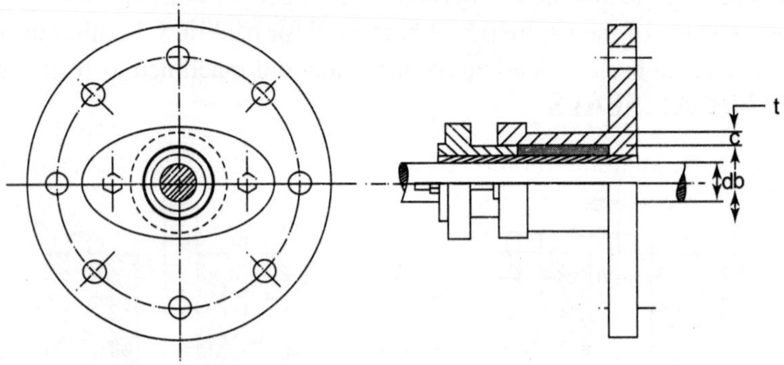

Fig. 4.20 Stuffing Box and Gland

$$b = d + \sqrt{d} \text{ (cm)} \tag{4.30}$$

$$t = \left(\frac{pb}{2f}\right) + 6 \text{ (mm)} \tag{4.31}$$

where, d — diameter of shaft (cm)
p — pressure (N/mm^2)
b — internal diameter of stuffing box (mm)
t — thickness (mm)
f — permissible stress in the material of stuffing box (N/mm^2).

Load on the gland $$F = \frac{\pi}{4} p\,(b^2 - d^2) \tag{4.32}$$

If the shaft is reciprocating, the load is increased by 30%.

The studs take the load F

$$F = \frac{\pi d_i^2}{4} \times n \times f_t \tag{4.33}$$

where, n – number of studs
f_t – permissible stress in studs
d_i – mean diameter of stud.

The diameter of stud is increased to allow for initial tightening. A minimum diameter may be 16 mm. The flange thickness may be about 1.75 times the stud diameter.

The stuffing box may be made of cast iron or mild steel, while the gland pusher may be made of cast iron, brass or gunmetal. The studs are of steel. For corrosive conditions, special materials may be selected as lining materials (see Figs. 14.16, 14.17 and 14.18). By providing a cooling arrangement on the stuffing box, one can increase significantly the service life of a gland packing (Fig. 14.19). If process fluid in non-hazardous (toxicity and flammability) a cooled gland arrangement can be used at pressure as high as 25 bar and even at temperatures exceeding 150°C.

4.14 MECHANICAL SEALS

Mechanical sealing arrangement, though costlier, is becoming popular among plant operators because it is easier to meet SHE (Safety, Health and Environment) compliance around rotating equipment, such as pumps, centrifuges, compressors, blowers, agitated reactors and so on. In its simplest form, a mechanical seal consists of two surfaces (rings), one on the rotating shaft and the other is stationary. The clearance between surfaces is in microns separated by a thin fluid (most often liquid but gas can also be used) acting as a sealant, to prevent process fluid from escaping into the surroundings.

Figure 4.21 indicates details of a mechanical seal, which consists of a rotary ring held on the shaft by a flexible O-ring. The stationary ring is also placed in a box through another O-ring. The flexible O-ring prevents leakage between the rotary seal ring and the shaft, as also between the stationary ring and the box, while allowing the rings sufficient freedom of movement to maintain a full free contact. The O-rings also compensate for the lack of alignment, run out, thermal expansion and shaft vibration. O-rings are made of natural, synthetic or silicone rubbers or Teflon. The sealing faces of the ring are finely lapped, so that they maintain close contact. The rotary ring is usually metallic, while the stationary ring is made of a non-metallic material. Due to low coefficients of friction, carbon or graphite is used for this purpose. A spring gripping the shaft at one end and the rotary seal at the other end provides an initial pressure on the sealing faces.

4.14.1 Mechanical design features

The detailed discussions on the design aspects are out of scope of this book. The recent articles by Boyson (2004) and Patil (2006) presents a refresher on this subject. The book by Mayers (1977) presents a comprehensive treatment on design aspects of mechanical seals.

It is necessary to provide two optically flat surfaces, which will maintain intimate contact and have a low coefficient of friction under all conditions. The parallel lapped surfaces must not be disturbed by distortion due to mechanical, hydraulic or thermal effects. Mountings for the seal and face should prevent distortion while allowing for flexibility of shaft movement relative to the casing.

The seals can be classified based on their application such as those used in high-speed applications such as pumps, blowers, compressors, etc., and those used at relatively low-speed, such

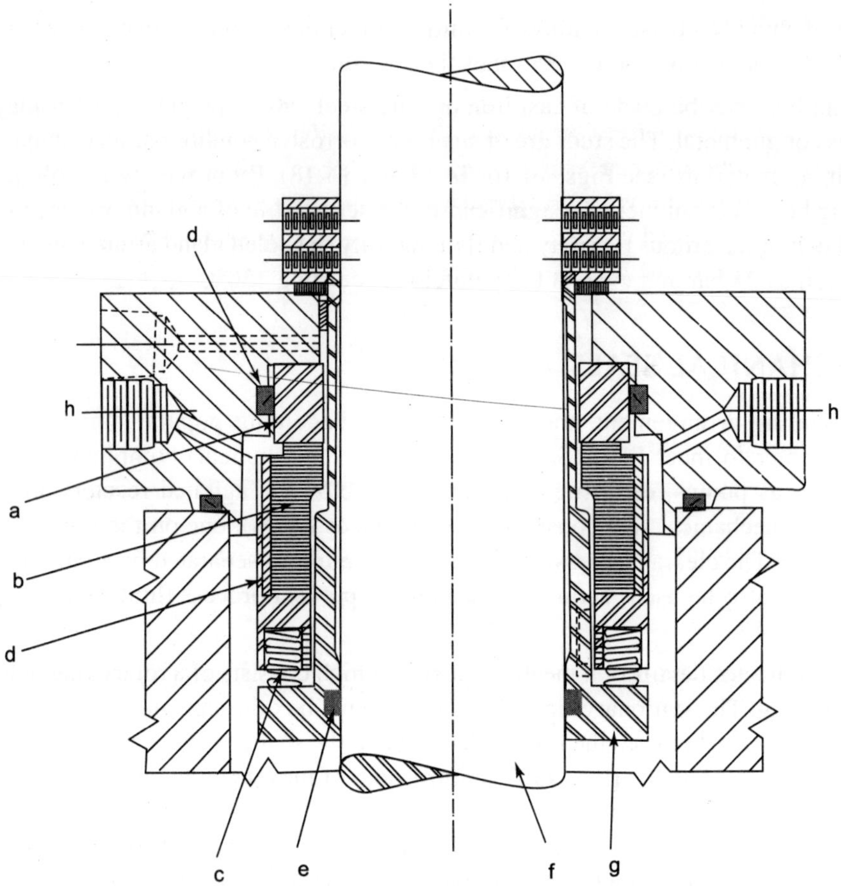

Fig. 4.21 Mechanical Seal: (a) Fixed Ring, (b) Rotating Ring, (c) Spring, (d) and (e) O-rings, (f) Shaft, (g) Shaft Sleeve, (h) Purge Fluid

as those on the shaft of a reactor. Some seals are for high temperature applications while some are for cryogenic service. Various seal arrangements such as spring type, pusher type having O-ring and non-pusher type with metallic or non-metallic bellows have been used in the chemical process industry.

The face material could be of tungsten carbide, silicone carbide, bronze, carbon, cast chrome steel, poly tetra fluro ethylene (PTFE) glass fiber filled or PTFE carbon filled. Natural rubber, synthetic rubber such as nitrile, neoprene, polyisoprene, Viton, EPDM could be materials for O-rings.

4.14.2 Heat Dissipation and Power Absorption

The operating conditions in many seals may be light and power consumption due to friction between ring faces may be ignored. However, for heavy-duty application, the vendor should be consulted for power consumed by the seal.

Many a time the seal becomes hot due to the heat generated by friction, due to the conduction of heat from the hot content of the reactor/vessel by the shaft. In such a case, a cooling arrangement is provided. The compatible liquid is used by thermosiphon action from the small tank for cooling purpose. Cooling water circulated through a coil immersed in the tank in turn cools the tank contents. The tank liquid is kept under pressure by instrument air or nitrogen, which is at a slightly higher pressure than that inside the vessel. Inert nitrogen atmosphere avoids oxidative degradation of the coolant. Alternately, a forced circulation arrangement can be used when thermosiphon arrangement is not adequate to remove the heat. The system consists of a pump, a heat exchanger, a tank, and an online filter.

4.14.3 Leakage Control

In order to reduce power losses and an excessive temperature rise by rubbing of seal faces, it is essential that a thin film of liquid be present between the faces for lubrication purposes. The loss of liquid through this gap constitutes leakage. The seal design must aim at keeping both leakage and loss to a minimum. There are three major factors, which influence the formation of a fluid film between seal faces namely,

(a) Wetting properties of the face and seal material

(b) Vapour pressure and boiling point of coolant

(c) Fluid pressure relative to face pressure loading.

The sealing liquid should be compatible with the reactor content. It should be noted that the mechanical seal is not a zero-leak device. There exists a possibility of a small leak of the sealing liquid into a reactor. This leak might result in unacceptable contamination for perfumery or pharma products or it might poison the noble metal catalyst used, as in hydrogenation. In such a situation preferably the product or a reactant is used as a sealing liquid.

(a) Double mechanical seals are commonly used for high-pressure service to minimize leaks.

(b) Gas sealed mechanical seals are special seals used for a specific application.

4.14.4 Seal Selection

In the selection of the seal, it is necessary to know the characteristics of the fluid for which the seal is to be used. Specifications must be available of pressure, temperature, vapour pressure, viscosity, corrosive properties, solid particles or abrasives to which the seal is likely to be exposed.

4.15 ZERO-LEAK DRIVE

While handling flammable and hazardous gases and chemicals, it becomes necessary to have a zero-leak drive. It is possible to have such an arrangement by having a magnetic drive often known as magne drive. In such a device, magnets are fixed on the shaft. The magnets having an opposite polarity with close clearance are mounted on the outside rotor. The motor then drives this rotor. As the outside field rotates, the inside shaft starts rotating due to the presence of the mounted magnets. This results in a leak proof drive assembly. The laboratory high-pressure reactors, and pumps are equipped with such drives. They are also used in plants.

References and Recommended Readings

Boyson, S., 'Mechanical Seals – Evaluating What's Right for You', *Chemical Engineering*, **111** (13) 36, (2004).

Busher, T.A., 'Bearing – Why They Fail', *Chemical Engineering Deskbook*, **56** Co., London (1977).

Fairs, V.M., *Design of Machine Elements*, Macmillan, New York (1965).

Greenwood, D.C., *Mechanical Power Transmission*, McGraw-Hill Book Co., UK (1962).

Hesse, H.G., and Rushton, I.H., *Process Equipment Design*, Affiliated East-West Press, New Delhi (1967).

Hoyle, R., 'How to Select and Use Mechanical Packings', *Chemical Engineering*, **85** (22) 103 (1978).

Mayer, E., *Mechanical Seals*, Third edition, English translation, Butterworth and Co., London (1977).

Patil, V., 'Learn Important Aspects of Mechanical Seals, Its Installation and Operations', *Chemical Industry Digest*, **XIX** 3, 49 (2006).

Ransden, I.H., 'How to Choose and Install Mechanical Seals', *Chemical Engineering*, **85** (22) 97 (1978).

Redford, G.D., *Mechanical Engineering Design*, Macmillan, London (1966).

Shigley, J.E., *Dynamic Analysis of Machines*, McGraw-Hill Book Company, Inc New York, Toronto, London (1961).

Spotts, M.F., *Design of Machine Elements*, Prentice-Hall, New Delhi, (1969).

Wright, J., 'Which Shaft Coupling is Best', *Hydrocarbon Processing*, **54** (4) 191 (1975).

5

Chapter

Pressure Vessels

5.1 INTRODUCTION

Several types of equipment which are used in the chemical industry have an unfired pressure vessel as a basic component. Such units are storage vessels, kettles, distillation columns, heat exchangers, evaporators, autoclaves, etc. Each one of these is covered in detail in subsequent chapters. The general design procedure applicable to the main parts of the pressure vessel is considered in this chapter.

5.2 OPERATING CONDITIONS

Pressure vessels are usually spherical or cylindrical with domed ends. They are provided with openings or nozzles with facilities for making threaded or flanged joints. Various methods are used for supporting the vessel. The operating conditions may be specified as those resulting from the operation during maximum or normal conditions, as well as those that exist during the start up or shut down or during a change in loading. The latter are known as transient conditions.

5.2.1 Normal Conditions

These include the following:

(a) Operating pressure (internal or external) existing during normal operation.

(b) Operating temperature is decided by the contained fluid. The maximum and the minimum temperatures have to be specified.

(c) Influence of environment, including possible corrosion from chemical attack of the fluid contained and of the atmosphere. Similarly, effects of erosion caused by a high velocity of flow and effects of irradiation have to be considered.

(d) External loading such as wind and snow. Other external loadings are that result from the reaction of the piping system attached to the pressure vessels, dead weight of agitator systems, pumps, valves, etc., supported by the vessel and in general all forms of local loading imposed during the service.

5.2.2 Transient Conditions

These may be repetitive, for example, those occurring during start up and shut down. It is necessary to know the anticipated modes of operation, including the rates of change of fluid temperature, procedure for starting up and shutting down and finally, possible emergency operations and loads due to earthquakes.

5.3 PRESSURE VESSEL CODES

A number of National Codes which specify requirements for the design, fabrication, inspection and testing of unfired pressure vessels are available (Table 5.1). The Indian Standards Institute has prepared a similar code IS-2825. The ASME is widely followed for fabrication. In most countries the National Codes have the force of law and strict adherence to their rules is required. It is, therefore, desirable that any design, regardless of the method adopted, should be checked with the standard code, as the codes have evolved with years of experience.

5.4 SELECTION OF MATERIAL

Pressure vessels from a major part of the equipment used in the chemical industry. It is, therefore, desirable to consider the suitability of the different materials for construction of pressure vessels operating under different conditions. Some of these conditions are: temperature in the range of 600°C to –200°C, pressure in range of vacuum conditions to as high as 3000 kg/cm^2, corrosive effects due to acid and alkalis, steady or cyclic loading, etc., (see Appendix E for strength properties). Apart from the mechanical properties and corrosion resistance of the material, fabrication problems, commercial availability of materials and cost will have to be critically assessed in the final selection of the material. A survey of the materials of construction and the methods of protective coatings are covered in Chapter 2. A review of the important materials accepted for construction of pressure vessels is indicated here. Metallic materials may be divided into three groups:

(a) *Low cost:* Cast iron, cast carbon and low alloy steel, wrought carbon and low alloy steel.

(b) *Medium cost:* High alloy steel (12% Cr and upwards) aluminium, copper, nickel, and their alloys, lead.

(c) *High cost:* Platinum, silver, tantalum, titanium, zirconium.

The materials in the second and third groups can be used in the form of cladding or bonding for materials in the first group. Similarly, non-metallic linings such as rubber, plastic, etc., may also be used.

Table 5.1 Principal National and International Codes

Country	*Code Title*	*Scope*
Australia	Standards Association of Australia Boiler Code, Parts I-V	B
		UFPV
Austria	Dampfkessel Veror Dnung (DKV) RGBI No. 83/1948 Werkstoff und	B
	Bauvorschriften (WBV) RGBI No. 264/1949	PV
Canada	C.S.A. Standard B 51-1957 incorporating A.S.M.E. Rules	B
		UFPV
Finland	Dimensioning, Materials and Welding of Steel Pressure Vessels	B
		UFPV
France	SNCT No. 1	UFPV
	Reglementaton des Appareils a Vepeur et a Perssion de Gaz	Government rules not strictly forming a design code
Germany	Workstoff und Bauvorschriften fur Dampfkessel und Dampfkessel Bestimmungen	B
	AD – Merkblatter DIN-2413	UFPV Pipes
Holland	Grondslagen Waarop de Becoordeling van de Constructie en het Material van	B
	Stoomtoestelen, Dampoestelenen Druck-houdersberust	UFPV
India	Indian Boiler Regulations 1950, Code for Unfired Pressure Vessels IS-2825-1969	UFPV
	Specification for Formed Ends for Tanks and Pressure Vessels IS-4049-1971	UFPV
	Specifications for Shell Flange for Vessels & Equipment IS-4864-4870-1968	UFPV
	Manholes & Inspection Openings for Chemical Equipment IS-3133	UFPV
	Code of Practice for Design Fabrication & Erection of Vertical Mild Steel Cylindrical welded Oil Storage Tanks IS-803-1976	Tanks
	Specification for Shell & Tube Type Heat Exchangers IS-4503	Heat exchanger
	Specifications for Mild Steel Tube IS-1239-1968	pipe, tube
International ISO-TC-11		
Italy	Controllo Della Combustione Aparecchi a Pressione	B
		UFPV
New Zealand	N.Z. Boiler Code	B
	N.Z. Pressure Vessel Code	UFPV
Sweden	Tryckkarlanormer	B
		UFPV
	Angpanneformer	B Weld
	Pannsvetnormer	Code
Switzerland	Regulations of the Swiss Association of Boiler Properties	B
		UFPV
Britain	Lloyd's Rules	B
		UFPV
		piping

(*Continued*)

Table 5.1 (*Contd.*)

Country	Code Title	Scope
	Rules of the Associated Office	B
	Technical Committee (AOTC)	UFPV
	B.S. 1500 : 1958 Pt. I	UFPV
	B.S. 1515 : 1965 Pt. I	
	B.S. 5500 : 1991	UFPV
	B.S. 1113 : 1958	B
	B.S. 806 : 1954	Pipes
	B.S. 1306 : 1955	Power
	B.S. 3351 : 1961	Pipes (oil)
	B.S. 2971 : 1961	
	B.S. 2654 : 1989 Pt. I	Vertical tanks
	: 1989 Pt. II	
	B.S. 3274 : 1960	Heat exchangers
	B.S. 3915 : 1965	Nuclear vessels
U.S.A.	A.S.M.E. Codes:	B
	SEC. I – Boilers	UFPV
	SEC. II – Materials A, B, C, D	
	SEC. III – Nuclear vessels	
	SEC. VIII Div 1 & 2	UFPV
	SEC. IX	Welding
	Tentative Structural	UFPV
	Basis for Reactor	Nuclear
	Pressure Vessels and Directly Associated Components	
	ASA – B31	Piping
	API – A.S.M.E. (similar to A.S.M.E. Codes)	
	TEMA	
	Tubular Heat Exchangers Manufacturers Association, 1978	Heat exchangers
	API 650 for Storage Tanks	

B – Boiler, UFPV – Unfired pressure vessel.

5.4.1 Steel

This is the most versatile and widely used material of construction in the pressure vessel industry. The carbon content in pressure vessel steels is usually limited to 0.3 %. Low carbon or mild steel is the cheapest and most commonly used amongst the pressure vessel steels. Depending on the degree of deoxidation, a steel may be rimmed, semi-killed or killed. Rimmed steels are seldom used in pressure vessel construction, due to their lack of chemical homogeneity. They may be used for light duty vessels. Semi-killed steels are cheapest steels used for general purpose light duty service. Almost all the plates used in the pressure vessels, up to 25 mm thickness for this type of service are semi-killed. Fully deoxidized silicon killed steels are more homogeneous. They are more expensive and are used for thicker vessels, or in all thicknesses for severe duty. These steels have 0.2% carbon and 0.7 to 0.9% manganese. Aluminium is usually added as a grain refiner, to improve notch toughness of the material. Mild steel is generally used in a normalized condition. This condition may be achieved by selecting the appropriate temperature and cooling rate during rolling and forming. Mild steel is readily attacked by most fluids and atmospheric environments. The corrosion resistance of the material depends to a large extent on the cleanliness and homogeneity of the steel and the fabrication process.

Low alloy steels with a carbon content of the order of 0.15%, manganese 1.0%, silicon about 0.3% and other alloying elements such as chromium (0.5 to 5%), molybdenum up to 0.5% are used for high temperature service and under mild conditions.

Chrome-molybdenum steel is widely used in the petroleum and petrochemical industry for hydrogen resistant applications. The most important pressure service is for steam drums particularly for large diameters and high pressure. The steels have 1 to 1.5% manganese with small additions of Cr, Mo, and/or Ni. High strength low alloy steels are also used for thick wall pressure vessels in the chemical and petroleum industry when the wall thicknesses are in the region of 75 mm or more. Quenched and tempered carbon and low alloy steels are used for pressure vessel construction when high strength is required. In the case of carbon steel the brittle transition temperature is reduced by quenching and tempering.

Low alloy steels, developed for high temperature application, have a lower elongation to rupture than ordinary mild steels. The means that cracks may initiate during a pressure test at points where there are high, local stresses. It is important to ensure that the maximum stress in low alloy pressure vessels does not exceed the yield point during the pressure test to any significant extent.

High alloy steels are used when it is necessary to have a corrosion resistant material and whenever contamination to the fluid by dissolved corrosion products or by the material itself has to be avoided. This is more so in the case of pharmaceuticals and speciality products. High alloy vessels are mainly used for severe duty. Such vessels can only be designed on the basis of a comprehensive knowledge of the material properties and the behaviour of the material during fabrication.

For pressure vessel construction the high alloy steels used, include ferrite steels containing 13, 17 and 27% chromium and the austenitic steels of the 18/8, 25/12, 25/20 and 32/22 chromium-nickel composition. The straight chromium ferrite steels are generally specified for corrosion resistant duties in the form of a clad plate. Only 13% Cr steel is employed for refinery vessels, while 17% Cr and 27% Cr steel are rarely employed. Austenitic chromium nickel steels are used for sub-zero temperatures, for corrosion resistance and for high temperature operation.

5.4.2 Aluminium and Aluminium Alloys

Aluminium alloys have been developed with mechanical properties comparable to those ' steel. They retain ductility at sub-zero temperatures, and are easy to fabricate. Alumir aluminium alloys are used for handling most organic solvents, foodstuffs, hydrogen pero' acid, sea water and inorganic salts. This material is seldom used for vessels operating a'

In all cases, care must be taken to prevent any contact between aluminium and nobler metals, such as copper and iron. The major disadvantages of aluminium are low strength and low melting point.

5.4.3 Copper and Copper Alloys

These are used in food processing plants, in the manufacture and recovery of organic solvents and in heat exchangers and evaporators for general purposes. Selection of various alloys is based on corrosion requirements, strength and weldability.

5.4.4 Nickel and Nickel Alloys

These are used for handling a variety of corrosive fluids at low or elevated temperatures and where it is essential to prevent the contamination of the contained fluid. These materials are very expensive. They are, therefore, used primarily as cladding materials in pressure vessel construction. The only exceptions are heat exchanger tubes and tube plates and small thin-walled vessels.

5.4.5 Titanium

The material is used as bonded titanium clad steel for pressure vessels. Though the density is almost half, strengthwise it is comparable to mild steel. Titanium heat exchangers are common in refinery and power plants where sea water is used for cooling purpose.

5.5 VESSELS OPERATING AT LOW TEMPERATURES

The ductility of some metals, including carbon and low alloy steel is significantly diminished when the operating temperature is reduced below some critical value (–28°C). This critical temperature, commonly described as the transition temperature, depends upon the material, method of manufacture, previous treatment and kind of stress system present. Fractures occurs at temperatures above the transition temperature only after considerable plastic strain or deformation, while below the transition temperature, fractures may take place in a brittle manner with little or no deformation. Brittle fractures are likely to be extensive and may lead to catastrophic fragmentation of a vessel.

Construction features producing a notch effect or sudden change of section are particularly objectionable in vessels designed for low temperature operation, since they may create a state of stress such that the material will be incapable of relaxing high localised stresses by plastic deformation. For this reason materials for low temperature service are tested for notch ductility.

Carbon steels are used down to –60°C. Control of notch ductility in such materials is obtained through proper composition, steel making practice, heat treatment and fabrication practice. They have low carbon content with increased manganese carbon ratio. Various additions are made to promote fine grain size and improve notch ductility, the most effective of which is aluminium. Notch ductility can also be improved by normalising. Between –50°C and –100°C low alloys steels are used. 1% Cr, 0.25% Mo steel and nickel steels containing between 1 to 5% Ni have been specified for this range. Below –100°C, the choice is made from among austenitic chromium-nickel steel, 9% Ni steel, aluminium or copper. All these materials are notch tough down to the lowest temperature.

5.6 VESSELS OPERATING AT ELEVATED TEMPERATURES

There are two main criteria in selecting a steel for elevated temperature use, namely, strength and metallurgical stability. All carbon steels are reduced in their strength properties due to rise in temperature and are liable to creep. Their use is, therefore, generally limited up to 500°C. High alloy steels can be used for higher temperature up to 1000°C with proper pre-conditioning. The other major requirement for satisfactory use at an elevated temperature is that the metal should not deteriorate significantly in service. Metallurgical faults must be avoided and corrosion rates must be controlled. Above 600°C, carbon steel is oxidised at a very high rate both in air and steam. Chromium in amounts of 3% and above significantly reduces corrosion rates and all heat resisting steels contain chromium for this reason. It is better to use austenitic chromium-nickel steel for unstressed parts operating in the range 500 to 600°C. For higher stressed parts stabilised grades of this steel are preferable due to their high tensile properties at higher temperature.

Embrittlement of carbon and alloy steel, may occur due to service at elevated temperature. In most instances brittleness is manifested only when the material is cooled to room temperature. Embrittlement is inhibited by the addition of molybdenum that also improves the tensile and creep properties.

5.7 DESIGN CONDITIONS AND STRESSES

The design conditions for pressure vessels are specified primarily in terms of pressure and temperature. Vessels are protected with safety devices such as safety valves or bursting discs (see Section 17.5). The design pressure is, therefore, based on the setting pressure of the safety devices, which is generally taken as equal to the maximum operating pressure or working pressure, including the static head plus 10%. Vessels subjected to vacuum and not provided with vacuum breaker valves are designed for 1 atmosphere gauge external pressure. When both external and internal pressure is operating, design is based on each pressure separately and then choosing the higher safe value of the thickness. For example, the reactor contents might be at a pressure 0.4 N/mm^2 (4 bar) but heat transfer fluid in the limpet coil or jacket might be at 1.5 N/mm^2 (15 bar). Further, in practice, safety considerations might overrule the rest. For example, in one instance, the owner of a polymerization reactor decided to have a design pressure 3 times more than the maximum operating pressure for safety, because in the system a sudden rise in pressure could occur. Normally, design pressure is 1.1 times the maximum operating pressure. The design temperature is also considered in an analogous way. Suppose we have a reactor wherein the maximum operating temperature is say, 220°C. The heat transfer fluid in the outside jacket (limpet coil) is at 260°C. The design temperature then would be 20 to 30°C higher than the higher of the above two temperatures, say in this case the design temperature could be 280°C. We would like to caution here, that as in the case of stainless steel, the allowable stress does depend on temperature. It means the reactor will have an extra thickness if the temperature is taken arbitrarily on the higher side resulting in a costlier reactor. Thus, a judicious choice of the design temperature is made. Please note that safety is never compromised. Here also, as a rule of thumb, design temperature is taken as 25°C higher than the highest operating temperature. This decision is subjective.

Appendix G presents comparison of different design features of different Pressure Vessel Codes.

5.7.1 Design Stress (Allowable Stress)

This is based on the ultimate tensile stress, the yield stress or the 0.2% proof stress or creep properties. The design stress is finally obtained by using a factor of safety. According to the basis accepted by different codes the design stress might vary. This may be observed from Table 5.2. The design stresses as per the IS, ASME and BS codes are based on the ultimate tensile stress (U.T.S.) with a factor of safety of 3, 3, or 2.35, whereas in AD Markbatter (DIN code) or in some of the continental codes, design stress is based on yield strength (Y.S.) with a factor of safety of 1.5 (Table 5.2). Recent developments in carbon and low alloy steels have led to a proportionally greater increase in yield strength than in ultimate tensile strength. A comparison of the allowable stresses for some ferrous materials, as recommended by codes are given in Table 5.2. The design stress value would be lower of U.T.S. and yield strength incorporating safety factor.

Table 5.2 Safety Factors by Different Codes for Design Stresses

National Code	*Country*	*a (U.T.S.)*	*b (Y.S.)*	*c (S_R)*	*d (S_C)*
ASME VIII Div. 1-'68	USA	3.0	1.5	—	1.0
ASME VIII Div. 2-'68	USA	3.0	1.5	—	—
BS-1500	UK	4.0	—	—	—
BS-1515 Part I-'65	UK	2.35	1.5	1.5	1.0
BS-1515 Part II-'65	UK	2.5	1.5	1.5	1.0
BS-5500-'76	UK	2.35	1.5	1.3	1.0
AD Markblatter	Germany	—	1.5	—	1.0
IS-2825-'69	India	3.0	1.5	1.5	1.0
RAP de Gaz	France	3.0	—	—	—
ANCC code	Italy	—	1.5	1.5	—
Pressure vessel code	Japan	—	1.7	—	—
ISO-TC-11	International	2.4	1.5	1.6	1.0

where,

a – factor of safety on U.T.S. at design temperature

b – factor of safety on Y.S.

c – factor of safety on S_R

d – factor of safety on S_C
S_R – Average stress of rupture after 1,00,000 hours at operating temperature (kg/cm^2)
S_C – Average stress for 1% creep in 1,00,000 hours at operating temperature (kg/cm^2)

Design stress values, at different temperatures for various ferrous and non-ferrous metals and alloys, as suggested by ISI, are given in Appendix E.

Design codes usually give rules for the design of vessel shell areas and components under steady pressure. In addition, it is necessary to assess the effects of other loads, such as support forces and temperature differences, some of which will be constant in the service life and others which may be cyclic. With the increase in vessel size and duty these loads may be more significant. In many cases the codes recommend relatively low design stress to cover approximately the effects of additional loads, thus, simplifying the design procedure. The present trend is to improve and rationalize component design and the codes are being revised accordingly.

5.7.2 Design Criteria

The methods of design are primarily based on elastic analysis. There are, however, other criteria such as stresses in the plastic region, fatigue, creep, etc., which sometimes are a consideration. Elastic analysis has been developed on the assumption that the material is isotropic and homogeneous and that it is loaded in the elastic range. In the plastic range, elastic analysis ceases to be applicable. However, elastic analysis can still be used if the plastic flow is limited to small regions in the neighbourhood of structural discontinuities or other stress raises, since the deformation of such regions is forced by that of the adjoining elastic material. Similarly, under cyclic variations of load, due to temperature or pressure causing the plastic flow, the material hardens and the behaviour of the material becomes purely elastic. This is known as the phenomenon of shakedown or cessation of plastic deformation under cyclic loading. Elastic analysis, therefore, is the most important method of designing pressure vessel shells and components. Beyond the elastic limit, the material yields and the plastic region spreads without increased value of load. The value of the load for which this occurs is called collapse load or bursting pressure. Limit analysis is concerned with calculating the load or pressure at which the flow of the structure material occurs due to yielding. This method of analysis is seldom applied to the design of pressure vessels. When vessels are subject to cyclic loading, it is necessary to consider the requirements for elastic cycling of the material and the implications of this on component behaviour. In the case of a discontinuity of shape, the load may give rise to a plastic yield, limited to a small area. This will be under elastic and plastic cycling. Under these conditions a shakedown will occur. The maximum shakedown load is twice that of the first yield load. Elastic analysis is, therefore, valid up to this range of load, under cyclic loading conditions. The fatigue strength is assessed by means of curves representing the stress range versus full reversal cycles to failure. A factor of safety of 2 is applied on the stress.

A low-stress brittle fracture of a pressure vessel is possible as a result of the following factors:

(a) Presence of a notch
(b) High localised stresses in the vicinity of a notch

(c) Operation at a sufficiently low temperature

(d) Wrong selection or treatment of material.

To guard against this contingency, the presence of a notch acting as a severe stress raiser or in a highly stressed region, at low temperatures, must be avoided. In addition an appropriate material must be selected for the specified service conditions and must not suffer any unexpected damage during fabrication or in operation. At elevated temperatures, in the so-called creep range for metals, creep may precipitate plastic instability or it may cause unacceptable deformations or ruptures. In such cases it is necessary to introduce new design stresses based on creep rather than on short time properties. Two criteria are usually specified, one for rupture and the other for deformation. According to the rupture criterion, the design stress is defined as 0.6 to 1.0 times of the minimum or average stress for rupture after 1,00,000 hours at the design temperature. In accordance with the deformation criterion, the design stress is usually defined as the average stress for 1% creep strain after 1,00,000 hours, or 0.01% per 1000 hour creep strain rate. In case of carbon steels the design stresses can be based solely on short time properties up to a design temperature of 350°C, while a knowledge of the creep behaviour of the material is required for design temperature above 400°C. These temperature limits are higher by about 75°C for alloy steels.

5.7.3 Corrosion Allowance

Corrosion of chemical equipment due to environmental conditions is a common feature. Various factors causing corrosion and the methods used to prevent it are indicated in Chapters 2 and 4. Every attempt should be made to avoid corrosion. However, this may not be always possible. In pressure vessels corrosion rates may be predictable, unpredictable or negligible. An allowance is, therefore made, to allow for expected corrosion, by providing additional material thickness over and above the thickness determined from design conditions. For carbon steel and cast iron parts, the corrosion allowance is 1.5 mm on all parts except tubes. In the case of those chemical industries where severe conditions are expected the corrosion allowance may be 3 mm. For high alloy steels and non-ferrous parts under pressure no corrosion allowance may be necessary.

5.8 DESIGN OF A SHELL AND ITS COMPONENTS

The simplest pressure vessel considered here is a single unit when fabricated. However, for convenience of design it is divided into following parts (see Plate I):

(1) Shell

(2) Head or cover

(3) Nozzle

(4) Flanged joint

(5) Support

Most of the above components are fabricated from sheets and plates. Seamless or welded pipes can also be used. Parts of vessel are formed and connected by welded joints. In designing these parts and connections between them, it is essential to take into account the efficiency of the joints. For welded joints, the efficiency may be taken as 100% if the joint is fully checked by radiograph. It is taken as 85% if it is checked at only a few spots. Efficiencies between 50 and 85% are taken when radiographic test is not carried out. The pressure vessels considered here are those operating at pressures less than 200 kg/cm^2. Vessels for higher pressures are considered in Chapter 6. The design procedure is primarily based on fabrication by welding.

5.8.1 Cylindrical and Spherical Shells

The following design procedure is applicable for thin wall shells having the ratio of outside diameter to thickness exceeding 4.

5.8.1.1 Shells Subject to Internal Pressure

(a) ***Cylindrical shell*** The internal pressure in the shell gives rise to stresses in the shell thickness, one in the circumferential and the other in the longitudinal direction. These are given by (see Eqs. 3.22 and 3.23).

$$\text{Circumferential stress, } f_p = \frac{pD}{2t} \tag{5.1}$$

$$\text{Longitudinal or axial stress, } f_a = \frac{pD}{4t} \tag{5.2}$$

where, p – internal pressure
D – mean diameter of the shell.

Both the above stresses are tensile. Since the circumferential stress is greater, 'this is taken as the design stress. The shell is generally formed by a joint in the longitudinal direction, which is considered in terms of joint efficiency. The thickness of the shell is, therefore, given by (IS2825).

$$t = \frac{pD}{2fJ} \tag{5.3a}$$

$$t = \frac{pD_i}{2fJ - p} \tag{5.3b}$$

and

$$t = \frac{pD_o}{2fJ + p} \tag{5.3c}$$

where, p – design pressure
D – mean diameter ($D_i \simeq D_o$)
J – joint efficiency

f – design or permissible stress at design temperature

D_i – internal diameter

D_o – outside diameter.

Values of the design or permissible stresses for different materials are given in Appendix E.

(b) *Spherical shell* In this case both stresses are equal to the longitudinal stress indicated above. The thickness is, therefore, given by (see Eq. 3.24)

$$t = \frac{pD}{4fJ} = \frac{pDi}{4fJ - p} = \frac{pDo}{4fJ + p} \tag{5.4}$$

5.8.1.2 Cylindrical Vessel under Combined Loadings

In addition to the internal pressure, the other loadings are the weight of the vessel with its contents and the wind. The effect of the connected piping can also be taken into account. The stresses created due to each of the above loadings can be stated as:

(1) Stress in the circumferential direction due to internal pressure as per Eq. (5.1). This is also known as tangential or hoop stress.

$$f_t = \frac{p(D_i + t)}{2t} \text{ (tensile)} \tag{5.5}$$

where t – thickness of shell calculated from Eq. (5.3).

(2) Stresses in the longitudinal or axial direction

(a) Due to internal pressure as per Eq. (5.2)

$$f_1 = \frac{pD_i}{4t} \text{(tensile)} \tag{5.6}$$

(b) Due to weight of vessel and contents (vertical vessels only)

$$f_2 = \frac{W}{\pi t(D_i + t)} \text{(compressive)} \tag{5.7}$$

where, W – weight of vessel and contents.

The weight W, includes weights of vessel shell, vessel fittings, internal fittings, external fittings (ladders, platforms, etc.) insulation, any auxiliary equipment (condensers, agitators) and the weight of liquid required to fill the vessel.

The vessel will be filled with water for hydraulic pressure tests, and may get filled with process liquid due to misoperation. The weight of the vessel may, therefore, be increased by 10 to 20 per cent to account for the above fittings and to obtain the weight W.

(c) Due to wind or piping, in the case of vertical vessels or due to weight of vessel in the case of horizontal vessels

$$f_3 = \pm\frac{M}{Z} = \frac{M}{\pi D_i^2 t} \text{ (tensile or compressive)} \tag{5.8}$$

where, M— bending moment, due to loads normal to the vessel axis (calculations of bending due to wind load are given in Chapter 13).

Z — modulus of section of the cylindrical vessel.

Total stress in the longitudinal or axial direction

$$f_a = f_1 + f_2 + f_3 \text{ (tensile or compressive)} \tag{5.9}$$

(3) Stress due to offset piping or wind

$$f_s = \frac{2T}{\pi t D_i (D_i + t)} \tag{5.10}$$

where, T — torque about the vessel axis.

Combining the above stresses on the basis of *shear strain* energy theory criteria, the equivalent stress is

$$f_R = [(f_t^2 - f_t f_a + f_a^2 + 3f_s^2)]^{1/2} \tag{5.11}$$

For satisfactory design, the following conditions must be satisfied:

f_R (tensile) ≤ f_t (permissible)

f_a (tensile) ≤ f_t (permissible)

f_a (compressive) ≤ f_c (permissible)

The axial compressive stress may cause wrinkling of the shell. The safe compressive stress which can be imposed without failure by wrinkling is given by (see Section 4.4)

$$f_e \text{ (permissible)} = \frac{1}{12} \cdot \frac{E}{\sqrt{3(1-\mu^2)}} \cdot \frac{t}{(D_o/2)} \tag{5.12}$$

where, E — modulus of elasticity

μ — Poisson's ratio

D_o — outside diameter of vessel.

If these conditions are not satisfied, the thickness t will have to be increased by employing the trial and error method. The final thickness should be determined by addition of corrosion allowance.

To ensure sufficient rigidity to withstand its own weight and the other attached loads, there is a need to provide a minimum wall thickness for any vessel. Various standards used in fabrication, specify minimum thickness to impart rigidity to equipment say, pressure vessels, tanks, etc. For instance, IS 803 specifies the minimum thickness, inclusive of corrosion allowance, as 5 mm for carbon steel tanks.

In the case of tall vertical shells, all the above stresses have to be considered. But in addition, the stresses resulting from seismic forces are also needed to be considered. The design procedure is indicated in Chapters 11 and 13.

The design procedure outlined above does not take into consideration certain loads such as those caused due to supports, eccentricity of loading, temperature, etc.

5.8.1.3 Shells Subjected to External Pressure

(a) *Cylindrical shell* Elastic buckling (Section 4.4) is usually the decisive criterion in the design of vessels operating under external pressure or vacuum. The critical buckling pressure for cylindrical vessels under external pressure is given by

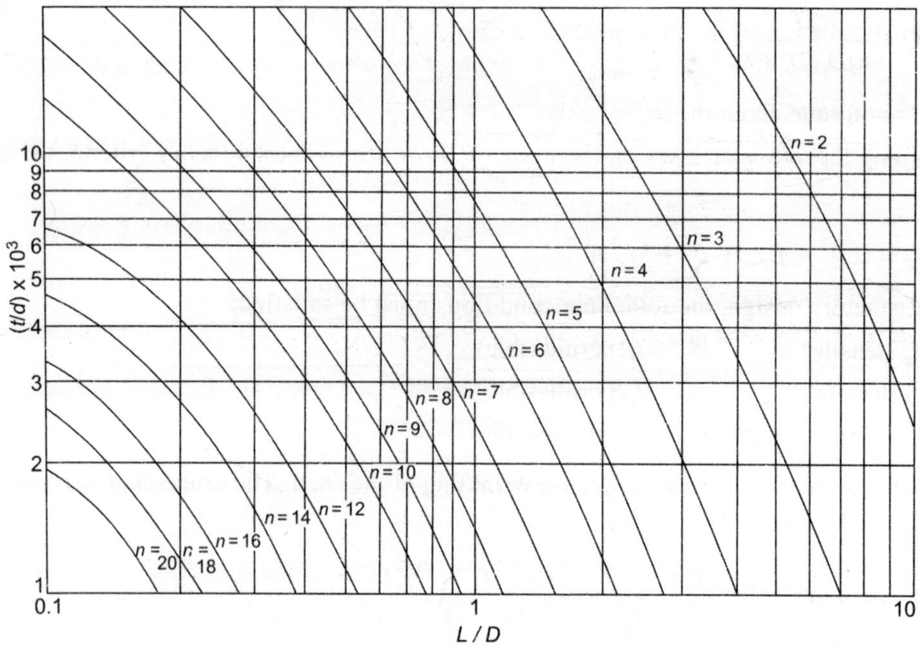

Fig. 5.1 Number of Lobes Formed at Buckling

$$p_c = \frac{1}{3}\left[n^2 - 1 + \frac{2n^2 - 1 - \mu}{n^2\left(\frac{2L}{\pi D_o}\right)^2 - 1}\right]\frac{2E}{(1-\mu^2)}\left(\frac{t}{D_o}\right)^3 + \frac{2E\frac{t}{D_o}}{(n^2-1)\left[n^2\left(\frac{2L}{\pi D_o}\right)+1\right]^2} \qquad (5.13)$$

where, L — unsupported length of the vessel

D_o — outside diameter

t — thickness without corrosion allowance

E — modulus of elasticity

μ — Poisson's ratio

n — number of lobes formed at buckling (Fig. 5.1).

In cases of shells where the ends are closed, the critical buckling pressure is given by

$$p_c = \left\{ \frac{1}{3}\left[n^2 + \left(\frac{\pi D_o}{2L}\right)^2\right]^2 \frac{2E}{1-\mu^2}\left(\frac{t}{D_o}\right)^3 + \frac{2E\left(\frac{t}{D_o}\right)}{\left[n^2\left(\frac{2L}{\pi D_o}\right)^2 + 1\right]^2} \right\} \times \frac{1}{\left(n^2 + \frac{1}{2}\right)\left(\frac{\pi D_o}{2L}\right)^2} \tag{5.14a}$$

An approximate equation is for quick computation of p_c is

$$p_c = \frac{2.42E}{(1-\mu^2)^{3/4}} \frac{\left(\frac{t}{D_o}\right)^{5/2}}{\left\{\left(\frac{L}{D_o}\right) - 0.45\left(\frac{t}{D_o}\right)^{1/2}\right\}} \tag{5.14b}$$

The distance L is measured as shown in Figs. 5.2(a) and 5.2(b). The other terms have the same meanings as in Eq. (5.13). When stiffening and reinforcing rings are used 'L' be considered as in Fig. 5.2(b).

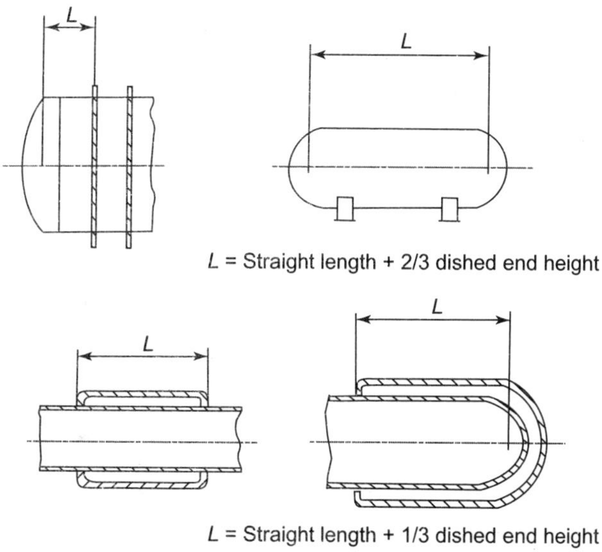

Fig. 5.2(a) Unsupported Length of Cylindrical Vessel

In practice, cylindrical vessels do not have a perfectly circular cross-section and it becomes necessary to calculate the reduction in the critical buckling pressure caused by this out of roundness in order to specify manufacturing tolerances and choose a safety factor. As a rough guide, if the maximum deviation from the true circular shape is equal to shell thickness, the critical buckling pressure will be less than 50% of the calculated value for the perfect shell. If it is one-tenth of the shell thickness, the reduction of critical buckling pressure is not more than 25%.

Considerable savings in weight and material can be made by use of *reinforcing rings or stiffening rings* attached to the inside or outside surface of the shell. These rings extend over the whole circumference and serve the same purpose as end supports. T-beams, flat plate rings welded with a continuous or intermittent seam are usually preferred as being simple, light and effective. I-beams, U-channels or angles bolted, or welded to the shell may also be used [Fig. 5.2(b)]. In these cases the length L is taken as the distance between stiffening rings. The required moment of inertia of the ring is given by

$$I = \frac{p_c D_o^3 L}{24E} \tag{5.15}$$

where, E — modulus of elasticity

D_o— outside diameter of shell

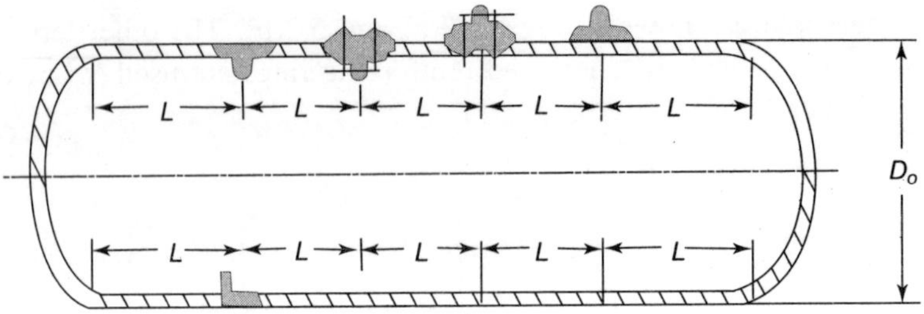

Fig. 5.2(b) Attachment of Different Types of Reinforcing Rings

The value of L may be reduced by about 30% to take into account the resistance of the shell. From the value of the critical pressure p_c, the critical compressive stress can be determined as

$$f_c = \frac{p_c D}{2t} \tag{5.16}$$

The permissible values of p_c and f_c can be worked out by using a factor of safety of 4.

It may be seen from Eq. (5.13) or Eq. (5.14) that to calculate the value of p_c, it is necessary to know the thickness t. To compute the critical pressure it is, therefore, necessary to select values of t/D and L/D. The working pressure is calculated with a factor of safety of 4. A trial and error procedure is essential to arrive at the requisite working pressure. Finally, it is necessary to check the maximum stress, which should not exceed the permissible value based on the ultimate stress or yield stress.

Indian Standard (IS-2825 – Appendix F) gives a procedure for calculating the allowable working external pressure based on the ratios L/D_o and D_o/t. Graphs have been supplied for different materials.

In addition to the external pressure the cylindrical shell may have to withstand axial loads, as well as wind loads. These may be considered as combined loadings and an equivalent stress may be determined from Eq. (5.11). The thickness t of the shell may be finally calculated by trial and error method to satisfy all the stress conditions. A suitable corrosion allowance may be added if required.

(b) *Spherical shell* The critical buckling pressure is given by

$$p_c = \frac{8E}{\sqrt{3(1-\mu^2)}}\left(\frac{t}{D}\right)^2 \tag{5.17}$$

and the corresponding critical stress is

$$f_c = \frac{2E}{\sqrt{3(1-\mu^2)}}\left(\frac{t}{D}\right) \tag{5.18}$$

The permissible value is obtained by assuming a factor of safety of 6.

The design procedure is to select a value of t/D and obtain the critical pressure and stress. These must satisfy the requisite external pressure and the permissible stress. A corrosion allowance may be added if necessary.

Indian standard (IS-2825 – Appendix F) gives a procedure for calculating the allowable external working pressure.

5.8.2 Head or Cover

To close either end of the cylindrical shell a cover or a closure is essential. This can be attached to the shell by a welded or bolted construction. The simplest cover is a flat plate of the same diameter as the shell. This plate can be firmly secured to the shell by a suitable method and must be capable of withstanding the working pressure. The more common types of closures are formed to a specific shape and are generally known as formed heads. The most economic formed head is that produced by simply forming a straight flange with a radius on a flat plate [Fig. 5.3(a)]. These heads are mainly used for horizontal cylindrical storage vessels at atmospheric pressure. The flat portion of such heads can be dished to form what are known as flanged and dished heads (Fig. 5.3). Depending on the depth of dishing and its shape these are classified as shallow dished, flared and dished, torispherical, elliptical, hemispherical and conical. Flared and shallow dished heads are used for low pressure vessels. Torispherical heads are used in the pressure range between 0.1 and 1.5 N/mm^2. Ellipsoidal heads (2:1) are preferable for pressures over 1.5 N/mm^2. Hemispherical heads are the strongest of the formed heads, but are expensive and are, therefore, used only in a few cases. In practice, the cost consideration decides the choice. Conical heads are widely used as bottom heads for a variety of equipment, where removal or draining of the material is facilitated. Cones having an angle of apex 60° are commonly used for the removal of solids.

See Appendix J for torispherical head dimensions.

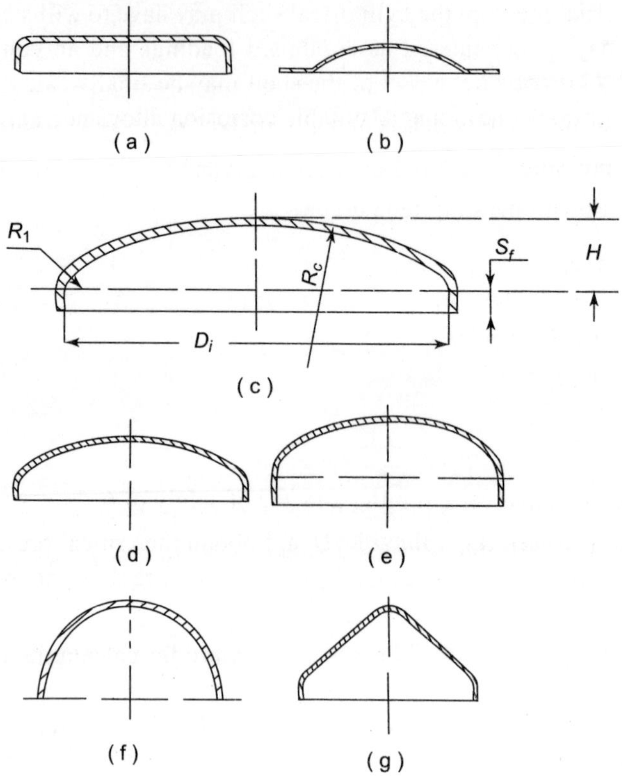

Fig. 5.3 Formed Heads: (a) Plain Formed Head, (b) Flared and Dished Head, (c) Shallow Dished Head, (d) Torispherical Dished Head, (e) Ellipsoidal Head, (f) Hemispherical Head, (g) Conical Head R_1: Knuckle Radius and R_c Crown Radius.

5.8.2.1 Flat Heads and Covers

Flat closures are often used as manhole covers in low-pressure vessels and small bore openings. Their use is generally limited to light duty service, since they are uneconomical for other applications. The stresses created in a flat plate, due to pressure, acting as a uniformly distributed load and with fixed edges are indicated under Section 3.9. The maximum stress is at the circumference and is given by (see Eq. 3.18).

$$f = \frac{3}{4}p\left(\frac{R}{t}\right)^2 = \frac{3}{16}p\left(\frac{D}{t}\right)^2 \tag{5.19}$$

The thickness of the plate can, therefore, be calculated as

$$t_h = CD\sqrt{\frac{p}{f}} \tag{5.20}$$

where, C— edge fixity constant

D— diameter of the plate which is actually under operating pressure

p — working pressure

f — design stress at operating temperature

R— $D/2$.

According to the method of attachment of the flat head to the shell the edge fixity changes. This is taken into account by the factor C, which varies from 0.4 to 0.7 (Fig. 5.4).

For the same working pressure and design stress, the thickness of a flat plate is several times the thickness of a shell. Further, due to the joint or the junction between the shell and the plate, severe localised stresses are created in the wall of the shell surrounding the junction. These are due to a sharp discontinuity in the shape at the junction. The substantial difference between the magnitude and direction of the deformation or dilations between the flat plate and the shell under pressure, and which are contained by the rigid junction results in localised shear and bending stresses. In low pressure service (below approximately 1.5 kg per cm^2), the magnitude of these stresses is small and can therefore, be ignored. For small diameter vessels slightly higher pressure may be possible. In general, for higher pressures it is preferable to use flanged and dished heads. Such heads are formed with a gradual change in shape (crown radius → knuckle radius → shell radius), to facilitate a smooth junction with the shell, resulting in a reduction of localised stresses at or near the junction.

5.8.2.2 Formed and Dished Heads

For a large number of cylindrical shells, dished heads with torispherical or elliptical shapes are commonly used. The hemispherical dished head is used in a few cases, due to excessive dish forming required. The thickness of these heads is calculated on the basis of the circumferential stresses, created either due to internal pressure (i.e., acting on the concave side) or external pressure (i.e., acting on the convex side).

Indian Standard-4049 gives specifications for formed heads for tanks and pressure vessels. These include different sizes of semi-ellipsoidal and torispherical heads for pressure vessels, tanks and non-pressure vessels. The length of the straight flange (S_f) on the head should not be less than three times the head thickness with a minimum of 20 mm (Fig. 5.3).

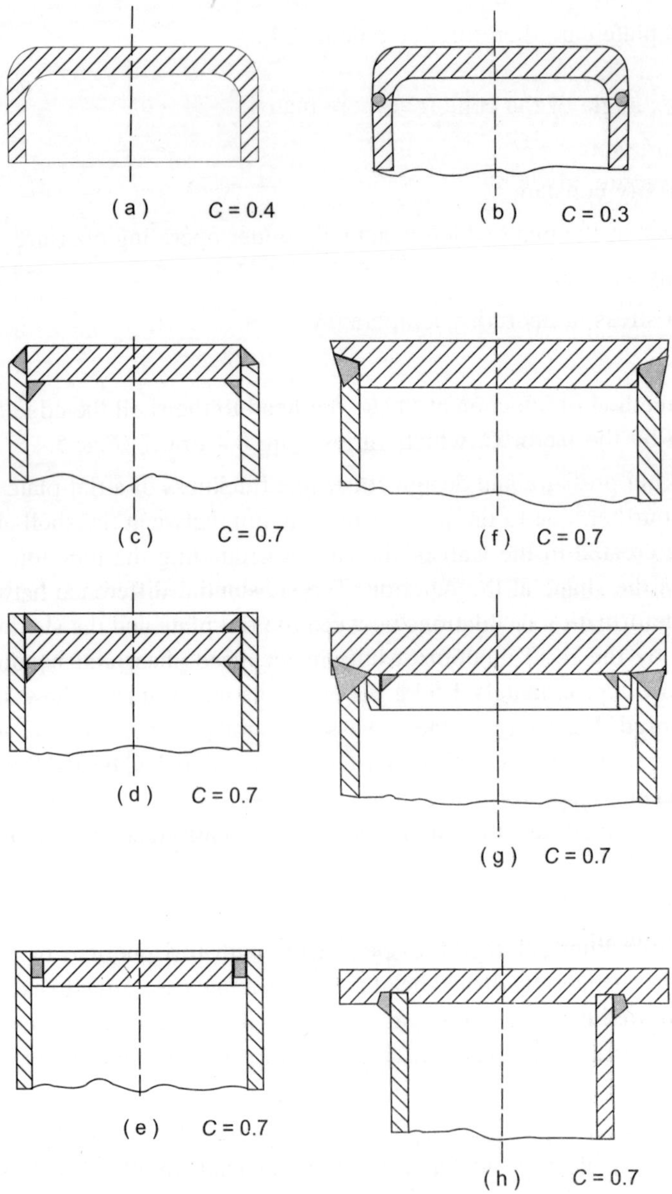

Fig. 5.4 Attachment of Flat Heads or Plates to Shell: (a) Forged Head, (b) Welded Head (c), (d), (e) Internal Plates, (f), (g), (h) External Plates

(a) *Internal Pressure*

(i) *Conical head* [Fig. 5.3(g)]: The circumferential stress in this type of formed head is given by

$$f = \frac{pD}{2t\cos\alpha} \tag{5.21}$$

where, α — half apex angle of the cone (not more than 45°)

p — design pressure

The thickness t_h is, therefore, given by

$$t_h = \frac{pD}{2fJ\cos\alpha} \tag{5.22}$$

At the junction of the conical head and the cylindrical shell, localized stresses are produced, which are likely to exceed the circumferential stress. In order to reduce the intensity of such localized stresses in the case of conical heads where the half apex angle exceeds 30°, provision has to be made to attach it to the shell, by a knuckle having an inside radius preferably not less than 10% of the internal diameter of the shell. In some cases, instead of providing a knuckle, a compression ring welded to the shell at the junction may be acceptable.

(ii) *Shallow dished and torispherical head* [Figs. 5.3(c) and (d)]: These heads have the dish radius or crown radius equal to or less than the diameter of the head. To reduce the stresses at the corner of the head, a knuckle is formed, with a radius of at least 6% of the inside diameter. The thickness of such heads is given by

$$t_h = \frac{pR_cW}{2fJ} \tag{5.23}$$

where, p — internal design pressure

R_c — crown radius

W — stress intensification factor $= \dfrac{1}{4}\left(3 + \sqrt{\dfrac{R_c}{R_1}}\right)$

R_1 — knuckle radius.

(iii) *Elliptical Head* (*Semi-ellipsoidal*) [Fig. 5.3(e)]: The thickness of such heads is given by

$$t_h = \frac{pDV}{2fJ} \tag{5.24}$$

where, p — internal design pressure

D_o — major axis of ellipse

V — stress intensification factor $= \dfrac{1}{6}(2 + k^2)$

$$k = \frac{\text{major axis}}{\text{minor axis}}$$

A common value of k is 2 making $V = 1.0$

(iv) *Hemispherical Head* [Fig. 5.3(f)]: The thickness of a hermispherical head is given by

$$t_h = \frac{pD_o}{4fJ + p} = \frac{pD_i}{4fJ - p} \tag{5.25}$$

(b) *External Pressure*

(i) *Conical head:* Depending on the apex angle these heads are designed either as a cylindrical shell under external pressure or as a flat plate. If the apex angle is less than 45°, the conical head is designed as a cylindrical shell with the same diameter as the large end of the cone and a length equal to the axial length of the cone. Heads with apex angle between 45° to 120° are designed as a shell, except that the diameter at the large end of the cone is also taken as the length of an equivalent cylinder. Heads with apex angles greater than 120°, are designed as a flat plate having a diameter equal to the largest diameter of the cone.

(ii) *Torispherical, elliptical and hemispherical heads:* These heads are designed for external pressure, with the application of the same equations, as those indicated for internal pressure, except that the internal pressure in these equations is to be taken as 1.67 times the external pressure.

In addition, these heads are designed on the basis of prevention of collapse due to buckling. They should have sufficient elastic stability.

Assuming a suitable factor of safety for the buckling pressure, the thickness is given by

$$t_h = \alpha R_c \left\{3(1-\mu^2)\right\}^{1/4} \sqrt{p/2E} + c \tag{5.26}$$

where, c — corrosion allowance

p — design external pressure

R_c — crown radius for torispherical and hemispherical heads and equivalent crown radius for elliptical head

E — modulus of elasticity

α — factor of safety may be = 4 to 4.4

In all the above cases of heads or covers, a corrosion allowance may be added to the thickness as required by the rate of corrosion.

IS 2825 presents a graphical calculation procedure.

5.8.2.3 Attachment of Head and Shell

Heads are attached to the shell by a welded or flanged joint. In the case of a welded joint, either a lap or butt construction is made according to the thickness of the shell and head. Figure 5.5 shows welded joints. Flanged joints are discussed under Section 5.8.4. These are detachable, unlike the welded joints which are permanent.

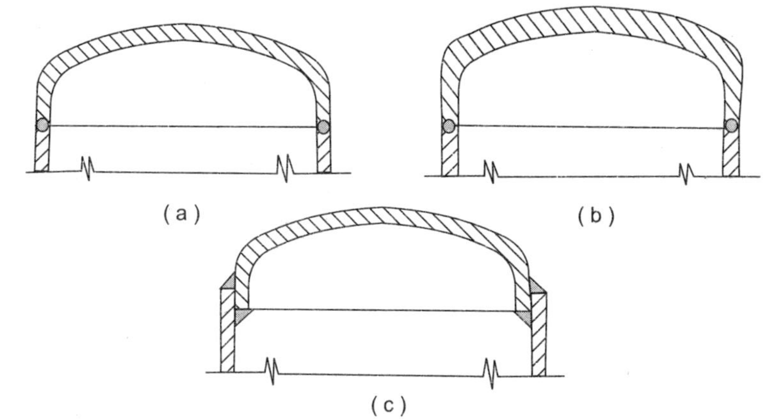

Fig. 5.5 Attachment of Formed Heads to Shell: (a) Butt Welded, (b) Butt Welded with Unequal Thickness of Shell and Head, (c) Lap Welded

5.8.3 Nozzles

Nozzles or openings are provided in the pressure vessels, to satisfy certain requirements such as inlet or outlet connections, manholes, handholes, vents and drains, etc. These may be located on the shell or head according to functional requirements and are circular or elliptical or round (with a ratio of major to minor inside diameter not exceeding 2). On the basis of the method of forming and attachment, nozzles can be classified as (a) integral, (b) fabricated and (c) formed.

(a) Integral nozzles are fabricated from a portion of the shell or head, by cutting and shaping the material to obtain the contour of the nozzle. This operation can be satisfactorily carried out only in cases where the material has sufficient ductility. These nozzles are, therefore, adopted only in a few cases.

(b) Fabricated nozzles are essentially short pieces of pipes, tubes and plates. They are cut to specific length from standard fittings and are welded to the vessel wall at an opening made in the wall. Such nozzles can be conveniently and satisfactorily used for relatively small openings and for low and medium pressures. Figures 5.9 to 5.15 show different types of fabricated nozzles welded to shell or head. Figure 5.6 shows a few methods of welding a piece of pipe to the vessel wall, while Fig. 5.7 shows three methods of welding a pad to the vessel wall. The pads are provided with tapped holes for attachment to other components. Figure 5.8 shows an extended nozzle welded to flanges and lined with a corrosion resistant material. Figure 5.9 shows a manhole formed out of a pad with a cover attached by studs. Figure 5.10 shows a nozzle attached to shell with a reinforcement pad. The nozzle is free from the vessel wall. A cover is also shown. Figure 5.11 shows a pipe length extending inside a vessel through a nozzle. At one end of the pipe a bend is welded while at the other end a flange is welded. The arrangement facilitates the removal of the pipe from the vessel. Figure 5.12 shows a similar arrangement with an extended nozzle and the bend which is screwed. Figures 5.13 and 5.14

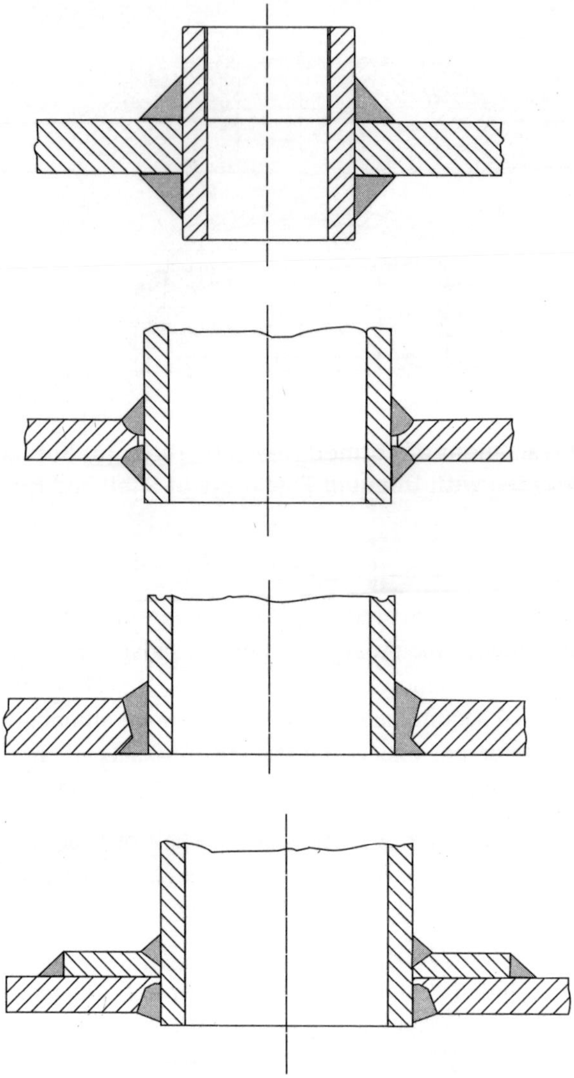

Fig. 5.6 Attachment of Pipes to Vessel Wall

show arrangements of fixing a sight glass. To prevent the fog formation on the surface of the glass, two glass pieces are provided as shown in Fig. 5.14. The other possible reason could be safety, particularly if the vessel contents are at high temperature. There is permissible Δt for glass. Between the two glass pieces electrical heating is provided. Figure 5.15 shows the sight glass. Generally, above 10 bar pressure sight glass is not used. As may be seen from the above figures, it is possible to provide nozzles in various ways to satisfy certain functional requirements. These nozzles may be connected to a pipe or valve by use of threads or flanges and bolts.

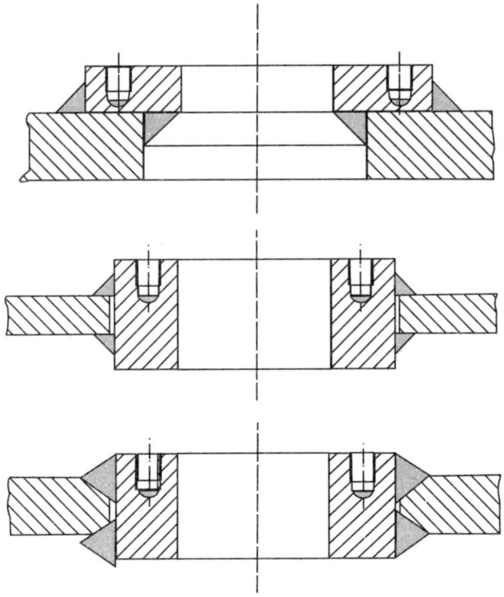

Fig. 5.7 Attachment of Pads to Vessel Wall

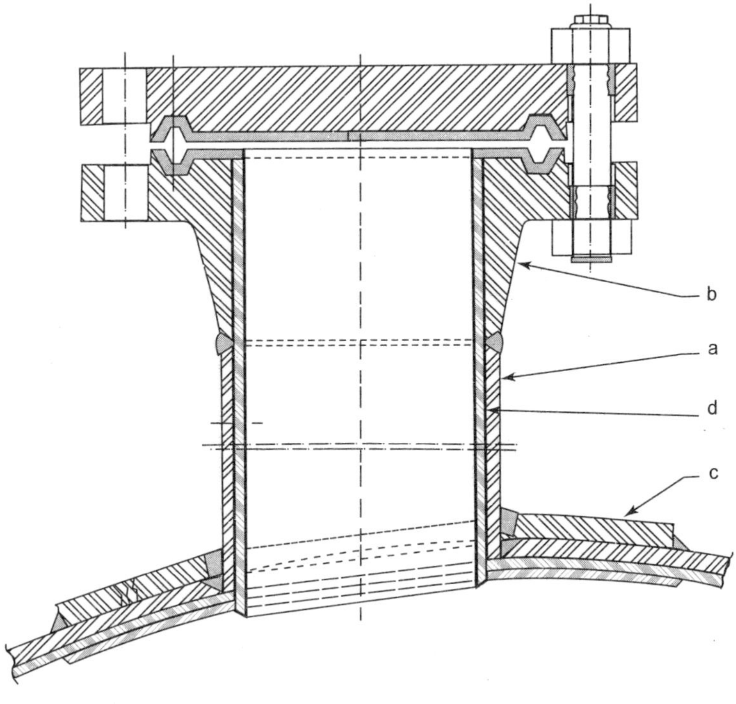

Fig. 5.8 Extended Nozzle with Flanged Joint: (a) Pipe, (b) Welding Neck Flange, (c) Reinforcement, (d) Lining

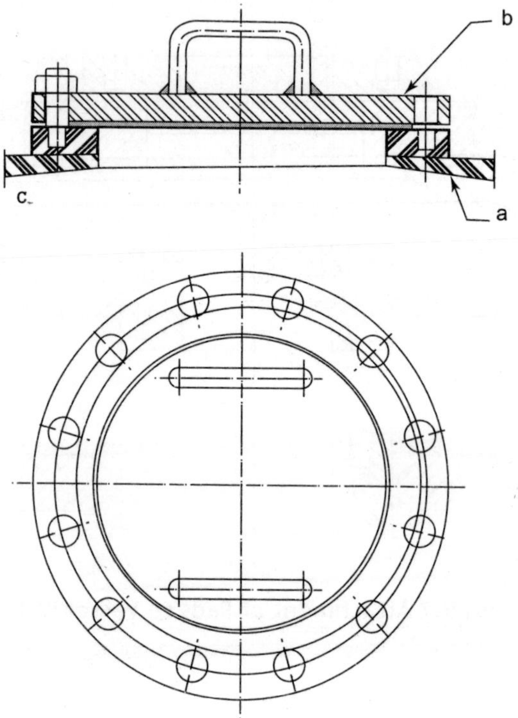

Fig. 5.9 Manhole: (a) Vessel Wall, (b) Cover

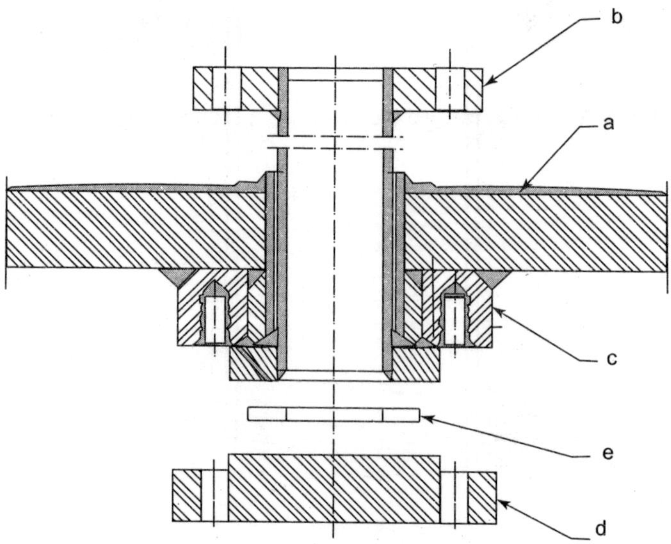

Fig. 5.10 Nozzle with Flanges at Either End: (a) Vessel Wall, (b) Flange, (c) Reinforcement, (d) Blank Flange, (e) Gasket

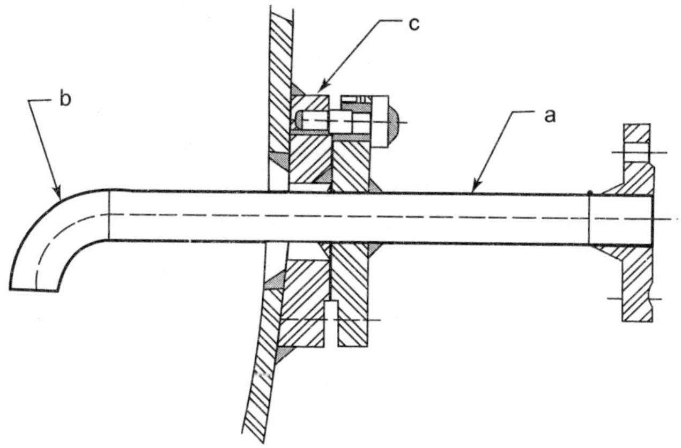

Fig. 5.11 **An Extended Pipe through a Nozzle: (a) Pipe, (b) Bend, (c) Pad for Nozzle**

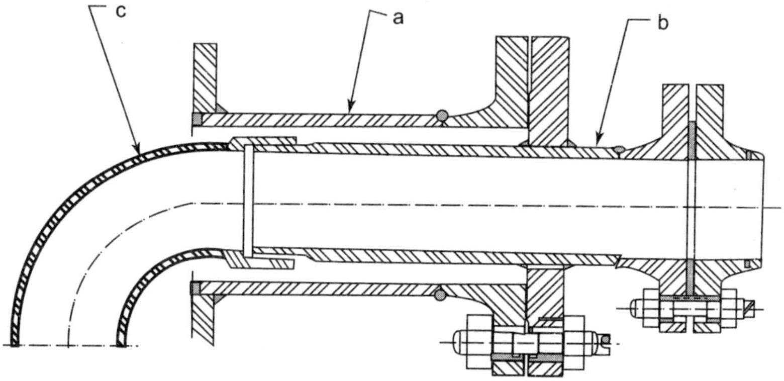

Fig. 5.12 **A Pipe through an Extended Nozzle: (a) Extended Nozzle, (b) Pipe, (c) Bend**

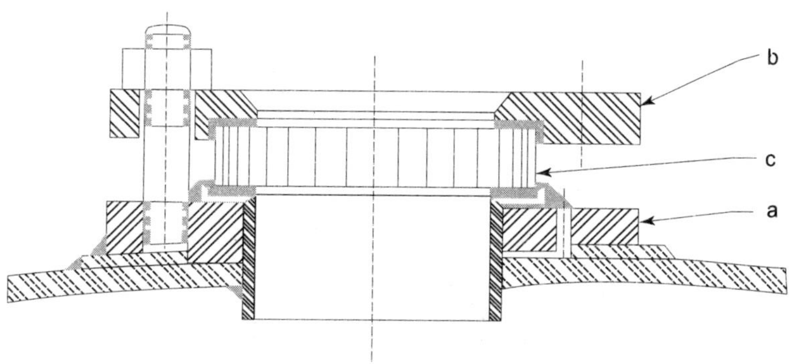

Fig. 5.13 **Sight Glass: (a) Pad, (b) Cover, (c) Glass**

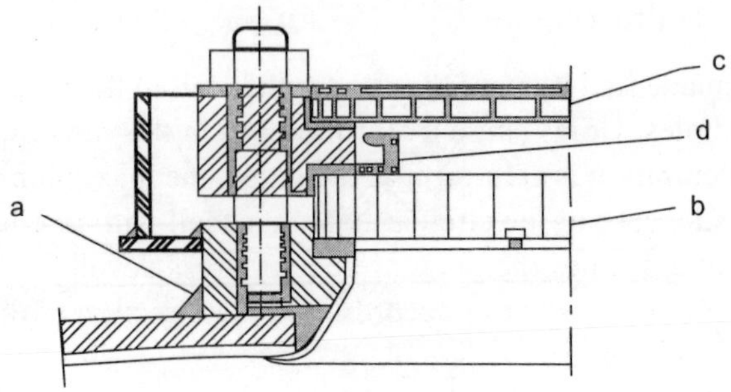

Fig. 5.14 Heated Sight Glass: (a) Pad, (b) Sight Glass, (c) Auxiliary Glass, (d) Heater

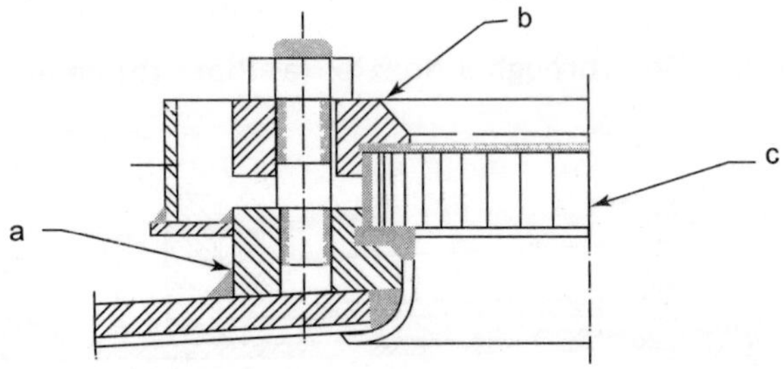

Fig. 5.15 Sight Glass: (a) Pad, (b) Cover, (c) Sight Glass

(c) Formed nozzles are fabricated to a specific shape and size, preferably with flanges. These are formed by rolling or forging and are attached to the vessel at an opening made in the wall. Standard formed nozzles of various sizes are manufactured to withstand different temperature and pressure ratings. IS-3133 gives the specifications for manhole and inspection openings for chemical equipment: Figure 5.16 shows different types of formed nozzles.

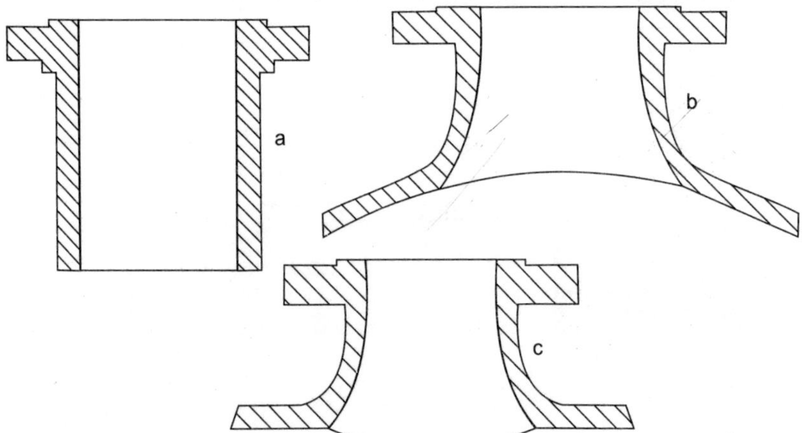

Fig. 5.16 Formed Nozzels: (a) Welding Neck, (b) and (c) Sweep Type

5.8.3.1 Stresses and Reinforcement

Openings in vessels are made first by making holes in the wall of the vessel. Nozzles are then formed or welded around these holes. Holes cause a discontinuity in the vessel wall. Due to the pressure or other loads a stress concentration is created near the holes, the maximum value of the stress being at the edge of the holes. In the case of the circular hole in a shell with internal or external pressure, the stress in the hole is about 2.5 times the circumferential stress developed in the shell. In the case of elliptical holes the maximum stress varies according to the direction of the axis and also the ratio of the major to minor axis of the ellipse.

The stress concentration at the hole can be reduced by increasing the thickness of the vessel in the vicinity of the nozzle. This can be done either by providing additional thickness to the vessel wall itself near the nozzle or by use of a separate reinforcing plate attached to the vessel wall covering an area surrounding the hole. Sometimes the nozzle wall at the base can be made sufficiently thick to act as a reinforcement. In all such cases the reinforcement is provided with a view to satisfy the following requirements:

(a) The weakening due to the hole made for the nozzle has to be compensated by sufficient additional material. However, this should not disturb the general strain pattern prevailing in the wall of the vessel.

(b) The reinforcing material should be placed immediately adjacent to the hole, but suitably disposed in profile and contour, so as not to introduce any stress concentration.

For a circular hole, considering the variation of stress in an area surrounding the hole, and the deflection characteristics of the nozzle, it has been observed that the satisfactory extent of reinforcement is up to a radius of four times the radius of the hole.

All openings up to 50 mm in size can be left uncompensated. If two openings in vessels have a distance equal to or more than the value L min, then the interaction between them can be neglected and they can be treated as isolated openings.

$$L_{\min} = \frac{d}{1 - 0.95(t_s'/t_s - C)} = 2d$$

whichever is larger.

However, if the clearance between any two openings is less then $L_{\min}$ then the conpensation required will be equal to that for a single opening of the sum of the two diameters of openings.

Figure 5.17 slows the reinforcement boundary limits, viz., ABCD. The hole are EFGH removed from the vessel wall is compensated partly by shaded nozzle area and additionally by compensating reinforcing material, located within the boundary ABCD. When the reinforcing material is placed within the above boundary, maximum reinforcement is obtained with the minimum material.

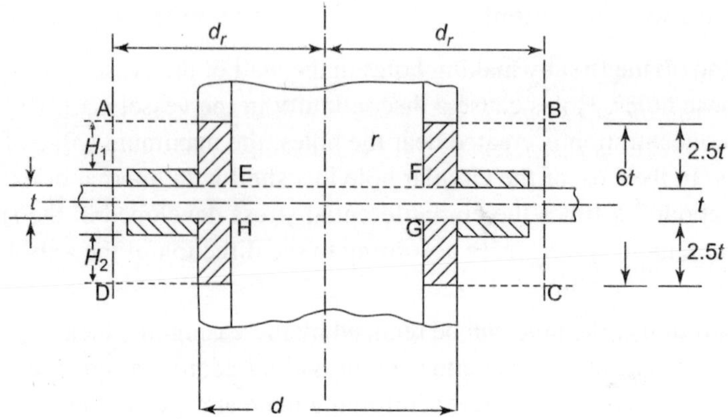

Fig 5.17 Reinforcing Pad Limits for Nozzle

The reinforcing material can be placed: (a) On both sides of the vessel wall (balanced) [Fig. 5.18(a)]. (b) On the inside or outside surfaces of the vessel wall (unbalanced)[Figs. 5.18(b) and (c)]. In the first case the reinforcement is balanced and, therefore, prevents creation of local bending moments and stresses. In the second case the reinforcement is unbalanced.

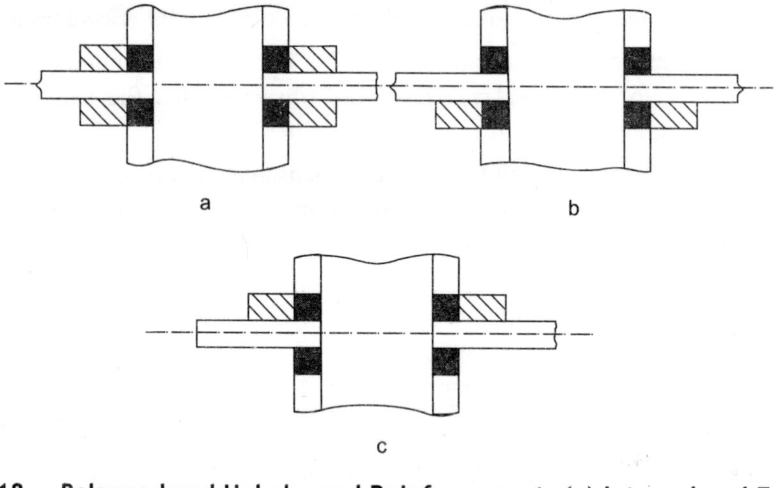

Fig. 5.18 Balanced and Unbalanced Reinforcement: (a) Internal and External Reinforcement, (b) Internal Reinforcement, (c) External Reinforcement

Reinforcement on the inside surface is preferable to the outside surface, in the second case. However, the recommended practice is to have reinforcement outside to keep it away from corrosive contents of the vessel.

5.8.3.2 Design of Reinforcement

There are three methods of designing reinforcement for a nozzle.

(a) *Area for area method:* This method has been suggested in codes. It has proved to be both simple in application and essentially reliable in service.

(b) *Controlled maximum stress method:* In this method the maximum stress in the vessel adjacent to the nozzle is restricted to a fixed multiple of the design stress. This multiple is usually 2.25.

(c) *Experimental yield method:* This method is based on pressure tests for a series of cylinder/cylinder junctions, and the pressure (P_{02}) to produce 0.2% residual strain at the junctions are used to derive 'weakening factors'. The design pressure equals (P_{02}) 1.5, so that the nozzle has the same reserve on yield as the membrane areas where the design stress equals the yield stress divided by 1.5.

(a) *Area for area method of compensation (internal pressure):* The area to be replaced and the effective area of compensation are shown in Fig. 5.17 by the limits of ABCD.

The maximum horizontal distance AB = $4r = 2d$

The maximum vertical distance AD = $6t$, or $(3.5t_s + 2.5t_n)$ whichever is smaller.

If the compensation is only provided by the nozzle then

$$H_1' = H_2 = 2.5t_s$$

If the compensation is to be provided by a combination of nozzle and a compensation ring, then

$$H_1 = 2.5t_n$$

The area for which compensation is required is given by

$$A = dt_s$$

Area available for compensation:

(i) The portion of the shell or head as excess thickness

$$A_s = d(t_s - t_s' - C)$$

(ii) The portion of the nozzle external to the vessel

$$A_0 = 2H_1(t_n - t_s' - C)$$

(iii) The portion of the nozzle inside the vessel

$$A_1 = 2H_2(t_n - 2C)$$

Where the calculated area A is greater than $(A_s + A_0 + A_1)$, additional compensation

$$= A - (A_s + A_0 + A_1),$$

is to be provided by use of a compensation ring welded to the nozzle and shell as shown in Fig. 5.17.

r — inner radius of nozzle

d — inner diameter of nozzle

t_s — actual thickness of the shell or head

t_s' — theoretical minimum thickness of the shell or head calculated with suitable, joint efficiency (Eqs. 5.3, 5.23, 5.24)

C — corrosion allowance

t_n — actual thickness of the nozzle

t_n' — theoretical minimum thickness of the nozzle calculated with suitable joint efficiency (Eq. 5.3)

H_1 — height of effective compensation in branch wall external to vessel measured from outside surface of compensation height ring or vessel wall

H_2 — height of any portion of the branch pipe projecting inside vessel and effective as compensation, measured from the inside surface of vessel or compensation ring.

If the permissible stress for the compensating ring material is different from that of the shell material, the compensation provided by the ring is to be modified accordingly.

Area of compensation (external pressure): This is only 50% of that required with internal pressure, where t_s is the required minimum thickness under external pressure.

5.8.4 Flange Joints

When heads of pressure vessels are to be of the detachable type, they must be provided with flanges, with an arrangement for bolting. Similarly, nozzles and piping systems may also be provided with flanges. The use of such bolted-flanged connections which are readily removable, facilitate disassembly of different components.

A flanged joint consists of a pair of flanges, one each attached to the two components to be joined, held securely together by a series of bolts or studs. A gasket is interposed between the two adjoining flange faces. A variety of flanges, gaskets and bolts are available for this purpose. *The joint must have structural integrity with negligible leakage during service.*

5.8.4.1 Flanges

Different types and sizes of flanges are fabricated either by casting, forging or formed from structural sections and plates. Flanges used for connections between shell section and between the shell and the head are standardized (IS-4864 to 4869). These are made of carbon steel and are welded to carbon steel or stainless steel pressure vessel shells. According to the type of flange and working temperature, these flanges can be used up to 25 kg/cm^2 pressure. Steel and cast iron flanges used for pipe joints are also standardized (IS-6392 and IS-6418). Pipe flanges are subject to bending and other stresses due to piping connections, unlike the shell flanges which are free from such external bending moments. Figures 5.19 and 5.20 show different types of standard flanges. For large diameter vessel subjected to high pressure, the flange neck will be thicker than vessel shell. Non-standard shell flanges can be formed from a structural section such as an angle iron or from a plate. These can be strengthened by gusset plates and may be used for low and medium pressure applications. Figure 5.21 shows two flanges one fabricated out of an angle iron (shell flange) and the other from a plate

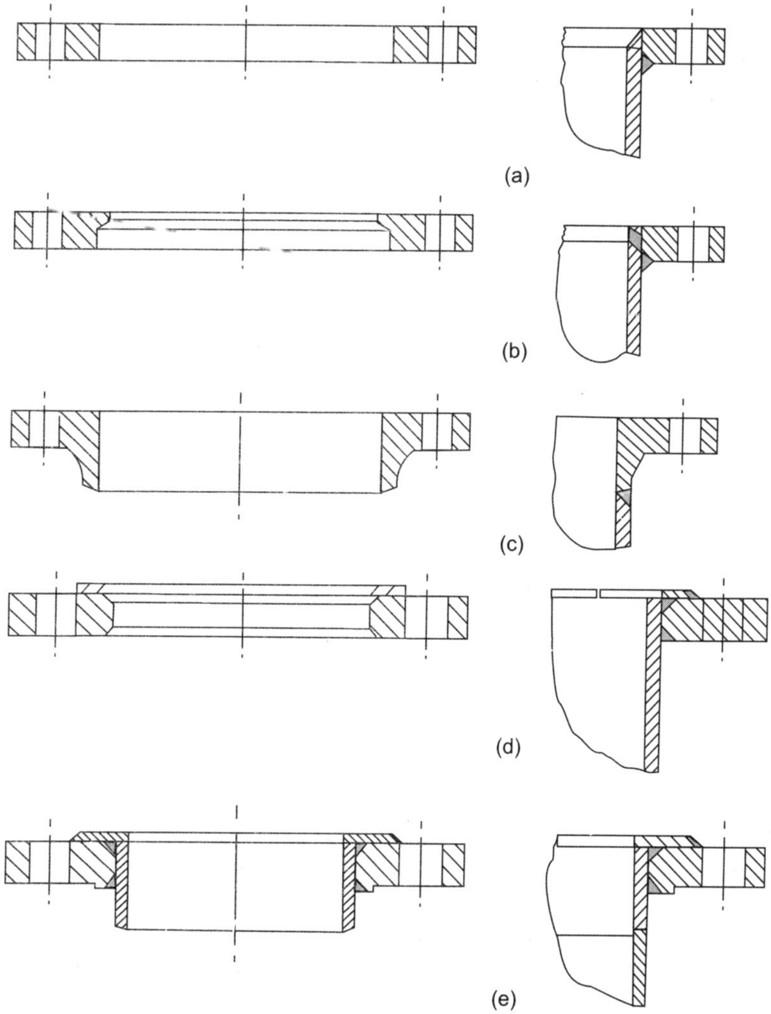

Fig. 5.19 Standard Flanges (Shell Flanges): (a) Ring Flange for Non-pressure Service, (b) Ring Flange for Carbon Steel Pressure Vessels, (c) Welded Neck Flange for Carbon Steel Pressure Vessels, (d) Ring Flange (with Stainless Steel Lining) for Stainless Steel Pressure Vessels, (e) Hub Type Flange (with Stainless Steel Lining) for Stainless Steel Pressure Vessels

(cover flange). The vessel shell and its conical had are made of special material such as copper, while the flanges are made of carbon steel. The arrangement is used for low pressure. A nozzle is also shown at the top of the head. Figure 5.22 shows a flared and dished head attached to the shell. The flange of the shell is fabricated from a plate and is strengthened by gusset plates. Figure 5.23 shows a similar arrangement except that the flange portion of the head is strengthened by an additional ring. A tongue and groove facing is provided. Figure 5.24 shows flanges made out of carbon steel plates,

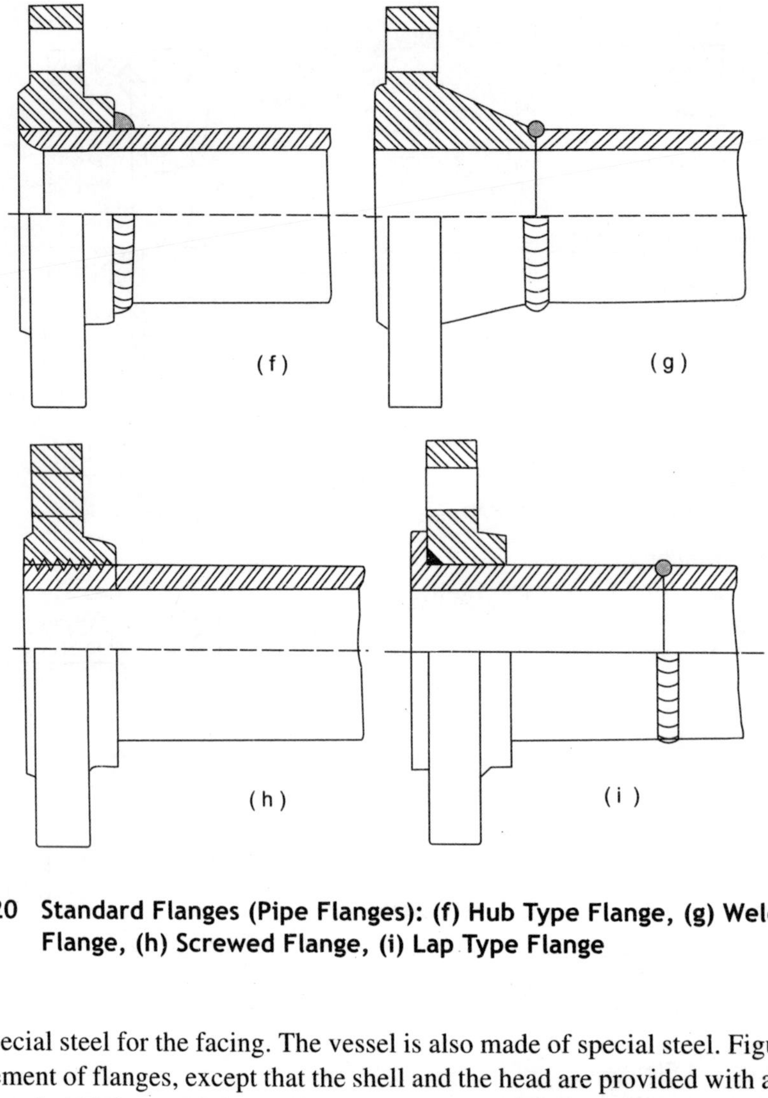

Fig. 5.20 Standard Flanges (Pipe Flanges): (f) Hub Type Flange, (g) Welding Neck Flange, (h) Screwed Flange, (i) Lap Type Flange

but lined with special steel for the facing. The vessel is also made of special steel. Figure 5.25 shows a similar arrangement of flanges, except that the shell and the head are provided with an outside ring, thereby increasing the thickness of the portion adjacent to the flange. This method is adopted when the vessel thickness is not enough to carry out a satisfactory welding of the carbon steel flanges to the shell or head. Figure 5.26 shows loose flanges held tight by bolts. These flanges are made to rest on the rings welded to shell and head. The flange facing the gasket is provided on the ring. Figure 5.27 shows special hub type flanges and stud with locking arrangement. This arrangement is particularly used in cases where the vessel has to sustain continuous vibrations, probably from certain rotating components. Figure 5.28 is an arrangement which combines the essential features of the earlier figures, but in addition it provides a jacket with a flange.

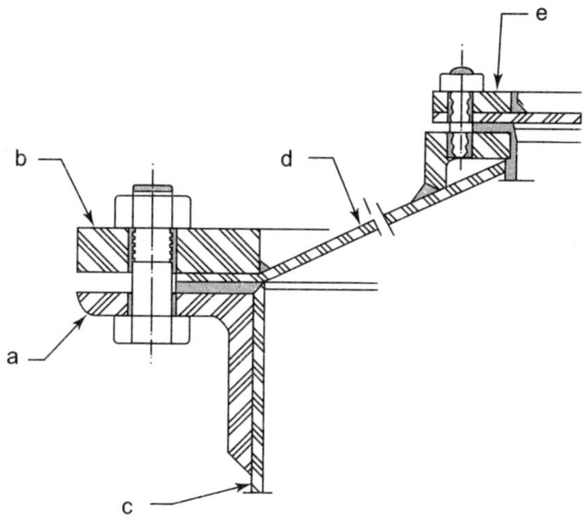

Fig. 5.21 Non-standard Flanges: (a) Angle Ring Flange for Shell, (b) Ring Flange for Head, (c) Shell, (d) Conical Head, (e) Nozzle

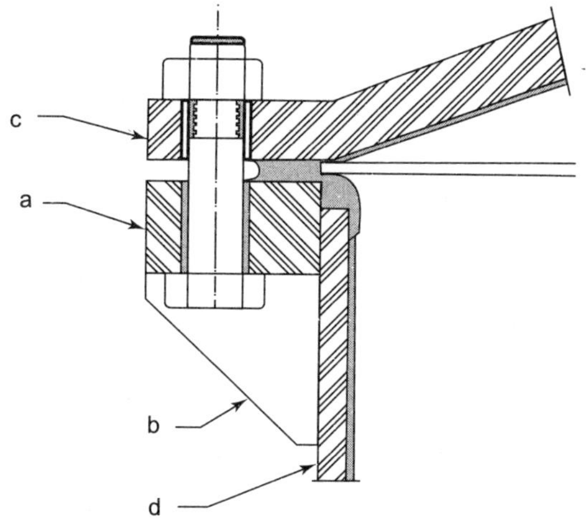

Fig. 5.22 Non-standard Flanges: (a) Flange for Shell, (b) Gusset plate, (c) Flared and Dished Head, (d) Shell

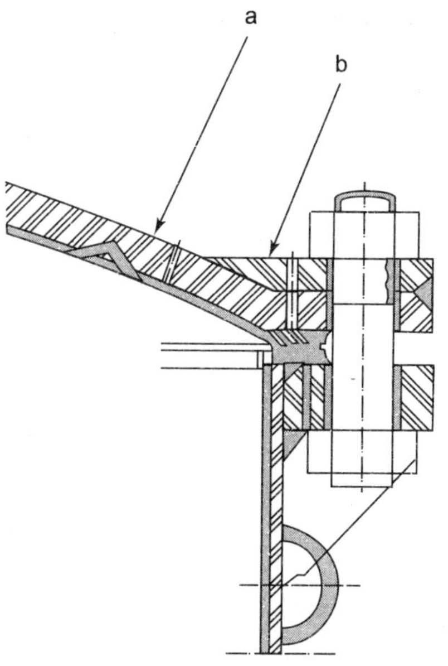

Fig. 5.23 Non-standard Flanges: (a) Flared and Dished Head, (b) Stiffening Ring

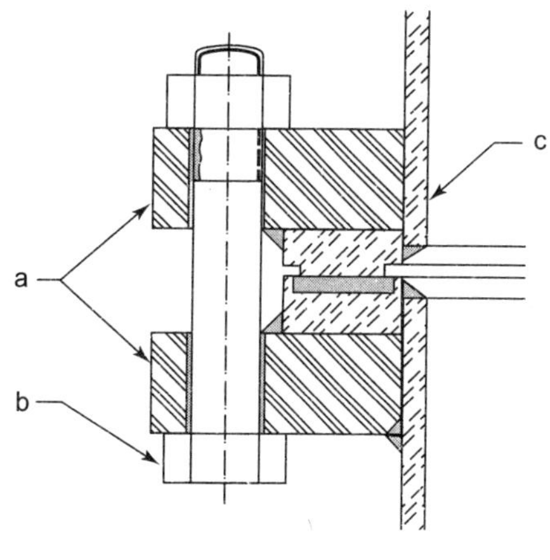

Fig. 5.24 Ring Flanges Lined with Special Steel: (a) Flanges, (b) Bolt, (c) Lining

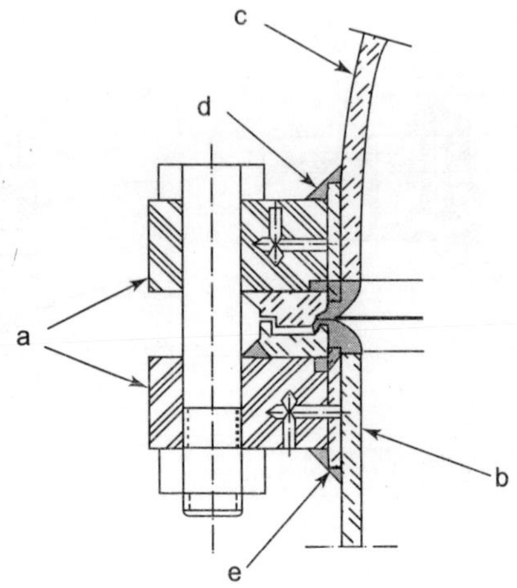

Fig. 5.25 Ring Flanges Lined with Special Steel: (a) Flanges, (b) Shell, (c) Head, (d) Ring for Head, (e) Ring for Shell

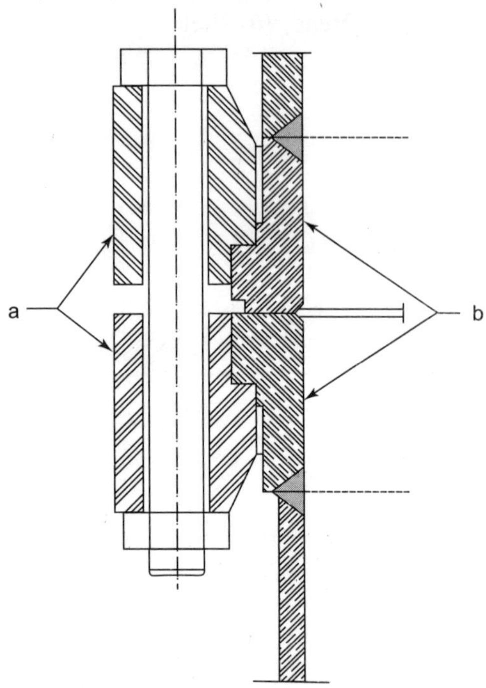

Fig. 5.26 Loose Flanges: (a) Flanges, (b) Shell and Head Rings

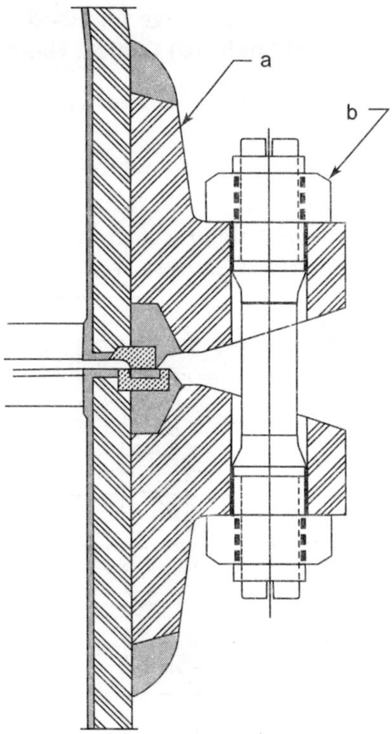

Fig. 5.27 Special Flanges with Studs: (a) Welding Neck Flange, (b) Stud with Nut

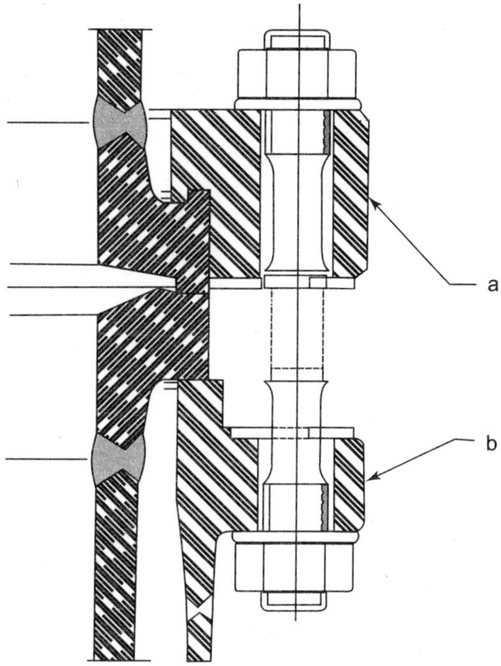

Fig. 5.28 Loose Flanges with Jacket Flange: (a) Loose Flanges, (b) Jacket Flange

Figures 5.29 and 5.30 show ring type flanges with eye bolts, which are hinged. This arrangement is used where the cover is to be dismantled frequently. Figure 5.31 shows special flanges connected by swing bolts and clamps, instead of normal bolts and nuts. No holes are made in the flange. The clamp can be either hinged or free. The arrangement facilitates dismantling.

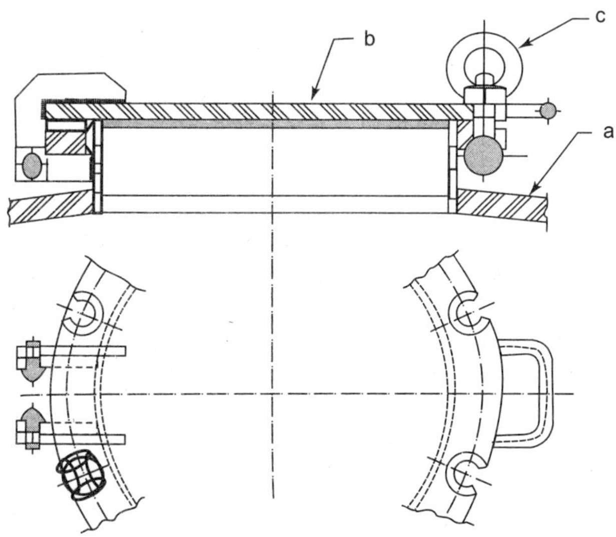

Fig. 5.29 Ring Flanges with Eye Bolts: (a) Vessel Wall, (b) Hinged Cover, (c) Eye Bolt

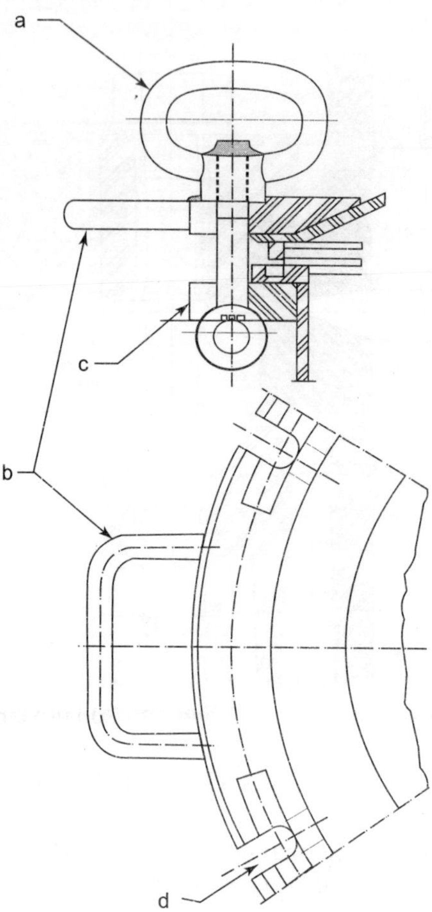

Fig. 5.30 Ring Flanges with Lining and with Eye Bolts: (a) Eye Bolt, (b) Handle for Cover Flange, (c) Shell Flange, (d) Slot for Eye Bolt

5.8.4.2 Flange Facings

In order to locate the gasket in an appropriate position on the flange faces, these faces are formed to specific shapes and finish. According to the width of gasket covering the flange facing, these can be divided into two broad types.

(A) *Wide-face flanges* The gasket in this case extends over the full width of the flange face. These are suitable only when used with comparatively soft gaskets and should not be used for pressures exceeding 20 kg/cm^2 or temperatures exceeding 220°C [Fig. 5.32 (a)].

(B) *Narrow-face flanges* The gasket in these cases does not extend beyond the inside of bolt holes. The facings are as follows (Fig. 5.32).

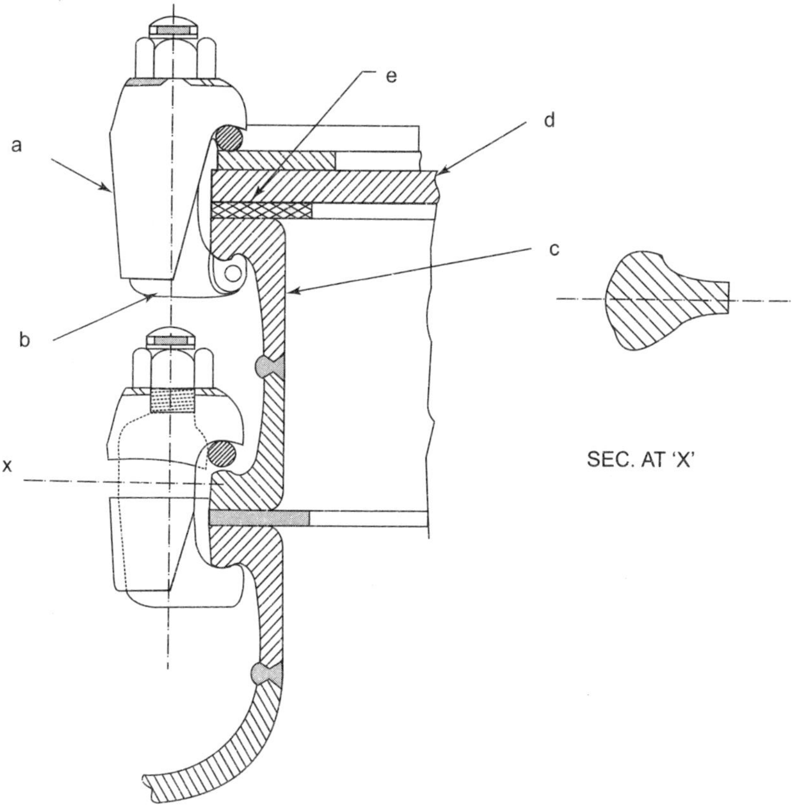

Fig. 5.31 Special Flange with Swing Clamps: (a) Clamp, (b) Swing Bolt, (c) Nozzle Wall, (d) Cover, (e) Gasket

(a) A portion of the flange face is raised to locate the gasket. The raised face is finished with spiral or concentric grooves to hold the gasket. This type of facing is commonly employed. The gasket used is in the form of a flat ring of some soft material and is usually equal in width to the width of the raised face [Fig. 5.32 (b)].

(b) Male and female facings are formed with one flange facing recessed and the other flange face having a corresponding raised face portion. The surface of the faces is smooth, or serrated. The outer diameter of the female face acts to locate and retain the gasket [Fig. 5.32 (c)].

(c) Tongue and groove facings have a groove formed on one flange facing and a raised ring on the other known as tongue. The gasket is held in position in the groove on both the inner and the outer diameter. The small tongue and groove construction requires only a narrow flat gasket [Fig. 5.32 (d)].

(d) Ring facing is formed by making identical grooves usually trapezoidal in shape on both the flange facings. The gasket in the form of a ring fits into the grooves. The ring is well protected, since it is retained on the grooves [Fig. 5.32 (e)].

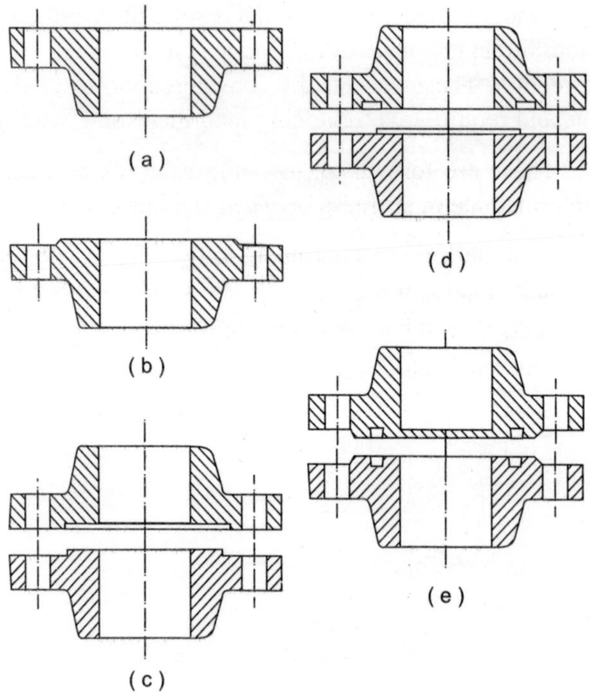

Fig. 5.32 Flange Facings: (a) Plain Face, (b) Raised Face, (c) Male and Female Facings, (d) Tongue and Groove Facing, (e) Ring Facing

5.8.4.3 Gaskets

Gaskets are interposed between two adjacent flange faces and are held tight by a series of bolts, which clamp the two flanges together. The gasket is, therefore, compressed, which causes a yielding of its surface, thus sealing the irregular surface of the flange faces. According to properties and the shape, different types of gaskets can be made (Tables 5.3 and 5.4).

(a) ***Flat ring gaskets*** These may be cut from sheets of different materials such as paper, cloth, rubber, compressed asbestos, soft iron, nickel, copper, aluminium, teflon, PVDF and teflon coated compressed asbestos, etc. Combinations of metals and non-metals can also be used. Their thickness normally range from 0.5 mm to 3 mm. Widths are usually 6 mm upwards. Narrow gaskets are preferable since they require less force for tightening, but they must not be too narrow otherwise they are likely to be crushed under the tightening force. Paper, cloth and rubber should not be used above 120°C. Asbestos may be employed up to 350°C, while ferrous or nickel base metal gaskets are satisfactory for higher temperatures. IS-4870 gives dimensions of flat gaskets for shell flanges.

(b) ***Serrated gaskets*** These are flat metal gaskets having concentric grooves or serrations machined on to their faces. With serrations the contact area is reduced to a few concentric lines, and the tightening force is greatly reduced and an efficient joint is obtained in cases where soft gaskets are unsuitable.

(c) ***Laminated gaskets*** These are made of metal and a soft filler, usually soft asbestos sheet. In some gaskets the laminations are in the plane of the flange face, in others the laminations are formed in an axial or edgewise direction. These gaskets are considered more suitable for high temperature than plain asbestos gaskets and require less tightening load than solid metal gaskets.

(d) ***Corrugated gaskets*** These are formed by inserting asbestos between the corrugations of a metallic sheet. They are intermediate in stiffness between flat non-metallic and metallic gaskets.

(e) ***Ring gaskets*** These are in the form of continuous rings with usually octagonal or oval cross-sections. Such rings are normally made for metals for severe service conditions, but for relatively low temperatures, rings made of plastic or rubber may be employed.

(f) ***Self-sealing gaskets*** Rings of different cross-section such as Lens, Delta, Grayloc wave are formed for use at high pressures. The are so shaped that a rise in pressure makes the sealing action more effective and hence the name self-sealing.

(g) ***Spinal wound gaskets*** This gasket is more suitable for high temperature and high pressure application. It is also used in the case where there is variation in stress due to frequent changes in temperature and pressure. It consists of a thin corrosion resistant metal strip wound spirally, with or without filler material, such as graphite, teflon, etc.

IS 2825 may be referred to for more details of gaskets.

5.8.4.4 Design of Flanged Joint

As indicated earlier, standard steel flanges (IS-4864-4867) for shell sections, heads and pipes are selected according to the pressure and temperature requirements. In such cases, no design calculations are necessary. However, where standard flanges cannot be used due to non-availability of specified material; special shape or size of flange, more elaborate bolting or clamping arrangements, etc., it becomes necessary to follow a design procedure which involves selection of a gasket, flange facing, bolting, hub proportions, flange width and flange thickness.

In designing a satisfactory flanges joint, the main considerations are: (1) To ensure a positive contact pressure at the gasket-flange interfaces under all service conditions and to prevent leakage. The sealing force or the tightening force must satisfy this requirement. (2) To create a tightening force by the use of bolts, without over-stressing the bolts. (3) To ensure structural integrity of the flange faces, and to minimise deflections of the flanges.

Under the action of the tightening force, the gasket is compressed and stressed. It, therefore, gives rise to a reaction. Three types of reactions are likely to be created.

(a) ***Elastic reaction*** When the gasket is not over-stressed by the tightening force and its mass remains within the elastic limit. Only the surface of the gasket yields, filling the irregularities of the flange faces, but the thickness or the mass of the gasket remains practically underformed, i.e., it is stressed within the elastic limit.

(b) ***Plastic reaction*** When the gasket is over-stressed and its mass becomes plastic.

(c) *Springy reaction* When the gasket is fully enclosed and resists compression like a confined fluid. Under varying pressure conditions plastic reaction is unsatisfactory, since once the mass of the gasket becomes plastic, it is unable to adjust itself to the required thickness and leakage starts.

The sizing of the gasket, and for the determination of the number and diameter of bolts necessary to create the tightening action, it is essential to evaluate the forces due to gasket reaction, both under atmospheric conditions and also under the operating pressure condition. These forces are resisted by the bolts, creating a load on the bolts.

Under atmospheric conditions, the bolt load due to gasket reaction is given by

$$W_a = A_g Y_a \tag{5.27}$$

where, A_g — area of gasket under compression

Y_a — stress in the mass of the gasket (seating stress).

Y_a should be less than the yield stress of the gasket if elastic reaction is expected.

After the internal pressure is applied, the gasket which is compressed earlier, is released to some extent and the bolt load is given by (Fig. 5.33):

$$W_a = A_g Y_p + A_h p \tag{5.28}$$

where, Y_p — residual stress in the mass of the gasket

A_h — area of the flange on which the pressure is exerted

p — operating pressure.

The change in the gasket mass stress is

$$\Delta Y = Y_a - Y_p \tag{5.29}$$

The initial tightening action causes a compression of the thickness of the gasket, extension of the bolt length and deflection of both the flanges. Under pressure the gasket is released and its thickness changes which is equal to the change in bolt length plus the change in the deflection of both the flanges. This is given by

$$\Delta T = \Delta L - 2\Delta f \tag{5.30}$$

where, ΔT — change in thickness of gasket

ΔL — change in bolt length

Δf — change in deflection of each flange.

Applying Hooke's law

$$\frac{T\Delta Y}{E_g} = \frac{L\Delta W}{E_b A_b} + 2C\Delta W \tag{5.31}$$

E_g, E_b — modulus of elasticity of the gasket and bolt material respectively

T — thickness of gasket

L — length of bolt

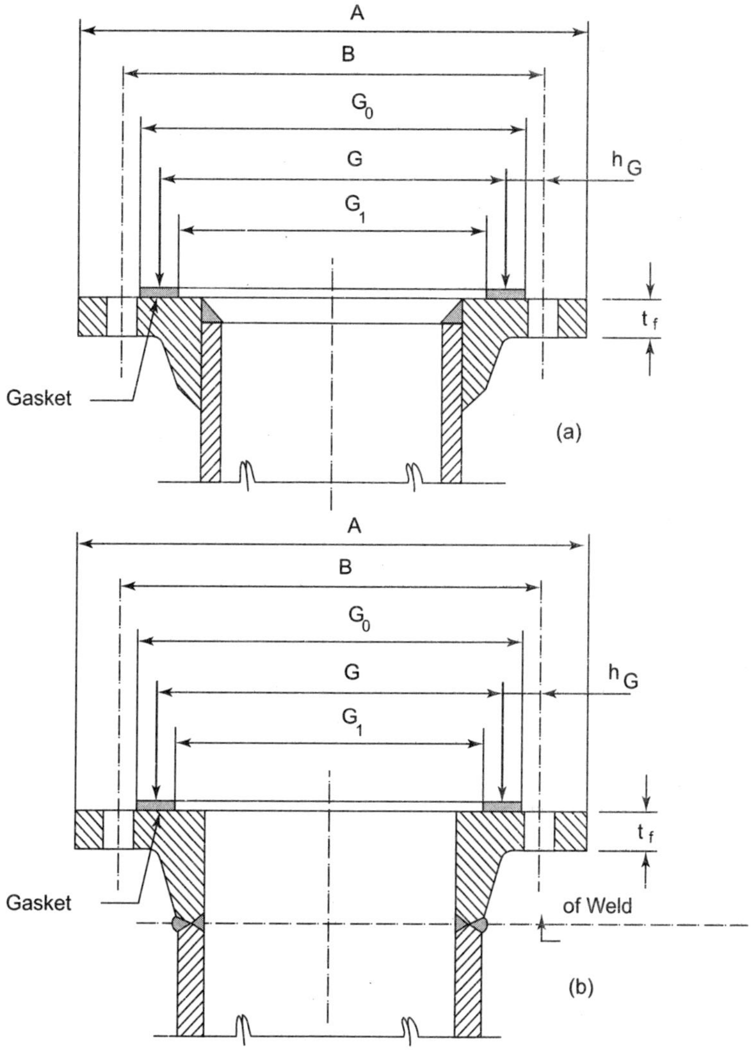

Fig. 5.33 Flange Dimensions: (a) Lap Welded, (b) Butt Welded

$\Delta W = (W_p - W_a)$ (change in bolt load)

C — deflection factor of the flange.

From Eqs 5.27, 5.28 and 5.29

$$\Delta W = (W_p - W_a) = A_h p + \Delta Y A_g \tag{5.32}$$

From Eqs 5.31 and 5.32

$$\Delta Y = \frac{BA_h p}{\dfrac{T}{E_g} + BA_g} \tag{5.32a}$$

where, $B = \dfrac{L}{E_b A_b} + 2C$

From Eqs 5.32 and 5.32a, change in bolt load

$$\Delta W = W_p - W_a = A_h p \left\{ 1 - \frac{BA_g}{\dfrac{T}{E_g} + BA_g} \right\} \tag{5.33}$$

From this equation it may be observed that in case of metallic gaskets whose modulus of elasticity (E_g) is very large, the term T/E_g is negligible. The value of the change in bolt load ΔW is almost zero. In such gaskets the bolt load during operating pressure conditions as well as atmospheric conditions remains practically constant. In the case of rubber or plastic materials, where E_g is less or in the case of thick gaskets, the term T/E_g is significant and, therefore, the bolt loads under both conditions must be assessed.

It is difficult to ascertain the value of the residual stress (Y_p) in the mass of the gasket, under pressurized conditions. However, it must be of sufficient magnitude to ensure that the joint remains tight, without leaking. In effect, a minimum compression of the gasket is necessary to seat the gasket properly. Tests on actual gaskets have shown that, if the joint is to remain pressure tight, when subjected to internal pressure, it is necessary to ensure that the residual gasket stress (Y_p), should be greater than m times the pressure (p). The term m (which is greater than 1) is known as gasket factor, which is dependent on the frictional resistance of the flange and the gasket contact surfaces. It is, therefore, a property of the gasket material and its construction.

Tables 5.3 and 5.4 show gaskets, materials with their properties, namely seating stress (Y_a), gasket factor (m) and basic seating widths (b_o). Gasket selection can be made from these tables, depending upon the temperature, pressure and corrosion properties of the confined material. For low temperature and low pressure service, rubber, fibre and asbestos sheet material are commonly used. For pressures up to 20 kg/cm^2, and temperature up to 250°C the most common materials used are compressed asbestos sheet and metallic laminated asbestos sheet. For higher temperatures, corrugated metallic gaskets and plain iron aluminium, copper and monel sheet gaskets are used. When pressure exceeds 20 kg/cm^2 and temperature exceeds 400°C plane metal gasket are preferred.

Gasket may be either wide, covering almost the whole face of the flange or may be narrow. In most applications, a narrow gasket is preferred. An approximate estimate of proportions of the gasket may be made as follows:

Gasket seating force – Hydrostatic pressure force = Residual gasket force

Table 5.3 Gaskets Types and Materials

Sketch	Dimension N (min) mm	Gasket Material	Gasket Factor	Minimum Design Seating Stress N/mm²	Refer to Table 6.8 Use Facing Sketch	Refer to Table 6.8 Use Column
	10	Rubber without fabric or a high percentage of asbestos fibre: Below 70 IR HD 70 IR HD or higher	 0.50 1.00	 0 1.4	1 (a, b, c, d) 4, 5	II
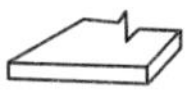		Asbestos with a suitable binder for the operating conditions: 3.2 mm thick 1.6 mm thick 0.8 mm thick	2.00 2.75 3.50	11.2 26 45.7		
		Rubber with cotton fabric insertion	1.25	2.8		
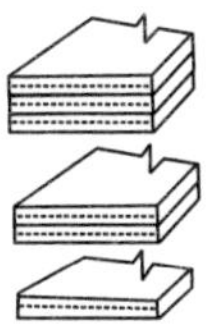		Rubber with asbestos fabric insertion, with or without wire reinforcement: 3-ply 2-ply 1-ply	2.25 2.50 2.75	15.5 4, 5 20.4 26	1 (a, b, c, d)	
		Vegetable fire	1.75	7.7		
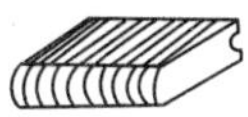		Spiral-wound metal, asbestos filled: Carbon steel stainless steel or monel metal	2.50 3.00	20.4 31.6		
 		Corrugated metal, asbestos inserted or Corrugated metal, jacketed asbestos filled: Soft aluminium Soft copper or brass Iron or soft steel Money metal or 4–6% chrome steel Stainless steel	2.50 2.75 3.00 3.25 3.50	20.4 26 31.6 38.7 45.7	1 (a, b)	II
		Corrugated metal: Soft aluminium Soft copper or brass Iron or soft steel Monel metal or 4–6% chrome steel Stainless steel	2.75 3.00 3.25 3.50 3.75	26 31 38.7 45.7 53.4	1 (a, b, c, d)	

(*Continued*)

Table 5.3 (*Contd.*)

Sketch	*Dimension N (min) mm*	*Gasket Material*		*Gasket Factor*	*Minimum Design Seating Stress N/mm²*	*Refer to Table 6.8* *Use Facing Sketch*	*Use Column*
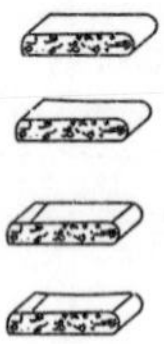		Flat metal jacketed asbestos filled	Soft aluminium	3.25	38.7	1a, 1b, 1c, 1d, 2	
			Soft copper or brass	3.50	45.7		
			Iron or soft steel	3.75	53.4		
			Monel metal or 4–6% chrome steel	3.50	56.2		
		Grooved metal	Stainless steel	3.75	63.3	1 (a, b, c, d) 2, 3	II
			Soft aluminium	3.25	38.7		
			Soft copper or brass	3.50	45.7		
			Iron or soft steel	3.75	53.4		
			Monel metal or 4–6% chrome steel	3.75	63.3		
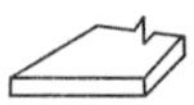	6	Solid flat metal	Stainless steel	4.25	71	1 (a, b, c, d) 2, 3, 4, 5	I
			Soft aluminium	4.00	61.9		
			Soft copper or brass	4.75	91.4		
			Iron or soft steel	5.50	126.6		
			Monel metal or 4–6% chrome steel	6.00	153.3		
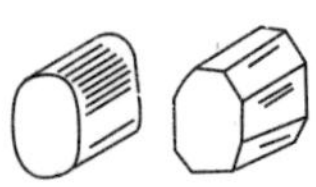		Ring joint	Stainless steel	6.50	182.8	6	
			Iron or soft steel	5.50	126.6		
			Money metal or 4–6% chrome steel	6.00	153.3		
			Stainless steel	6.50	182.8		
		Rubber o-rings					
		Below 75 IRHD		3	0.0		
		Between 75 and 85 IRHD		6	1.5	7 only	
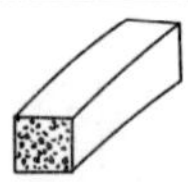		Rubber square section rings;					
		Below 75 IRHD		4	1.0		
		Between 75 and 85 IRHD		9	2.8	8 only	II
		Rubber T-section rings:					
		Below 75 IRHD		4	1.0		
		Between 75 and 85 IRHD		9	2.8	9 only	

Table 5.4 Basic Gasket Seating Widths b_o

Facing Sketch — Exaggerated	*Column I*	*Column II*
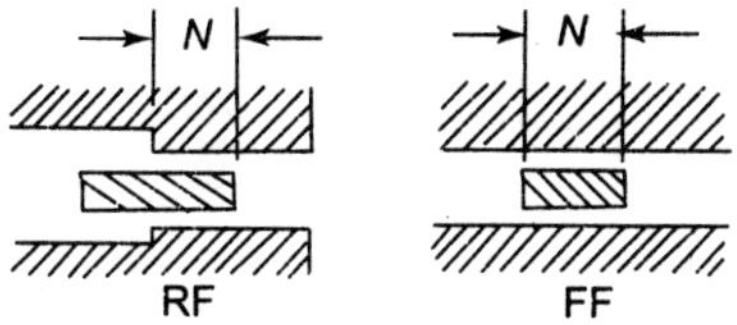	$\frac{N}{2}$	$\frac{N}{2}$
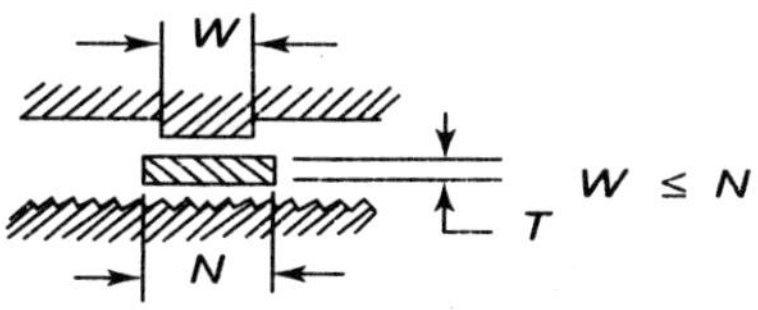	—	—
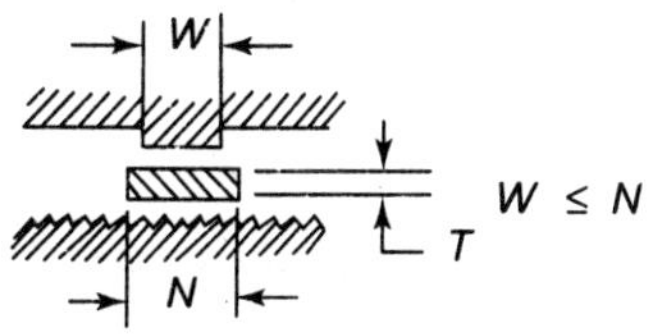	$\frac{w+25T}{2}$; $\frac{w+N}{4}$ Max	$\frac{w+25T}{2}$; $\frac{w+N}{4}$ Max
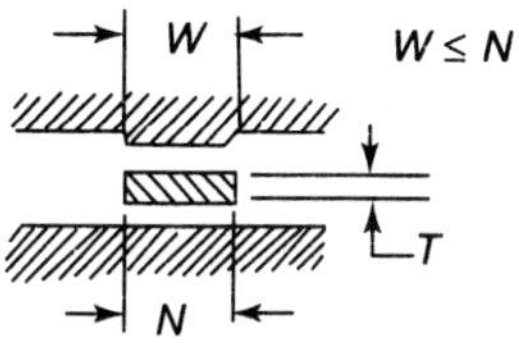	—	—
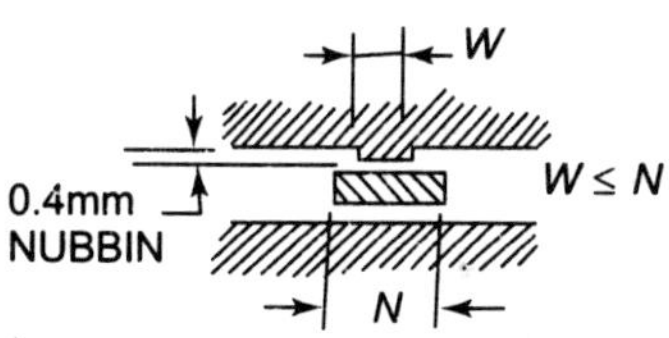	$\frac{w+N}{4}$	$\frac{w+3N}{8}$
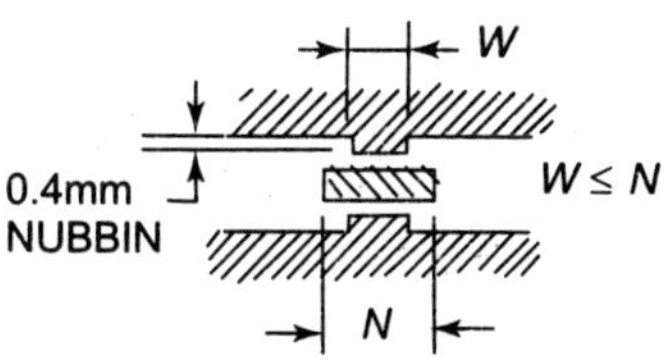	$\frac{w}{2}$; $\frac{N}{4}$ Min	$\frac{w+N}{4}$; $\frac{3N}{8}$ Min

(*Continued*)

Table 5.4 (*Contd.*)

Facing Sketch — Exaggerated	*Column I*	*Column II*
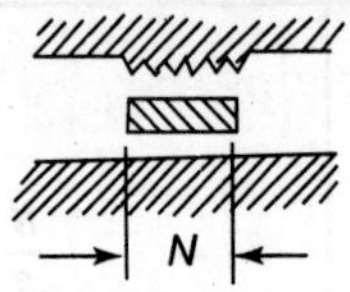	$\frac{3N}{8}$	$\frac{7N}{16}$
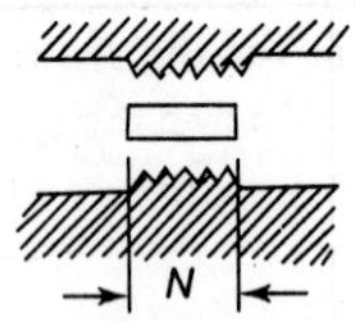	$\frac{N}{4}$	$\frac{3N}{8}$
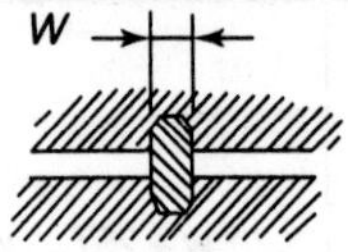	$\frac{w}{8}$	—
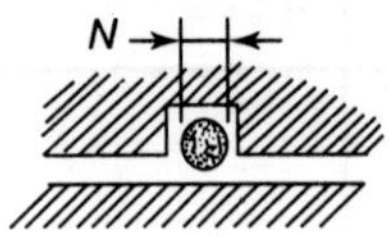	—	$\frac{N}{2}$
	—	$\frac{N}{2}$
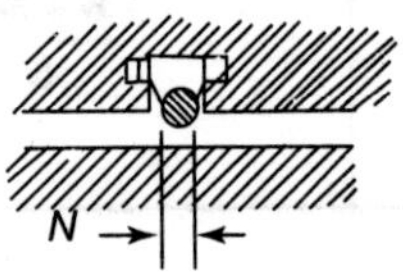	—	$\frac{N}{2}$

$$\left[\frac{\pi}{4}(G_o^2 - G_i^2)Y_a\right] - \left[\frac{\pi G_o^2}{4}p\right] = \frac{\pi}{4}(G_o^2 - G_i^2)Y_p = \frac{\pi}{4}(G_o^2 - G_i^2)mp \tag{5.34}$$

G_o and G_i are the outside and the inside diameters respectively of the gasket ring (Fig. 5.33), m is the gasket factor and p the design pressure.

$$\frac{G_o}{G_i} = \sqrt{\frac{Y_a - pm}{Y_a - p(m+1)}} \tag{5.35}$$

Based on the above equations, an appropriate size of the gasket, suitable for the requisite flange facing may be chosen. Gaskets having different values of the seating stress (Y_a) and the gasket factor (m) may be tried to arrive at a suitable width, and to satisfy the operating conditions.

5.8.4.5 Bolts

The minimum bolt load and the number and size of the bolts can be determined from Eqs. 5.27 and 5.28, which can now be written in terms of the size of the gasket (Figs 5.33 and 5.34).

Atmospheric condition

$$W_{m_1} = A_g Y_a = \pi b G Y_a \tag{5.36}$$

where b — effective gasket or joint contact surface width (from Table 5.4 and Fig. 5.34)

G — diameter of gasket load reaction

Y_a — gasket seating stress

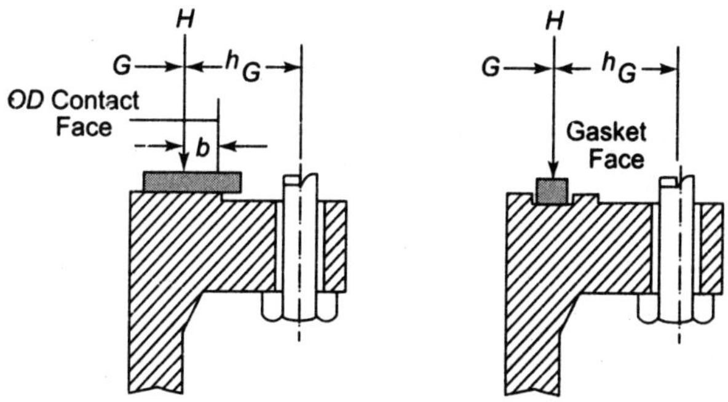

For $b_o > 6.3$ mm For $b_o \leq 6.3$ mm

b = Effective gasket seating width

b_o = Basic gasket seating width (From Table 5.3)

$b = b_o$ when, $b_o \leq 6.3$ mm

$b = 2.5\sqrt{b_o}$ when, $b_o > 6.3$ mm

Fig. 5.34 Effective Gasket Seating Width

Operating condition

$$W_{m_2} = A_g Y_p + A_h p = \pi\,(2b)Gmp + (\,\pi/4)G^2 p \tag{5.37}$$

m — gasket factor (Table 5.3)

$2b$ — effective gasket or joint contact surface width under pressure

p — design pressure.

The bolt loads either W_{m_1} and W_{m_2} will create a tensile stress in the cross-section of the bolt.

$$A_{m_1} = \frac{W_{m_1}}{f_a} \quad \text{and} \quad A_{m_2} = \frac{W_{m_2}}{f_b} \tag{5.38}$$

where, W_{m_1} and W_{m_2} — cross-section of bolt

f_a — permissible tensile stress in bolts under atmospheric conditions

f_b — permissible stress in bolts under operating conditions.

Further,

$$A_{m_1} \text{ or } A_{m_2} = A_m \times \text{Number of bolts} \tag{5.39}$$

where, A_m = area of bolt.

The actual bolt are provided (A_b) should not exceed the value given by

$$A_b = \frac{2\pi Y_a G N}{f_a} \tag{5.40}$$

where, N — width of gasket (Table 5.4)

The actual bolt area (A_b) should be greater than A_{m_1} or A_{m_2}. To determine the number and size of bolts, the larger of the above two areas (A_{m_1} or A_{m_2}) should be considered. A large number of small sized bolts or a small number of large sized bolts can be selected. The bolt spacing must be enough to facilitate the use of wrenches. However, a large bolt spacing would cause a deflection of the flange. If this deflection is excessive, the joint will leak. The pitch between two bolts should be decreased with pressure. It may generally be between 3.5 and 5 times the diameter of the bolt. The total number of bolts may be approximately equal to the mean diameter of gasket in centimetres divided by 2.5 and should be a multiple of four. The final choice of the number of bolts will depend on the proper bolt spacing. Bolts smaller than 15 mm should only be used when considerable care is taken during tightening up to prevent yielding. The bolt stress area for the chosen bolt nominal size is given in IS 1367 part III, 1979. However, one can quickly calculate the bolt stress area with the help of expression (Eq. 5.41), without sacrificing engineering accuracy. After all, one computes the stress and then ensures that bolts are adequate.

$$Y = a\,(X)^b \tag{5.41}$$

where, Y = bolt area, mm^2

X = nominal diameter, mm

$a = 0.51$ and $b = 2.09$

The next step is to determine the bolt circle diameter, consistent with the gasket diameter and bolt spacing. The final choice will be made only when the diameter of the flange hub is determined. The hub can be short and straight as in the case of a hub flange, or tapered as in the case of a welding neck flange. The slope of the tapered of the hub may be equal to or less than 1 : 4. Assuming a suitable outside diameter of the hub at the base, the bolt circle diameter and the outside diameter of the flange can be determined. Within the practical limitations, the bolt circle, the gasket (centre line) circle, and the radial distance between them should be kept as small as possible.

5.8.4.6 Flange Thickness

An approximate value of the flange thickness may be given by (Fig. 5.33):

$$t_f = G\sqrt{\frac{p}{k \cdot f}} + c \tag{5.42}$$

where, $$k = \frac{1}{\left[0.3 + \dfrac{1.5\,W_m h_G}{HG}\right]}$$

G — diameter of gasket load reaction

p — design pressure

f — permissible stress

B — bolt circle diameter

c — corrosion allowance

W_m — total bolt load

h_G — radial distance from gasket load reaction to bolt circle

$$= \left(\frac{B - G}{2}\right)$$

H — total hydrostatic end force $= \dfrac{\pi}{4} G^2 p$.

With low pressures the thickness of flange may work out to be negligible. It is necessary to provide a minimum thickness. The minimum flange thickness for carbon steel is between 20 mm and 40 mm depending on the diameter of flange.

A rigorous approach to the determination of stresses created in the flange and the calculation of thickness is given in Indian Standard-2825 (paras 4.6, 4.7 and 4.8).

5.9 SUPPORTS

Details and design considerations of supports used for pressure vessels are indicated in Chapter 13. These are non-pressure parts and are, therefore, designed essentially as structural members. They are attached to the vessel by welding and produce local moments and stresses in the vessel wall area surrounding the attachment.

5.10 STRESSES FROM LOCAL LOADS AND THERMAL GRADIENT

As has been pointed out earlier under Sections 5.2.1 and 5.8.1.2, apart from the three important loads namely, pressure, weight of vessel with its contents and wind load, there are a number of other forces which give rise to localized stresses in the vessel parts. These forces are due to

(a) Local loads arising from
   (1) supports for the vessel
   (2) structures supported by the vessel
   (3) loads imposed on branches by the piping system, etc.
(b) Steady and transient thermal gradients
(c) Local areas and lines of stiffening or thickening of the shell.

It is difficult to determine the effect of each of the forces. Most of the above forces give rise to bending stresses in the shell or head, which decrease rapidly with distance from the area of application of the load. Thermal gradients, however, give rise to not only bending stresses, but longitudinal thermal gradients give rise to a combination of direct stresses and bending stresses.

The design criteria depends on whether the local bending stress system extends over a small or a large proportion of the circumference of the vessel. In the former case the criterion adopted is that the bending stresses are limited to the value which will just cause a plastic hinge to develop, i.e., the localized material looses elasticity; the criterion is considered safe because a very small amount of distortion, consequent upon the formation of a plastic hinge will cause a local increase in direct stresses which will prevent further distortion. The stress intensification factor corresponding to this condition is of the same order as that occurring around nozzles under the action of pressure. If, however, the highly stressed area extends over considerable fraction of the circumference of the vessel, loads approaching the plastic limit are liable to cause a kink to form in the shell. Under these conditions special precautions should be taken to reduce the bending stresses. For example, one can provide stiffening rings above and below lug support. The limpet coil also serves the purpose of a stiffening ring.

5.11 THERMAL STRESSES IN CYLINDRICAL SHELL

In addition to the stresses created in the wall of a vessel, due to pressure and other loadings, stresses are also produced due to the temperature variations through the wall. Thermal gradients through the wall are encountered when the contents of the vessels are heated from the outside or when heat from a reaction in the vessel is removed through the wall. Temperature gradient will cause expansion at any radius in three directions, namely, radial, circumferential and longitudinal. Portions of the wall having lower temperature will expand less and will resist expansion of the high temperature regions. Thermal stresses are set up due to uneven expansion.

If the heat flows from the inner to the outer wall of the shell, compressive tangential stresses are set up. When pressure is applied to the inside of this shell, tensile stresses are superimposed on the compressive thermal stresses. Thus, in a shell subjected to internal pressure, a temperature gradient with high temperature inside equalizes the stresses throughout the wall. When the shell is heated from the outside, the inside layer is put under tension and these thermal stresses are added to the tensile stresses due to internal pressure. At extreme temperature gradients thermal stresses may reach the yield point of the material, even in the absence of an internal pressure.

In the case of thin cylinders, the determination of thermal stresses is simplified and they are given by

$$f_t = f_a = \frac{\alpha E (t_2 - t_1)}{2(1 - \mu)} \tag{5.43}$$

t_1 and t_2 = temperature at inner and outer radius

μ = Possion's ratio

α = coefficient of expansion

E = modulus of elasticity

The stresses determined are due to a steady static temperature distribution throughout the wall. While warming up and cooling down, additional stresses are induced in the wall of the vessel due to the transient temperature gradient. Theories in respect of these have not been worked out. However, in order to reduce these stresses, it is necessary to heat or cool the vessels gradually.

5.12 FABRICATION

The methods used for fabrication of the components of vessels may differ to some extent depending on the materials of construction. However, in a majority of the ferrous and non-ferrous vessels, the methods of fabrication are essentially the same.

The most common methods used are welding, brazing, forging, flame cutting, plasma cutting, riveting and other machine operations. Though riveting is used in the aerospace industry, it has become obsolete in the Chemical Process Industry (CPI). Brazing is used to a limited extent in copper and brass alloy applications. Fusion welding is applicable to the rest of metal fabrication, for example, joining of the dished ends to the vessel shell, flange nozzle to the top dished end and so on.

The process of fusion welding can be classified as below:

- *Gas welding* suitable for carbon steel. The flame used is oxy-acetylene and oxy-LPG.
- *Shielded Metal Arc* suitable for carbon steel. Stainless steel.
- *Metal Inert Gas (MIG)* suitable for carbon steel, The gas used is carbon dioxide or argon.
- *Tungsten Inert Gas (TIG)* suitable for carbon steel, stainless steel and other exotic metals (for example, Titanium) and alloys like hastelloy, It is the most versatile. The inert gas used is argon.

The welding electrode plays a key role in the welding process. The electrode is selected in such a way that the weld deposits have near identical composition to the parent parts to ensure homogeneity and mechanical properties superior to that of the parent materials.

The electrode coating plays an important role too.

- Rutile coated electrodes of mild steel are common for general purpose welding jobs. They can be used in any position, Rutile coated stainless steel electrodes are suitable for various grades of stainless steel. Special electrodes such as Niobium stabilized are also available. They are suitable for welding subjected to radiography test.
- Rutile and iron powder electrodes are used for special jobs of welding mild steel to material having higher thickness.
- Cellulose coated electrodes are used in piping welding jobs. The weld has sound mechanical strength and can be suitable for radiography quality finish.
- Nickel containing electrodes are good for cast iron parts, casting welding of cast iron to mild steel, and other non-ferrous meterial.
- Basic coated electrodes also known as low hydrogen content weld. They are common in CPI for boilers, pressure vessels, reactors, storage tanks, etc. They are used for radiography finish.

The electrodes can be operated on AC or DC power supply.

It is absolutely essential that the electrode prior to use must be dry. There is an Indian Standard-IS 815, for electrodes wherein electrode coding is given based on the method of electrode manufacturing, type of coating, welding position, welding current, strength, special property such as hydrogen controlled, etc.

Plates of various sizes and thicknesses are available for forming the shell and the heads. These plates are cut to suitable sizes by oxy-acetylene flame or by shearing in the case of thin plates. In the case of stainless steel, plasma arc or abrasive wheels may be used for the cutting operations. The edges of these plates are then bevelled by an oxy-acetylene torch or can be machined to make them square, or ground by abrasive wheels.

The plates are then rolled to form a shell. In the case of thick plates the edges need crimping, prior to the rolling operation, since the edges cannot be formed to the desired curvature by the method of rolling alone. It is advisable to clamp together the two adjoining edges of the shell. The clamps can be removed as soon as these edges are tack welded. Finally, the longitudinal seam of the shell is welded by any one of the fusion welding methods. For large diameter vessels and heavy sections, the submerged arc method is most suitable. This method is automatic, and gives a neat and regular weld which is free from any defects. In the case of copper and some special steel shells, after the initial rolling and welding, the shells are rolled again to give a uniform circular shape. The rollers on such rolling machines are removable for insertion of the original shell. Vessels having a wall thickness above 50 mm may be fabricated by forging or welding.

The various types of dished heads are produced from blanks by either pressing them in a hydraulic press or by spinning them to the desired shape. In the second method the blank is made to rotate, while a pair of rollers are made to exert pressure on it. The roller is gradually moved in the radial direction simultaneously. The pressure of the rollers is, therefore, gradually transferred to concentric areas on the blank. This pressure gives rise to the dished shape as required. It may be necessary to form a bevel on the edge of the head for welding it to the shell. This can be produced by oxy-acetylene torch, which can be attached to the tool post of a vertical boring machine. The head is slowly rotated with the table of the machine, while the torch is being operated. The head is connected to the shell by welding and the assembly may be heat treated depending on the thickness.

Holes in the shell or the head can be made by an oxy-acetylene torch which is attached to a compass or a pantograph mechanism. The edges of the holes may be upset by a hydraulic press, by heating the area around the holes locally. The nozzles are welded to these openings as required. It is desirable to tack weld these nozzles and check their correct position and alignments prior to the final welding.

Nozzles may be fabricated from plates by the method of rolling and welding or may be formed to the desired shape by forging. Flanges are produced by forging with the help of a steam or hydraulic press. After suitable heat treatment, these are machined to give the final shape and size.

Post weld heat treatment is carried out in the case of carbon and low alloy ferritic steels either by normalizing or by stress relieving. Special heat treatment is necessary in case of alloy steels and non-ferrous metals. Thermal stress relieving is necessary when the vessel is: (a) intended for containing toxic or inflammable material; (b) intended for operation below 0°C; (c) intended for use with media liable to cause stress corrosion cracking; (d) subject to excessive stress concentration or risk of fatigue due to changing loads; (e) under risk or brittle fracture; (f) to have dimensional accuracy and shape in service; (g) to be made of a plate thickness (including corrosion allowance) at any welded joint in the vessel shell or head, which exceeds the values given in Table 5.5.

The vessel is stress relieved as a whole in an enclosed furnace wherever practicable. Where it is not practicable to do so, special precautions will have to be taken to heat the vessel in sections. To obtain good finish the vessel is sand blasted, ground and polished.

5.13 REVIEW OF UNFIRED PRESSURE VESSEL CODES (APPENDIX G)

A detailed procedure of the design of pressure vessel components and an outline of the fabrication technique has been given so far. A reference to the pressure vessel codes is made in Section 5.3.

Table 5.1 gives a list of codes pertaining to pressure vessels and other components, which are applicable to equipment used in the chemical industry. Pressure vessel codes specify the minimum construction requirements for design, fabrication, inspection and testing. In general, the basic concepts in the various codes are somewhat similar. Even so, there are differences in the approach arising out of historical background, experience, raw material availability and design philosophy adopted in various countries where these codes have evolved. A comparison of the design procedure suggested by some of the important codes have been presented in Appendix G.

In the selection of material the most important consideration is the mechanical strength. The design or the allowable stress values are obtained by dividing the ultimate stress or yield stress by a safety factor. Table 5.2 indicates the safety factors for different codes. Allowable stress values for some materials are presented in Table 5.6. It may be observed that the ASME code, with a higher factor of safety on ultimate stress, gives lower values of allowable stresses. On the other hand, AD-specification considers yield stress values. It is, therefore, advantageous to use high yield strength materials, if the design is to be based on AD-specification. This will result in reduction of thickness for a component and consequently a reduction in weight and material handling. If a higher yield strength material is chosen, it is essential to assure the required quality of fabrication. Further, simple reduction of weight should not be the sole criterion, since the cost of the material is likely to be higher.

From a study of the design procedure of the various codes (Appendix G), it may be observed that the methods of calculation of thickness of items such as shell, head, etc., are somewhat similar. If, therefore, the allowable stress values for a given material of construction are identical, the thickness values should work out to be almost the same. For the design of nozzle compensation, it is usual to provide a pad in the form of a ring. The sizes of the rings obtained by codes are comparable. If, however, formed nozzles are used, major differences are observed in the code values. The efficiency of welded joints in all cases is linked with radiographic examination.

It may be mentioned that apart from the design procedure, it is essential to compare different codes in the requirements for welding procedure, fabrication tolerances, heat treatment requirements, etc., before making a choice in favour of any particular code. Even so, significant savings in raw material cost can be achieved by the selection of a particular code.

5.14 INSPECTION AND TESTS

It is essential to carry out inspection of the pressure vessels. Such inspections are carried out during vessel manufacture and after completion of fabrication. Inspection is carried out of the materials used, components as fabricated and welded joints. Finally, the vessel is tested for pressure.

5.14.1 Welded Joints

Radiographic Examination: Crack, porosity, lack of fusion, slag inclusion, inadequate penetration and other interfections decrease the joint efficiency of a weld. Radiographic examination has proven highly successful for inspecting plate and pipe welding and for measuring corrosion damage. The examination can locate the defects, and repairs can be carried out to bring joint efficiency up to specification. Not only weld joints but shell and dished ends are also subjected to radiography or ultrasonic testing to bring out defects developed during the rolling and fabrication respectively.

5.14.2 Hydrostatic and Acoustic – Emissions Testing

Hydrostatic testing simply consists of pressurising of a vessel with a fluid to check for any leaks.

5.14.2.1 Testing with Incompressible Fluids

Water is pumped in a vessel, with vent.

$$\text{The test pressure} = 1.3 \times \text{design pressure} \times \frac{f_1}{f_2} \quad \text{(IS 2825)}$$

f_1 = allowable stress value of the material at test temperature.

f_2 = allowable stress value of the material at the design temperature.

The vessel shall be maintained at a specified test pressure for a sufficient length of time to permit a thorough examination to be made of all seams and joints but in no case less than 10 minutes.

Table 5.5 Maximum Plate Thickness above which Post Weld Heat Treatment is Required

Percentage Chemical Composition and Mechanical Properties	*Plate Thickness*
Carbon 0.20 Max Manganese 1.00 Max Residual elements 0.80 Max Specified minimum yield strength 260 N/mm^2 Max Carbon 0.25 Max Manganese 1.60 Max Residual elements 0.80 Max Specified minimum yield strength 380 N/mm^2 Max Carbon 0.30 Max Manganese 1.20 Max Residual elements 0.40 Max Specified minimum yield strength 380 N/mm^2 Max	38 mm
Carbon 0.25 Max Manganese 1.60 Max Chromium 0.60 Max Molybdenum 0.60 Max Vanadium 0.12 Max Residual or other elements 0.80 Max Specified minimum yield strength 440 N/mm^2 Max Carbon 0.35 Max Manganese 1.20 Max Residual or other elements 0.40 Max Specified minimum yield strength 440 N/mm^2 Max	16 mm
Carbon 0.25 Max Manganese 1.60 Max Chromium 1.50 Max Molybdenum 1.50 Max Vanadium 0.16 Max Residual or other elements 0.80 Specified minimum yield strength 440 N/mm^2	All thicknesses
All other ferritic steels	All thicknesses

Table 5.6 Minimum Ultimate Tensile Stress, Minimum Yield Stress, and Allowable Stress as per Different Codes for Ferrous Materials

Material	Ultimate Tensile Stress Minimum N/mm^2	Yield Stress Minimum N/mm^2	Allowable Stress N/mm^2			
			ASME-VIII, DIV.I	IS-2825	AD Merkblatter	BS-5500
Carbon steel IS-2002 GR2A at ambient temperature	420.0	210.0	104.8	140.6	140.6	140.6
At 515 Gr 70 at ambient temperature	492.2	267.2	137.9	163.8	187.6	178.6
At 515 Gr 70 at 400°C	413.4	186.3	111.0	124.4	124.4	124.4
Stainless steel A 240 type 304 at ambient temperature	527.3	210.9	137.9	140.6	140.6	140.6
Stainless steel A 240 type 304 at 400°C	106.7	121.6	106.8	81.1	81.1	81.1

The values reported above are within engineering accuracy and during conversion from different unit to SI some rounding of figures has taken place.

5.14.2.2 Testing with Compressible Fluids

Pressure testing with air or gas may be carried out in lieu of hydraulic test in the following cases:

(i) Vessels that are also designed, constructed or supported that they cannot safely be filled with water or liquid.

(ii) Vessels that are used in services where even small traces of water cannot be tolerated.

The pneumatic test pressure shall not be less than the design pressure but need normally less than hydraulic test pressure.

5.14.2.3 Acoustic - Emissions Testing

When vessels are hydraulically tested, stresses in the specimen cause discontinuities to emit high-frequency sounds that can be picked up by piezoelectric sensors positioned on the specimen's surface before the start of the test. Sound waves received by the sensors are then fed as an electrical signal into an analyzer system, that counts the total number of acoustic events occurring in the specimen. The systems of testing are capable of indicating over-pressurization and imminent catastrophic failure. Results from the test can be displayed on a CRT screen or on a printout sheet.

5.14.2.4 Halogen Leak Inspection of Pressurized Gas Systems

Following the preliminary hydraulic test, the halogen-Freon gas—is introduced in the vessel. After the introduction of the Feron gas, the vessel is pressurized with an inert gas or clean dry air to approximately, 3 kg/cm^2 pressure or the design pressure whichever is less. The vessel should remain under pressure for a maximum of one-half hour. The presence of leaks may be detected with either a diode electrical-detector or a halide torch.

5.15 ILLUSTRATION

The following numerical problem will give a feel of the above theoretical considerations. You may refer to Plate I.

Data

(a) *Shell*

Diameter (o.d.)	:	1200 mm
Material of construction (MOC)	:	Stainless Steel Grade 304
Allowable stress at 150 °C, f	:	130 N / mm^2
Internal pressure *p*	:	0.3 N / mm^2
Joint efficiency	:	0.85 (spot radiography of the main welds)
Corrosion allowance *C*	:	nil

(b) *Head-flanged and shallow dished end*

External diameter	:	1200 mm
Crown radius	:	1200 mm
Knuckle radius	:	72 mm
Material of construction	:	SS (304)

(c) *Flanges*

Material	:	Carbon steel (IS 2002) grade I
Permissible stress	:	95 N/mm^2
Gasket	:	asbestos

(d) *Bolts*

Material	:	Hot rolled carbon steel
Permissible stress up to 50°C	:	58.7 Nmm^2
Up to 200°C	:	54.5 N/mm^2

(e) *Nozzle*

Welded to head

Diameter, i.d.	:	150 mm
Thickness	:	3 mm
Material of construction	:	SS

5.15.1 Design of Shell

The design pressure is 10% excess than the pressure inside. Therefore, the design pressure is = 1.1 × 0.3 = 0.33 N/mm^2

From Eqn. (5.3.c) we have,

$$t = p\, D_0 / (2\, fJ + p) + C$$

The joint efficiency, J = 0.85 (spot radiography)

$$t = 0.33 \times 1200 / (2 \times 130 \times 0.85 + 0.33)$$

$$= 1.79 \text{ mm}$$

In reality, considering the rigidity of the vessel, we have, t = 4 mm. Please note no corrosion allowance is required as the vessel is fabricated in stainless steel (SS).

Check for loadings

The calculations are based on 4-mm thickness. (If corrosion allowance is used then the thickness for this purpose is the actual thickness less corrosion allowance).

(a) *Tangential or hoop stress*

From Eqn. (5.5) we have,

$$f_t = p\,(D_i + t)\,/\,2\,t$$

$$D_i = 1200 - 2 \times 4 = 1192 \text{ mm}$$

$$f_t = 0.33\,(1192 + 4)\,/\,2 \times 4$$

$$= 49.34 \text{ N/mm}^2 \text{ within permissible limit.}$$

(b) *Stress in axial direction due to internal pressure* (Eqn. 5.6)

$$f_1 = p\,D_i/4 \times t$$

$$= 0.33 \times 1192/4 \times 4$$

$$= 24.59 \text{ N/mm}^2 \text{ within permissible limit.}$$

(c) *Stress due to weight of the vessel:*

As an approximation we calculate the weight of the cylindrical part of vessel and multiply by 1.25 to get the total weight. During hydrostatic testing water is used to detect any leaks. We take into consideration weight due to water when the vessel is full. If a heavier liquid material is likely to be processed, we have to take that weight instead of weight due to water.

The volume of a cylindrical portion $\left(v = \frac{\pi}{4} D^2 \cdot L\right)$ of the vessel assuming L / D ratio to be 1.2, is 1.6 m^3. The volume of two dished ends can be calculated as follows:

The volume of the head for knuckle radius to be min 6% of the internal diameter of the vessel is given by

$$V = 0.081\,D_i^3$$

$$= 0.081\,(1.192)^3$$

$$= 0137 \text{ m}^3$$

The straight length of the dished end is 40 mm. The volume of this cylindrical portion is 0.045 m^3.

There are two dished ends. Therefore, the total volume of the dished ends are $= 2\,(0.137 + 0.045)$

$$= 0.364 \text{ m}^3.$$

$$\text{Total volume of the vessel} = 1.6 + 0.364 + \text{volume of nozzles}$$

$$\text{say} = 2 \text{ m}^3$$

The density of material is assumed to be same as that of water. Therefore total weight of the liquid filled in vessel is = 2000 kg. Please note volumes of all top and bottom nozzles are included.

The shell is fabricated out of SS plate of thickness 4 mm.

The estimated weight is = $\pi \times D \times H \times t \times \rho$

We have mean D = 1196 mm; H = 1.2 × 1196 mm

and t = 4 mm; density = 8020 kg/m^3

The weight w = 172.4 kg, say 173 kg

Weight of the dished ends is calculated as follows:

From Appendix J, we have,

$$W = (\pi/4)\, B^2 \times t \times \rho$$

$$B = 1.024\,(1.2) + 0.67 \times 0.072 + 2 \times 0.040 = 1.357 \text{ m}$$

$$\rho = 8020 \text{ kg/m}^3$$

$$W = (\pi/4) \times (1.357)^2 \times 0.004 \times 8020 = 46.4 \text{ kg}$$

There are two dished ends and hence total weight = 2 × 46.4

Say = 93 kg

Weight of the carbon steel body flanges (Number 2)

$$= 2 \times (\pi/4)\,[(1.315)^2 - (1.2)^2] \times 0.045 \times 7850 = 160.5 \text{ kg}$$

Therefore, the weight of the vessel including nozzles, manhole and support is considered to be 10 to 20% higher than the sum of weights of vessel shell, dished ends and flanges.

Thus weight of the vessel = 1.15 (173 + 93 + 160.5) = 490.5 kg

Say = 491 kg

The total weight, W = vessel + content

= 491 + 2000

= 2491 kg

say, for the design purpose W = 2500 kg

From Eqn. (5.7), we have,

$$f_2 = W/(\pi \times t\,(D_i + t))$$

$$= 2500/(\pi \times 4\,(1192 + 4))$$

$$= 0.166 \text{ kg/mm}^2$$

$$= 1.63 \text{ N/mm}^2 \text{ (compressive)}$$

The reaction vessel is inside the plant building and, therefore, wind load is not there. As a result f_3 is = 0.0.

From Eqn. (5.9),

$$f_a = f_1 + f_2 + f_3$$
$$= 24.59 + (-1.63) = 22.96 \text{ N / mm}^2$$

We assume torque, T = 500,000 N mm, due to the offset piping.

The stress due to offset piping is given by Eqn. (5.10),

$$f_s = 2 \times T/(\pi t D_i (D_i + t))$$
$$= 2 \times 500{,}000 / (\pi \times 4 \times 1192 (1192 + 4))$$
$$= 0.0558 \text{ N / mm}^2.$$

Combining above stresses on the basis of shear strain energy theory we get equivalent stress from Eqn. (5.11),

$$f_R = \left[f_t^2 - f_t \times f_a + f_a^2 + 3f_s^2\right]^{0.5}$$

Substituting values of respective stresses computed above, we get

$$f_R = [(49.34)^2 - (49.34)(22.96) + (22.96)^2 + 3(0.0558)^2]^{0.5}$$
$$= 42.76 \text{ N/mm}^2$$

It is thus seen that both f_R and f_a are less than permissible stress 130 N/mm^2. Therefore, the thickness of 4 mm is acceptable.

5.15.2 Head Design - Flanged and Shallow Dished Head

We use the expression Eqn. 5.23 to calculate thickness for the internal design pressure 0.33 N/mm^2.

$$t_h = p.R_c.W/2f.J$$

We assume J = 1.

The stress intensification factor, W, is given by

$$W = 0.25\left\{3 + \sqrt{(R_c / R_1)}\right\}$$
$$= 0.25\left\{3 + \sqrt{(1200/72)}\right\} = 1.77$$

We then have

$$t_h = 0.33 \times 1200 \times 1.77/(2 \times 130 \times 1)$$
$$= 2.7 \text{ mm}$$

Use the plate of 4 mm thickness (no corrosion allowance is necessary). Straight flange length = 3 × 4 = 12 mm. However, minimum specified by code is 20 mm. Use the straight length of the dished end 40 mm.

5.15.3 Flanges for Top Head and Shell

A flanged joint connects the shell and the top head. The vessel is made of stainless steel for which the flange may be selected from Indian Standard Specification (IS-4868 – Welded shell flanges for stainless steel pressure vessels and equipment). Table 1 of the specification gives a flange of 1200 mm nominal diameter. This requires a minimum shell thickness of 6 mm. The shell thickness as per design is only 4 mm. It is, therefore, necessary to weld an additional ring between the shell and the flange (see Fig. 5.25). The particulars of flange and bolts are as follows:

Nominal diameter of flange	:	1200 mm
Inside diameter of flange	:	1202 mm
Outside diameter of flange	:	1315 mm
Outside diameter of stainless steel lining ring (raised face)	:	1240 mm
Bolt circle diameter	:	1270 mm
Thickness of flange	:	45 mm
Number of bolts	:	48
Nominal diameter	:	18 mm (M 18)

If a flange is a non-standard flange, the design procedure is illustrated below:

(a) *Design of gasket and bolt*

The flange is made of carbon steel with a stainless steel lining (raised face) in the form of a ring. A flat asbestos gasket of 1200 mm inside diameter and 1240 mm external diameter with 3 mm thickness is used to cover the raised face.

From Tables 5.3 and 5.4 and Fig. 5.34, we have,

Gasket factor $m = 2.0$

Minimum design seating stress, $Y_a = 11.2 \text{ N/mm}^2$

Basic gasket seating width, $b_0 = 1/2\ [(1240 - 1200)/2]$

$= 10$ mm

The effective gasket width, $b = 2.5\sqrt{(b_0)} = 7.9$ mm

The minimum bolt load for atmospheric condition is obtained from Eqn. (5.36) as:

$$W_{m_1} = \pi \times b \times G \times Y_a \qquad G = (1240 + 1200)/2 = 1220 \text{ mm}$$

$$= \pi \times 7.9 \times 1220 \times 11.2 \qquad = 339121.104$$

say $= 340{,}000$ N

Under the operating condition we have the bolt load from Eqn. (5.37) as:

$$W_{m_2} = \pi\ (2b)\ mGp + (\pi/4)\ G^2\ p$$

$$= \pi\ (2 \times 7.9) \times 2 \times 1220 \times 0.33 + (\pi/4)\ (1220)^2 \times 0.33$$

$$= 39967.8 + 385765.6 = 425733.4 \text{ N}$$

say $= 426{,}000$ N

We consider a higher load for the design purpose.

The number of bolts = (mean gasket dia) / 2.54

= (1220) / 2.54 = 48.03

The bolts are always in multiple of 4.

Therefore, number of bolts is = 48.

The bolt area is obtained from Eqn. (5.38) with the permissible stress 54.5 N/mm^2,

$$A_{\text{bolt}} = 426{,}000/54.5 = 7816.5 \text{ mm}^2$$

Area per bolt = 7816.5/48 = 162.8 mm^2

Let us consider a M16 bolt with nominal diameter 16 mm.

From Eqn. (5.41) we have, bolt area for stress as

$A_b = 0.51\ (16)^{2.09} = 167.56$ mm^2, which is just matching.

However, we recommend M 18 bolt with nominal diameter 18 mm

$A_b = 0.51\ (18)^{2.09} = 214.33$ mm^2, which is about 31% higher area than the required bolt area.

Therefore, 48 number of bolts with nominal diameter 18 mm (M18) are adequate for the service.

The total bolt area, $A_{bt} = 48 \times 214.13 = 10287.8$ mm^2

From Eqn. (5.40), we have,

$$A_{bt} = 2 \times \pi \times Y_a \times G \times N/f_a = 2 \times \pi \times 11.2 \times 1220 \times 20/58.7$$

$$= 29251.6 \text{ mm}^2$$

The total bolt area provided is more than the above value. Bolt size and number of bolts are thus adequate.

Pitch of bolts = 4.6 × 18 = 82.8 mm (pitch of bolts is in between 3.5 and 5 times bolt dia.)

Bolt pitch circle diameter,

$$B = 82.8 \times 48/\pi = 1265.09 \text{ mm}$$

or

$$B = \text{outside dia. of the gasket} + 2 \times \text{bolt dia.} + 12$$

$$= 1240 + 2 \times 18 + 12 = 1288 \text{ mm.}$$

This matches well with the data wherein pitch circle diameter is given as 1270 mm.

The outside diameter of the flange is $= B + 2 \times \text{bolt dia}$

$$= 1270 + 2 \times 18$$

$$= 1306 \text{ mm}$$

This figure matches well with the standard flange of 1315 mm diameter.

(b) ***Flange thickness***

From Eqn. (5.42), we have,

$$t_f = G\sqrt{p/kf} + c;$$

where,

$$k = 1/[0.3 + 1.5W_m\, h_G / HG]$$

we have, $W_m = 426{,}000$

$$h_G = (1288 - 1220) / 2 = 34$$

$$H = (\pi/4)\,(1220)^2\,(0.33)$$

$$G = 1220$$

Substituting above values, we get $k = 2.88$.

Therefore,

$$t_f = 1220\,(0.33/(2.88 \times 95))^{0.5} = 42.4 \text{ mm.}$$

The flange thickness thus computed is 42.4 mm and it matches with that of 45 mm given in the data. This is a trial and error method, for more detailed treatment, refer to ASME section VIII or Indian Standard IS-2825.

During fabrication, the plate choice depends upon the machining allowance to obtain the desired surface finish. In this case, one may use the plate of a higher thickness, say 50 mm.

5.15.4 Nozzle Reinforcement

The computation of the reinforcement will be exemplified with the help of one calculation. Similar procedure should be adopted for other nozzles. The nozzle is provided on the head. It has an internal diameter of 150 mm. The design pressure will be the same as that of the vessel. We have from Eqn. (5.3b)

$t_n = pD_i/(2\,fJ - p)$, we assume $J = 1.0$

$$= (0.33) \times 150/(2 \times 130 \times 1 - 0.33)$$

$$= 0.19 \text{ mm}$$

There is no corrosion allowance as the material of construction is stainless steel. For the sake of rigidity we have a nozzle with thickness of 3 mm which is the standard size. The reinforcement design procedure will be that outlined in Section 5.8.3.2.

The maximum nozzle distance = $2 \times d = 2 \times 150 = 300$ mm.

The maximum vertical distance for compensation provided outside the vessel.

$$AD = 2.5 \times t_h$$
$$= 2.5 \times 2.7 = 6.75 \text{ mm}$$

Area for which compensation is required,

$$A = dt_s$$
$$= 150 \times 2.7 = 405 \text{ mm}^2$$

The area available for compensation is computer as follows:

(i) Portion of the head having excess thickness

$$A_s = 150 \times (4 - 2.7) = 195 \text{ mm}^2$$

(ii) Portion of the nozzle external to the vessel having excess thickness

$$A_0 = 2 \times 2.5 \times 4\ (3 - 0.191) = 56.18 \text{ mm}^2$$

(iii) Since the nozzle is not projected inside $H_2 = 0$

Therefore, area available for compensation = $195 + 56.18 = 251.18$ mm^2

The extra area to be given as compensation = $402 - 251.18 = 150.82$ mm^2

This can be provided by a ring of outside diameter 225 (which is less than $2d = 300$) and inside diameter 156 mm (150 + 6) and 3 mm thick (<6.75 mm) which is within the requirement as shown in Fig. 5.17.

$$\text{Area of compensation} = (225 - 156) \times 3$$
$$= 207 \text{ mm}^2$$

This is satisfactory as it is 37% in excess.

(For simplicity the area of weld is ignored.)

In reality, the top dished ends might have many nozzles (refer to the reactor sketch at the end). One has to calculate the reinforcement required, if any, for each nozzle.

References and Recommended Readings

Bickell, M.B., and Ruiz C., *Pressure Vessel Design and Analysis*, Macmillan, London, (1967).

Brownell, L.E., and Young E.H., *Process Equipment Design*, John Wiley, New York, (1968).

Chuse, R., *Unfired Pressure Vessels: The ASME Code Simplified*, 4th Edition, F.W. Dodge Corporation, New York, (1960).

Czernik, D.E., *Gaskets Design, Selection and Testing*, McGraw-Hill, New York, (1996).

Dimoplon, W. Jr., 'How to Determine the Geometry of Pressure Vessel Heads, *Hydrocarbon Processing*', 53 (8)71(1974).

Earland, S., Nash, D. and Garden B., *Guide to European Pressure Equipment*, Professional Engineering Publishing. Bury St. Edmonds and London (2003).

Fort, H., Ben, K.E., Roggen, J. and Gordner H.E., Design Criteria for Pressure Vessels: Safety in High Pressure Polyethylene Plants, CEP Technical Manual, 15 (1973).

Gill, S.S., *Stress Analysis of Pressure Vessels and Pressure Vessel Components*, Pergamon, Oxford, (1970).

Hammer, J., '*Pressure Vessel Guide*', British Chemical Engineering, **15**(7) 869 (1970).

Harvey, J.F., *Pressure Vessel Design; Nuclear and Chemical Applications*, D. Van Nostrand Company Inc., Princeton, New Jersey, (1967).

Jethanandani, H.H., 'Design and Fabrication Codes for Chemical Plants and Equipment', *Chemical Age of India*, **27**(9) 789 (1976) and **27**(11) 953 (1976).

———, 'Pressure Vessel Industry in India', *Chemical Age of India*, **22**(6) 351 (1971).

Krishnamurti, S., 'Application of Stainless Steel in Design and Fabrication of Chemical and Process Industries', *Chemical Age of India*, **22**(11) 794 (1971).

Laithwate, J., 'Pressure Vessel Standardization and Design', *Chemical Process Engineering*, **51**(10) 61 (1970).

Lancaster, J.F., 'Recent Developments in Pressure Vessel Steel: U.S. vs. Europe', *Hydrocarbon Processing*, **43**(10) 161 (1964).

Mahajan, K.K., *Design of Process Equipment (Selected Topics)*, Pressure Vessel Handbook Publishing Co., USA, (1979).

Markovitz, R.E., 'Choosing the Most Economical Vessel Head', *Chemical Engineering*, **78**(16) 102 (1971).

Mead, W.J., *Encyclopedia of Chemical Process Equipment*, Reinhold, New York, (1964).

Megyesy, E.E., 'Size Vessel Nozzles Faster', *Hydrocarbon Processing*, **57**(12) 100 (1978).

Nichols, R.W., *Pressure Vessel Engineering Technology*, Elsevier Pub. Co., Amsterdam, (1971).

Perry, R.H., Green D.W. and Maloney J.O., *Perry's Chemical Engineers' Handbook*, 6th Edition International Edition McGraw-Hill Book Co., Singapore, (1984).

———, 'Pressure Vessel', *Hydrocarbon Processing*, **57**(12) 89 (1978).

———, 'Proceedings of a Symposium on Pressure Vessel Technology', *Chemical Age of India*, **23**(5) 329 (1972).

Raghavan, K.V., Pressure Vessel Design, *Chemical Process Engineering*, **51**(10) 81 (1970).

Riegel, E.R., *Chemical Process Machinery*, Reinhold, New York, (1953).

Spring, H.M., *Pressure Vessels for Industry*, McGraw-Hill, New York, (1947).

Thandavan, N. and Fogelman J.A., 'Fabrication Stainless Steel Pressure Vessels', *Chemical Age of India*, **22**(11) 826 (1971).

Thielsh, H., *Defects and Failures in Pressure Vessels and Piping*, Reinhold Publishing Corporation New York, (1965).

Vardarajulu, R., Blank Diameters for Formed Heads of Pressure Vessel', *Chemical Age of India*, **27**(5) 488 (1976).

6

Chapter

High Pressure Vessels

6.1 INTRODUCTION

These vessels are generally used as reactors, separators or receivers and heat exchangers. Most of the batch reactors used for high pressure work are termed as autoclaves. They are vessels with an integral or welded bottom and a removable or welded top head and are generally provided with an inlet, heating and cooling systems and very often an agitator system. Separators for high pressure work are also vessels with an inlet and in some cases a centrifugal device for separating the gas from the liquid.

6.2 CONSTRUCTIONAL FEATURES

High pressure vessels are used for a pressure range of 10 N/mm^2 to say 300 N/mm^2. These are essentially thick walled cylindrical vessels, ranging in size from small tubes to several metres in diameter. Both the size of the vessel and the pressure involved will dictate the type of construction used.

Following are the various types of construction:

(A) A solid walled vessel produced by forging or boring a solid rod of metal (Fig. 6.1).

(B) A cylinder formed by bending a sheet of metal with a longitudinal weld.

(C) A multilayered vessel built up by wrapping a series of sheets of relatively thin metal tightly round one another over a core tube, and holding each sheet with a longitudinal weld. Rings are inserted in the ends to hold the inner shell round while subsequent layers are added. The inner cylinder is 6 to 12 mm thick, while the subsequent layers are generally less than 6 mm thick (Fig. 6.2).

(D) Shrink-fit construction (Fig. 6.3): It consists of concentric cylindrical shells, fitted together by shrinking one cylinder over the other. The method involves heating the outer shell in order

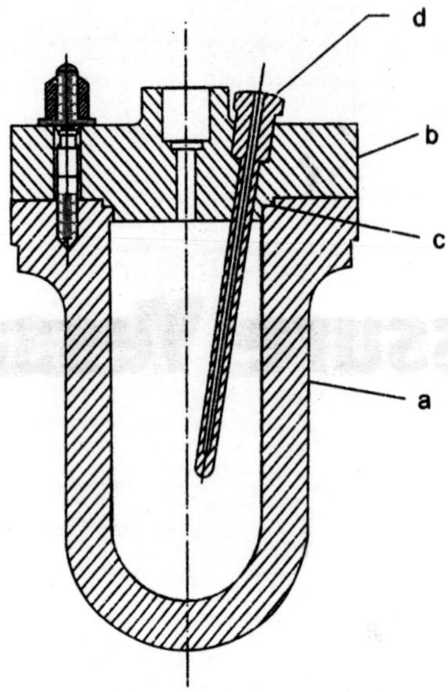

Fig. 6.1 Solid Walled Vessel: (a) Shell, (b) Cover, (c) Gasket, (d) Thermometer Pocket

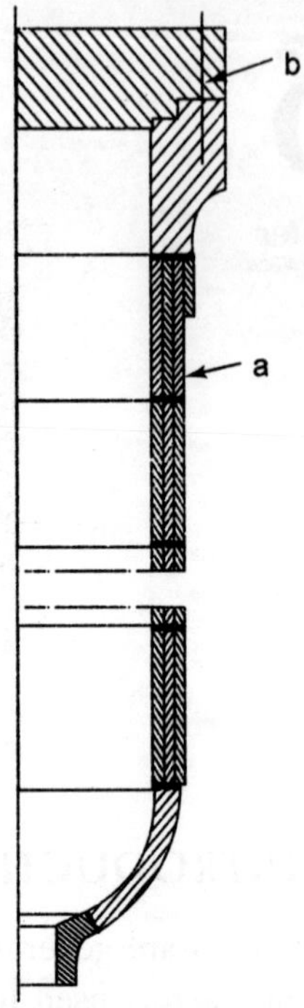

Fig. 6.2 Vessel Formed by Wrapping of Sheets: (a) Shell, (b) Cover

to expand it beyond the interference on the diameter, and then slipping it on the inner shell while expanded. In practice, it is often desirable to make the innermost component of some special material, for instance, a corrosion-resistant alloy.

(E) A vessel built up by wire winding around a central cylinder. The wire is wound under tension around a cylinder of about 6 to 10 mm thick (Fig. 6.4).

(F) A vessel built up by wrapping successive layers of interlocking strips around a central cylinder made from a 25 to 40 mm thick plate. Spiral grooves

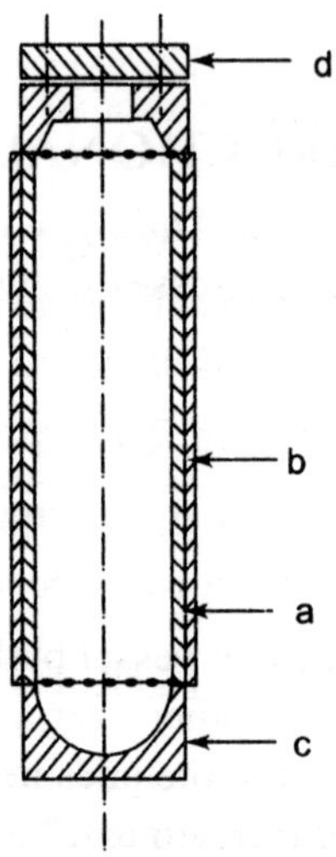

Fig. 6.3 Vessel with Shrink-fit Construction: (a) Shell, (b) Outer-shell, (c) Head, (d) Cover

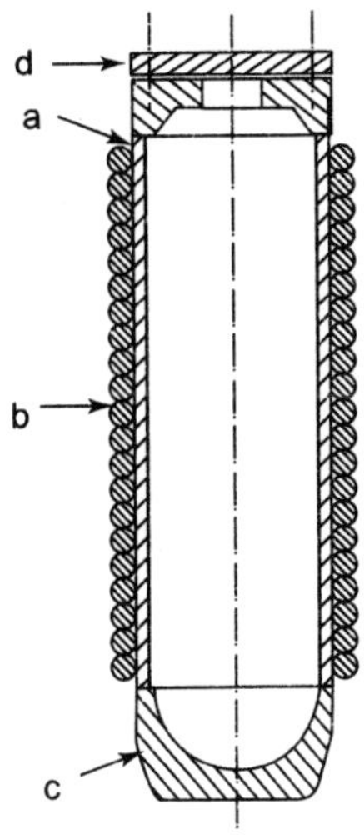

Fig. 6.4 Wire Wound Vessel: (a) Shell, (b) Wire, (c) Head, (d) Cover

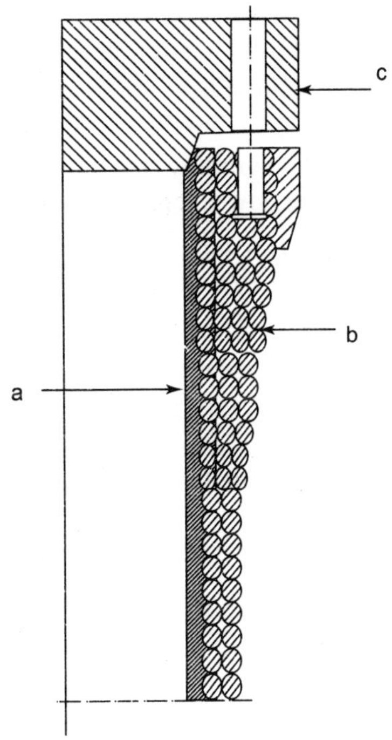

Fig. 6.5 Tape Wound Vessel: (a) Shell, (b) Tape, (c) Cover

are cut on the surface, which is then wrapped with a tape, 65 to 85 mm wide and 6 mm thick. The tape is also grooved so that it will interlock with succeeding layers. The succeeding layers are offset from the preceding ones to get a proper interlock and produce axial strength. During winding of tape, it is heated, so that on cooling it will produce compressive stresses in the inner layer, thus prestressing the vessel (Fig. 6.5).

(G) Autofrettage: It involves the use of a solid forged vessel, to which is applied a large internal pressure, so as to expand the internal diameter, causing overstrain. When the pressure is released, the outer material will try to contract the inner material. As a result, residual compressive stresses are developed in the inner material and tensile stresses in the outer material.

6.3 MATERIALS FOR HIGH PRESSURE VESSELS

Ductility is an essential requirement in steel for use in a pressure vessel and should guide the choice of steel and its heat treatment.

(a) *Ni-Cr-Mo steels:* These are the most common steels used for high pressure vessels. Heat treatment followed is generally by oil quenching from 850°C with subsequent tempering between 600°C and 650°C (Table 6.1).

(b) *Maraging steels:* These steels have been developed recently and their main asset is that they develop their strength as a result of age hardening which is initiated by relatively low temperature heat treatment. A typical steel of this type has a composition of Ni 18.5%, Mo 4.9%, Co 8.7%, Ti 0.7%, Al 0.1%. These steels can be machine finished to final dimensions in the solution – annealed condition before heat treatment due to negligible changes in length and the low temperatures involved.

(c) *Creep resistance steels:* Special creep resistance steels which have the property of retaining their strength at these temperatures are used above 300°C.

(d) *Non-ferrous alloys:* For temperatures above 500°C, non-ferrous alloys such as nimonic or hastelloy are required, the latter giving superior creep resistance. These are essentially Ni-Cr-Co alloys. Creep, however, cannot be wholly avoided.

Table 6.1 High Tensile Steels for Heavy-Wall Pressure Vessels in Quenched and Tempered Condition

Standard	*C*	*N*	*Cr*	*Ni*	*Mo*	*V*	*YP* N/mm^2	*UTS* N/mm^2	*Elongation* %
A 10	0.10...0.15	0.50	11.5...13.5	0.50			550	600 – 750	18
43 OH	0.37...0.45	0.6...0.95	0.65...0.95	1.5...2.0	0.2...0.3		750	800 – 900	14
En 15	0.27...035	0.5...0.7	0.5...0.8	2.3...2.8	0.4...0.7		750	800 – 1000	14
En 26	0.35...0.44	0.5...0.7	0.5...0.8	2.3...2.8	0.4...0.7		850	900 – 1000	11
En 29	0.25...0.35	0.65	2.5...3.5	0.4	0.3...0.7		900	1000 – 1150	11
Cr MoV 135	0.17...0.23	0.3...0.5	3.0...3.3		0.5...0.6	0.45...0.55	750	800 – 950	14
30 Cr MoV 9	0.26...0.34	0.4...0.7	2.3...2.7		0.15...2.25	0.10...0.20	1050	1250 – 1450	9

6.4 SOLID WALLED VESSEL

This type of vessel consists of a single cylindrical shell, with closed ends. Due to the high internal pressure and large thickness the shell is considered as a 'thick' cylinder. In general, the physical criteria are governed by the ratio of diameter to wall thickness and the shell is designed as a thick cylinder, if its wall thickness exceeds one-tenth of the inside diameter.

6.4.1 Stresses in Shell

The main loading in the shell is due to internal or external pressure. The stresses across the shell thickness due to pressure are not uniform. Stresses due to other loads, such as dead loads, are comparatively smaller in magnitude and may, therefore, be ignored.

Three principal stresses are produced in the wall of the shell due to pressure. These stresses are: (a) tangential (circumferential or hoop), (b) radial and (c) longitudinal (axial). The tangential stresses are of high magnitude. According to Lame's analysis (assuming a fully elastic cylinder) the variation of tangential and radial stresses along the radius of the shell is given by (Eqs. 3.25, 3.26 and 3.27)

Tangential stress $f_p = A + \frac{B}{R^2}$ (6.1)

Radial stress $f_R = A - \frac{B}{R^2}$ (6.2)

Longitudinal stress $f_a = A$ (6.3)

where, R – any radius

A, B – constant, given by Eqns 6.4 and 6.5 below.

The tangential stress may be either tensile or compressive, depending on the relative magnitude of the internal and external pressure. The radial stress is always compressive and the axial stress is tensile. It is assumed that the axial force due to pressure is wholly carried by the wall.

If at the inner radius R_1, the pressure is p_1 and at the outer radius R_2, the pressure is p_o, the constants A and B can be determined as

$$A = \frac{p_i R_1^2 - p_o R_2^2}{R_2^2 - R_1^2} = \frac{p_i - p_o K^2}{(K^2 - 1)} \tag{6.4}$$

where, $K = \frac{R_2}{R_1}$

$$B = \frac{(p_i - p_o) R_1^2 R_2^2}{(R_2^2 - R_1^2)} = \frac{(p_i - p_o) K^2 R_1^2}{(K^2 - 1)} \tag{6.5}$$

These relations show some important features. (1) The tangential and radial stresses consist of a pure shear stress $\left(q = -\frac{B}{R^2}\right)$ with a superimposed hydrostatic stress (A). (2) The maximum values of the stresses occur at the inner radius. (3) The maximum values depend only on the ratio of the radii and not on their absolute magnitude. In many cases, only the internal pressure p_i is present and the above equations are, therefore, simplified. The stresses at the inner surface are (at $R = R_1$) (Fig. 6.6).

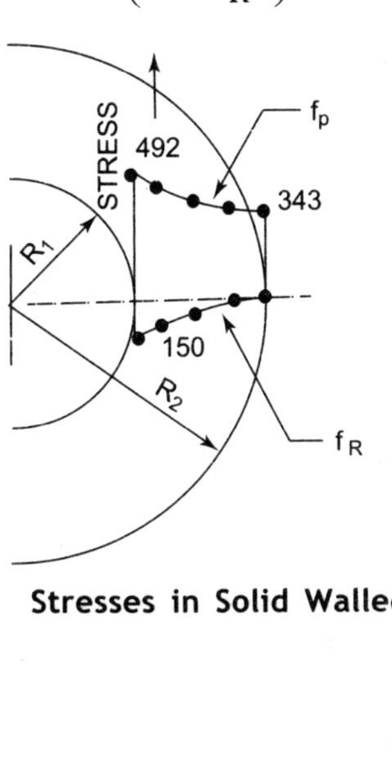

Fig. 6.6 Stresses in Solid Walled Shell

When the vessel is subjected to very high internal pressures, one can simplify the calculation of f_p, f_R and f_a by assuming $p_o = 0$ although it is atmospheric pressure.

For instance, if internal pressure is 150 N/mm^2 and the outside pressure is atmospheric pressure (= 0.101 N/mm^2) one can see that

$$p_i - p_o = 150 - 0.101$$
$$= 149.9 \text{ N/mm}^2$$
$$\simeq 150 \text{ N/mm}^2$$

and, therefore, assuming $p_o = 0$

We do not have major error and expressions yield answers within acceptable engineering accuracy.

Therefore, we have for $p_o = 0$

At radius R_2, i.e., outside wall of the vessel

$$f_p = A + \frac{B}{R_2^2} = \frac{p_i(K^2+1)}{(K^2-1)} \tag{6.6}$$

$$f_R = A - \frac{B}{R_2^2} = \frac{p_i}{(K^2-1)}(1-K^2) = -p_i \tag{6.7}$$

and (always compressive)

$$f_a = \frac{p_i}{(K^2-1)} \tag{6.8}$$

The factor $\left(\dfrac{K^2+1}{K^2-1}\right)$ will tend to unity as the value K increases. This means that beyond a radius ratio of generally $K = 5$, the increase in permissible pressure will be marginal even with a substantial increase in the vessel wall thickness.

6.4.2 Application to Shell Design

As mentioned earlier we assume $p_o = 0$ for ease of computation without sacrificing engineering accuracy. The internal pressure is p_i.

6.4.2.1 Design Based on Tangential Stress

The thickness of the cylinder can be determined on the basis of the maximum tangential stress at inner surface of the cylinder. Equation 6.6 can be written in terms of thickness $[t = (R_2 - R_1)]$ as

$$f_{(\max)} = \frac{p_i}{t}\left(\frac{R_1^2 + R_2^2}{R_1 + R_2}\right) \tag{6.9}$$

If f_p is taken as the allowable tensile stress f then

$$t = R_i\left(\sqrt{\frac{fJ + p_i}{fJ - p_i}} - 1\right) \tag{6.10}$$

where, J – joint efficiency.

Corrosion allowance is neglected if the thickness is more than 30 mm.

6.4.2.2 Design Based on Shear Stress

Another approach to design is based on the maximum shear stress produced at any point on the wall of the cylinder. This is given by

$$f_{s(\max)} = \frac{1}{2}(f_p - f_R) = p_i \frac{K^2}{(K^2 - 1)} \tag{6.11}$$

The limit of elastic action occurs when

$$f_s \text{ (at yield point)} = \frac{1}{\sqrt{3}} f_{yp} \text{ (direct tensile stress at yield point).}$$

If the cylindrical shell is to be designed within the limits of elastic action, the maximum internal pressure at the inner radius is given by

$$p_{\max} = \frac{1}{\sqrt{3}} f \left(\frac{K^2 - 1}{K^2} \right) \tag{6.12}$$

For purposes of design a factor of safety should be used to get the allowable stress. The value of K can then be determined as

$$K = \sqrt{\frac{f}{f - p_i\sqrt{3}}} \tag{6.13}$$

where, f is the allowable stress from which the thickness t can be obtained.

6.4.2.3 Design Based on Maximum Energy of Distortion

A third criterion of design is based on the maximum energy distortion theory, also known as Von Mises or Hencky criterion. The limit of an elastic action is reached according to this criterion, when the relation between f_p, f_R and f_a is (see Eq. 3.44)

$$(f_p - f_R)^2 + (f_R - f_a)^2 + (f_a - f_p)^2 = 2(f_{yp})^2 \tag{6.14}$$

where, f_{yp} is the yield point of the material in simple tension. The three stress can be written for the internal pressure alone. The values of f_p by Eq. f_R and f_a are given by Eqns 6.6, 6.7 and 6.8.

The thickness t of the cylinder can be determined, by substituting respective values in Eq. 6.14 in terms of 'K' and then solving for K by trail and error method.

The maximum shear stress criterion of design and the maximum energy of the distortion theory mentioned above are based on the allowable yield stress of the material. The thickness of the vessel can be reduced if the yield stress of the material can be increased. This is made possible by use of alloy steels and by heat treatment. The alloying elements are chromium, nickel, molybdenum and vanadium. The three criteria for design of the thickness generally give widely differing values of the required thickness. Experimental results satisfy one criterion for one material and another criterion for another material. When working with a new material, it is important to make a preliminary test with a vessel constructed of the material. The experimental vessel may be hydrostatically tested until a permanent set is obtained. The pressure measured at the beginning of the permanent set can be compared with that predicted by the various criteria and the most appropriate equation. Steels with high tensile strength would be expected to fail by shear. A few typical compositions and their strength properties are indicated in Table 6.1.

6.4.2.4 Strains in Shell

In many designs of 'thick walled' vessels, a knowledge of the magnitude of the change in dimensions of the vessel is pertinent, where a more critical analysis is desired. With internal pressure p_i, the change in inside radius ΔR_1 and outside radius ΔR_2 is as follows.

$$\Delta R_1 = p_i \frac{R_1}{E}\left[\frac{R_2^2 + R_1^2}{R_2^2 - R_1^2} - \mu\left(\frac{R_1^2}{R_2^2 - R_1^2} - 1\right)\right] \tag{6.15}$$

$$\Delta R_2 = p_i \frac{R_2}{E}\left[\frac{R_2^2}{R_2^2 - R_1^2}(2 - \mu)\right] \tag{6.16}$$

where μ is Poisson's ratio, and E, the modulus of elasticity.

6.4.2.5 Vessels with Thermal Stresses

If the temperature of the vessel wall is not uniform, thermal stresses must be taken into account in determining the safe internal pressure or thickness as desired. The nature of the thermal stresses created in the wall has been stated previously (Eqs. 5.42 and 5.43). Two cases may be considered, (a) in which the vessel is subjected to internal pressure and heated internally and (b) the vessel is subjected to internal pressure, but heated externally. In the former case the design is based on the larger of the two stresses, namely, stress due to internal pressure or thermal stress. In the latter case the design is based on the combined stress due to pressure and thermal gradients.

6.4.3 Shells Formed by Bending a Sheet of Metal with a Single Longitudinal Weld

This type of vessel can be designed on the same basis as a solid walled vessel except that the joint efficiency must be taken into consideration in all calculations of thickness determination. The permissible stress may be multiplied by the joint efficiency.

6.4.4 Multi-shell and Prestressing

So far the design of a single shell formed out of a solid block of material or shell formed from a sheet by welding was considered. As the internal pressure in the shell increases, the required shell thickness also increases. With higher pressure the problems of economic use of material as well as those of fabrication become critical. The remaining methods of construction involve use of additional shells or layers of materials or autofrettage. In all these methods of construction, a better utilisation of the material is attempted by setting up an initial stress in the shell, known as prestressing.

As may be seen from Fig. 6.6 in a solid walled vessel, the internal pressure causes a tangential stress, which is maximum on the inside of the cylinder. The thickness of the wall is determined by the allowable maximum stress. To utilize the material more economically, it would be desirable to create a uniform stress in the entire thickness of the vessel wall. It is possible to satisfy this condition, by prestressing. The extent of prestressing required to create uniform stress is assessed as follows. The average stress in the vessel wall can be calculated as

$$f_{av} = \frac{p_i R_1}{t} = \frac{p_i R_1}{(R_2 - R_1)} \tag{6.17}$$

$$= p_i \left(\frac{1}{K-1} \right)$$

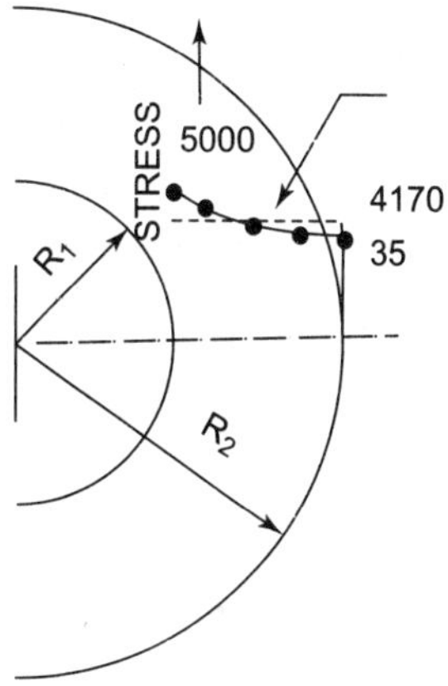

Fig. 6.7 Prestressed Shell

If the tangential stresses due to internal pressure are f_i and f_o at the inner and outer radius, then, the extent of prestressing would be, $(f_i - f_{av})$ compressive stress in the part of the wall thickness and $(f_{av} - f_o)$ tensile stress in the remaining wall thickness (Fig. 6.7). The various methods of construction satisfy these conditions only partially. Ideally prestressing to create a uniform stress is difficult to achieve.

6.5 MULTI-SHELL CONSTRUCTION

These vessels are built up by wrapping a series of sheets over a core tube. The construction involves the use of several layers of material, usually for the purpose of quality control and optimum properties. Each layer may be sufficiently thick, and is considered as a thick walled cylinder. Between the two cylinders in contact there is an interface pressure p_f which acts on the outer surface of the inner cylinder and on the inner surface of the outer cylinder. Thus, with only two layers the inner cylinder is subjected to an internal pressure p_i and an external pressure p_f; the outer cylinder is subjected to an internal pressure p_f only. The value of the interface pressure p_f is determined from the deformation at the interface of both cylinders.

Multi-shell construction is used for higher pressures. It provides inbuilt safety, utilises the material economically, no stress relief is required, fabrication and assembly cost is lesser, and the vessels can be easily repaired. For corrosive applications the inner liner is made up of special material and is not considered for strength criteria. The outer load bearing shells can be made of high tensile carbon steel. These vessels are constructed up to 233 N/mm^2 pressure, 400°C to –20°C temperature range and a 300 mm wall thickness.

6.5.1 Shrink Fit Construction

In this construction the vessel is built up of two or more concentric shells, each shell progressively shrunk on from the inside outward. As a result, the inner shells are subjected to an initial compressive circumferential stress so that when the internal pressure is applied, the resultant tensile circumferential stress on the shell layer does not exceed the allowable safe working stress. The magnitude of the compressive stress due to shrinkage is determined by the initial difference between the bore of the outer and the external diameter of the inner cylinder. The permissible extent of shrinkage involves a consideration of the maximum temperature to which the outer cylinder can conveniently be heated without impairing the properties of the material. From economic and fabrication considerations, the number of shells should be limited to two.

6.5.2 Stresses in Multi-Shell or Shrink Fit Construction

In the general case of two shell construction the two components have different properties and are held together by an interface pressure p_f. The radial deformations of the two shells can be given as follows:

For the inner shell

$$\xi_i = \frac{p_f R_2}{E_o}\left(\frac{R_1^2 + R_2^2}{R_2^2 - R_1^2} - \mu_o\right) \tag{6.18}$$

For the outer shell

$$\xi_0 = \frac{p_f R_2}{E_o}\left(\frac{R_2^2 + R_3^2}{R_3^2 - R_2^2} + \mu_o\right) \tag{6.19}$$

where, R_1, R_2 and R_3 are the radii as shown in Fig. 6.8, E_i and E_o – moduli of elasticity of respective shells and μ_i and μ_o – Poisson's ratio for respective shells.

The total shrinkage is the absolute sum of the individual deformations.

$$\Delta = \xi_o + \xi_i \tag{6.20}$$

If the two shells are of the same material then $E_i = E_o$ and $\mu_i = \mu_o$, and the interface pressure p_f is given by

$$p_f = \frac{E\Delta}{R_2}\left[\frac{(R_2^2 - R_1^2)(R_3^2 - R_2^2)}{2R_2^2(R_3^2 - R_1^2)}\right] \tag{6.21}$$

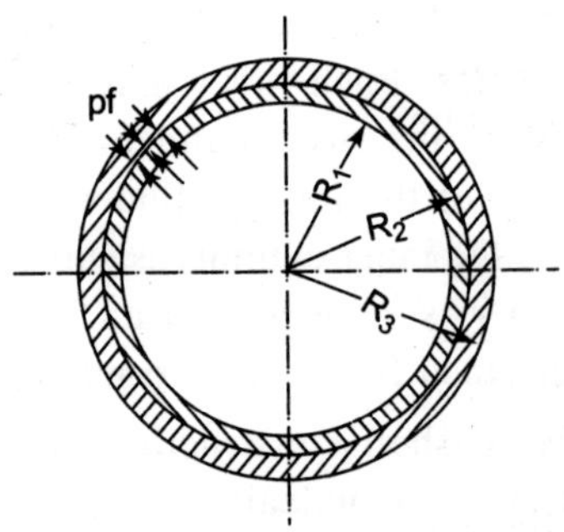

Fig. 6.8 Two Shell Construction

In shrink fit construction the radial deformation of the cylinder is obtained by heating the outer cylinder. The temperature required for shrinking operation is given as

$$t_f - t_i = \frac{R_f - R_i}{\alpha R_i} = \frac{\xi}{\alpha R_1} \tag{6.22}$$

where R_f, R_i = final and initial radii, α = coefficient of thermal expansion and t_f, t_i = final and initial temperatures. In either method, it is necessary to calculate the interface pressure. If such a built up cylinder is now subjected to internal pressure, the stresses produced by this pressure are the same as those in a solid wall cylinder of thickness equal to the sum of those of the individual cylinders, with radii R_1 and R_3. These are superimposed on the stresses produced due to p_f. Thus, the maximum tangential tensile stress is reduced, creating a more favourable stress distribution (Fig. 6.9).

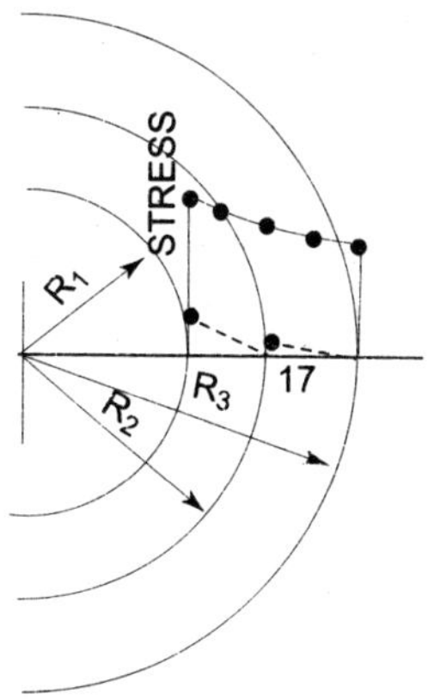

Fig. 6.9 Stresses in Single and Two Shell Construction

6.5.3 Optimum Design of Multi-shell Construction

For certain special applications where weight is critical and maximum utilization of the material is desired it is necessary to design for optimum conditions. Thus, if all the shells are of similar strength, the completed vessel is designed so that when there is an internal pressure and no external pressure, each component carries the same maximum shear stress, which is given by

$$f_{s\ \max} = \frac{p_i K^{2/m}}{m(K^{2/m} - 1)} \tag{6.23}$$

where, m – number of shells in the compound cylinder

K – ratio of radii.

If the maximum shear stress is taken as equal to the shear yield strength, then the maximum internal pressure is given by

$$p_{max} = \frac{mf_{yp}(K^{2/m}-1)}{K^{2/m}} = \frac{mf_{yp}(K^{2/m}-1)}{\sqrt{3}K^{2/m}} \tag{6.24}$$

Compare Eq. 6.12

The allowable stress value for f_{yp} will give the design value of K, if internal pressure is known. There is a limit to the maximum allowable pressure in a compound cylinder regardless of dimensions, which is given by

$$p = \frac{2f_{yp}(K^2-1)}{\sqrt{3}K^2} \tag{6.25}$$

There is also a limit on the temperature difference which can be used for shrinkage operation. The optimum shrinkage per unit radial length is $2p_i\mu/E$.

In an actual shrinkage operation the outer cylinder must be heated so that the thermal expansion difference is equal to the shrinkage required plus the radial clearance needed for construction. The optimum value of the interface radius is given by

$$R_2 = \sqrt{R_1 R_3} \tag{6.26}$$

6.6 VESSEL CLOSURES

6.6.1 Permanent Closures

It is the usual practice to close off one end of the cylindrical vessel permanently by a head which is forged or machined from a solid bar to form the shell and the head as the single piece. Alternately, a closure in the form of a cup or dish may be welded to the shell. For vessels of small diameter, a simple flat disk may be welded to the shell. This type of closure is used for moderate pressures.

6.6.2 Removable Closures

The type of the closure and the method of sealing and fixing the closure to the shell depend on a number of factors such as the size of vessel, the nature of its contents, the working temperature and pressure and the frequency with which the closure is removed. The joints made between the body of the vessel or shell and the closure are of different types depending on whether the joint can be made without a gasket or with the help of a gasket. For the latter a gasket joint can be either of the compression type or of the unsupported type. Compression type gasket joints are discussed under Section 5.8.4.2 and are either of the unconfined type, such as a full face gasket and a ring gasket on raised face, or confined gaskets located on male and female or tongue and groove facings. Unsupported gasket joints, also known as self-sealing types are especially applicable to high pressure vessel closures.

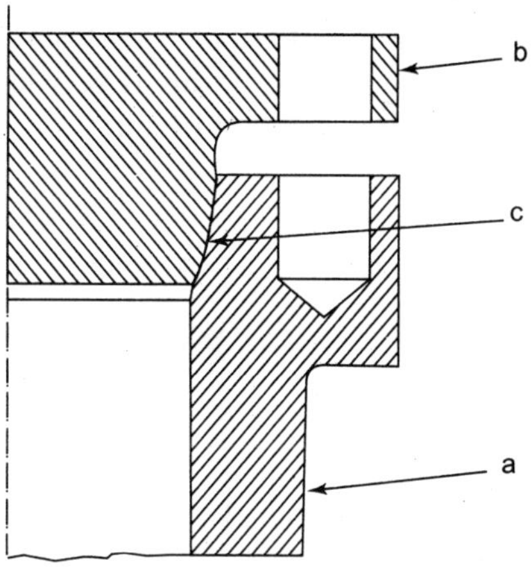

Fig. 6.10 Single Cone Joint: (a) Shell, (b) Cover, (c) Cone

The plain joint which is made without a gasket, does not require excessively large clamping forces since the two components, namely, the vessel and its cover make contact over a comparatively small area. Figure 6.10 illustrates the single cone joint which is one of the most common. The maximum pressure the joint can withstand is determined by the initial compressive stress that can be developed across the sealing surfaces. With these joints the effect of over-tightening is to increase the area of contact between the two components and, thus, increase the difficulty of obtaining the required initial stress across the seating. Such joints are used for small vessels operating at a moderate pressure. Compressive type of gasket joints are made with gaskets which are comparatively soft and ductile. They are able to sustain an initial high compressive stress. The gasket diameter and width are kept as small as possible. For high pressure work the gasket is confined and supported and is made of copper, silver, iron or aluminium (Table 5.4). Unsupported or self-sealing gaskets are so constructed that the stress between the adjoining faces is automatically maintained at a higher value than the pressure in the vessel. Several designs of the self-sealing type joints have been developed; the most widely adopted forms are indicated below.

6.6.2.1 Delta-ring Closures

These are used for vessels operating under wide range of pressures up to 100 N/mm^2 (Fig. 6.11). The ring has a triangular cross-section with the longest side of the triangle facing the centre of the ring. V-shaped grooves are cut in the flange of the vessel and in the underside of the cover. These grooves match the delta-ring, so that when the cover is tightened down on the flange, it will form a seal. The delta-ring is made of steel, machined, polished and copper plated. The copper plating serves as a gasket. The ring is designed to function below its elastic limit under bolt operating and hydrostatic testing conditions.

Delta-ring closures are simple to fabricate and install. The closure may be opened and closed with minimum time. The disadvantage is that when the closure is very large, the delta-ring lacks rigidity during machining and dimensional control becomes difficult.

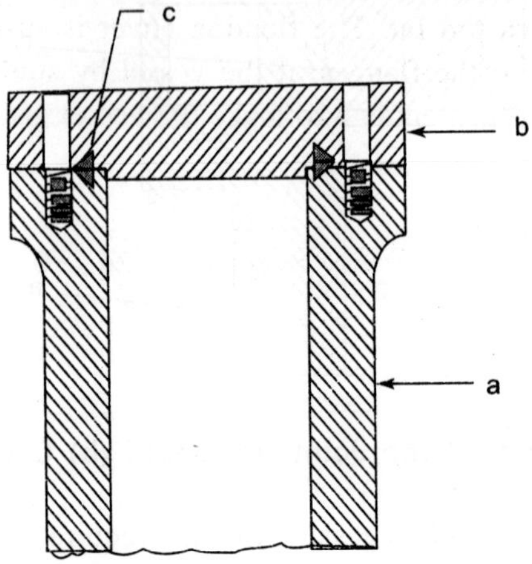

Fig. 6.11 Delta-Ring Closure: (a) Shell, (b) Cover, (c) Delta-Ring

6.6.2.2 Double-Cone Seal Ring (Lens Ring)

In this type of closure a ring of soft steel with a plain vertical face towards the inside, and two bevelled edges towards the outside are used (Fig. 6.12).

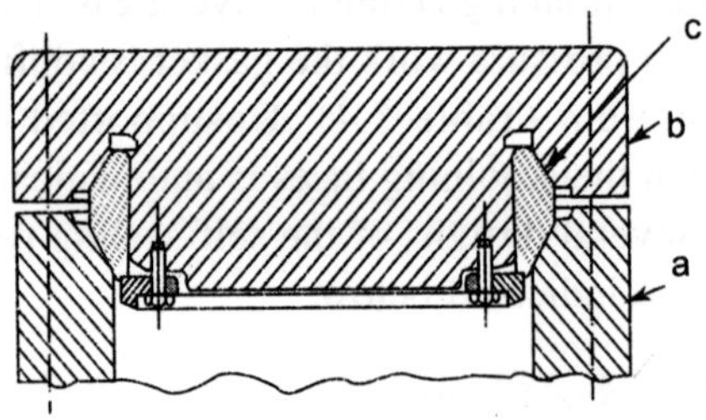

Fig. 6.12 Double-Cone Seal Ring Closure: (a) Shell, (b) Cover, (c) Double-Cone Ring

The bevels extend nearly to the middle line of the ring and each bevel makes an angle of 30° with the axis of the vessel. The ring is made from an endless ribbon formed from an aluminium sheet about 1 mm thick. The length of the ribbon is made equal to the inside circumference of the double-cone ring and a width equal to 2½ times the height of the ring. This ribbon fits in the inside circumference of the ring, wraps around and covers the conical surfaces of the ring, thus forming a gasket. No groove is required in the flange of the shell. The studs can be placed on a smaller pitch circle, thus reducing the size of the closure. These seal rings are used up to 50 N/mm^2 pressure. The above gaskets are unsupported in the radial direction.

6.6.2.3 The Bridgeman Closure

This consists of a partly triangular gasket of a relatively soft metal like copper or aluminium backed by a steel seal-ring. These two rings are accurately machined to fill the space between the mouth of the vessel and a floating cover (Fig. 6.13). A narrow edge in the mouth of the vessel prevents the floating cover from going down too far. The floating cover is suspended by screws from a ring-shaped cap, which is fastened to the flanges of the vessel by studs. In making the closure, these suspension screws are tightened just enough to make the vessel gas-tight without internal pressure.

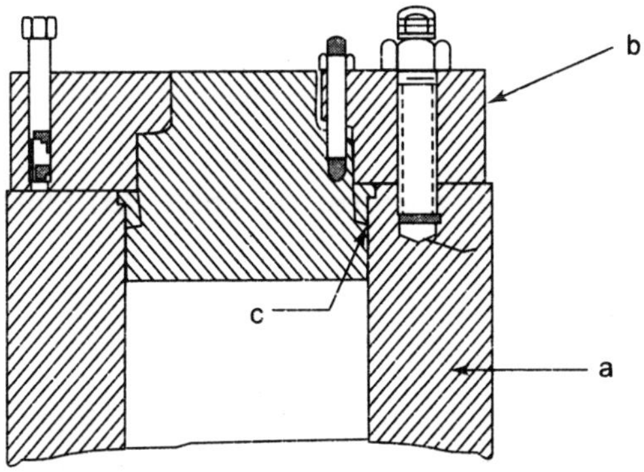

Fig. 6.13 Bridgeman Closure: (a) Shell, (b) Cap, (c) Seal-Ring

When the internal pressure on the vessel is applied, the axial pressure on the underside of floating cover pushes it against the gasket. A component of the pressure then pushes the gasket against the mouth of the vessel making a seal, the tightness of which is proportional to the internal pressure.

The chief advantages of this closure are: (a) the gasket is totally confined, and is in no danger of being blown out even at very high pressure, (b) the gasket is unsupported in the axial direction and, hence, the function of taking the end load and of prestressing the joint between the vessel and its cover can be separated and (c) the closure is simple to fabricate, easy to install and can be used for any diameter.

6.6.2.4 The Uhde-Bredtschneider Closure

It consists of a floating cap with spherical shoulder and a seal-ring with a wedge-shaped cross-section (Fig. 6.14), which fits into the slightly tapered throat of the vessel. This feature facilitates removal of the seal-ring at any time after the vessel has been under testing or under normal operation. The seal-ring is designed to function within the elastic limit under operating and testing conditions and is, therefore, made of steel.

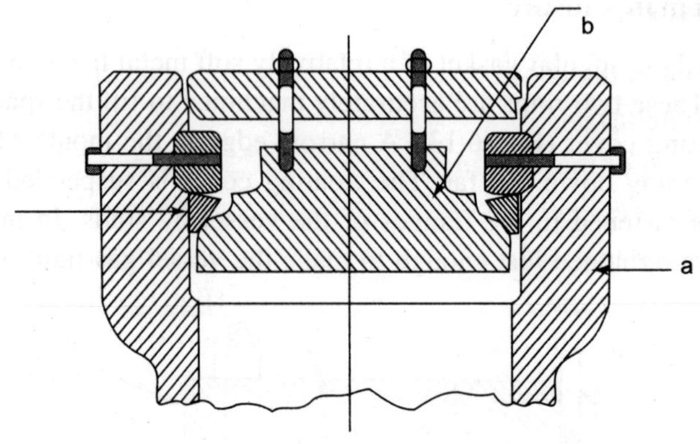

Fig. 6.14 Uhde-Bredtschneider Closure: (a) Shell, (b) Floating Cap, (c) Seal-Ring

6.7 JACKET FOR VESSELS

Some high pressure vessels need heating and cooling devices. Figure 6.15 shows two methods of providing jackets, one for a solid-wall vessel and the other for a strip-wound vessel. In both the cases a half coil is used, which (see Section 8.6.2) is welded to the wall or to the outer strip in strip-wound vessel. Water or steam with a pressure of up to 10 N/mm^2 can be used in these coils.

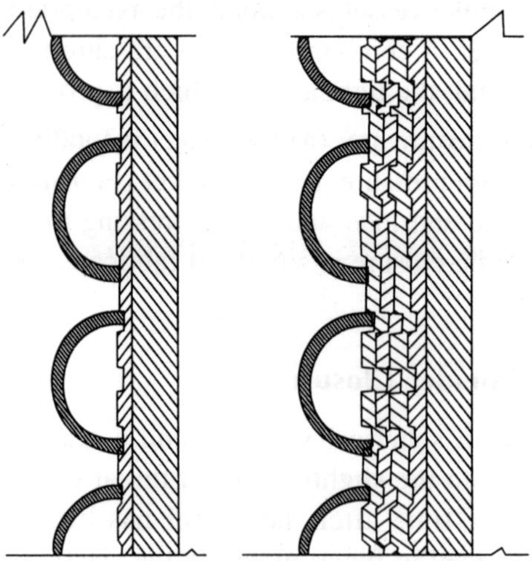

Fig. 6.15 Jackets

6.8 ILLUSTRATION

Following example makes the above discussion more comprehensive.

High Pressure Vessel Shell Data:

Internal diameter of the shell, p_i	: 300 mm
Internal design pressure	: 150 N/mm^2
External pressure, p_o	: 0.1 N/mm^2
Material of construction	: High tensile steel (Cr, Mo, V)
Permissible Internal stress (based on UTS)	: 500 N/mm^2
Permissible stress (based on YS), f	: 700 N/mm^2
Modulus of elasticity	: 2×10^5 N/mm^2

6.8.1 Vessel Shell Thickness

We have from Eq. (6.10)

$$t = R_i\ [\{(fJ + P_i)/(fJ - P_i)\}^{0.5} - 1]$$

The joint efficiency J = 1 (forged shell)

$$t = (300/2)\ [\{(500 \times 1 + 150)/(500 \times 1 - 150)\}^{0.5} - 1]$$

$$= 54.4 \text{ mm},$$

say, $t = 55$ mm

Therefore, we have $R_2 = 150 + 55 = 205$ mm.

By definition we have $K = (R_2/R_1) = 205/150 = 1.37$

6.8.2 Variation of Stress Along the Thickness

Tangential and radial stresses can be computed as follows:

From Eq. (6.4), we have,

$$A = (p_i - p_o K^2)/(K^2 - 1) = (150 - 0.1\ (1.37)^2)/((1.37)^2 - 1) = 170.84$$

From Eq. (6.5)

$$B = \frac{(p_i - p_o)K^2 R_1^2}{(K^2 - 1)}$$

$$= \frac{(150 - 0.1)(1.37)^2 (150)^2}{((1.37)^2 - 1)}$$

$$= 7.22 \times 10^6$$

At R_1 = 150 mm, the tangential stress is given by Eq. (6.1)

$$f_p = A + B/R_1^2;\ \text{N/mm}^2\ (\text{Tensile}) = 491.7$$

say, = 492 N/mm^2

The radial stress, f_R, is given by Eq. (6.2)

$$f_R = A - B/R_1^2;\ \text{N/mm}^2$$
$$= 170.84 - 7.22 \times 10^6/(150)^2$$
$$= -150.04$$

say, = –150; N/mm^2 (–ve sign indicates compressive stress)

Thus, we can have at various values of R_1 stress values as listed below:

R_1	f_p N/mm^2	f_R N/mm^2
150	492	–150
160	453	–111
170	421	–79
190	371	–29
205	343	00

(Please note all above values have been rounded off)

From Eq. (6.8) we have longitudinal stress, f_a as:

$$f_a = p_i/(K^2 - 1)$$
$$= 150/((1.37)^2 - 1)$$

say, = 171; N/mm^2

From Eq. (6.17) we have the average stress, f_{av}

$$f_{av} = p_i/(K - 1)$$
$$= 150/(1.37 - 1)$$

say, = 405 N/mm^2

The extent of prestressing would be:

$$f_i - f_{av} = 492 - 405 = 87\ \text{N/mm}^2;\quad (\text{compressive})$$

$$f_{av} - f_o = 405 - 343 = 62\ \text{N/mm}^2;\quad (\text{tensile})$$

6.8.3 Maximum Shear Stress Theory

Here we assume p_o = 0 which is acceptable within engineering accuracy as inside pressure is very high.

$$\Delta P = 150 - 0.1 = 149.9 \quad \text{say} = 150\ \text{N/mm}^2$$

According to the maximum stress theory the value of K can be determined via Eq. (6.13)

$$K = \left(f/(f - p_i\sqrt{3})\right)^{0.5}$$

$$= \left(700/(700 - 150\sqrt{3})\right)^{0.5}$$

$$= 1.26$$

By definition of K, we have,

$$R_2 = 1.26\ (150) = 189.15 \text{ mm}$$

Thickness, $t = 189.15 - 150 = 39.15$ mm

6.8.4 Maximum Energy of Distortion Theory

From Eq. (6.14) one can solve by trial and error to obtain the thickness, which is around 39.2 mm.

It is thus seen that the thickness based on the maximum tangential stress (hoop stress) is 55 mm and is higher than that by both theories. The design is, therefore, acceptable.

References and Recommended Readings

For more detailed and comprehensive treatment of this topic the following books are recommended:

Brownell, L.E., and Young, E.H., *Equipment Design*, John Wiley and Sons, (1968).

Harvey, J.F., *Process Vessel Design: Nuclear and Chemical Applications*, D. Van Nostrand Company Inc. Princetone, New Jersey, (1963).

Taganov, N.I., *Fundamentals of Process Equipment Design,* Academic Books, Bombay, (1968).

7

Chapter

Storage Vessels

7.1 INTRODUCTION

Liquid and gaseous products must be stored during intervals between production, transportation, refining, blending and marketing. The object of storage at each of these stages is firstly to supply a sufficient balance of each stock to ensure continuity of operations, and secondly, to ensure that the product is conserved and maintained at an acceptable level of quality.

In general, the storage of liquid or gaseous products is done in large-sized vessels. These consist of units either completely fabricated in the shop or vessels so large that they must be manufactured on site. Various parts of the vessels are made in sizes which can be conveniently transported. The most common material used is mild steel. Vessels are also made of stainless steel, aluminium and clad steel, where the cladding may be either stainless steel, nickel, monel or inconel.

7.2 STORAGE OF FLUIDS

Fluids normally stored in bulk can be classified in three groups namely, non-volatile liquids, volatile liquids and gases.

7.3 STORAGE OF NON-VOLATILE LIQUIDS

These liquids are usually stored in standard cylindrical tanks in sizes in excess of 60 metre diameter and with heights up to 30 metres. Very small tanks may be rectangular, made of flat plates. These tanks require stiffening arrangements, since flat plates are structurally weak, and are unable to withstand pressures. Small size tanks may also be horizontal cylindrical vessels with flat or slightly dished ends, supported on two saddles.

7.3.1 Standard Free Roof Cylindrical Tanks

Vertical cylindrical tanks consist of a vertical shell usually butt welded, with a flat bottom and roof sheets, supported on an umbrella type roof truss. The bottom and the roof are fabricated from sheets by welding. The roof sheets are not usually attached to the roof truss, and the tank is thus free to breathe with the changes in temperature and pressure, the roof sheet acting like a membrane (rupture disk) leaves the truss when subject to internal pressure (resulting even due to fire). Typically, the capacity to hold pressure is 500 mm of water.

7.4 STORAGE OF VOLATILE LIQUIDS

These liquids are stored in specially designed tanks which are capable of conserving the contents. If these are stored in normal fixed volume tanks, a loss of liquid takes place due to the following causes:

(a) ***Breathing losses*** A space above the liquid level is filled with air and vapour mixture. Vaporization increases with rise in temperature during the day. Increase in temperature will, therefore, result in the expansion of the mixture and a loss of a certain volume through the vent. To minimize the loss often a vent condenser is provided. The temperature drops during the night and some condensation takes place, the air-vapour mixture contracts and fresh air is sucked into the tank.

(b) ***Filling losses*** When the liquid is discharged from a fixed volume tank, the tank will be filled with air-vapour mixture. If the tanks is then filled again with a fresh batch of liquid, certain volume of air-vapour mixture will be vented and lost. This loss can be substantial, if such filings take place very frequently.

(c) ***Boiling losses*** If the vapour pressure of the liquid stored is higher than the atmospheric pressure, it boils and the vapour thus produced is lost. In such cases, the liquid should be stored in a vessel operating under a pressure not less than the vapour pressure of the liquid at the maximum temperature of storage.

7.4.1 Standard Fixed Roof Storage Tanks

Depending on the volatility of the liquid, the tank is designed to have a pressure-tight roof. Normally, the pressure variation in the vapour space of the tank roof may be from + 200 mm to – 25 mm w.g. In some cases, the tanks are built with a higher pressure up to 625 mm. A vent is fitted on such tanks with a pressure relief valve set to open at an internal pressure of 200 mm or 625 mm, as the case may be. An anti-vacuum valve should also be provided and also connected separately to the vent, so that air or an inert gas like N_2 can enter the tank whenever the tank develops a vacuum of 25 mm, or higher as specified. When the temperature of air-vapour mixture in the tank drops, or when the liquid is discharged from the tank a vacuum higher than 25 mm is developed on account of the fact that the vent is connected to the atmosphere through a breather valve and not directly. Fixed roof tanks (Fig. 7.1), or any closed tank should be designed to withstand the vacuum conditions. With proper adjustment, the breather valve losses can be completely eliminated and the filling losses reduced to some extent.

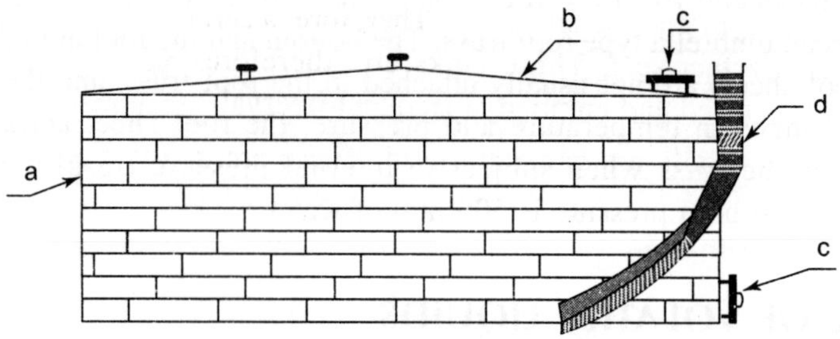

Fig. 7.1 Standard Fixed-roof Storage Tank: (a) Shell, (b) Roof, (c) Manhole, (d) Ladder

7.4.2 Variable Volume Tanks

7.4.2.1 Vapour-Lift Roof Type (Gas Holder)

In this construction, the roof normally rests on the usual roof supports as in a fixed volume cone-roof tank. When the volume of the vapour increases due to rise in temperature or when the liquid is pumped, the roof rises vertically increasing the vapour space (Fig. 7.2). A liquid seal prevents the loss of vapour at the junction of the roof and shell. A simple mechanism, which directs and stabilizes the upward and the downward movement of the roof, consists of tracks placed at regular intervals at an angle of 45°, roller guides ride in this tracks and other roller guides place at 90° to the rollers in the tracks, ride against the side of the tank and take any lateral load such as from winds. The liquid seal and stabilizers are fully protected from the weather by a wind apron. Suitable pressure or vacuum relief valves are used to prevent accidental excess pressure or vacuum. The liquid seal is designed to relieve excess pressure in case the relief valve fails to function. These tanks will eliminate all breathing losses and also the filling losses to a certain extent.

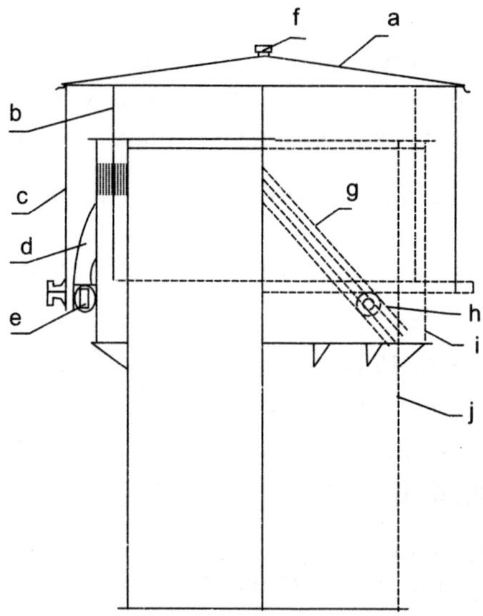

Fig. 7.2 Vapour-Lift Roof Storage Tank: (a) Vapour Lift Roof, (b) Moving Shell, (c) Wind Apron, (d) Guide Track, (e) Guide Roller, (f) Emergency Relief Valve, (g) Guide Track, (h) Guide Roller, (i) Seal Tank, (j) Fixed Tank Shell

7.4.2.2 Floating Roof Type

In this type of variable tanks, the roof (Fig. 7.3) is not supported either by the shell or by columns, but is made to rest on the stored liquid and is free to move with the level of the liquid, the vapour space is thereby practically eliminated. No breathing losses are possible, since there is little space in which the air-vapour mixture can form. It is difficult to create a perfect seal between the floating roof and the tank shell. Therefore, a certain amount of liquid is expected to be lost through the imperfect seal. These tanks are, therefore, most economical when used as working tanks which are filled and emptied at frequent intervals, than merely as storage tanks.

Three types of floating roofs are generally used. These are: (a) Pan, (b) Pontoon, (c) Double-deck.

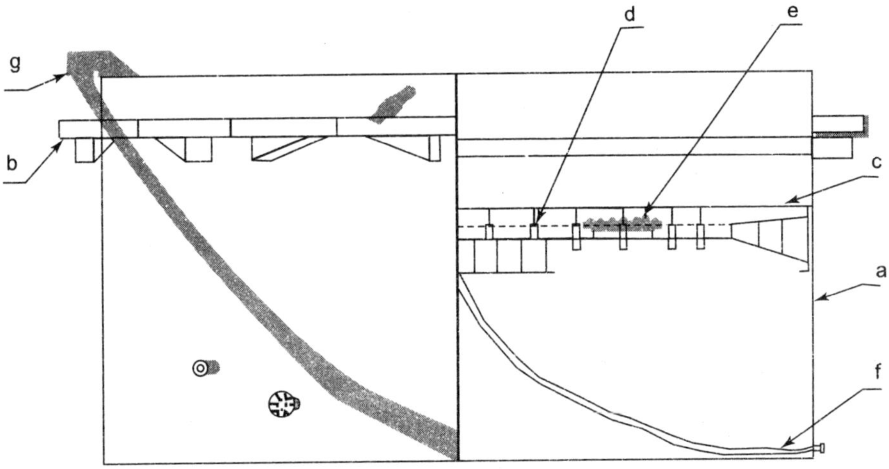

Fig. 7.3 Floating Roof Storage Tank: (a) Shell, (b) Shell Reinforcement (Wind Girder), (c) Pontoon Type Floating Roof, (d) Adjustable Pipe Support, (e) Manhole, (f) Hose Drain or Swivel Joint, (g) Ladder

The pan type of roof is the cheapest. The double-deck type will give the greatest structural protection as well as hold the evaporation losses to a minimum

(a) *Pan roof* This type of roof has an upstanding rim which gives the roof floating characteristics. Various types of pan roofs are shown in Fig. 7.4. Since the pan is a single-deck roof, an unchecked leak through this deck will eventually cause it to sink. It has no other buoyancy other than that provided by the deck and rim. If the roof is loaded by rain or snow, the rim gets compressed and begins to deform. Excessive distortion can cause an imperfect seal.

(b) *Pontoon roof* This represents a significant improvement over the pan roof. Buoyancy and stability are increased. Also, the insulation of the product from the sun is enhanced. As a result, these roofs have been widely used in both cold and hot climates and on corrosive and non-corrosive products with vapour pressures up to 1 kg/cm^2. Some of the pontoon roofs are shown in Fig. 7.5. It is common to use a pontoon with 35 cm to 45 cm depth, and a pontoon area of about 20% to 40%. These roofs are able to carry a considerable water load.

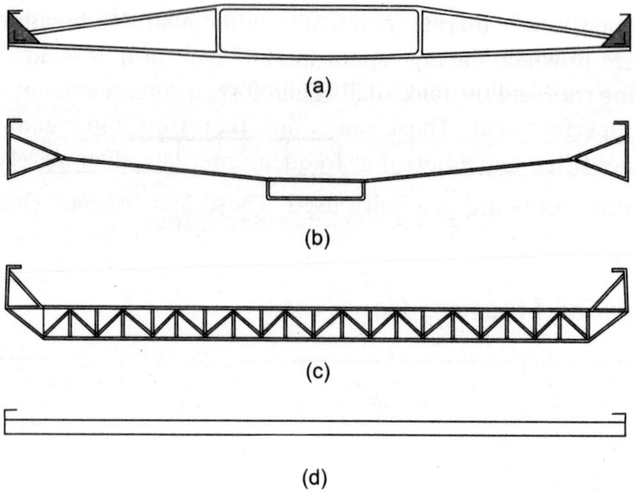

Fig. 7.4 Pan Type Roofs: (a) Trussed Pan Roof, (b) Pan Roof with Centre Weight, (c) Pan Roof with Underdeck Truss, (d) Pan Roof with Underdeck Girder

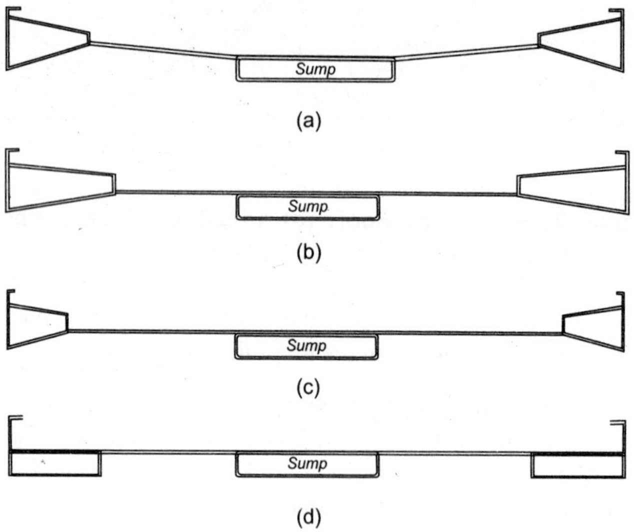

Fig. 7.5 Pontoon Type Roofs: (a) Pontoon Roof with Centre Weight, (b) Pontoon Covering 30 Per cent of Roof Area, (c) Standard Pontoon Roof, (d) Pontoon Roof with Deck Raised

(c) *Double-deck roof* This roof comprises upper and lower decks separated by bulkheads and trusses (Fig. 7.6). Circular bulkheads divide the entire roof into annular liquid-tight compartments and some of these compartments are in turn divided into liquid-tight sectors by means of radial bulkheads. Radial stresses are dispersed throughout to provide the required strength and stability. Because of its obviously superior load capacity and insulating characteristics, the double-deck roof is used extensively for all types of products in all climates.

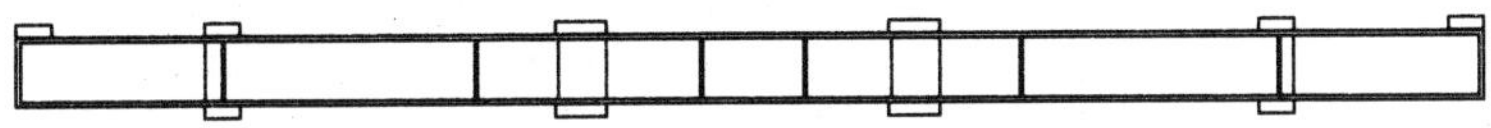

Fig. 7.6 Double-deck Roof

7.4.3 Accessories of Floating Roof Tank

Seal A flexible seal between the floating roof and the tank shell is an essential feature. Two types of seals are shown in Fig. 7.7.

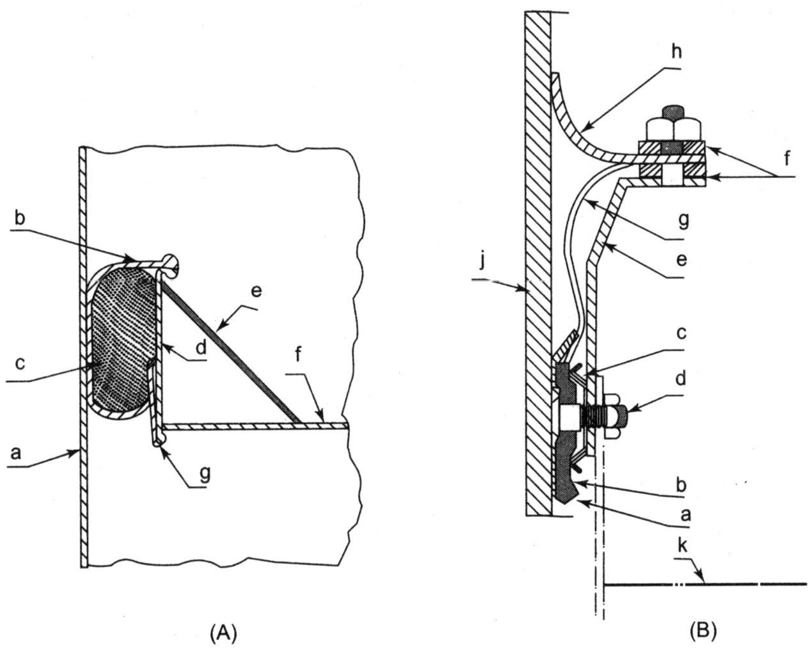

Fig. 7.7 Seals for Floating Roof Tank: (A) (a) Tank Shell, (b) Endless Fabric, (c) Flexible Foam Ring, (d) Rim, (e) Stiffener, (f) Deck Plate, (g) Hold Down Bar, (B) (a) Sliding Shoe, (b) Continuous Seal, (c) Clamp, (d) Bolt, (e) Apron Plate, (f) Washer, (g) Curtain Seal, (h) Secondary Wiper Seal, (j) Shell, (k) Floating Roof

(i) The space between the pan type of roof and the shell, is filled by a seal consisting of a resilient foam doughnut covered by a polyurethane-coated nylon seal envelope [Fig. 7.7(A)]

(ii) Figure 7.7(B): This consists of a sliding shoe: (a) of extruded section formed, so as to contact the tank shell along two horizontal sections. It is covered on the inside by a continuous seal,

(b) of flexible material, which is held in position by clamps, (c) and large number of bolts, (d) A series of segmental apron plates, (e) are held in position by bolts (d) at one end and washer bars (f) at the other end. The gaps at the top, bottom and sides of the apron are sealed by narrow strips of flexible material, known as curtain seals (g). A secondary wiper seal (h) consisting of flexible material is provided to exclude air currents. The assembly is connected to the floating roof.

Drains Drains are necessary to remove accumulated water from the floating roof. Open drains are needed on Double-deck roof. They allow the water to flow into the stored product, where it settles to the bottom to be drawn off later. A siphon drain is an open pipe with screened top, having a bottom reservoir. The water reservoir seals against the product and prevents its flow on to the deck. A pipe with a swivel joint conducts water from a central sump on the roof to the outside of the tank through a shell nozzle.

In addition to the drain, it is necessary to provide certain other facilities for the floating roofs such as an adjustable pipe support to keep the roof from resting on the tank bottom when the tank is empty or nearly so. These pipe supports hold the roof about one metre off the bottom during normal operation, and two metres for cleaning and inspection. Vents are required to provide air ingrace under the roof as the tank is emptied or to relieve pressure as the tank is filled when the roof is resting on the bottom. Manholes or openings are placed in the deck or pontoons for inspection and cleaning operations and a rolling ladder for getting on top of the roof. Anti-rotation device is needed for keeping the roof from rotating during product level changes and, thereby, prevent damage to rolling ladder, pipe drain, swing line and gauging equipment. A liquid level gauge platform and foam guns for fire fighting are usually placed at the top of the tank.

7.5 STORAGE OF GASES

7.5.1 Spherical Vessels or Hortonspheres

These types of vessels (Fig. 7.8) are extensively used for storage of gases and volatile liquids in the pressure range of 1 to 10 atmospheres or higher. With higher pressure the increased plate thickness necessitates stress relieving of the vessel.

Under similar operating pressure conditions, the thickness of the plates required to form a spherical vessel is half of that required for a cylindrical shell of the same diameter. Moreover, the ratio of surface area to volume is less for a sphere than any other shape of a vessel. This is an advantage and reduces the plates required. In case of insulated vessels, this reduces the amount of insulation required.

Hortonspheres usually range from 6 metres diameter up to about 25 metres diameter. They are supported by 6 to 12 tubular or rolled section columns attached to the shell at the equator. The upper section of each column is welded to the shell. The columns are braced with rods and turnbuckles to carry the wind load. Each column rests on a concrete pier. The upper half of the sphere acts like a hemispherical dome roof and the lower half of the sphere may be considered as a hemispherical suspended bottom.

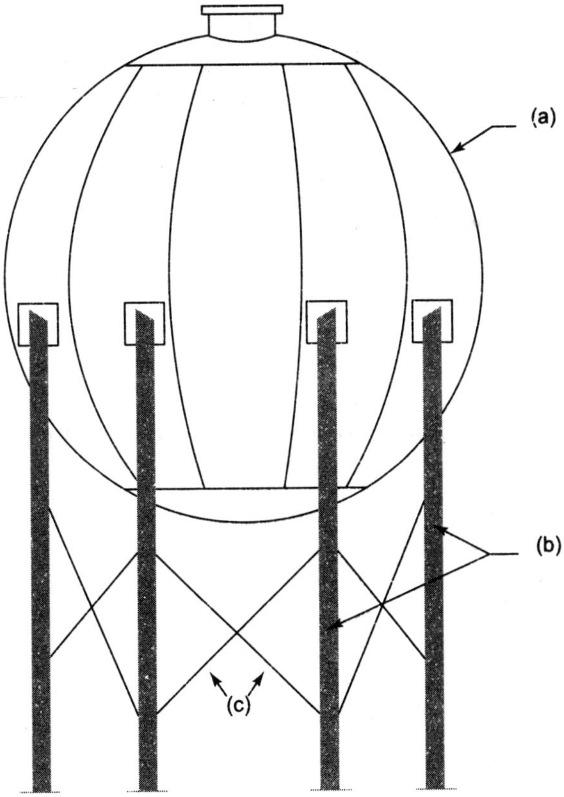

Fig. 7.8 Hortonsphere; (a) Shell, (b) Column for Support, (c) Tie Rods

7.6 DESIGN OF RECTANGULAR TANKS

Rectangular tanks are preferred in certain processes. These are particularly suitable for storing liquids without any pressure. They are easily fabricated and are limited to smaller capacities than the cylindrical tanks. Design of rectangular tanks is based on the use of rectangular plates of uniform thickness and an allowable deflection of the plate to about half the plate thickness. Rigidity can be obtained by stiffening the top edge and the walls of the tank.

7.6.1 Design without Stiffeners

The maximum pressure on the side wall due to liquid level is given by

$$p = \rho \cdot H \tag{7.1}$$

where, H — height of the tank
ρ — density
p – pressure.

If the longer side of the tank is a, then the maximum bending stress on the side plate is given by

$$f = \frac{\beta p a^2}{(t-c)^2} \tag{7.2}$$

where,

β — constant (from Table 7.1)
t — thickness of wall plate
c — corrosion allowance.

The maximum deflection is given by

$$Y_{max} = \frac{\alpha p a^4}{E(t-c)^3} \tag{7.3}$$

Maximum deflection should be limited to one half of the corroded-plate thickness. The value of α can be obtained from Table 7.1 and E is the modulus of elasticity.

Table 7.1

Ratio (H/a)	0.5	0.667	1.0	1.5	2.0	2.5	3.0
β	0.11	0.16	0.2	0.28	0.32	0.35	0.36
α	0.026	0.033	0.04	0.06	0.058	0.064	0.067

The following correlations may be used:

$$\beta = 0.2127 + 0.1444 \ln (H/a); \quad r^2 = 0.99$$

and

$$\alpha = 0.0429 + 0.0237 \ln (H/a); \quad r^2 = 0.96$$

7.6.2 Design with Top-Edge Stiffeners

All edges of the tank may be considered supported if a top-edge stiffener of sufficient size is fixed. The plate can be considered as a beam, supported at the top and bottom and with a load that increases uniformly from top of the tank to the bottom. The reaction at the bottom edge is

$$R = \frac{Hm}{6} \tag{7.4}$$

where, m — unit width of plate.

The maximum deflection of the beam with this type of loading is

$$Y_{max} = \frac{Ra^4}{384\,EI} \tag{7.5}$$

where, I — moment of inertia of the stiffener cross-section and a-area of plate. Neglecting the moment of inertia of the plate, the required moment of inertia of the top stiffener can be calculated, by limiting the maximum deflection to half plate thickness.

$$\left(\frac{t-c}{2}\right) = \frac{Ra^4}{384\,EI} \tag{7.6}$$

$$I = \frac{Ra^4}{192E(t-c)} \tag{7.7}$$

A minimum angle of 65 × 65 × 6 mm should be used. The plate thickness is calculated from Eqs. (7.2) and (7.3), with values of α and β taken from Table 7.2.

Table 7.2

Ratio (H/a)	0.25	0.286	0.333	0.4	0.5	0.667	1	1.5	2
β	0.024	0.031	0.041	0.056	0.08	0.116	0.16	0.26	0.34
α	0.00027	0.00046	0.00083	0.0016	0.0035	0.0083	0.022	0.043	0.06

The above data could be approximately correlated as:

$$\alpha = 0.0157\ (H/a)^{2.64};\quad r^2 = 0.972$$

and

$$\beta = 0.1614\ (H/a)^{1.247};\quad r^2 = 0.98$$

7.6.3 Horizontal and Vertical Stiffeners

Stiffeners increase the rigidity of the entire plate, and hence the plate thickness can be considerably reduced. A large stiffener can be attached horizontally all round the tank and located at a distance $H/\sqrt{3}$ from the bottom of the tank. Similarly, vertical stiffeners can be attached. This will give higher H/a ratios. By obtaining values of β and α from Table 7.2, thickness and deflection can be calculated by the use of Eqs. 7.2 and 7.3, with the reduced value of a.

7.6.4 Bottom Plate

When the entire surface of the bottom plate is supported, a minimum thickness of 6 mm is sufficient in almost all cases. If the bottom is supported by angels or beams, a special design analysis should be made to ensure rigidity.

7.7 DESIGN OF TANKS

The design and construction features of the tanks are given in detail in the following codes:

(1) Indian Standard-803

(2) API Standard-650

The above codes cover requirements for mild and alloy steel, vertical cylindrical tanks with conical or domed covers and flat bottoms, in various sizes, for erection above ground. Some of the important design features are indicated in the following. For design purposes, these tanks can be divided into three categories:

(a) Fixed roof tanks designed to operate only at a nominal pressure/vacuum, designated briefly as non-pressure tanks (internal pressure 75 mm water gauge and vacuum of 25 mm water gauge).

(b) Fixed roof tanks designed to withstand small pressures/vacuum.

   (i) Class 'A' tanks – For an internal pressure of 200 mm and vacuum of 65 mm water gauge.

   (ii) Class 'B' tanks – For an internal pressure of 550 mm and a vacuum of 65 mm water gauge for tanks normally not exceeding 20 metres in diameters.

(c) Open-top tanks.

7.7.1 Materials

Structural mild steel plates and sections are used for tank construction. These steels are based on IS-2062 specification for structural steel. Suitable sizes of plates or strips are indicated in IS-1730.

7.7.2 Bottom Design

The flat bottom of the tank is constructed of rectangular plates as shown in Fig. 7.9. On the periphery full rectangular plates cannot be used. The plates have to be cut to form the circular shape. These are known as sketch plates. For tanks over 12 metre diameter, an annular ring of segmental plates is provided. All joints in the plates are lap welded, with an overlap of not less than 5 times the thickness of the thinner plate, but need not exceed 25 mm.

It is assumed that the bottom plates are fully supported, either on a concrete slab or on a well consolidated and leveled ground. The bottom plates are, therefore, subjected to direct pressure of liquid and not to any bending moment. However, the bottom plates near the shell bottom joint are highly stressed due to the bending moment caused by the pressure of liquid on shell. Further, although the bottom plates are supposed to be uniformly supported by the ground, it is possible, particularly in the case of large tanks, that certain portions of the ground may settle unevenly and the support may not be uniform. All the bottom plates should have, therefore, a minimum nominal thickness of 6 mm. All sketch plates for tanks greater than 12 metre diameter should have a minimum thickness of 8 mm.

In the case of tanks which area supported on beams for providing a clearance under the tank, the bottom plate is subjected to a bending moment by the liquid height. The unsupported part of the bottom can be considered as a rectangular plate, uniformly loaded. The minimum thickness of the bottom plate in mm is given by

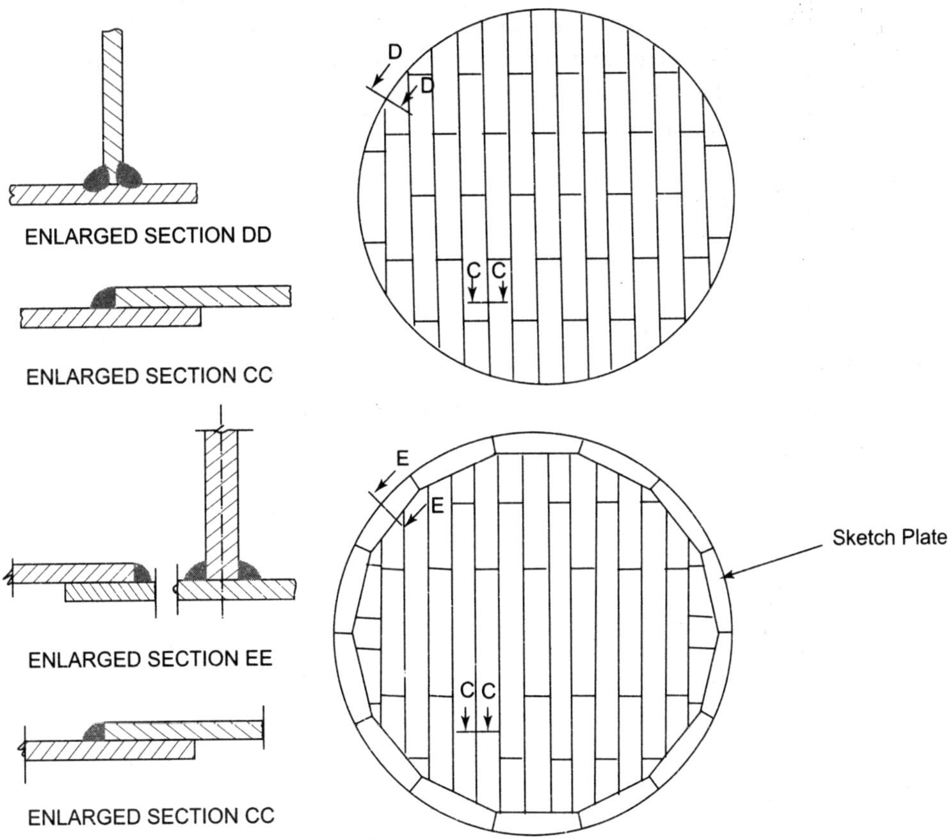

Fig. 7.9 Tank Bottom Details

$$t_b = 10^2 \sqrt{\frac{3\rho H l^2}{4f}} \tag{7.8}$$

where, ρ — density of the liquid stored in kg/cm^3

H — maximum liquid height in tank in metres

l — length of bottom plate freely supported between successive beams in cm

f — allowable bending stress in kg/cm^2.

7.7.3 Shell Design

7.7.3.1 Internal Loading

The internal pressure in the tank-shell is computer on the assumption that the tank is filled to the full height by the liquid stored. The pressure varies from maximum at the lowest point to a minimum at top. The thickness of the tank has to be determined according to the pressure variation. The shell is constructed by welding sheets of required thickness. It is assumed that, the bottom plate gives some support to the lowest shell course, and for other courses the lower courses would give some restraint to the top courses. Therefore, the tension in each course is computed at 300 mm above the centre line of the lower horizontal joint in question.

7.7.3.2 External Loading

(a) Where tanks are subjected to high wind velocities, the shell may be stiffened to prevent failure by buckling when empty.

(b) Heavy loads from platforms and elevated walkways should be distributed appropriately.

7.7.3.3 Shell Thickness

The minimum thickness of the shell plate is determined as follows:

The internal pressure is calculated from the liquid height ($H - 0.3$) as discussed above.

$$p = \rho(H - 0.3) \times 9.807 \times 10^{-6} \tag{7.9}$$

where, p — pressure in N/mm^2

ρ — density in kg/m^3

H — height from bottom of the course under consideration to the top of the roof-curb angle in m.

$$t = \frac{pD}{2fJ} + c \tag{7.10}$$

where, t — thickness of shell in mm

D — nominal diameter of tank in mm

J — joint efficiency (varies between 0.7 to 1.0 depending upon the level of radiography)

c — corrosion allowance in mm

f — permissible stress in N/mm^2.

The above equation is applicable to non-pressure and class 'A' tanks only. For class 'B' tanks, the pressure p, is calculated by an addition of the internal pressure to the pressure due to liquid height as calculated from Eq. 7.9.

In no case should the thickness of shell plates be less than given in Table 7.3.

Table 7.3 Tank Diameter and Minimum Thickness

Nominal Tank Diameter (metres)	*Minimum Nominal Thickness (mm)*
Smaller than 15	5.0
Over 15 up to and including 36	6.0
Over 36 up to and including 60	8.0
Over 60	10.0

The above thickness includes corrosion allowance (according to IS-803). Maximum plate thickness is 40 mm.

7.7.3.4 Shell Joints

Horizontal and vertical joints between shell plates are of the butt type. The type of butt welds are, square butt, single bevel or double bevel for horizontal joints according to thickness of plates (Fig. 7.10). Vertical joints are square butt, single V or U-butt, double V or U-butt type, according to the thicknesses of plates.

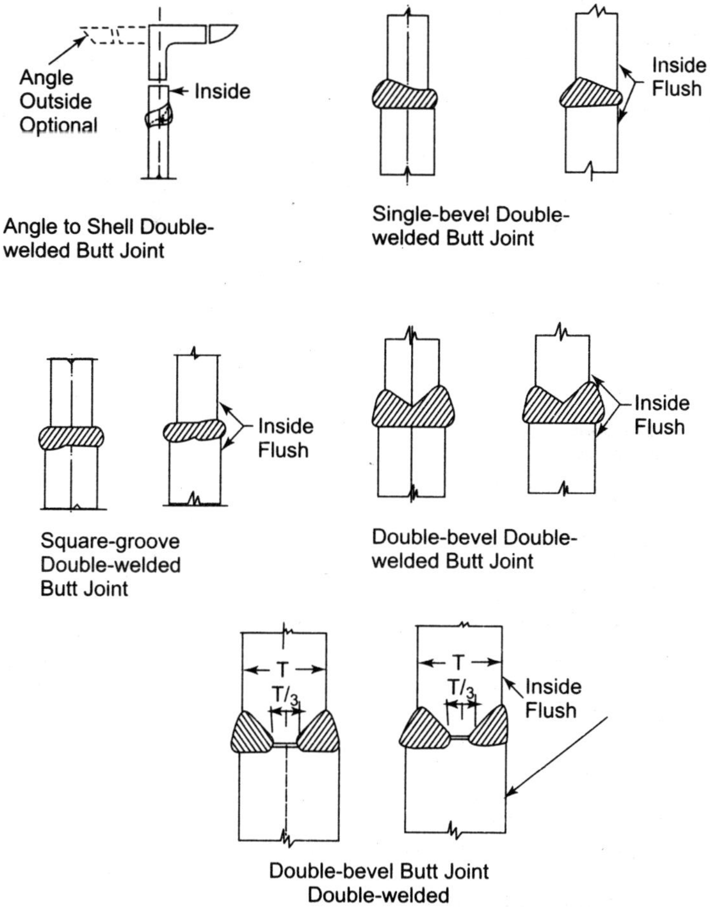

Fig. 7.10 Horizontal Joints for Shell Plates

7.7.4 Wind Girders for Open-Top Tanks

These tanks are provided with stiffening rings to maintain roundness when the tank is subjected to a wind load. Such rings, known as wind girders, are located near the top course, preferably on the outside of the tank-shell. The required minimum section modulus of the stiffening ring is given by

$$Z = 0.059D^2H \qquad (7.11)$$

where, Z — section modulus in cubic centimetres
D — normal diameter of tank in metres
H — height of tank in metres.

The section modulus of the stiffening ring of the shell is based upon the properties of the applied members and may include a portion of the tank-shell for a distance of 16 plate thicknesses below, and if applicable, above the ring shell attachment. Stiffening rings are made up of either structural, formed plate sections, or sections built up by welding. Attachments of wind girder sections to shell are shown in Fig. 7.11.

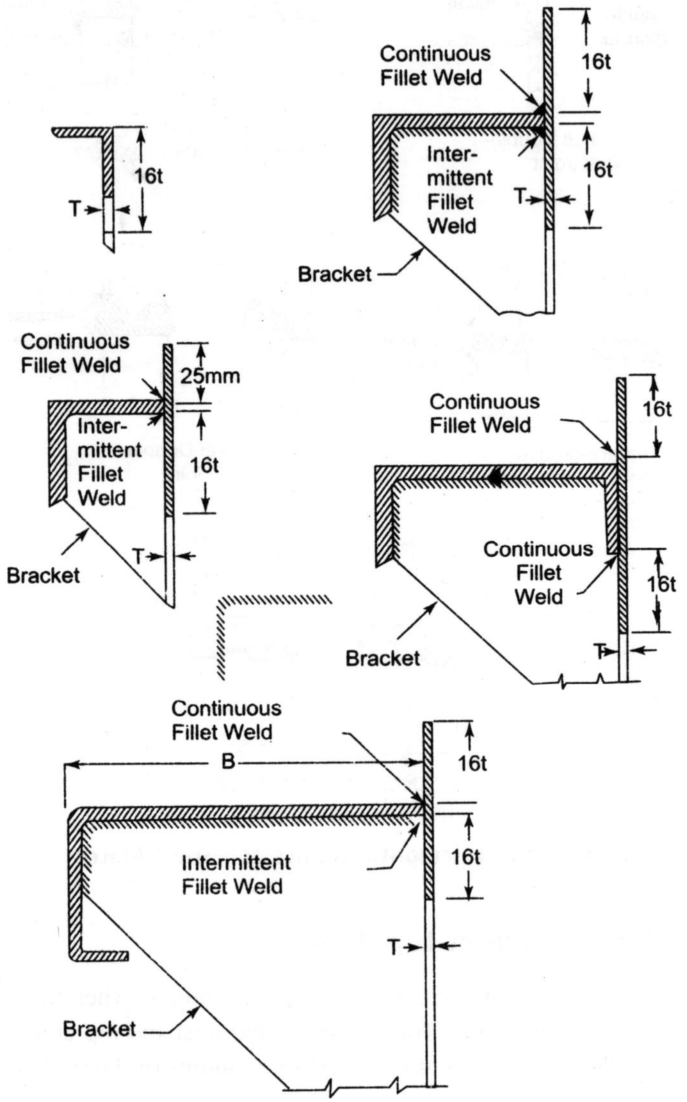

Fig. 7.11 Wind Girder Attached to Shell Top

7.7.5 Roof-Curb Angles

For closed top tanks, the tank-shells are provided with top-curb angles, which are attached to the upper edge of the shell plate by a continuous double-welded square butt joint or continuous double-fillet lap joint (Fig. 7.12).

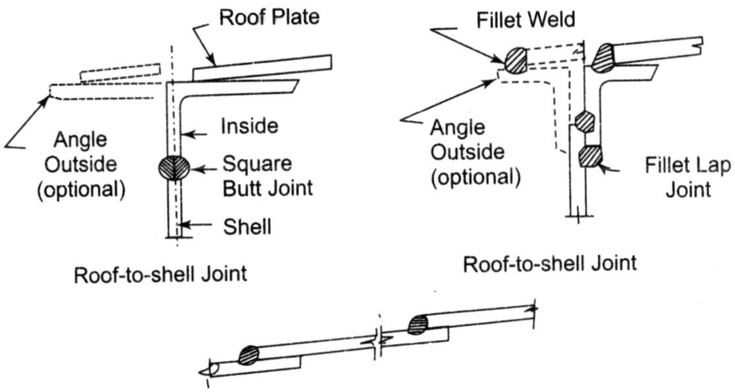

Fig. 7.12 Roof-Curb Angles

(a) ***Non-pressure tanks*** For these tanks, the top-curb angles are of the following minimum sizes:

(1) Tanks up to 10 metre diameter: 65 × 65 × 6.0 mm

(2) Tanks over 10 metre diameter and up to 18 metre diameter: 65 × 65 × 8.0 mm

(3) Tanks over 18 metre diameter and up to 36 metre: 75 × 75 × 10.0 mm

(4) Tanks over 36 metre diameter: 100 × 100 × 10.0 mm

(b) ***Pressure tanks (class A and class B)*** The top-curb member in such tanks will be required to take up the roof load, if the roof is of the self-supporting type. In such cases the top-curb angle has to be increased in size. The cross-sectional area required is indicated later under self-supporting roof design (Eq. 7.18).

7.7.6 Self-Supporting Roof Design

In this type of roof, the entire roof load is supported by the tank periphery. The roof shape may have the following forms:

(a) *Cone roof*: Appropriate to the surface of a right cone. The slope of the cone is 1 in 5 or 1 in 6.

(b) *Dome roof:* The radius of curvature is a spherical radius.

(c) *Umbrella roof:* A modified dome roof, so formed that any horizontal section is a regular polygon with as many sides as there are plates.

7.7.6.1 Roof Loading

Roofs shall be designed to support the following loads and pressures:

(a) A superimposed load of not less than 125 kg/m^2 measured on the horizontal plane in addition to dead load of roof sheets and supporting structure. This load usually consists of snow, wind and men walking on the roof. This may be reduced if there is absolutely no possibility of snow fall in the area.

(b) An internal pressure equivalent to

(i) 75 mm water gauge or 75 kg/m^2 or non-pressure tanks

(ii) 200 mm water gauge or 200 kg/m^2 for class 'A' tanks

(iii) 550 mm water gauge or 550 kg/m^2 for class 'B' tanks.

7.7.6.2 Stresses in Cone Roof

The stresses in the cone roof plates are determined by considering the above loads. The stress in the cone roof plates under either external or internal pressure is given by

$$f = \frac{pD}{2t\sin\theta} \tag{7.12}$$

where, f — stress in roof plate

p — pressure due to the roof loadings

D — diameter of cone roof

t — thickness of cone roof plates

θ — angle between cone roof and horizontal

In the case of large diameter conical roofs with an external load, the wrinkling of the roof or the elastic instability may cause a failure of the plates. The theoretical critical compressive stress that can cause failure of a curved plate due to wrinkling is given by

$$f_c \text{ (critical)} = \frac{E}{\sqrt{3(1-\mu^2)}}\left(\frac{t}{r}\right) \tag{7.13}$$

where, E — modulus of elasticity

μ — Poisson's ratio

r — radius of curvature of roof $= \dfrac{D/2}{\sin\theta}$ (Fig. 7.13)

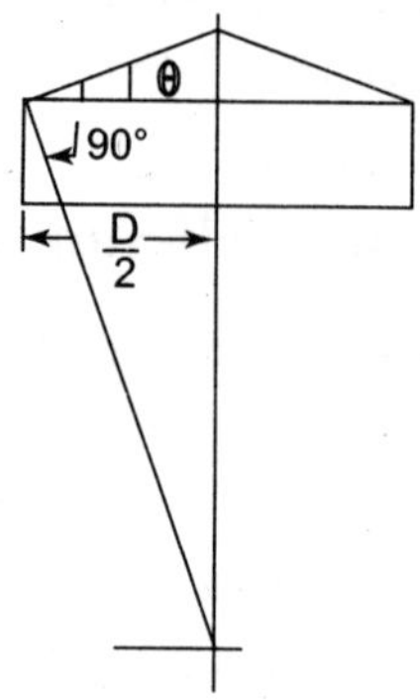

Fig. 7.13 Conical Roof Geometry

The safe compressive stress is taken with factor of safety of 12 and with Poisson's ratio of 0.33 for steel.

$$f_c \text{ (permissible)} = \frac{1}{12}\frac{E}{\sqrt{3(1-0.33^2)}} \times \frac{1}{D/2}\sin\theta \tag{7.14}$$

$$f_c \text{ (permissible)} = 0.102\,E\left(\frac{t}{D}\right)\sin\theta \tag{7.15}$$

Considering the external load on the roof, the stress in Eq. (7.12) is compressive. Equating stresses in Eqs. (7.12) and (7.15)

$$(\sin\theta)^2 = \frac{pD^2}{0.204\,Et^2} \tag{7.16}$$

$$\sin\theta = \left(\frac{D}{t}\right)\sqrt{\frac{p}{0.204\,E}} \tag{7.17}$$

The slope of the cone is limited to 1 in 5 or 1 in 6, i.e., $\tan\theta < \frac{1}{5}$ or $\frac{1}{6}$, from which the plate thickness may be determined.

The minimum thickness of the steel roof plates in 5 mm and the maximum is 12 mm. The roof plates are lapped and welded with a continuous fillet weld on the top side only. They are supported on purlins. The spacing of roof purlins shall be such, that the distance between them should not exceed 1.2 metres. In addition, the purlin frame is provided with bracings (or stay bars).

7.7.6.3 Area of Curb Angles

In a self-supporting roof, the load on the roof is transferred to the top of the shell and to the top curb angle. The curb angle must have enough cross-sectional area to withstand this compressive load. The total area required should not be less than

$$A = \frac{W\cos\theta}{2\pi f} \tag{7.18}$$

where, W — total load on roof in kg

θ — cone roof angle (Fig. 7.13)

f — allowable stress in kg/cm^2

The area A consists of

$$A = A_c + A_s + A_r \tag{7.19}$$

where, A_c — area of the curb member

A_s — area of shell plates effective = $1.5t_s\sqrt{Rt_s}$ (cm^2)

A_r — area of roof plates effective = $0.75t_s\sqrt{R_1 t_s}$ (cm^2)

t_s — thickness of shell plate in cm

t_r — thickness of roof plate in cm

R — radius of tank in cm

R_1 — radius of curvature of roof in cm.

7.7.7 Column Supported Roof

This type of roof (Fig. 7.14) has columns transmitting the roof load to the tank bottom. The roof plates are placed on a structural framework and are designed to prevent the plates from bending. The roof plates are assumed to act as flat continuous plates with a uniform roof load. The rafters and girders are assumed to act as uniformly loaded beams with freely supported ends. The plates are designed as rectangular plates, uniformly loaded and freely supported at the edges (Eq. 7.8). If 5 mm thick plates are used the rafter spacing has to be limited to 2 metres.

Rafters and girders are designed as uniformly loaded beams with the freely supported ends. Each rafter is considered to support the roof plates and roof load over an area extending on either side of the rafter and bounded by the centre lobe of the adjacent rafter. The length of rafters is limited to about 6 metres.

Column supports are used at the centers of the tank along with each ring of girders. Usually 5 or more straight girders are joined end-to-end to form a polygonal support for the ends of rafters. Column supports are designed on the basis of Rankine-Gordon formula (Eq. 3.16).

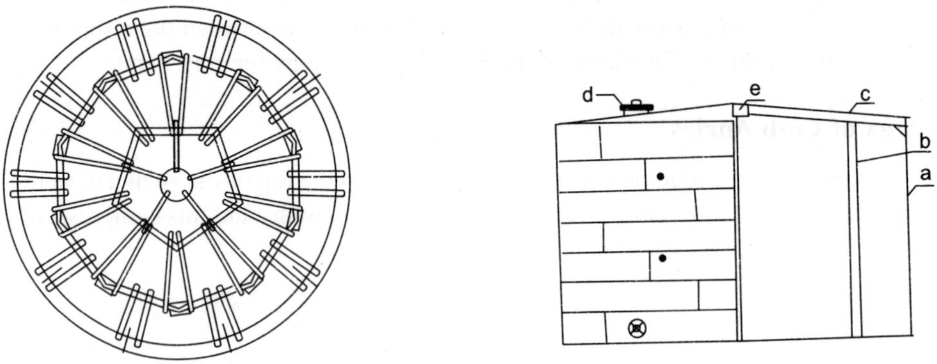

Fig. 7.14 Columns Supported Roof: (a) Shell, (b) Column, (c) Roof, (d) Manhole, (e) Centre Column Cap

7.8 NOZZLES AND MOUNTINGS

The following nozzles are provided in tanks:

(1) Shell manholes (500, 600, 700 and 900 mm diameters)

(2) Inlet and outlet connections

(3) Level gauge connection

(4) Overflow connection

(5) Drain

(6) Roof manhole (500 and 600 mm diameter)

(7) Vent connection

  (a) Breather vent

  (b) Emergency vent

(8) Combined water draw-off and clean out sump.

Details of the nozzles are shown in Figs. 7.15 and 7.16. Openings in tank-shells larger than 60 mm in diameter should be reinforced.

Mountings are provided as (1) Stairways, (2) Gangways, (3) Handrailing, and (4) Ladders.

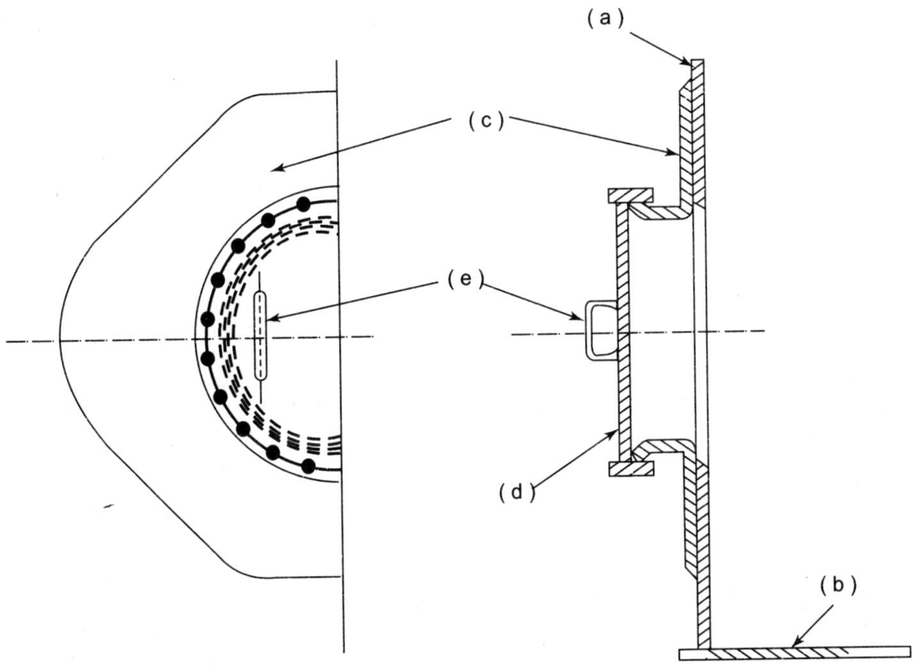

Fig. 7.15 Shell Manhole: (a) Shell, (b) Bottom, (c) Flange (Manhole), (d) Cover, (e) Handle

7.8.1 Breather Vents

The size of the vent (Table 7.4) should be adequate to prevent a pressure or vacuum to build up beyond the specified range.

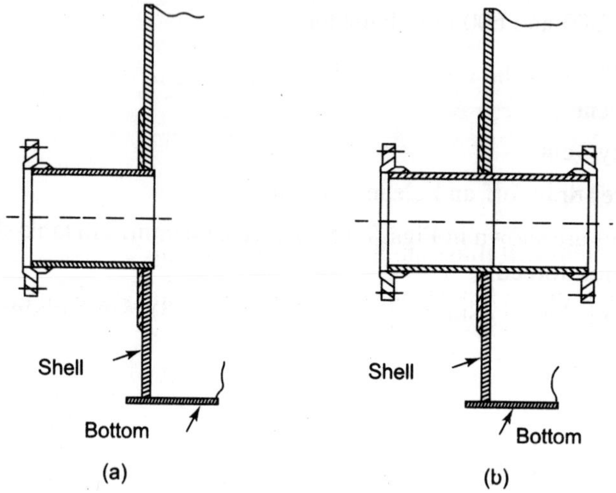

Fig. 7.16 Flanged Nozzles: (a) Single Flange, (b) Double Flange

Table 7.4 Breather Vent Size

Capacity of Tank (litres)	*Breather Vent Size (mm)*
4,000 – 8,000	35
8,000 – 40,000	50
40,000 – 2,00,000	75
6,00,000 – 16,00,000	100 and 75 used together

7.9 LARGE CAPACITY STORAGE TANKS

Designing of large capacity tanks is not covered by the codes mentioned in Section 7.7. For instance Indian Standard-803 has stated that the maximum tank diameter of 82 metres and thickness of the shell-plate is up to 40 mm. Tanks with much larger diameters and capacities require special consideration. Large diameter tanks require more ground area. The cost of the tank bottom and roof is increased substantially. Usually the tank area is limited and the shell heights are established at the maximum permitted by the permissible soil bearing pressure.

The stress in the shell of the large diameter tanks are reduced by the use of the circumferential steel band around the shell at an effective location. A close contact between the shell and the band should be provided, so that the initial radial expansion of the shell under hydrostatic pressure can immediately be resisted by the band. This will help in reducing the thickness of the shell within the limits specified by the codes. Another method of reducing stress and thickness in the lower portion of the tank shell is by the application of external pressure. For this purpose an additional shell of a limited height is provided surrounding the main shell. The tank bottom is also extended to support this additional shell. The liquid is maintained under pressure in the outer tank. The net result is a uniform internal pressure against the lower part of the inner shell. An alternate arrangement is to provide a concrete tank around the main tank. The outer tank is filled with water up to a suitable height.

7.10 ILLUSTRATION

Following numerical problem will illustrate the above discussions.

Detailed mechanical design of a fixed Conical Roof Cylindrical Tank with the help of the following data:

Tanks diameter (inside)	:	20 m
Tank height	:	12 m
Density of liquid	:	1000 kg/m^3
Superimposed load	:	1250 N/m^2
Conical roof slope	:	1 in 5
Material of construction (MOC)	:	carbon steel (structural) IS 2062
Permissible stress, f	:	165 N/mm^2
Density of the MOC	:	7850 kg/m^3
Modulus of elasticity	:	2×10^5 N/mm^2
Corrosion allowance	:	1.5 mm

7.10.1 Shell

From Eq. (7.9), the pressure at the bottom of the tank is given by,

$$p = \rho(H - 0.3) \times 9.807 \times 10^{-6}; \quad \text{N/mm}^2$$

$$= 1000(12 - 0.3) \times 9.807 \times 10^{-6}$$

$$= 0.1147 \quad \text{N/mm}^2$$

From Eq. (7.10), the thickness of the shell at the bottom:

$$t = \frac{pD}{2fJ} + c$$

The joint efficiency, $J = 0.85$ (spot radiography)

$$= (0.1147) \times 20{,}000/(2 \times 165 \times 0.85) + c$$

$$= 8.17 + 1.5, \text{ mm}$$

say, $t_1 = 10$ mm

Butt joints are provided for welding the vertical joints between plates to form the lowest layer of the shell. A gap of 2 mm is made between adjacent plates to facilitate welding. Total circumference with assumed 10 plates is

$$= (\pi \times 20.01 - 10 \times (2 \times 10^{-3})) = 62.84 \text{ m}$$

Therefore, the length of each plate is = 62.84/10 = 6.284 m.

The size of the plate selected from IS-1730 is 6300 (L) × 1800 (W).

To determine the number of layers to be used with 10 mm thick plate it is necessary to determine the height up to which a plate of lesser thickness can be used.

Assume the plate thickness to be 8 mm. Since this is inclusive of corrosion allowance of 1.5 mm, the actual thickness for stress calculation is = 8 – 1.5 = 6.5 mm.

From Eq. (7.10), and using joint efficiency to be 0.85, and given allowable stress (165 N/mm^2), we get

$$p = 6.5 \times 2 \times 165 \times 0.85/20 \times 10^3$$

$$= 0.091 \text{ N/mm}^2$$

Also from Eq. (7.9), we have

$$p = \rho(H - 0.3) \times 9.807 \times 10^{-6}; \quad \text{N/mm}^2$$

Therefore,

$$H = 0.091/(1000 \times 9.807 \times 10^{-6}) + 0.3 = 9.58 \text{ m}$$

say, H = 9.6 m

The total height is 12 m and a 8 mm plate can sustain a height of up to 9.6 m.

Therefore, the balance 12 – 9.6 = 2.4 m height is from the plate of thickness 10 mm. We do not want to waste the plate and, therefore, two lower most plates will be 10 mm thick. Since the width of each plate is 1800 mm we consider two such plates. This will cover the height of 2 × 1800 = 3600 mm = 3.6 m.

The remaining height of the tank can be fabricated out of a plate of thickness 8 mm. In order to lower the cost of the tank one should check up to what level one can use a 6 mm thick plate.

Assume the plate thickness to be 6 mm. Since this is inclusive of the corrosion allowance of 1.5 mm, the actual thickness for stress calculation is = 6 – 1.5 = 4.5 mm.

From Eq. (7.10)

$$p = 4.5 \times 2 \times 165 \times 0.85/20 \times 10^3 = 0.0631 \text{ N/mm}^2$$

From Eq. (7.9),

$$H = 0.0631/(1000 \times 9.807 \times 10^{-6}) + 0.3 = 6.74 \text{m}.$$

We can use one 8 mm thick plate and two 10 mm thick plates. Therefore, the balance height = 12 – 2 × 1.8 – 1 × 1.8 = 6.6 m can be fabricated out of the 6 mm plate as this is less than $H = 6.74$ m computed above. Alternately, to be on safe side, we can use two plates each of thickness 8 mm and 10 mm.

The height of 6 mm plate = 12 – 2 × 1.8 (10 mm thick) – 2 × 1.8 (8 mm thick)

= 4.8 m

Therefore, we can have 3 layers of 6 mm plate covering the balance height of 4.8 m. The extra plate height 0.6 m (3 × 1.8 – 4.8 = 0.6 m) can be adjusted by cutting the plate. Thus, 600 mm plate of thickness 6 mm is wasted.

Please note that depending upon the availability and for minimizing wastage in the tank fabrication one may adjust the tank diameter and height by iterative calculations as indicated above.

7.10.2 Tank Bottom

The diameter of the bottom tank extends beyond the shell by say 65 mm. As specified in IS-803.

Therefore, the bottom diameter, D_b, is given as:

$$D_b = 20 + 2 \times 10/1000 + 2 \times 65/1000 = 20.15 \text{ m}$$

Thickness of plates to be used to form bottom is:

Annular plate = 10 mm

Sketch plate = 6 mm

The bottom plates will be lap welded. The overlap between the sketch plates and annular plates is 65 mm and between the sketch plates, it is 5 times thickness of the plate, i.e., 30 mm. Please note that the annular plate thickness is the same as that of the lower most shell thickness, in this case it is 10 mm. The sketch plates are always 6 mm thick irrespective of the tank diameter. The joints between the plates will be staggered, so that no more than three plates are overlapping within 300 mm of each other or of the shell.

Annular plates

The minimum width of the annular plate = 65 + 300 + 10 + 65 = 440 mm

Say 500 mm. (Please note many fabricators use 1000 mm width).

The circumference of the bottom is = $\pi \times D_b = p \times 20.15 = 63.3$ m.

7.10.3 Self Supporting Conical Roof

The plate thickness for the roof plates is assumed to be 6 mm.

The weight per unit area is then $= 6 \times 10^{-3} \times 7850$

$= 47.1\ \text{kg/m}^2$

The force acting due to load is $= 47.1 \times 9.807 = 461.9\ \text{N/m}^2$

say, $= 462\ \text{N/m}^2$

The total load = superimposed load + load due to weight

$= 1250 + 462 = 1712\ \text{N/m}^2$

$p = 1712 \times 10^{-6}\ \text{N/mm}^2$

With the help of Eq. 7.17, we have

$$\sin\theta = (D/t)\sqrt{p/(0.204 \times 2 \times 10^5)}$$

$$= (20/6 \times 10^{-3})\sqrt{1712 \times 10^{-6}/(0.204 \times 2 \times 10^5)}$$

$$= 0.683$$

Therefore, $\tan\theta = 0.935$

We then have slope = 1 in 1.07.

However, the slope is limited to 1 in 6, i.e., $\tan\theta = 0.167$. The slope being greater than permissible value let us have thickness = 18 mm. Following above procedure we get $\tan\theta = 0.294$.

Even with a thickness of 18 mm, the slope is not within acceptable limit. Therefore, it is recommended not to use a self-supporting roof.

In case a column support type roof is used, it is possible to use 6 mm thick plate and provide for rafters under the roof plates (Fig. 7.14).

References and Recommended Readings

Binder, G., and Sutton N., *Large Plastic Tanks*, British Chemical Engineering, **15** (10) 1332 (1970).

Blenkin, B., and Wardale, J.K.S., *Storage Vessel – Design, Fabrication and Use,* Chemical Process Engineering, **44** (7) 344 (1963).

Bodley, R.W., *Choose a Right Floating Roof Tank*, Petroleum Refiner, **40** (3) 185 (1961).

Brownell, L.E., and Young E.H., *Process Equipment Design,* John Wiley, New York, (1968).

Forbes, K., McGregor, A., and Turner, S., 'Design of Fluid Storage Tanks from Polypropylene', British Chemical Engineering, **15** (10) 1333 (1970).

Karcher, G.G., *New Design Calculations for High Temperature Storage Tanks*, Hydrocarbon Processing, **57** (10) 137 (1978).

'Liquid Storage in the CPI', *Chemical Engineering Desk book Issue* (9), (1978).

Mahajan, K.K., *Design of Process Equipment (Selected Topics),* Pressure Vessel Handbook Publishing Co., U.S.A, (1979).

———., 'A Method for Designing Rectangular Storage Tank', *Chemical Engineering*, **84** (7) 107 (1977).

McGrath R.V., *New Concepts for Storage Tank Design, Hydrocarbon Processing*, **55** (5) 97 (1976).

———., *Highlights of Revised API Tank Standards, Hydrocarbon Processing*, **54** (5), 89 (1975).

Mead, W. J., *Encyclopedia of Chemical Process Equipment,* Reinhold, New York, (1964).

Perry, R.H., Green, D.W. and Maloney, J.O., *Perry's Chemical Engineers' Handbook,* 6th Edition, International Edition, McGraw-Hill Book Co., Singapore, (1984).

Riegel, E.R., *Chemical Process Machinery,* Reinhold, New York, (1953).

8

Chapter

Reaction Vessels

8.1 INTRODUCTION

As the name suggests, the reaction vessels are used not only to carry out unit processes (for example, hydrogenation, esterification, halogenations, alkylation and so on), but also used to conduct unit operations such as mixing, extraction, distillation, dissolution, crystallization and so on. The operating conditions might have high pressure, reduced pressure (vacuum) or near atmospheric pressure. The reaction vessel operating under high-pressure of say, more than 20 bar, is often refereed as an 'autoclave'. However, this classification is subjective in nature.

By and large, the reaction vessels are often indicated by their capacity expressed in volume, say litre or m^3. For a pilot plant the size can be from 10 to 1000 litres. For a commercial plant, the size could be between 1 and 20 m^3. Fermenters having capacities of 100 m^3 are common in bioprocesses.

The vessel may require heating, cooling and good mixing depending upon the process conditions. The utilities used, particularly, for heating could be under pressure. For example, as a rule of thumb, steam up to saturation temperature of 200°C, i.e., 16-bar (1.6 N/mm^2) pressure, is used for heating purposes.

The reaction vessel due to its own operation is under pressure (or vacuum) or may be subjected to an external pressure of utility. Therefore, for the purpose of mechanical design, it is considered as a pressure vessel. A closed vessel wherein vapours are condensing is designed for full vacuum.

The type of heads used for such vessels are usually shallow-dished, torispherical or ellipsoidal. When it is not necessary to remove the cover frequently, a normal flanged head secured by studs or by nut and bolt connection is very suitable. If, however, the cover is to be opened frequently a special type of flange is provided and swing type bolts are necessary (for details refer to Section 5.8.4). A common weakness of the covers of this kind is the wear and tear of the gasket sealing the cover. These gaskets must be protected and either a tongue and groove or male and female facings should be provided. Several nozzles, such as inlet, oulet, sight and light glasses, thermowell, handholes and manholes are provided as the case may be in line with standard practices to satisfy the process requirements. Bracket or column supports are generally satisfactory (see Chapter 13).

8.2 MATERIALS OF CONSTRUCTION

A number of metals can be used for construction of reaction vessels. The more common metals are low carbon steel, stainless steel and other alloy steels such as hastelloy. In special cases non-ferrous metals such as copper, nickel, aluminium, titanium are also used. At moderate pressures and temperatures, vessels can be made of glass reinforced polyesters, glass filled furons, phenolics, polypropylene, linear low density polyethylene and polyvinyl chloride reinforced with FRP. A saving in cost can be achieved by fabricating vessels from a low cost metal, with a cladding of corrosion resistant metals like stainless steel, nickel, inconel, monel, copper, etc. Vessels can also be lined with lead, rubber, glass and plastics to prevent corrosion. (For details see Section 2.4).

8.3 AGITATION

Agitation of the reaction vessel contents is a requirement in a number of processing operations. Both heat and mass transfer are greatly influenced by agitation or mixing. An agitator, also known as a stirrer, produces high velocity liquid streams, which move through the vessel. As the high velocity streams come into contact with the stagnant or the slower moving liquid, momentum transfer occurs. The classification of the agitation or mixing equipment is usually made on the basis of liquid viscosity, since viscosity is a major contribution to the forces tending to dampen flow through a mixing system. Details regarding agitation equipment and their applications are given in Chapter 14.

In a vertical cylindrical reactor the ratio of liquid depth to tank diameter (filling ratio) is normally between 0.5 and 1.5 and a value approximately equal to 1.0 is recommended for most purposes. When dispersing gas in a liquid, a filling ratio of about 2.0 is recommended in order to maintain a sufficiently long period of contact between the gas and the liquid. Flat-bottomed and cone-bottomed vessels have the disadvantage of low agitation efficiency in the corners formed between the walls and the bottom. This is particularly significant when mixing involves the suspension of heavy solids in the liquid. A vessel with a dished bottom is always preferable, as its power consumption is low.

8.4 CLASSIFICATION OF REACTION VESSELS

Agitated reaction vessels may be divided into three main modes of operation: batch reactors, continuous flow reactors and semi-batch reactors.

(a) *Batch reactors* These are almost exclusively used for liquid phase reactions. The reactants are added to the empty vessel and the contents are removed after completion of the reaction. In this system, temperature and pressure as well as composition may vary with time.

(b) *Continuous flow reactors* In these reactors the reactants flow continuously into the reactor and the products flow out continuously. Under ideal conditions, in a well-agitated system, a uniform concentration is maintained throughout the vessel.

(c) *Semi-batch reactors* In these reactors one of the reactants is initially charged batchwise, while the other reactant is fed into the reactor continuously.

For the mechanical design of the reaction vessel, it does not make much difference in which mode it is operating except for heating and cooling, wherein thermal cycles may lead to fatigue of the material.

8.5 HEATING SYSTEMS

Chemical reactions are accompanied by the absorption or liberation of heat. The reaction vessel must, therefore, be provided with the means of supplying or removing the heat of reaction. The rate of heat transfer is a function of the physical properties of the agitated liquid and the heating and cooling medium, the vessel geometry, the material and the thickness of the vessel wall, and the degree of agitation.

Heating systems for reaction vessel need special consideration. The devices used are either the direct or the indirect type. There are various electrical methods, which heat the vessels directly. The two most common methods are resistance heating and induction heating. These systems have a low overall efficiency and high operating costs due to ever increasing power tariff. Indirect heat transfer systems are the most widely used. The heat is received from fluids such as steam, hot oil, hot water, fine gas or air, molten salt mixtures, and special organic compounds such as downtherm, marlotherm and therminol. These organic compounds are high boiling point materials, so that the heat transfer systems can be operated at low pressures. Both liquid and vapour phase heat transfer system are used. In liquid systems, the heat transferred is from sensible heat. In vapour systems, the heat transferred is from latent heat; the entire heat transfer surface is, therefore, at a uniform temperature. In general, heat transfer coefficients are higher for condensing vapour systems than for liquid systems.

In indirect cooling systems, fluids employed are air, an evaporant such as liquid ammonia, water or brine, oil and organic compounds.

In the indirect heat transfer system, the fluid is supplied in either a jacket, which surrounds the vessel wall, or internal equipment which is placed inside the vessel. The most commonly used internal equipment includes helical coils, spiral coils, plate baffles and tube baffles. Heat transfer coefficients can be increased by increasing fluid velocities. Similarly, heat transfer coefficients can be increased by creating agitation of the contents of the vessel.

There is no specific choice between a jacket or coil for a vessel carrying out an exothermic or endothermic reaction, although generally a jacket is installed when it is necessary to supply heat and a coil to remove heat. The reasons for this selection are that in the majority of cases heat is supplied by condensation of some vapour, and for a given heat transfer area there is a greater space for condensation in a jacket than in a coil. The greater the space the jacket provides, the greater is the ease of drainage of the condensate. On the other hand, a cooling coil is generally more suitable than a cooling jacket because the rate of heat transfer is greater under forced convection conditions and greater turbulence can be achieved in the coolant liquid when it is pumped through a coil than when pumped through a jacket. The jacket can be constructed out of cheaper materials like mild steel even when the vessel has to be of an expensive alloy.

Cooling coils are basically non-removable; they have large a thermal expansion, requiring adequate support without constraining thermal movement. Plate baffles are good from the point of view of cleaning and provide a large heat transfer surface. They are excellent for low-volume flows of either heating or cooling liquid. Harp coils are more efficient than helical coil or plate baffles. They are easier to clean, adapt easily to thermal expansion, are less likely to interfere with the flow pattern of an agitator and can be made removable. They are excellent for high volumes of heating and cooling fluid.

For highly exothermic reactions one can insert a tube bundle in the reactor. The cooling utility flows through the tube bundle. Depending upon the heat transfer area required, multiple tube bundles (say, 4) could be inserted having a common header. Such tube bundles also act as baffles.

8.5.1 Jackets

A plain jacket [Fig. 8.1 (a)] is used for steam. The space between the vessel wall and the jacket shell should be narrow, for obtaining good heat transfer rates. The dimpled construction shown in Fig. 8.1 (g) helps to reduce the thickness of the shell. This construction is particularly suitable for large vessels, because of the reduction in weight. The dimpled jacket is generally used for condensing vapours such as steam or downtherm. High velocities of circulating fluids can be obtained by use of different types of jacket constructions, such as the channel or coiled jackets [Fig. 8.1 (b to g)].

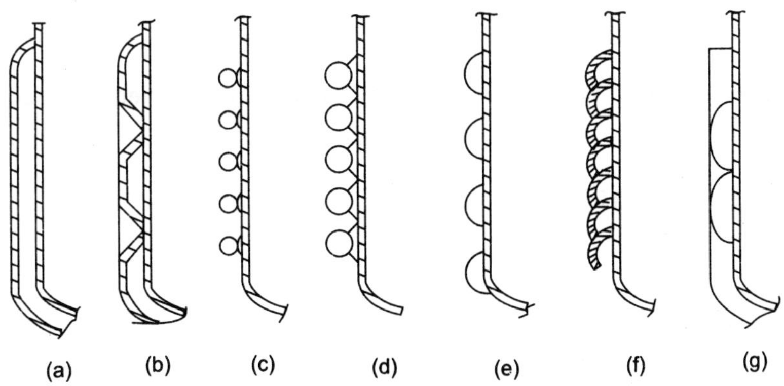

Fig. 8.1 Types of Jackets: (a) Plain, (b) Channel, (c) Tube, (d) Tube with Copper Strip Support, (e) Half Coil (Limpet), (f) Overlapped Cut Coil, (g) Dimpled Construction

A better way to increase the turbulence in the jacket is to have a spiral plate with some clearance to the jacket wall welded to the vessel shell. This also works as a stiffener and a finned heat exchanger. The spiral flow not only facilitates drainage of condensate but also improves heat transfer coefficient during cooling by cooling water.

8.5.2 Coils

Coils are used for heating or cooling by immersing them in the contents of the vessel (Fig. 8.2). Such coils are formed from a tubing by shaping them in the form of a helical or double helical coil. A pancake type of coil, which is a spiral rolled in a single plane so as to lie horizontally near the bottom of the vessel is also used to transfer heat by free convection. Tubes can also be arranged vertically in the vessel, which serve the dual purpose of heat transfer and baffling.

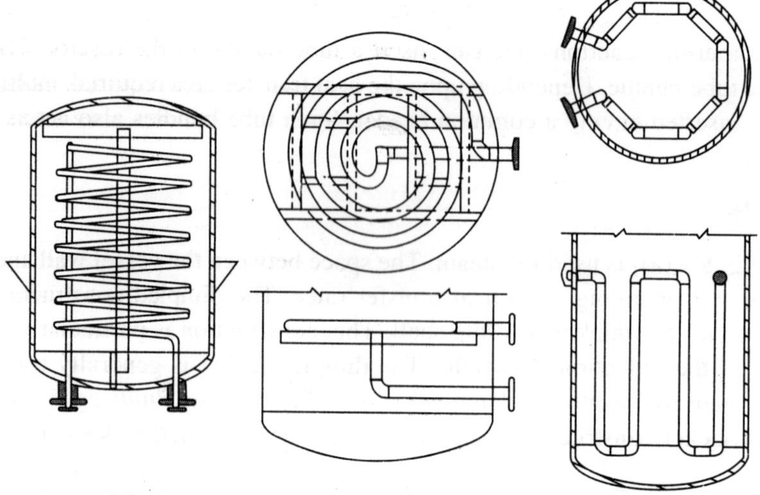

(a) (b) (c)

Fig. 8.2 Coils: (a) Helical Coil, (b) Spiral Coil, (c) Vertical Tubes

8.6 DESIGN CONSIDERATIONS

Design of the reaction vessel is based on pressure and temperature conditions. It is essentially a pressure vessel with heads and nozzles. A jacket construction is an additional feature, involving operation of external pressure on the vessel wall. The reaction vessel is, therefore, designed for both internal and external pressures operating independently and the higher value of the wall thickness is accepted as satisfactory. The vessel shell is subjected to external pressure and care must be taken in the design to prevent a collapse (see Section 5.8.1.3).

The methods adopted are:

(a) Making the shell thick enough in proportion to its diameter and length so that it is self-supporting.

(b) Using stay bolts for attaching the inner shell to the outer jacket.

(c) Using stiffening rings or corrugations in the shell of the vessel.

The procedure indicated in Chapter 5 is followed to size the various components of the vessel. The overall dimensions of the vessel, the size and position of nozzles are determined by the process conditions.

8.6.1 Jacket Design

A plain jacket is the simplest arrangement for heating or cooling. The jacket is generally made of low carbon steel and is designed for internal operating pressure of the heating fluid at the appropriate temperature. Various methods are adopted to attach the jacket to the vessel wall. A common method is to use two rings of square or rectangular section (Fig. 8.3), one at the top and the other at the bottom to which the jacket and the vessel shells are securely welded. Figures 8.4 and 8.5 show an alternate method of attachment of jacket to shell. Figures 8.6 and 8.7 show attachment of jacket for the top head. Figure 8.8 shows a detachable jacket. In the case of cast iron vessel the body flange is integrally casted. Most frequently, the jacket extends over the base of the reactor and in such cases provision must be made for emptying the vessel as well as for removing the heating fluid from the jacket. The design of the drainage connection of the vessel is more complicated than the jacket,

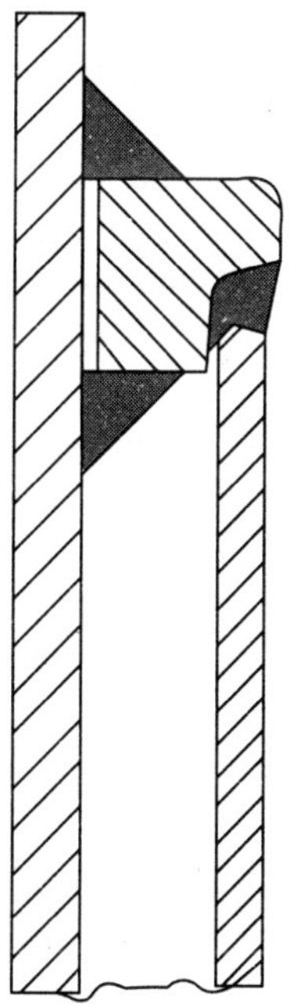

Fig. 8.3 Plain Jacket Welded to Shell by Ring

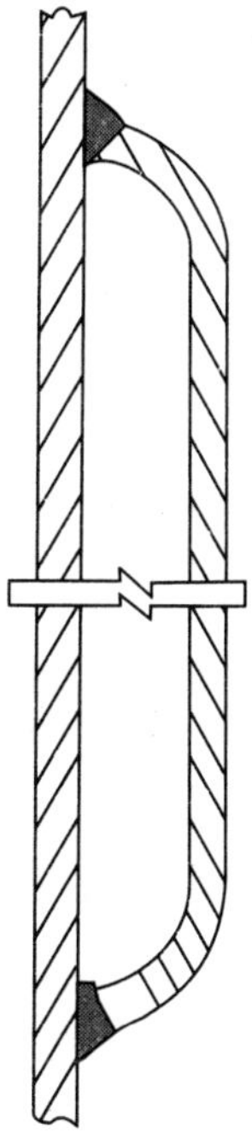

Fig. 8.4 Plain Jacket Welded to Shell

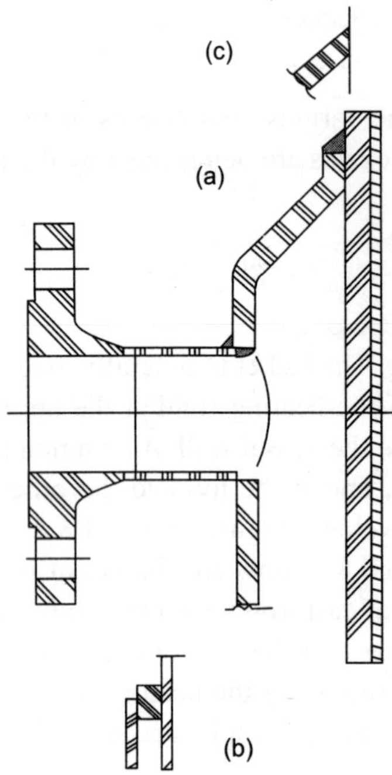

Fig. 8.5 Jacket Welded to Pressure Vessel Shell: (a) For Low Fluid Pressure, (b) For High Fluid Pressure, (c) For Medium Fluid Pressure

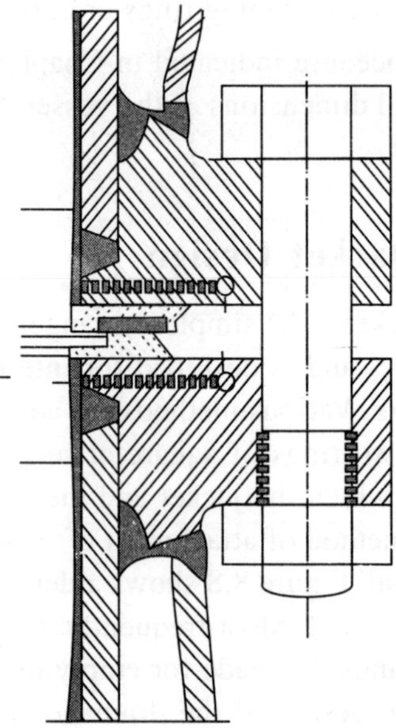

Fig. 8.6 Stainless Steel Lined Flanged Joint of a Reaction Vessel with Jacket for Shell and Head Welded to Flanges

because there will be differential expansion between jacket and vessel which, if not provided for, could lead to rupture of the vessel. Figures (8.9 to 8.13) show a few examples of the different methods suggested for jacket connections for the bottom head of the reactor. The thick vessel wall required for high jacket pressure makes the vessel expensive. The heat transfer through the thick wall will also be less. For higher jacket pressures it is, therefore, preferable to use other types of jacket designs. The dimpled jacket construction is formed by plug welds, which act as 'stays' and help to reduce the vessel wall thickness. Special precautions will have to be taken to prevent fouling, stress and cell corrosion at the welds.

8.6.2 Coil and Channel Design

The half coil (limpet coil) or channel construction shown in Figs. 8.1(e) and (b) is formed by a continuous spiral of a half pipe or channel section, attached to the vessel wall by continuous fillet welding with full penetration. Figure 8.14 shows details of portion of the vessel shell with a half coil attached to it. While Fig. 8.15 shows a channel jacket.

These methods reduce the shell thickness considerably. Heat transfer coefficients are also higher than that for jacket construction. The half coil or part circular coil is also known as limpet coil.

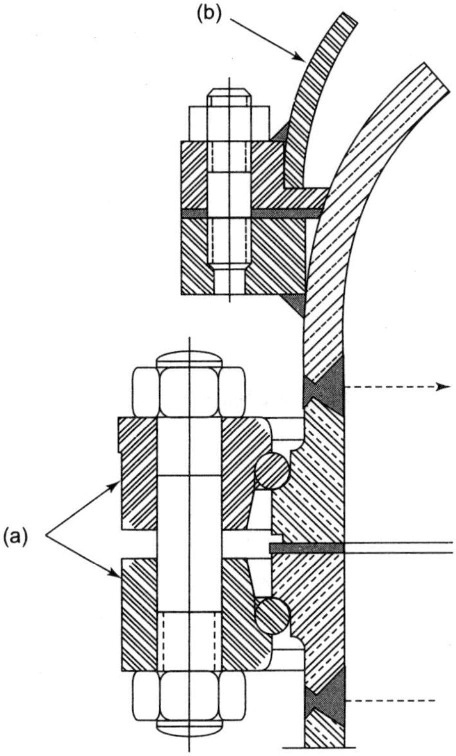

Fig. 8.7 **Detachable Joint for Head Jacket: (a) Loose Flanges, (b) Head Jacket**

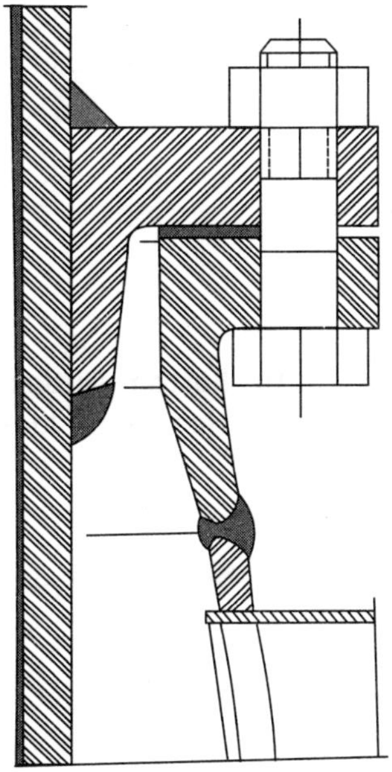

Fig. 8.8 **Detachable Jacket for Vessel Shell**

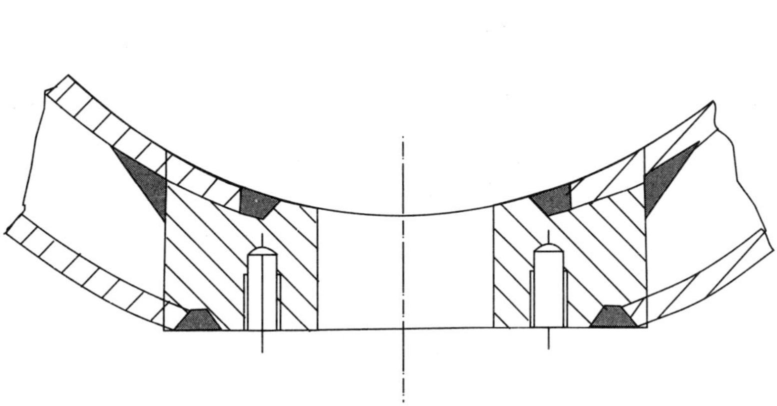

Fig. 8.9 **Bottom Outlet in a Pad Welded to Head and Jacket**

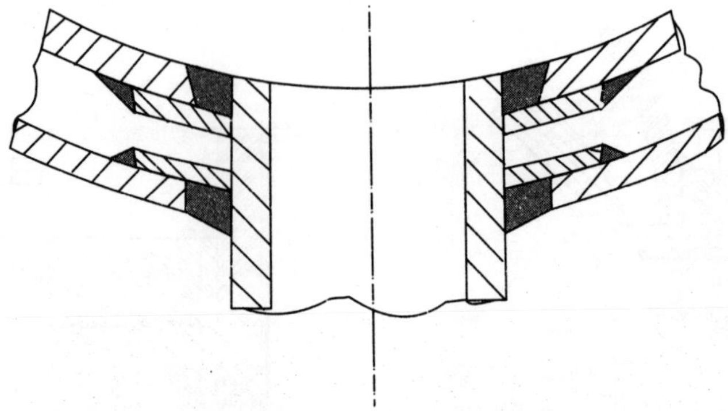

Fig. 8.10 Bottom Outlet Pipe with Reinforcing Ring Welded to Head and Jacket

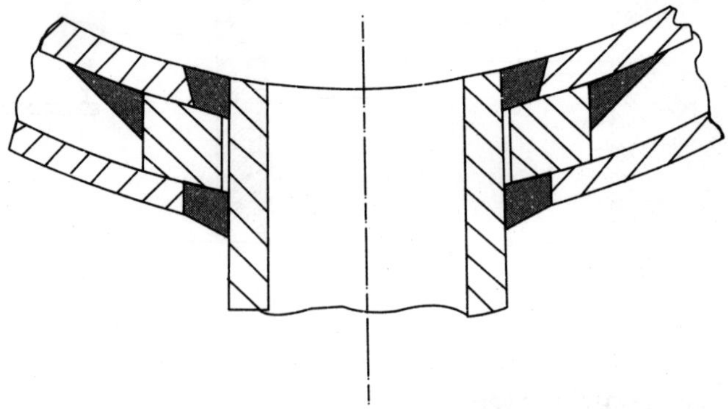

Fig. 8.11 Bottom Outlet Pipe with Separate Reinforcements for Head and Jacket

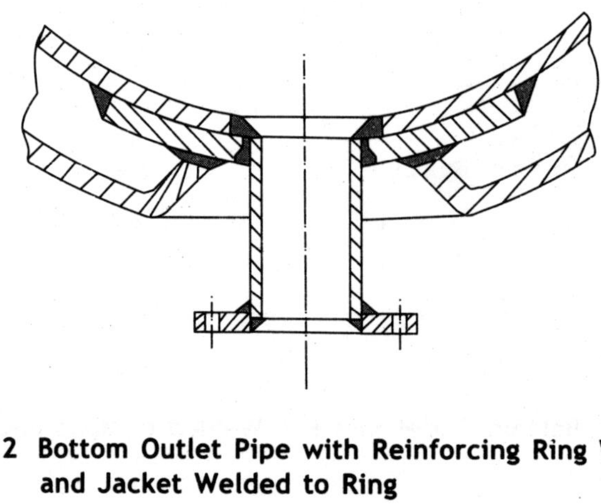

Fig. 8.12 Bottom Outlet Pipe with Reinforcing Ring Welded to Head, and Jacket Welded to Ring

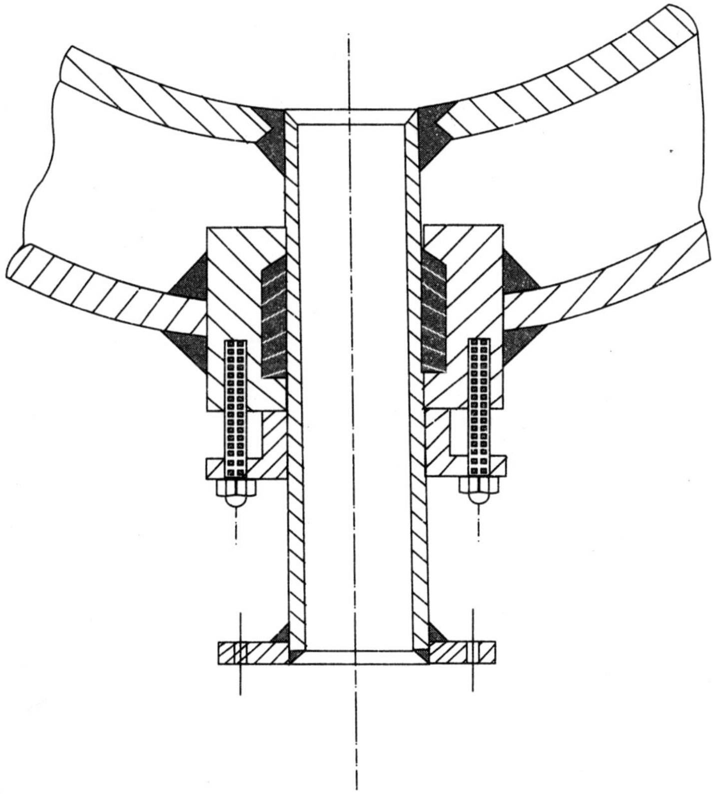

Fig. 8.13 Bottom Outlet Pipe with Stuffing Box Type of Packing Gland for Differential Expansion between Head and Jacket

8.6.2.1 Design of Vessel Shell with Half Coil

A half coil is a part of a taurus, for which the circumferential stress occurs at the junction with the shell, and is given by

$$f_{pc} = \frac{pd_i}{2t_c} \tag{8.1}$$

and the longitudinal (axial) stress is given by

$$f_{ac} = \frac{pd_i}{4t_c + 2.5t_s} \tag{8.2}$$

where, p — design pressure inside the half coil

d_i — internal diameter of the half coil

t_c — thickness of half coil

t_s — thickness of shell.

The thickness of the half coil is calculated from Eq. 8.1 above and by addition of the usual corrosion allowance.

The total circumferential stress f_{ps} in the shell is the sum of the circumferential stress f_p due to the pressure in the vessel and the longitudinal stress f_{ac} in the coil caused by coil pressure.

$$f_{pc} = f_p + f_{ac}$$

$$= \frac{pD_i}{2t_s} + \frac{pd_i}{4t_c + 2.5t_s} \tag{8.3}$$

The total longitudinal (axial) stress f_{ac}, in the vessel shell is made up of

(1) Longitudinal stress due to pressure in the vessel

(2) Longitudinal stress f_{ac} due to coil pressure

(3) The bending stress caused by the distortion of the shell at the junction with the coil

$$f_{as} = f_a + f_{ac} + f_b$$

$$= \frac{pD_i}{4t_s} + \frac{pd_i}{2t_s} + \frac{2\Delta pd_o^2}{3t_s^2} \tag{8.4}$$

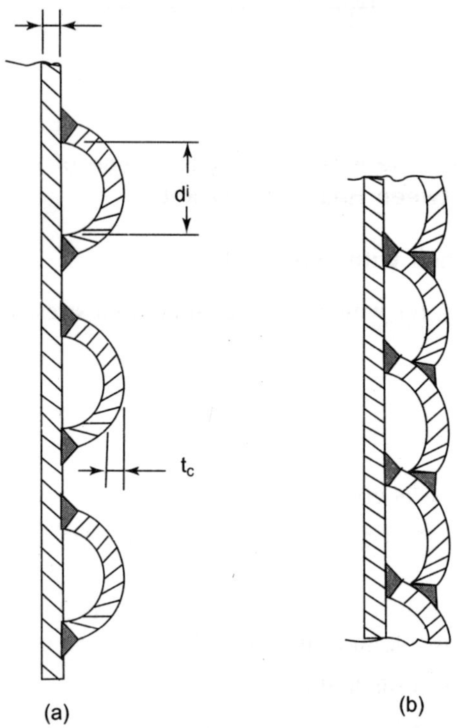

Fig. 8.14 (a) Half Coil Welded to Shell, (b) Overlapped Cut Coil Welded to Shell

where, p — internal pressure in the vessel

D_i — internal diameter of shell

Δp — maximum differential design pressure between coil and shell

d_o — external diameter of half coil.

The vessel shell thickness can be calculated from Eq. 5.3 and by the addition of corrosion allowance. The longitudinal stress may then be checked. It is further necessary to check for the buckling for the shell portion closed off by a coil with maximum differential pressure Δp.

Both from the elastic and thermal stress aspects, there should not be too great a disparity between the half coil thickness and vessel shell thickness. The preferred range of half coil thickness (t_c) is 0.3 to 1.2 times that of the shell thickness (t_s). A minimum half coil thickness of 3 mm is recommended for ease of fabrication and containing distortion.

8.6.2.2 Design of Vessel Shell with Channel Jacket

Figure 8.15 shows a part of a channel jacket. The thickness t_s of the vessel shell wall or t_c of the channel can be determined on the basis of uniformly loaded flat plate restrained at the ends. The thickness may be given as

$$t_s = d.\sqrt{\frac{k_1 p}{f_1}} + c \tag{8.5}$$

$$t_s = d.\sqrt{\frac{k_2 p}{f_2}} + c \tag{8.6}$$

where, d — channel width

p — design jacket pressure

f_1, f_2 — stresses in the material at the appropriate temperature

$k_1 = 0.167$; $k_2 = 0.12$

c — corrosion allowance.

For higher jacket pressures the width d can be reduced to obtain a reduction in thickness t.

It is desirable to ensure a sound weld between the shell and the surrounding half coil or channel. A minimum thickness (t_s or t_c) of at least 2 mm is recommended for making a satisfactory joint.

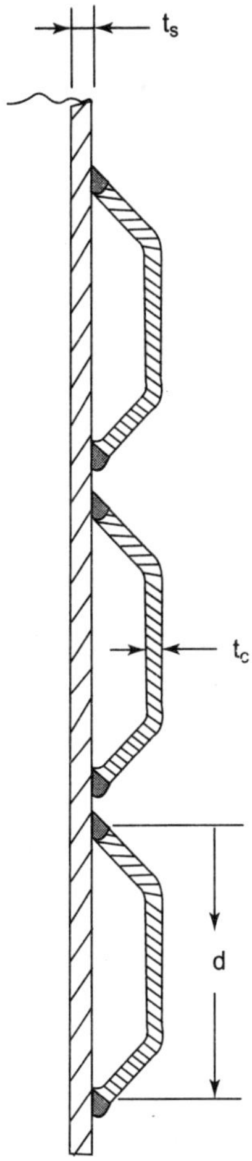

Fig. 8.15 Channel Jacket Welded to Shell

It is difficult to cut a half coil of the channel to suit the vessel bottom dished end. The possible ways are to have either a pattern of coils radially like spokes of a wheel or to have a spiral. The spiral would result in a higher heat transfer coefficient if a liquid coolant is used.

8.7 ILLUSTRATION

The design procedure is illustrated below with a numerical problem. We consider two situations wherein heat transfer fluid is in a plain jacket and in the second case it is in a limpet coil.

Reaction vessel

Data

Vessel shell internal diameter	:	2130 mm
Jacket internal diameter	:	2260 mm
Jacket length	:	2500 mm
Diameter of half coil or width of channel jacket	:	100 mm
Flanged dished head		
Internal diameter	:	2130 mm
Crown radius	:	2130 mm
Knuckle radius	:	128 mm
Straight flange length	:	60 mm
Internal pressure (shell)	:	0.55 N/mm^2
Internal pressure (jacket)	:	0.35 N/mm^2
Temperature	:	150°C
Material		
Allowable stress	:	96 N/mm^2
Modulus of elasticity	:	190×10^3 N/mm^2
Poisson's ratio	:	0.3

8.7.1 Shell with Plain Jacket

We consider here that the heat transfer fluid is in the plain jacket.

Thickness of Shell

We shall first find out the thickness by considering internal pressure and then check its adequacy for external pressure by fluid in the jacket.

(a) *Internal pressure*

Design pressure is 10% more than the maximum working pressure.

Design pressure = 1.1 (0.55) = 0.605 N/mm^2

The vessel thickness is given by Eq. (5.3 b)

$t = pD_i/(2fJ - p) + c$(corrosion allowance)

$= 0.605 \times 2130/(2 \times 96 \times 0.85 - 0.605) + c$

$= 7.9$ mm + 1.5 mm = 9.4 mm

Use a plate of thickness 10 mm inclusive of corrosion allowance.

(b) *External pressure*

External design pressure = 1.1 (0.35) = 0.385 N/mm^2

We have from Eq. (5.14 b) the critical buckling pressure, P_c, as

$$P_c = \frac{2.42 \times E \times (t/D_0)^{5/2}}{(1-\mu^2)^{3/4}\left\{(L/D_0) - 0.45(t/D_0)^{1/2}\right\}}$$

Use t = thickness calculated above without corrosion allowance.

= 7.9 mm

L = effective jacket length

= Jacket length + 1/3 dished end length (Appendix J)

= 2500 + 60 (st.portion) + 1/3(360)

= 2680 mm.

= 2700 (for our illustration purpose we consider)

D_o = external diameter = $D_i + 2t = 2130 + 2 \times 7.9$

= 2145.8 say, = 2146 mm

Substituting all these in the above, we get

$P_c = 0.33$ N/mm^2

The critical stress can be calculated using Eq. (5.16)

$f_c = P_c D_o/(2t) = 0.33 \times 2146/(2 \times 7.9)$

= 44.82 N/mm^2

By having a factor of safety, 4, we can compute the allowable stress, f_a

f_a = 11.64 N/mm^2,

which is far less than the given allowable stress of 96 N/mm^2

The other approach is to have an allowable pressure

$P_a = (1/4)\ P_c$

Therefore,

$$P_a = 0.33/4 = 0.083 \text{ N/mm}^2.$$

This is far less than the design pressure $P_i = 0.385$ N/mm^2

It is, therefore, concluded that the computed thickness 7.9 mm is not acceptable based on external pressure.

Therefore, let us try thickness $t = 16$ mm.

$$P_c = 2.0 \text{ N/mm}^2$$

Therefore,

$$P_a = 2/4 = 0.5 \text{ N/mm}^2$$

The design pressure, 0.385 N/mm^2 is less than allowable pressure, 0.5 N/mm^2. Therefore the thickness of 16 mm is acceptable based on external pressure. This conclusion can also be drawn by the rigorous method of IS-2825 Pressure vessel Code Appendix F.

The plate thickness 16 mm will increase the equipment cost. Therefore, let us work out other option of having stiffening or reinforcing rings.

OPTION I

It is proposed to have shell thickness of 10 mm and 4 stiffening rings each at a distance of 540 mm. Therefore, we have

$$P_c = \frac{2.42 \times E \times (t/D_o)^{5/2}}{(1-\mu^2)^{3/4}\left\{(L/D_o) - 0.45(t/D_o)^{1/2}\right\}}$$

$$= \frac{2.42 \times 190 \times 10^3 \times (10/2150)^{5/2}}{(1-(0.3)^2)^{3/4}\left\{(540/2150) - 0.45(10/2150)^{1/2}\right\}}$$

(please note, $D_o = 2130 + 2 \times 10 = 2150$ mm)

$$= 3.30 \text{ N/mm}^2$$

Therefore,

$$P_a = 3.30/4 = 0.83 \text{ N/mm}^2.$$

The design pressure is far less than the allowable pressure computed above.

OPTION II

It is proposed to have shell thickness of 10 mm with 3 stiffening rings each at a distance of 2700/4 = 675 mm. Therefore, we have from the identical calculation procedure, $P_c = 2.57$ N/mm^2. The allowable pressure $P_a = 0.64$ N/mm^2, which is higher than the design pressure 0.385 N/mm^2.

This option is acceptable. The actual plate of thickness = 12 mm is adequate including corrosion allowance of 1.5 mm.

Stiffening ring

The required moment of inertia for the stiffening ring is obtained via Eq. (5.15).

$$I = P_c D_o^3 L/(24E)$$

We take $P_c = 4 \times 0.385 = 1.54$ N/mm^2. Substituting appropriate values we get

$$I = \frac{1.54 \times (2150)^3 \times 675}{24 \times 190 \times 10^3}$$

$$= 2.27 \times 10^6 \text{ mm}^4$$

The value of I is derated by 30% to accommodate the resistance offered by the shell. Therefore, the desired I is $I_d = 0.7 \times 2.27 \times 10^6 = 1.59 \times 10^6$ mm^4.

The angle ISA $100 \times 100 \times 10$ has $I = 1.77 \times 10^6$ mm^4, which is more than required $I = 1.59 \times 10^6$ mm^4. Therefore, this angle is adequate for the stiffening ring (refer IS-808).

8.7.2 Thickness of Jacket

For the jacket internal design pressure is $1.1 \times 0.35 = 0.385$ N/mm^2. For Eq. (5.3.b) we have thickness, t_j

$$t_j = \frac{pD_i}{(2fj - p)} + c$$

$$= \frac{0.385 \times 2260}{(2 \times 96 \times 0.85 - 0.385)} + c$$

$$= 5.34 \text{ mm} + c$$

Considering the corrosion allowance of 1.5 mm we have plate of the standard thickness 8 mm.

8.7.3 Head Thickness

We have top and bottom dished ends. While there is no external pressure acting (only atmospheric pressure) on the top dished end, there is a jacket pressure acting on the bottom dished end.

(i) *Top head* (internal pressure only)

The thickness is given by Eq. (5.23).

The stress intensification factor, W, is given by

$$W = (1/4)\left\{3 + \sqrt{R_c / R_1}\right\}$$

$$= (1/4)\left\{3 + \sqrt{2130/128}\right\} = 1.77$$

We then have,

$$t_h = \frac{p_i R_c W}{2fJ}$$

$$= \frac{0.605 \times 2130 \times 1.77}{2 \times 96 \times 0.85}$$

$$= 13.98 \text{ mm}$$

The corrosion allowance is 1.5 mm.

Therefore, the plate thickness, t_h = 13.98 + 1.5 = 15.48 mm

Use the plate of thickness 18 mm for fabrication, to allow for thinning while forming the dish.

(ii) *The bottom dished end*

(a) As a rule of thumb stated in Section 5.8.2.2 item (ii) torispherical head, we use the design pressure, p_i = 1.67 × 0.385 = 0.643 N/mm^2.

$$t_h = \frac{0.643 \times 2130 \times 1.77}{2 \times 96 \times 0.85} + \text{corrosion allowance}$$

$$= 14.85 \text{ mm} + 1.5 \text{ mm} = 16.35 \text{ mm}$$

Therefore, for plate of thickness 18 mm is adequate, including thinning allowance.

(b) By considering the buckling pressure 2 times the design external pressure we calculate thickness by Eq. (5.26):

$$t_h = a \times R_c \times (3 \times (1 - \mu^2)^{1/4}) \times \sqrt{p/(2E)}$$

$$= 4 \times 2130 \times (3 \times (1 - (0.3)^2)^{1/4}) \times \sqrt{0.385/(2 \times 190 \times 10^3)}$$

$$= 11 \text{ mm}$$

Therefore, the plate thickness of 18 mm is adequate.

8.7.4 Vessel Shell with Limpet Coil

We illustrate here how the use of limpet coil becomes advantageous to reduce the shell thickness when compare to a plain jacket.

(a) *The design based on circumferential stress*

The diameter of limpet coil d_i = 100 mm (Fig. 8.4).

The design pressure = 1.1 × (0.35) = 0.385 N/mm^2.

From Eq. (8.1) we have the thickness, t_c, as

$$t_c = \frac{p_i d_i}{2f_c}$$

$$= \frac{0.385 \times 100}{2 \times 96}$$

$$= 0.20 \text{ mm}$$

Use minimum thickness 3 mm.

The circumferential stress, f_{ps}, in the shell is given by Eq. (8.3)

$$f_{ps} = \frac{pD_i}{2t_s} + \frac{pd_i}{(4t_c + 2.5t_s)}$$

$$= \frac{0.605 \times 2130}{2 \times 7.9} + \frac{0.385 \times 100}{4 \times 0.2 + 2.5 \times 7.9}$$

$$= 81.56 + 1.88$$

$$= 83.44 \text{ N/mm}^2$$

The allowable stress 96 N/mm^2 is higher than the above value. Therefore, design is acceptable.

(b) *The design based on axial or longitudinal stress*

We have total axial or longitudinal stress given by Eq. (8.4):

$$f_{as} = \frac{pD_i}{4t_s} + \frac{pd_i}{2t_s} + \frac{2 \times \Delta P d_o^2}{3t_s^2}$$

We calculate f_{as} for the shell thickness 7.9 mm. (Please note corrosion allowance of 1.5 mm and the rounding off to available plate thickness 10 mm are not to be considered in this calculation).

$$f_{as} = \frac{0.605 \times 2130}{4 \times 7.9} + \frac{0.385 \times 100}{2 \times 7.9} + \frac{2 \times (0.605 - 0.385) \times (106)^2}{3 \times (7.9)^2}$$

$$= 69.6 \text{ N/mm}^2$$

This value of f_{as} is less than allowable stress value.

Therefore, we use shell of 10 mm thickness.

References and Recommended Readings

Biswas, D. K., and Panthaki K.M., *Design and Construction of Limpet Coil, Pressure Vessel, Chemical Age of India,* **28** (6) 733 (1977).

Feichtinger, C.A., 'New Idea in Jacketed Vessels', *Chemical Engineering,* **67** (18) 127 (1960).

Kern, D.Q., *Process Heat Transfer,* McGraw-Hill, New York, (1950).

Kneale, M., 'Design of Vessels with Half Coils', *Trans. Institution of Chemical Engineers*, **47** T279 (1969).

Markovitz, R.E., *Picking Best Vessel Jacket, Chemical Engineering,* **78** (26) 156 (1971).

Markovitz, R.E., *Process and Project data Pertinent to Vessel Design, Chemical Engineering,* **84** (21) 123 (1977).

Perry, R.H., Green, D.W. and Malonoey, J.O. *Perry's Chemical Engineers Handbook* 6th Edition, *International Edition,* McGraw-Hill Book Co., Singapore, (1984).

Riegel, E.R., *Chemical Process Machinery,* Reinhold Publishing Corporation, New York, (1953).

9

Chapter

Heat Exchangers

9.1 INTRODUCTION

A device used for transfering or exchanging energy in the form of heat between two fluids is known as a heat exchanger. This heat exchange can be either by direct or indirect method. In the indirect method two fluids are separated by a surface, metallic or non-metallic – through which heat transfer takes place generally in the direction perpendicular to flow direction.

Broadly speaking, heat exchangers could be classified as fired or unfired. In the case of a fired heater, heat is transferred from a flame to the fluid mainly by radiation in the radiant zone and by convection in convection zone. In the case of unfired heaters, the mode of heat transfer is by convection in the fluids and by conduction across the solid surface separating the two fluids.

In the present chapter design of unfired and indirect type of heat exchangers are being considered.

9.2 TYPES OF HEAT EXCHANGERS

Heat exchangers can be classified by different methods:

(1) *Function:* Process function such as reboiler, condensers, cooler, heat recovery, steam generators, etc.

(2) *Materials of Construction:* Metallic, non-metallic like graphite, etc.

(3) *Mechanical Construction:* Double pipe, shell and tube, plate heat exchanger, spiral heat exchanger, finned tube, etc.

(4) *Flow Paths:* Co-current, counter current, cross flow, etc.

Amongst the different types of heat exchangers the design and construction features of the double pipe and the shell and tube type, being the most commonly used, are considered in details.

The plate and spiral heat exchangers are proprietary in nature and hence only their salient features are presented.

9.2.1 Double Pipe Heat Exchanger

This (Fig. 9.1) is used particularly when the flow rates are low, and when the temperature range is relatively high. In this type, one fluid flows inside a pipe, while a second fluid flows either in the same direction (co-current) or opposite direction (counter current) in the annulus between a larger pipe and the outer side of the inner pipe carrying the first fluid. The components of the heat exchanger unit are concentric pipes, connecting tees and return bends. Being truly in counter current mode of operation, heat transfer area required is minimum. As a rule of thumb, double pipe heat exchanger is considered for heat transfer area of less than 5 m^2.

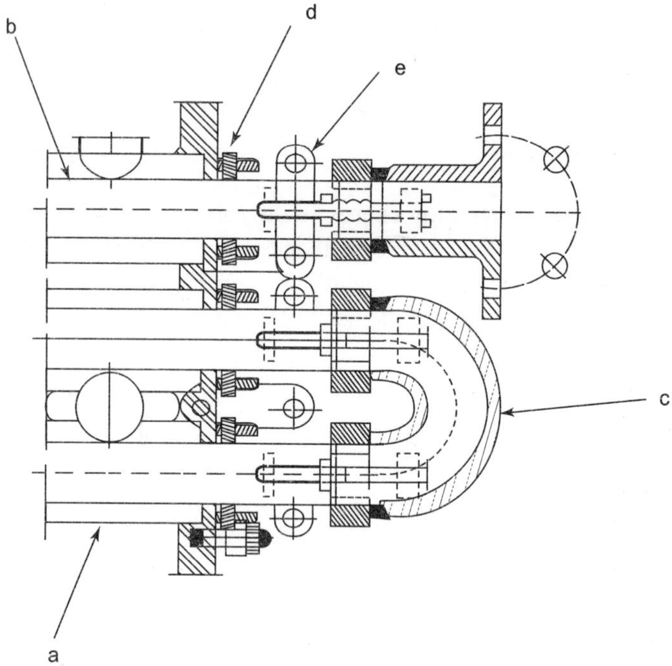

Fig. 9.1 Double Pipe Heat Exchanger: (a) Outer Pipe, (b) Inner Pipe, (c) Return Bend, (d) Gland with Packing, (e) Support Bracket

9.2.2 Shell and Tube Heat Exchangers

These are the most widely used types of heat exchangers (Fig. 9.2). The equipment consists of a number of parallel tubes enclosed in a relatively close fitting cylindrical shell. One fluid flows inside the tube and is called the tube side fluid, the other flows outside the tubes and is called the shell side

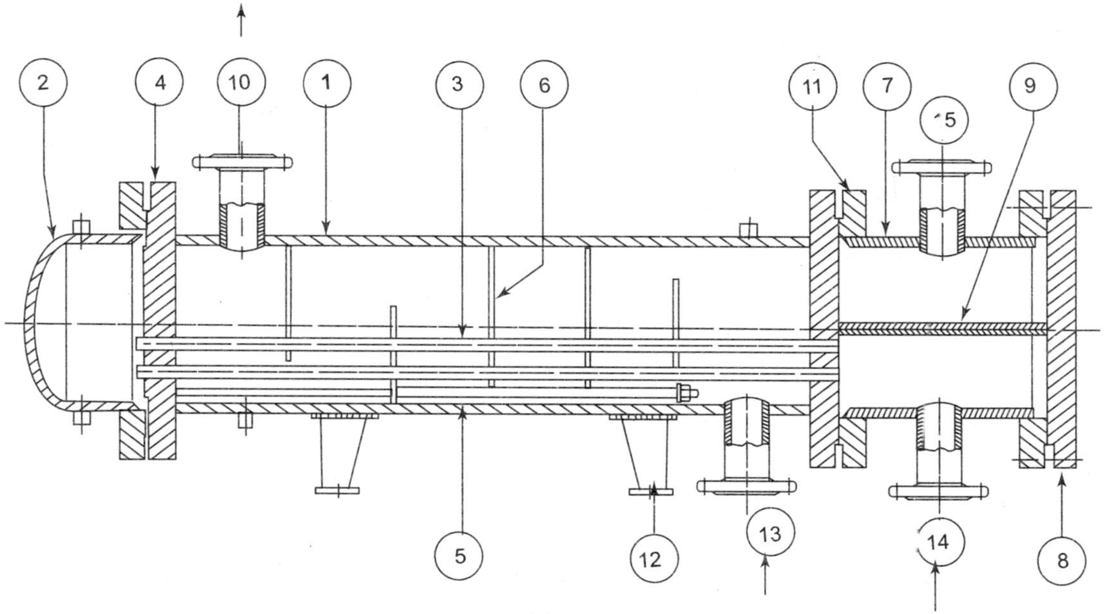

(a)

Fig. 9.2 Types of Shell and Tube Heat Exchangers: (a) Fixed Tube Sheet

Legend: (1) Shell, (2) Bonnet Cover, (3) Tube, (4) Stationary Tube Sheet, (5) Tie Rod, (6) Baffle (7) Stationary Head or Channel, (8) Channel Cover, (9) Pass partition, (10) Outlet Nozzle Shell Side Fluid, (11) Flange, (12) Support, (13) Inlet Nozzle Shell Side Fluid, (14) Inlet Nozzle Tube Side Fluid, (15) Outlet Nozzle Tube Side Fluid

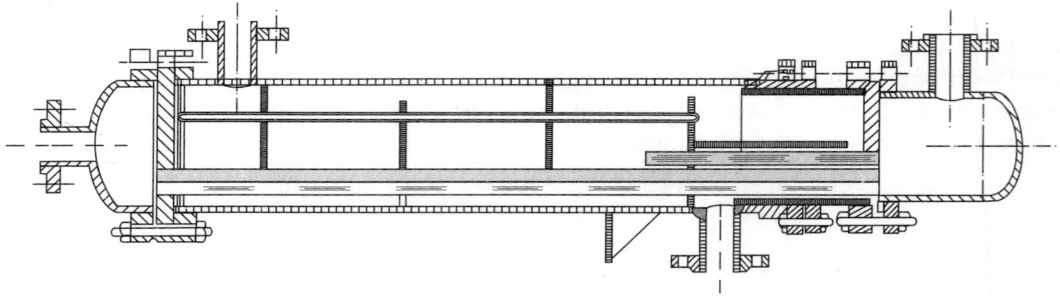

(b)

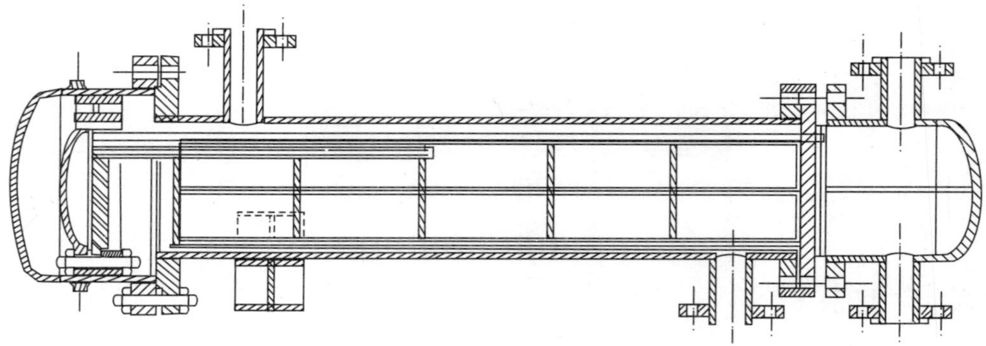

(c)

Fig. 9.2 Types of Shell and Tube Heat Exchangers: (b) Outside Packed Floating Head, (c) Internal Floating Head

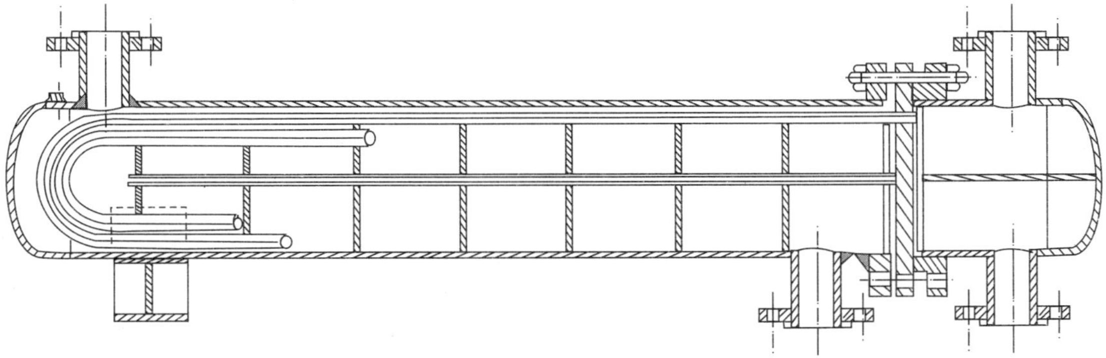

(d)

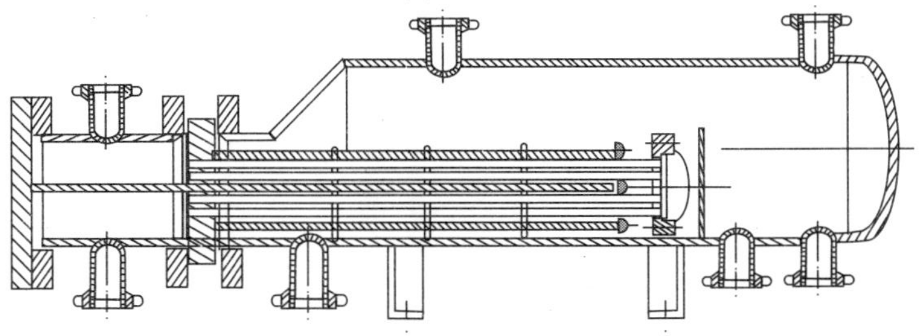

(e)

Fig. 9.2 Types of Shell and Tube Heat Exchangers; (d) U-tube Type, (e) Reboiler (with internal floating head)

fluid. If none of the fluids condense or evaporate, the unit is known as a 'heat exchanger'. If one of the fluids, either in the tube or the shell condenses, the unit is known as a 'condenser' or as a 'heater', depending on whether the primary purpose of the unit is to condense one fluid or to heat the other. Similarly, if one of the fluids evaporates (vaporizes) the unit is designated as an evaporator, a reboiler/vapourizer or as a cooler, depending on whether the primary purpose is to evaporate one fluid or to cool the other. In the case of vapour-in-tube condensers, the vapour distribution is uniform, pressure drop is reasonable and vapour sub-cooling is readily accomplished. For service at high vacuum conditions, horizontal vapour-in-shell condenser (direct column mounted condenser) is a preferred design, because at high vacuum the vapour lines are large, vapour temperatures are usually high and condensate sub-cooling is not too important. Such directly mounted condensers have a very low pressure drop which is very important in high vacuum distillation service such as in fatty acid recovery.

9.2.3 Special Types of Heat Exchangers

These are used in special cases where the standard forms of tubular exchangers are unsatisfactory.

(a) *Pipe coils* These coils of spiral or helical shape are used for cooling purposes, especially at high temperature and pressure applications. Such coils are immersed in a tank of water or sprayed with water. The pressure drop in the coils is high, but the construction is simple and less expensive.

(b) *Spiral heat exchanger* (Fig. 9.3) This consists of two plates, spirally wound one inside the other on a split mandrel, each turn being separated by spacer studs, flash welded to the plates before winding. After winding, alternate pairs of plates are edge welded to form two channels, each open at one side and welded at the other. The last turn of the spiral forms the shell of the cylindrical spiral body, the ends of which are closed by end covers. The hot fluid enters at the centre, flows through the inner passage and leaves at the periphery. The cold fluid enters at the periphery, flows through the outer passage, and leaves at the centre. The two fluids, thus, flow in true counter-current in between the curved surfaces, giving excellent flow characteristics. The curved channel gives a constantly changing angle of incidence to the particles that may, in conventional exchangers, cause fouling. The spiral heat exchanger is used for highly fouling and corrosive liquids.

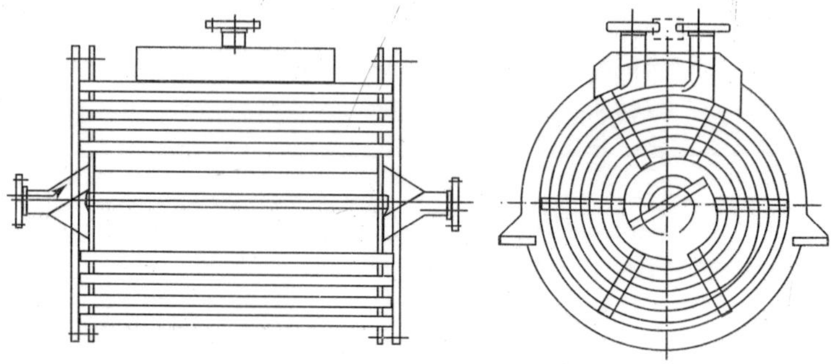

Fig. 9.3 Spiral Heat Exchanger

(c) *Plate type exchanger* This is constructed by a series of corrugated parallel plates held firmly together between frames. The heat transfer surfaces consist of the adjacent plates. Gaskets cemented to the plates help to form separate channels for the two fluids. The two liquids travel in counter-current directions, and heat transfer takes place through the plates. Headers connecting the two liquids connect the alternate plate compartments through corner plates. The whole assembly is bolted together with stay bolts and end plates. This has the advantage of being able to increase the area by adding extra plates. The advantages are: high rate of heat transfer, low pressure drop and easy cleaning and replacement of plates. This type of heat exchanger is used in the food, dairy and other industries where even the slightest contamination of the products due to a leakage cannot be tolerated. Recently, for the same application a double walled plate heat exchanger has been introduced in the market. They are very useful for the pharmaceutical industry as well, where even the slight contamination is not allowable.

The brazed/welded fabrication avoids a gasket, thereby making it suitable also for high temperature applications.

The introduction of the wide gap plate heat exchangers make them more advantageous over the conventional shell and tube heat exchangers in the food processing industry, paper and pulp industry and paint industry.

(d) *Finned tube exchanger* In case of heating air or gas, where the heat transfer coefficients are low, the surface area can be increased by use of fins either in transverse or longitudinal form. Transverse finned tubes are used for cooling process streams by air. Longitudinal finned tubes units are used at high temperature and high pressure service, where heat transfer rates are low.

(e) *Graphite heat exchanger* Carbon and graphite are used in many applications in the construction of chemical equipment, since these materials are stable over a wide range of temperatures and are chemically resistant to most corrosive agents. Graphite has a high thermal conductivity. Due to their low tensile strength and brittleness, techniques in design and fabrication of these materials differ from those employed for metals. Several types of graphite exchangers are made. In the normal types, the exchanger is made of the shell, and in the tube type, with graphite tubes and tube sheets. The cubic or rectangular block heat-exchanger consists of a graphite block perforated with rows of parallel holes for conveying the two fluids. The graphite blocks are made from a series of accurately machined plates of graphite laminated together and bonded with thermosetting resin. Headers are bolted to the four faces of the graphite block for supply and discharge of the fluids. In a cylindrical block of graphite, holes can be drilled and these blocks can be multi-stocked in a cylindrical steel shell that has gland fittings. These units have been designed for use as evaporators and reboilers. In another design, known as the cartridge block exchanger, a graphite core and process headers are housed in a cylindrical steel shell. The graphite core is perforated with holes for the process fluid, opening into headers at each end. Even graphite plate heat exchangers are now available.

9.3 DESIGN OF SHELL AND TUBE HEAT EXCHANGERS

In the chemical process industry, design of any equipment involves, detailed working of the process design and the mechanical design. The process design of a heat exchanger gives the data in respect of the following items:

(a) Type of heat exchanger.

(b) Inside diameter and internals of shell, shell passes, etc., in a shell and tube heat exchanger.

(c) Tube size, number of tubes and tube passes, (heat transfer area) in a shell and tube heat exchanger.

(d) Terminal temperatures of fluid streams.

(e) Mode of operation namely, counter-current, co-current and cross-flow.

(f) Location of fluid streams.

Table 9.1 gives the Heat Exchanger Process sheet which indicates the various specifications formulated from process design and mechanical design of shell and tube heat exchanger.

Due to the wide usage, the design of these heat exchangers and their construction features are standardized. A number of codes such as the Indian Standard-IS-4503, Standards of Tubular Exchanger Manufactures Association (TEMA) and BS–3274 specify the design procedure and details of construction. It is expected that the design of the heat exchanger will satisfy the requirements of the code.

In general, TEMA classifies heat exchangers in three classes. TEMA, R Class – Heat exchangers used in severe conditions as in Petroleum and Petrochemical Industry, for example, heat exchangers used in methanol plants in a synthesis loop and heat recovery system.

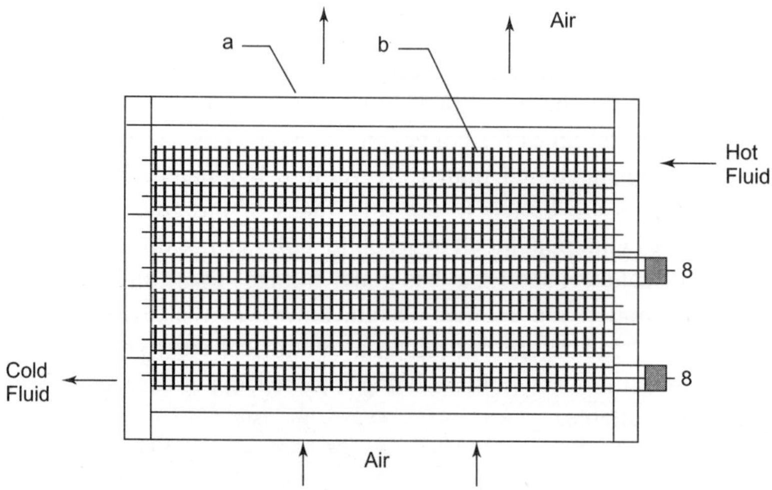

Fig. 9.4 Finned Tube Heat Exchanger: (a) Box, (b) Finned Tubes

Table 9.1 Heat Exchanger Process Data Sheet

SERVICE				Item Number		1
Orientation: ☐ Horizontal ☐ Vertical ☐ No Preference		Type:		Number of Units		2
PERFORMANCE OF ONE UNIT						3
	Measurement	FLUID A	Shell/Tube	FLUID B		4
	Units	In	Out	In	Out	5
Fluid Name						6
Total Fluid Quantity	kg/h					7
Liquid	kg/h					8
Vapour	kg/h | MW					9
Non-Condensables	kg/h | MW					10
Steam	kg/h					11
Water	kg/h					12
Temperature	°C					13
Mol. Wt. of Vapour + NC/Comp. Factor		/	/	/	/	14
Specific Gravity of Liquid						15
Viscosity of Liquid	cP					16
Viscosity of Vapour + NC	cP					17
Specific Heat of Liquid	kcal/kg°C					18
Specific Heat of Vapour + NC	kcal/kg°C					19
Thermal Conductivity of Liquid	kcal/h.m°C					20
Thermal Conductivity of Vapour + NC	kcal/h.m°C					21
Latent Heat	kcal/kg					22
Surface Tension	dyn/cm^2					23
Dew Point/Bubble Point	°C					24
Freeze Point/Pour Point	°C					25
Heat Release Curve or Tabulated Date		☐ Yes	☐ No	☐ Yes	☐ No	26
Inlet Pressure/Allowance Pressure Drop	kg/cm^{2}g/kg/cm^2					27
Total Allowable Stream Pressure Drop	kg/cm^2					28
Fouling Resistance	h.m^2°C/kcal					29
Heat Exchanged	kcal/h					30
Operating Temperature Max./Min.	°C					31
Maximum Operating Pressure	kg/cm^2 g					32
Minimum Design Pressure Allowed	kg/cm^2 g					33
Relief Valve Set Pressure	kg/cm^2 g					34
Design Pressure (Process)	kg/cm^2 g					35
Design Vacuum (Process)	kg/cm^2 a					36
Cyclic Service		☐ Yes	☐ No	☐ Yes	☐ No	37
Vapour Quality (Kettle Reboiler)						38
Materials						39
						40
Corrosion Allowance	mm					41
Line Size	mm	In	Out	In	Out	42
Heat Treatment for Process Reasons		☐ Yes	☐ No	☐ Yes	☐ No	43
Hazard Considerations						44
Insulation						45
Cleaning Requirement		☐H ☐C ☐P ☐W ☐S ☐No		☐H ☐C ☐P ☐W ☐S ☐No		46
Overdesign Multiplier to be Applied to Area		☐ Chemical	☐ Mechanical	☐ Chemical	☐ Mechanical	47

Wherever desired, corresponding SI units may be used.

TEMA, C Class – Heat exchangers for moderately severe conditions.

TEMA, B Class – Heat exchangers used for general purpose for example, those used in food and pharmaceutical industry.

There are certain limitations to the application of TEMA standards. These are: (1) Shell diameter not to exceed 1500 mm, (2) Design pressure of less than 200 bar.

9.3.1 Fluid Flow Arrangement

One of the methods of classification of heat exchangers is on the basis of the configuration of the fluid flow paths through the heat exchangers. The four most common type of flow paths are:

(a) Co-current or parallel flow units, where the two fluid streams enter together at one end, flow through in the same direction and leave together at the other end.

(b) Counter-current or counter flow units in which the two fluid streams move in opposite directions.

(c) Single path cross-flow units in which one fluid moves through the heat exchanger matrix at right angles to the flow path of the other fluid.

(d) Multipass cross-flow units, in which one fluid stream shuffles back and forth across the flow path of the other fluid stream usually giving a cross-flow approximating to counter-flow.

9.3.2 Classification of Shell and Tube Heat Exchangers

These types of heat exchangers are built of a shell in which a number of round tubes are mounted by means of tube plates. Many variations of this basic type are available. The difference lies mainly in the detailed features of construction and provisions for differential thermal expansion between the tubes and the shells. According to the mechanical configuration these exchangers can be classified as (Fig. 9.2):

(a) Fixed tube sheet

(b) Outside packed floating head

(c) Internal floating head

(d) U-tube type

(e) Reboiler (with internal floating head or U-tube type).

(a) *Fixed tube sheet exchanger* [Figs 9.2. (a) and 9.15] These types of exchangers are more often used than others. The essential part is a shell, welded to two tube sheets, one each at either end which also serve as flanges for the attachment of the channels or bonnets. It is often necessary to provide expansion bellows on the shell to eliminate excessive stresses in the shell and tubes due to differential thermal expansion. This type of exchanger is used only where the shell side fluid is clean and non-fouling and where the temperature difference is below 95°C.

In these exchangers, the shell is always welded to tube sheets, resulting in no gasketed or packed joints on the shell side. There is no possibility of intermixing between the hot and cold side fluids through internal gasket failures. The tubes are straight and easily cleanable on the inside. However, mechanical cleaning and usual inspection of the outside of the tubes is not possible with non-removable tube bundle.

(b) *Outside packed floating head exchanger* [Figs 9.2 (b) and 9.16] The shell of this heat exchanger is made up of two parts. The main shell has a flange welded at one end, and provision is made at the other end for the attachment of a short length shell piece, with a stuffing box and gland type joint. The short length of the shell is provided with a flange to which the tube sheet is held by a snap ring. The channel flange is fixed to the tube sheet by a gasketed flanged joint. The second tube sheet is clamped between the main shell flange and the bonnet flange by the use of gaskets on either side of the tube sheet. As the tubes expand, the short length shell piece along with the channel are free to move. This exchanger is fairly common. It is used as a vertical condenser with vapour inside the tubes, and is of a single pass type. The tube bundle is removable for inspection and mechanical cleaning of the outside surface of the tubes. Only the shell side fluid is in contact with the gland. In general the construction is rather elaborate and there are limitations as to the use of high temperatures and pressures for the shell packing gland.

(c) *Internal floating head exchanger* [Figs 9.2 (c) and 9.17] This exchanger has straight tubes fixed at both ends in the floating and stationary tube sheets. The stationary tube sheet is clamped between the channel and shell flanges. The floating head is fitted to the sheet at the other end and the floating head cover is held in position by a split backing ring to permit dismantling. The floating tube sheet is kept slightly smaller in diameter than the bore of the shell, so as to withdraw the entire tube bundle from the channel end. The shell is closed by a bonnet on the floating head side. The tube sheet along with the floating head is free to move and take the differential thermal expansion between the shell and the tube bundle. This exchanger is widely used in chemical industries. It is quite suitable for rigorous duties associated with high temperatures and pressures and also with dirty fluids.

The tube bundle of the exchangers is removable for inspection and the mechanical cleaning of the outside of the tubes. The tubes are replaceable and can be cleaned easily from the inside. The exchanger is costly because of its construction having several gasketed joints.

(d) *U-tube type exchanger* [Fig 9.2 (d)] The tube bundle in this type of exchanger consists of U-shaped tubes, both ends of which are fixed to a single stationary tube sheet. The differential thermal expansion between the shell and the tubes is absorbed by the U-bends which can expand and contract freely. The number of joints in this type of exchanger are reduced.

The exchanger requires only one tube sheet. The tube bundle can be removed for cleaning of the outside surface of the tubes. The inside of tubes can be cleaned only by chemical means. Bending and replacement of tubes is rather difficult. However, the overall cost is reduced. The exchangers are usually employed for high pressure and temperature applications, where a single tube sheet results in considerable reduction in cost.

(e) ***Reboiler exchanger*** [Fig. 9.2 (e)] This exchanger is either provided with an internal floating head arrangement or a U-tube arrangement. The shell is made up of a larger diameter to provide for vapour space, above the tube bundle. This exchanger is located at the bottom of a distillation column for converting the liquid at the bottom of the column into vapour.

9.3.3 Materials of Construction

Exchangers are generally constructed of carbon and low alloy steels. Carbon steels are recommended up to 540°C, while low alloy steels up to 590°C. In such exchangers the tubes may be of non-ferrous metal such as brass, copper, nickel, etc. For handling corrosive fluids non-ferrous alloys or high alloy steels are used. Such metals are copper, cupro-nickel, nickel, monel, inconel, aluminium, stainless steel, etc. In special cases heat exchangers are made of graphite, teflon and glass. For low temperature service, austenitic stainless steel or non-ferrous alloys are used. For high temperature service, stabilized grade of low carbon steel or alloys steel are used. Table 9.2 gives fluid temperature limitations for pressure parts of different materials.

Table 9.2 Fluid Temperature Limitations for Pressure Parts

Material	*Maximum Permissible Temperature °C*
Carbon steel	450
C-M_0 steel	590
Cr-M_0 steel	650
Low alloy steel (less then 6% chromium)	590
Alloy steel (less then 17% chromium)	590
Austenitic Cr-Ni steel	650
Cast iron	150
Brass	200

9.3.4 Design Pressure

For determining the minimum thickness of various components, the design pressure is obtained normally by adding (5% but generally 10%) to the maximum working pressure. This is finally decided by the severity of the working conditions.

9.3.5 Design Temperature

It is taken as 10°C higher than the maximum temperature that any part of the exchanger is likely to attain in course of its operation.

9.3.6 Corrosion Allowance

For carbon and cast iron pressure parts, the corrosion allowance is 1.5 mm except for tubes. For severe conditions the allowance is 3 mm. For non-ferrous and stainless steel parts no allowance may be required. For internal covers and tube sheets, corrosion allowance is provided on both sides. Non-pressure parts such as tie rods, spacers, baffles, supports need not have any allowance for corrosion.

9.3.7 Shell

The internal diameter of the shell is determined from the total area occupied by tubes. The procedure for calculation of shell diameter is indicated later. The ratio of tube length to shell diameter for liquid exchangers is in the range of 4 : 1 to 8 : 1. For gas to gas exchangers this ratio may be even less than 4 : 1. The shell may be cut to the required length from a standard pipe up to 600 mm diameter or fabricated from a plate. The thickness of the shell may be calculated as a cylinder with internal pressure or vacuum depending on the working pressure in the shell [Eqs 5.3 and 5.14 (b)]. The minimum thickness including corrosion allowance of the shell, made of carbon steel varies from 5 mm to 11 mm depending on the diameter. Table 9.3 gives minimum thickness, for different materials, when the shell is fabricated out of pipe. For a larger diameter, the shell is fabricated out of standard plate. The shell diameter is decided by the number of tubes, tube pitch diameter, and the tube arrangement, for instance on a triangular pitch or square pitch.

Table 9.3 Minimum Shell Thicknesses when Severe Conditions are not Expected

	Minimum Thickness (in mm)						
Nominal Diameter	*Cast Iron*	*Carbon Steel (including Corrosion Allowance)*	*Copper and Copper Alloys*	*Aluminium and Aluminium Alloys*	*Austenitic Stainless Steel*	*Nickel*	*Monel Inconel*
150	10	5	3.2	5	3.2	3.2	3.2
200	10	6.3	3.2	5	3.2	3.2	3.2
250	10	6.3	3.2	5	3.2	3.2	3.2
300	13	6.3	3.2	5	3.2	3.2	3.2
350	13	6.3	5	5	3.2	5	3.2
400	13	6.3	5	6.3	3.2	5	3.2
500	13	8	6.3	8	3.2	6.3	3.2
600	16	8	6.3	8	5	6.3	5
700	16	10	8.3	10	5	8	5
800	16	10	10	11.2	6.3	8	6.3
900	19	10	10	11.2	6.3	10	6.3
1000	19	10	11.2	12.5	6.3	11.2	6.3
1100	22	11.2	11.2	14	6.3	11.2	6.3

Note: The thickness values are exclusive of the corrosion allowance.

For determination of shell thickness, nozzle thickness and flanges, the equations applicable to pressure vessel design are used. Joint efficiency is also to be taken into account on the same basis (see Chapter 5).

9.3.8 Tubes

The two basic types of tubes are plain and finned, external or internal. The plain tube is used in the normal heat exchanger units. Finned tubes give an increased heat transfer surface. The duplex tube is a tube within a tube snugly fitted by drawing the outer tube on to the inner tube. This tube is useful in conditions where the shell side fluid is not compatible with the material for the tube side fluid. The outside diameters of the tubes vary from 6 mm to 40 mm, with thickness (1 to 3 mm) depending on the materials of construction and diameters. Tube lengths used are 0.5, 1, 1.5, 2, 2.5, 3, 4, 5, and 6 metres. The tubes are laid out either on an equilateral triangular pitch or a square pitch. The minimum pitch is 1.25 times the outside diameter of tubes. When tubes are on a square pitch a minimum cleaning lane of 6.5 mm is provided (Fig. 9.5).

The use of finned tubes is widely accepted due to their high heat transfer rates. These are most commonly transverse low fin type. The fin height is about 1.5 mm, with about 7 to 8 fins per cm tube length. It is usual to reduce the tube diameter during the finning process, so that the fin diameter is confined within the limits of the initial stock-tube diameter; the tube can thus be fed through holes in tube sheets and baffles, just as easily as the plain tube. The low fin tube is, therefore, ideally suited for shell and tube heat exchangers and can replace the plain tube in existing equipment as well as in the new units. In some applications, such as condensers and evaporators, a roped tube surface with appropriate groove depth, pitch and helix angle appear more suitable due to good drainage characteristics. Similarly, a fluted tube is preferred for vertical condensers. Tubes could be seamless or welded type, the seamless tubes are recommended for high pressure application. They are also considered for hostile (corrosive) environments.

9.3.9 Tube Sheet

It is essentially a flat plate with a provision for making a gasketed joint, around the periphery. A large number of holes are drilled in the tube sheet according to the pitch requirements of the tubes (Fig. 9.5). The common methods of fixing the tubes in these holes consist of expanding the ends of the tubes located in tube sheet holes by means of a rotating, roller expanding tool. The holes in the tube sheet are drilled undersize and are then reamed to a diameter slightly larger than the outside diameter of the tube. One or two grooves are cut inside the holes in order to prevent leakage and to increase the strength of the joints (Fig. 9.6).

Tubes are sometimes soldered or brazed into the tube sheet holes, particularly when the tubes have a small diameter. Tubes can also be welded in tube sheet holes, but this method makes the tube replacements very difficult. A minimum tube sheet thickness is necessary in order to fix the tubes to the tube sheet. The various methods of joining tubes to tube sheets do not entirely eliminate the possibility of leakage of shell and tube side fluids. For those cases, where mixing of these fluids

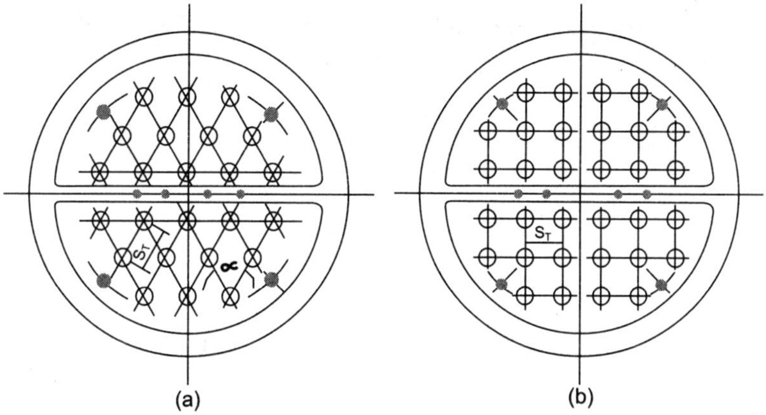

Fig. 9.5 Tube Sheet: (a) Triangular Pitch of Tubes, (b) Square Pitch of Tubes

cannot be tolerated, a double-tube sheet construction should be considered. This construction does not eliminate leakages; but it does eliminate leakage of shell side fluid into tube side fluid. Fig. 9.7 shows a conventional double tube sheet unit. One shell side tube sheet is bolted to the shell flange and channel side tube sheet is bolted to the channel flange. Adjacent tube sheets are joined to each other only by tubes. Shrouding of thin metal sheet may be used to seal the gap between adjacent tube sheets for collecting leakage. The gap between the tube sheets is determined by

$$G = \left[\frac{(1.5\, E d_o \Delta)}{f}\right]^{1/2} \tag{9.1}$$

where, d_o — outside tube diameter

E — modulus of elasticity for the tube material

f — permissible stress in tubes

Δ — deflection of tube

$$= \frac{d_t}{2}\left[\alpha_i (t - t_a) - \alpha_s (T - t_a)\right]$$

d_t — outside tube limit circle diameter

t — channel side tube sheet metal temperature

T — shellside tube sheet metal temperature

α_i, α_s — coefficient of expansion of the channel side and shell side tube sheets respectively

t_a — atmospheric temperature.

$$a = n \times 0.866\, S_T^2 \tag{9.2}$$

where, S_T — pitch of tube

n — number of tubes

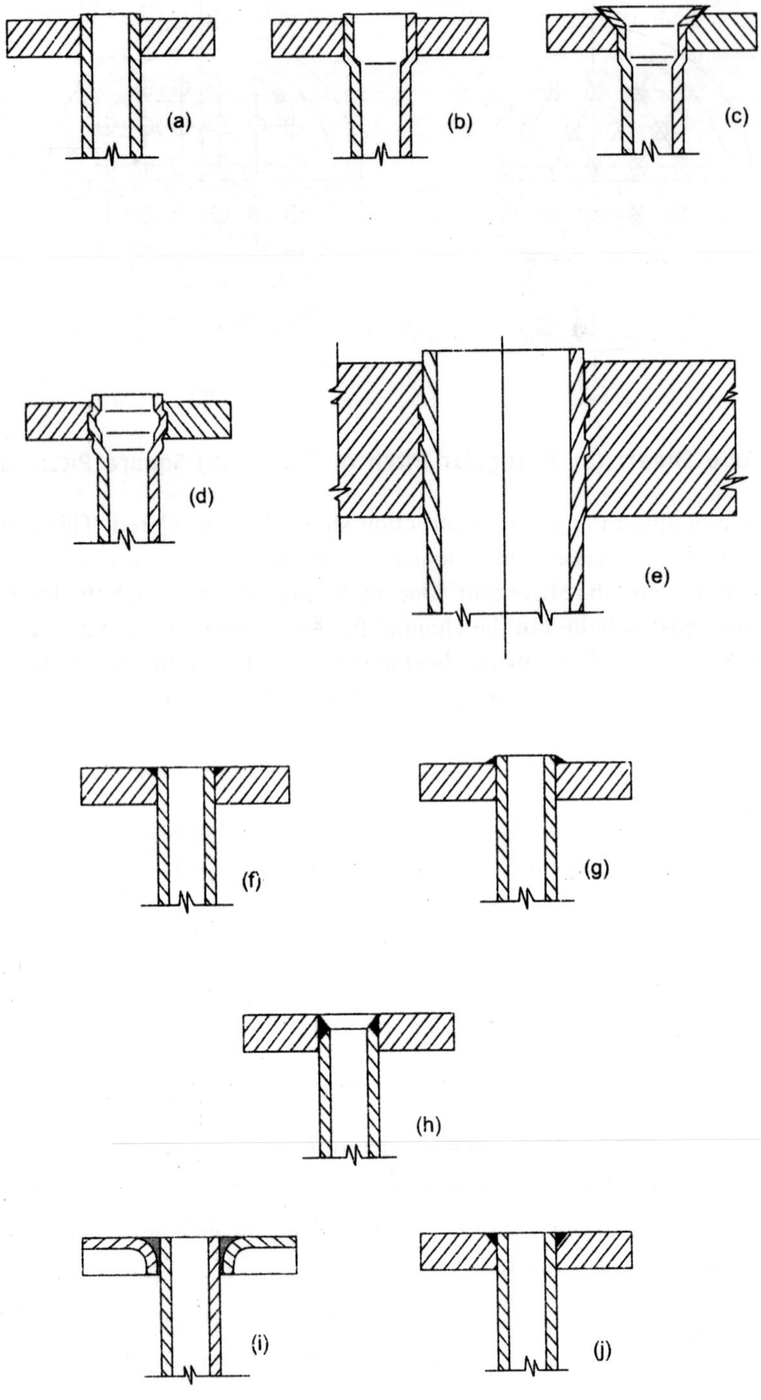

Fig. 9.6 Methods of Fixing Tubes to Tube Sheet: (a) Screwed Tube, (b), (c), (d) and (e) Rolled Tubes, (f), (g) and (h) Welded Tubes, (i) and (j) Brazed Tubes

On the basis of the type of tube arrangement and pitch (Fig. 9.5) it is possible to determine the tube sheet diameter and also the inner shell diameter. Assuming a triangular spacing, the area occupied by the tube is given in

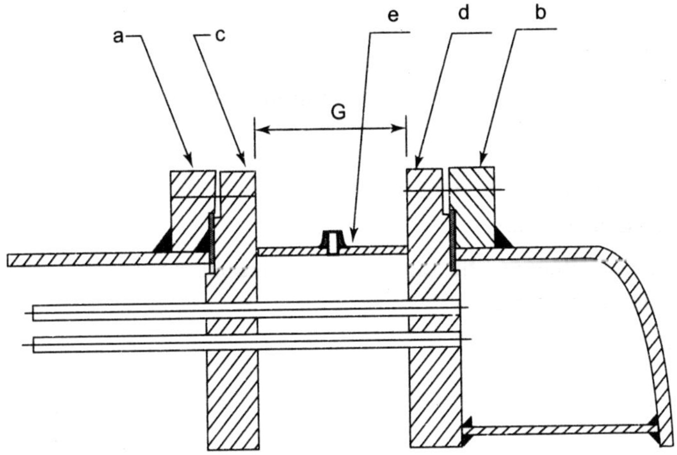

Fig. 9.7 Double Tube Sheet: (a) Shell , (b) Bonnet Flange, (c) Shell Side Tube Sheet, (d) Bonnet Side Tube Sheet, (e) Shroud

For square pitch, area is given by

$$a_n = n \times S_T^2 \tag{9.3}$$

To provide for the pass partitions, tie rods, etc., the actual area of the tube sheet for locating the tubes will be greater. This is assessed by dividing the area a_n by a proportionality factor β, whose values vary from 0.8 to 1 for single tube side pass, 0.7 to 0.85 for double pass and 0.6 to 0.8 for multipass heat exchangers. This area also corresponds to the area of the shell

$$A_s = \pi \frac{D^2}{4} = \frac{n \times 0.866 S_T^2}{\beta} \tag{9.4}$$

where, A_s— area of shell

D— inside diameter of shell.

The overall tube sheet diameter can be determined by providing an additional area required for positioning of the gasket. A similar procedure may be evolved for a square pitch.

The effective thickness of the tube sheet, is the thickness measured at the bottom of the pass partition groove minus the shell side corrosion allowance and the corrosion allowance on the tube side, in excess of the groove depth. It is determined by taking into consideration the effect of the working pressure on a flat plate with holes, as also the effect of expansion of the tube and shell due to a rise in temperature. The detailed calculations are rather complex. A simpler equation for effective tube sheet thickness is as follows:

$$t = F.G.\sqrt{\frac{0.25\,p}{f}} \tag{9.5}$$

where, t — effective thickness of tube sheet

p — design pressure

f — allowable stress at appropriate temperature

G — mean diameter of gasket.

The value of F varies according to the type of heat exchanger. In most cases it is taken as 1. For a U-tube exchanger it is taken as 1.25.

For a fixed tube sheet exchanger, the thickness is calculated for both tube side and shell side conditions, using whichever is greater.

For the tube side pressure

$$F = \sqrt{\frac{2+k}{2+3k}} \tag{9.6}$$

and for shell side pressure

$$F = \sqrt{\frac{k}{2+3k}} \tag{9.7}$$

where, $$k = \frac{E_s t_s (D_o - t_s)}{E_t N (d_o - t_t) \cdot t_t} \tag{9.8}$$

E_s — elastic modulus of shell

E_t — elastic modulus of tube

D_o — outside diameter of shell

d_o — outside diameter of tube

t_s — shell thickness

t_t – tube wall thickness

N – number of tubes in shell

G in Eq. 9.5 depends on the fixity of the tube.

(a) *For tube side pressure:* G is the mean diameter of the channel tube sheet gasket or if integral with channel, the inside diameter of the channel.

(b) *For shell side pressure:* G is the mean diameter of the shell tube sheet gasket or if integral with shell, the inside diameter of the shell.

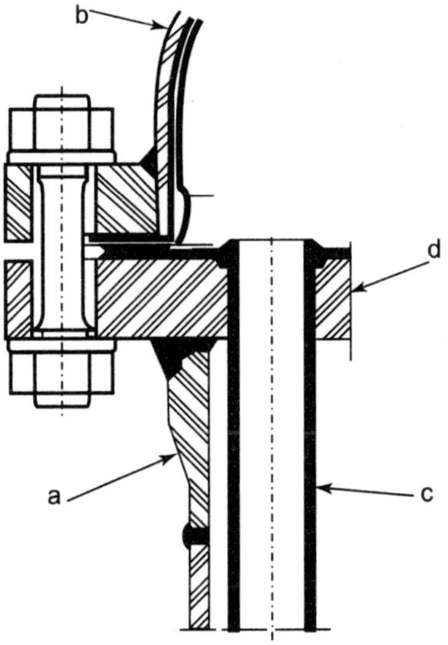

Fig. 9.8 Lined Fixed Tube Sheet: (a) Shell, (b) Channel, (c) Tube, (d) Lined Tube Sheet

Figure 9.8 shows a fixed tube sheet arrangement with an increased shell thickness to facilitate welding of the tube sheet to the shell.

A minimum tube sheet thickness must be provided. It is usually equal to the outside diameter of the tube up to 15 mm tube diameter. For higher diameter tubes, the thickness is smaller than the tube diameter. Table 9.4 gives the minimum tube sheet thicknesses.

The method of calculating the tube sheet thickness indicated above is based on the specification to Tubular Exchanger Manufacturers Association (TEMA). Tube sheet thickness can also be calculated based on IS-4503, British Standard (BS)-3274 and the German specification AS-Merkblatt. The procedures suggested in these standards are rather complex. It may be seen that the primary consideration in IS, BS, and TEMA codes has been the bending stress created in the tube sheet. The German code has no provision for considering the differential expansion of tubes and shell. The IS and BS code values generally lead to a greater economy in equipment cost without impairing the reliability.

Table 9.4 Minimum Tube Sheet Thicknesses

Tube Outer Diameter mm	*Thickness* mm
6	6
10	10
12	12
16	13
18, 19, 20	15
25, 25.4	19
31.8, 32	22.4
38, 40	25.4

9.3.10 Shell Side and Tube Side Passes

The direction of flow in the shell and tubes can be changed by use of different flow paths, known as passes. Such arrangements are used to obtain higher velocities and longer paths for the fluid to travel without increasing the length of the exchanger. The passes on the shell side are single pass, two pass, single split pass, and double split pass (Fig. 9.9). Passes on the tube side are one, two, four, six up to twelve (Fig. 9.10). Shell side passes are formed by use of baffles placed in the shell, while those on the tube side are formed by partitions placed in the shell cover and channels. These partitions are fixed between the channel covers and tube sheets by providing approximately 5 mm deep grooves, for higher pressure say more than 20 bar.

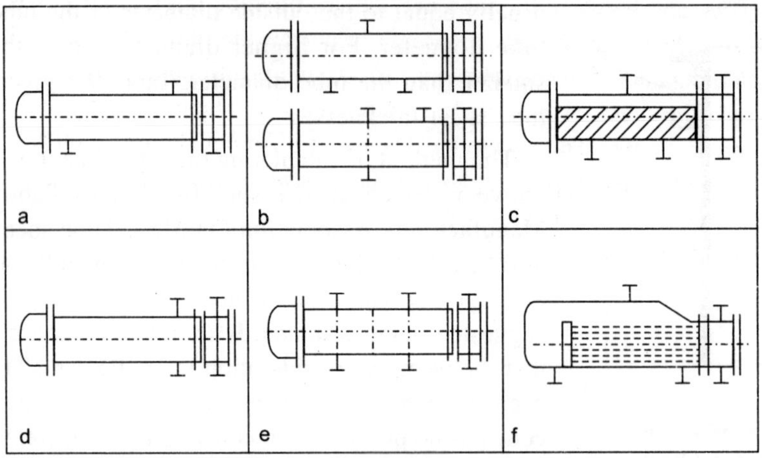

Fig. 9.9 Shell Side Passes: (a) Single Pass, (b) Single Split Pass, (c) Evaporator, (d) Two Pass, (e) Double Split Pass, (f) Kettle Type

The single pass type is by far the most commonly used. The split pass types are used to reduce the shell-side pressure drop where pressure drop rather than heat transfer is the controlling factor in the design.

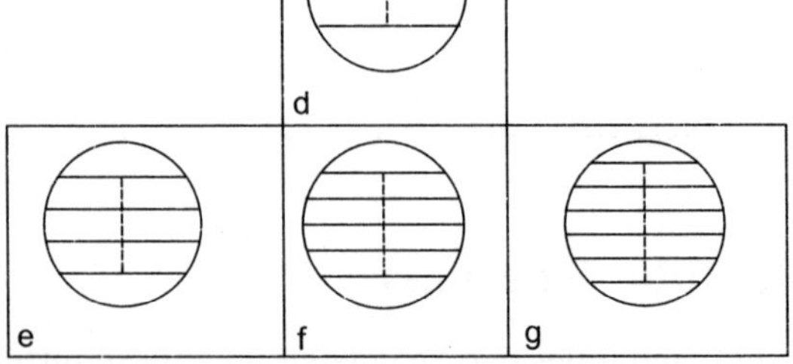

Fig. 9.10 Tube Side Passes: (a) Single Pass, (b) Two Passes, (c) Four Passes, (d) Six Passes, (e) Eight Passes, (f) Ten Passes, (g) Twelve Passes

9.3.11 Baffles and Tie Rods

Baffles are of two types, namely, transverse and longitudinal. These are used to increase the rate of heat transfer by increasing the velocity and turbulence of the shell side fluid. The clearance between the shell and the baffles and between the tube and baffles must be minimum to avoid bypassing of the fluid. However, the clearance should be enough to permit the removal of the tube bundle. The baffles are supported independent of the tubes by tie rods and positioned by spacers. The tie rods are fixed at one end in the tube sheet, by making blind tapped holes. The minimum number of tie rods is four, of at least 10 mm diameter. Transverse baffles are of different shapes, such as segmental, disc and ring, orifice, vane and helical types (Fig. 9.11). The common baffle shape is the segmental type. Longitudinal baffles consist of flat plates, which extend across the shell. These are used to obtain counter flow of the tube side and the shell side fluids.

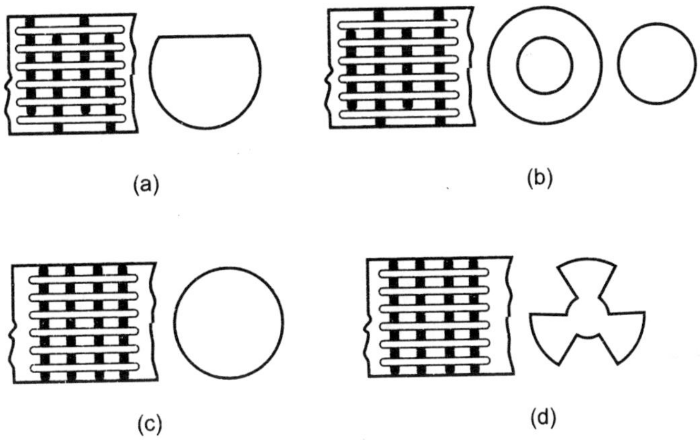

Fig. 9.11 Transverse Baffles: (a) Segmental, (b) Disc and Ring, (c) Orifice, (d) Vane

Baffle cuts from 15 to 45 per cent are used. Generally, a baffle cut of 20 to 25 per cent will be optimum, giving good heat transfer rates without an excessive pressure drop. Baffle spacings are used in the range from 0.2 to 1.0 shell diameters, the optimum spacing being 0.3 to 0.5 times shell diameter.

9.3.12 Channel and Channel Cover

Channels may either be fabricated or be of cast construction and may be of one piece 'bonnet' type or the 'straight' type with a separate bolted cover. The thickness of the channel shall be greater of the following two values: (a) shell thickness, (b) thickness calculated on the basis of pressure in a cylindrical channel. Under severe operating conditions, the depth in the case of multipass channels shall be such that the minimum cross cover area for flow between successive tube passes is at least 1.3 times the flow area through the tubes in one pass. Where severe conditions are not expected, the minimum cross over area shall be at least equal to the flow area through the tubes in one pass.

The effective thickness of the flat channel cover is calculated from the formula

$$t = G_c \sqrt{\frac{k \cdot p}{f}} \tag{9.9}$$

where, G_c — mean gasket diameter for cover

p — design pressure

f — permissible stress at design temperature

k is a factor which is 0.25 when the cover is bolted with full faced gaskets and 0.3 when bolted with narrow faced or ring types gaskets. Dished heads can also be used as channel covers. The final thickness is determined by the addition of corrosion allowance as required. A special type of channel is shown in Fig. 9.12. It shows a vertical exchanger with a dome shaped channel cover with tube sheet fixed to the channel. This arrangement facilitates complete draining of tube side liquid.

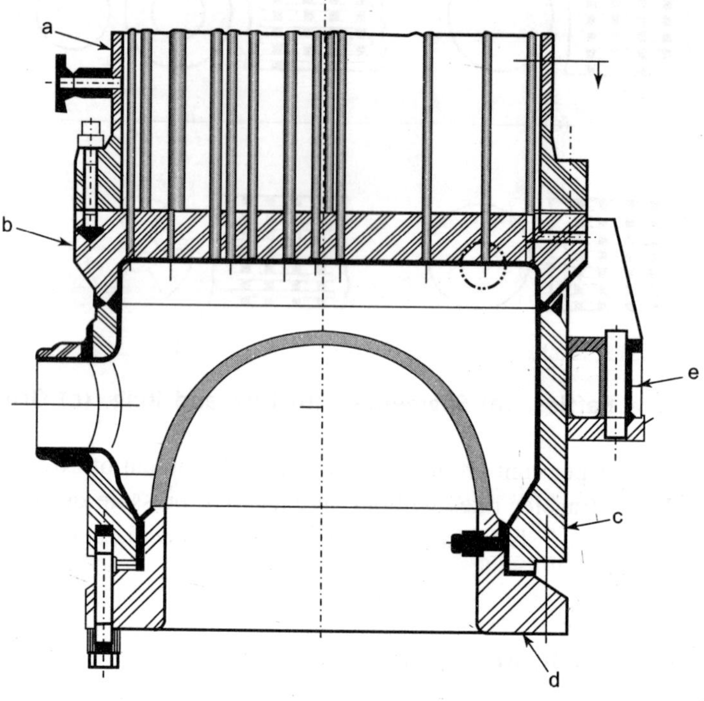

Fig. 9.12 Channel with Tube Sheet: (a) Shell, (b) Tube Sheet, (c) Channel, (d) Domed Channel Cover, (e) Support

9.3.13 Tube Sheet, Channel and Shell Joints

Flanged joints are used to hold the tube sheet between shell and channel, as also for channel covers. These flanges are of ring type or welding neck type. Ring flanges thicker than 100 mm are not recommended. Above this thickness, welding neck flanges are preferred. Flange facings may be of, raised face, male and female or tongue and groove (Fig. 9.13). A double tube sheet arrangement is also shown in the figure.

Metal jacketed or solid metal gaskets are used for all joints in contact with oil or oil vapour and also for pressure above 1.6 N/mm^2. For pressure below this value and when there is no contact with oil and oil vapour, compressed asbestos fibre, natural or synthetic rubber or other suitable gasket and packing materials having appropriate mechanical and corrosion resisting properties may be used.

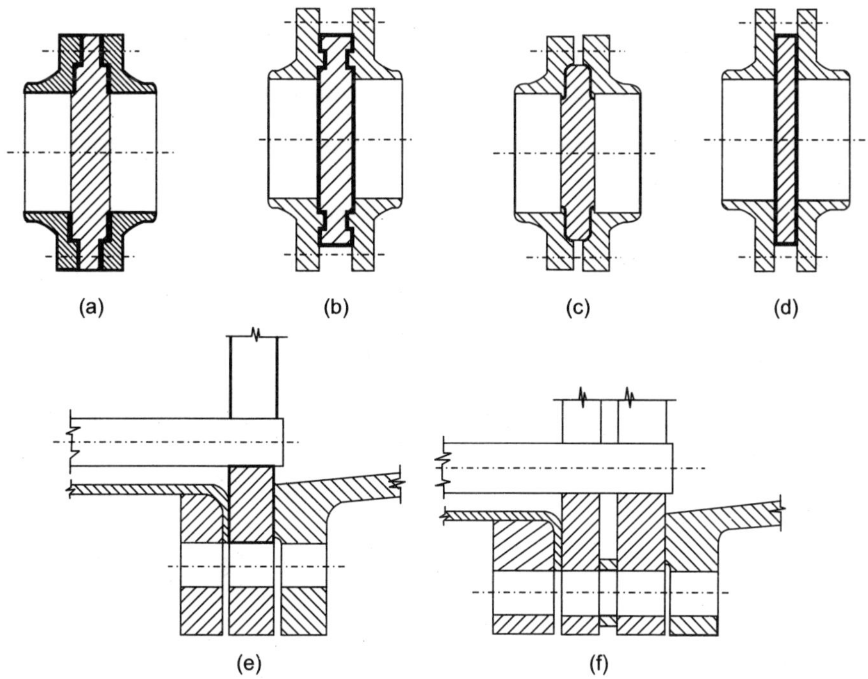

Fig. 9.13 Flanged Joint with Tube Sheet: (a) Male and Female (with Full Tube Sheet), (b) Male and Female, (c) Tongue and Groove, (d) Plain Face, (e) Single Tube Sheet (Removable), (f) Double Tube Sheet (Removable)

Detailed information on the gaskets, bolt loads type of flanges and the flange thickness is indicated under Chapter 5.

9.3.14 Expansion Provision

In heat exchangers, it is necessary to make a provision to allow for differences in the expansion of the tube and shell due to rise in temperature. If the shell and the tubes are fixed together, as in the case of fixed tube sheet exchangers, the two components will not be free to expand. Lack of provision for free expansion is likely to induce considerable stresses in the shell and tubes. There is also a likelihood of damage to the joints made for holding the tube sheet.

If free expansion due to temperature rise is prevented, the stresses in the shell and the tubes can be assessed as follows:

Elongation of tubes due to rise in temperature

$$\Delta l_t = \alpha_t L_t T_t \tag{9.10}$$

Elongation of shell due to rise in temperature

$$\Delta l_s = \alpha_s L_s T_s \tag{9.11}$$

where, α— coefficient of linear expansion

L— length of tube or shell

T— rise in temperature above atmospheric temperature.

If $T_t > T_s$, the tubes will elongate more than the shell. (Assuming same material of construction. In the case of different materials the relative elongation will depend on the coefficient of linear expansion for each material and temperature rise). The shell will, therefore, resist the free expansion of the tubes, while the tubes will create a greater expansion of the shell. If the mean elongation is Δl

$$\Delta l_s < \Delta l < \Delta l_t$$

The force on the shell will be tensile, and is given by

$$p = \frac{\Delta l - \Delta l_s}{L_s}(A_s E_s) \tag{9.12}$$

The force on the tubes will be compressive and is given by

$$p = \frac{\Delta l - \Delta l_t}{L_t}(A_t E_t) \tag{9.13}$$

$$\text{Stress in the shell} = \frac{p}{A_s} \tag{9.14}$$

$$\text{Stress in the tubes} = \frac{p}{A_t} \tag{9.15}$$

where, A — area of cross-section of shell or tubes

E — modulus of elasticity.

(a) ***Expansion bellow or diaphragm*** One of the methods of reducing the thermal stresses in fixed tube sheet heat exchangers is the interposition of a fairly flexible element in the comparatively stiff shell of the heat exchanger. This flexible element called an expansion bellow or diaphragm, it has sufficient flexibility and permits for a constrained expansion or contraction of the shell. However, it has a certain amount of rigidity and, thus, the heat exchanger shell behaves as a partially constrained member. A expansion bellow can thus never reduce the thermal stresses to zero. Figure 9.14 shows an expansion diaphragm consisting of two dish-shaped discs each welded to the shell. The discs are welded together of form a diaphragm. Fig. 9.15 shows the attachment of the diaphragm to the shell

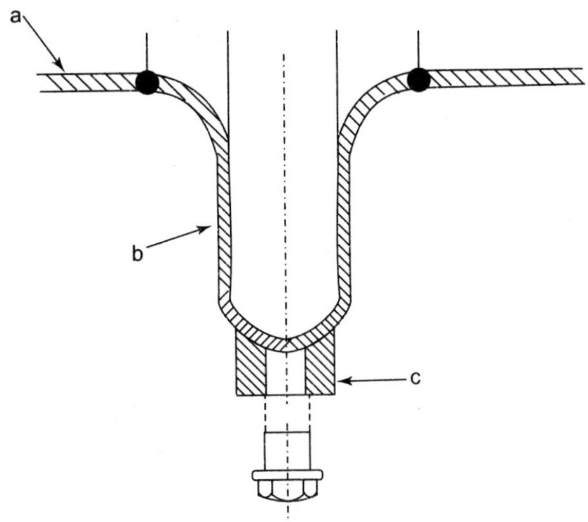

Fig. 9.14 Expansion Bellow: (a) Shell, (b) Bellow, (c) Drain

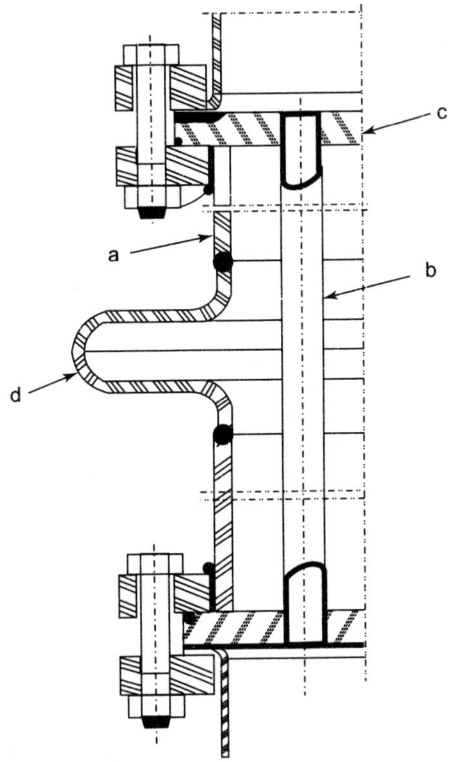

Fig. 9.15 Fixed Tube Sheet Heat Exchanger with Bellow: (a) Shell, (b) Tube, (c) Tube Sheet, (d) Diaphragm

(b) *Outside packet floating head* One end of the shell is formed into a stuffing box. An extension piece of the shell is provided with a gland. The two components of the shell are assembled together with a packing in between. The shell is, therefore, free to expand. It will slide against the gland packing and allow the extension piece to adjust its position accordingly. The tube sheet forms a part of the extension piece, which is closed by a separate cover.

Figure 9.16 shows details of outside packing arrangement.

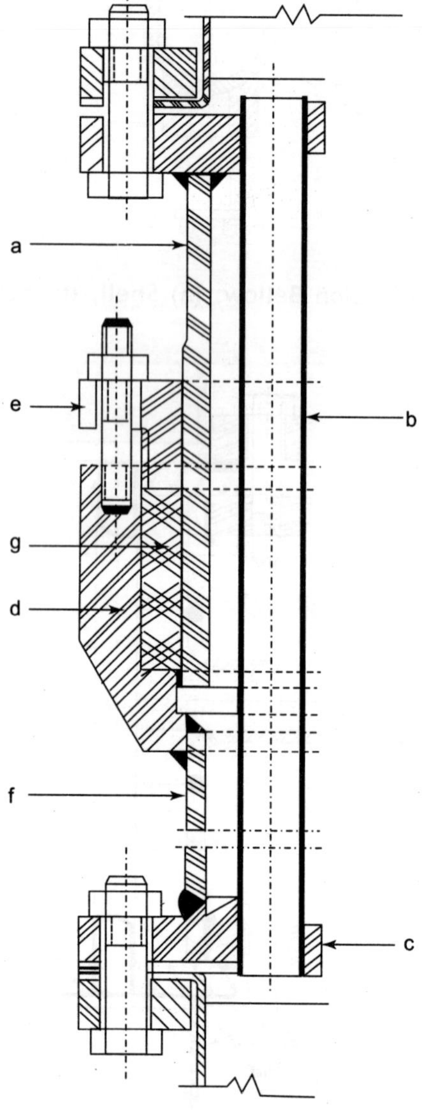

Fig. 9.16 Outside Packed Floating Heat: (a) Shell, (b) Tube, (c) Tube Sheet, (d) Stuffing Box, (e) Gland, (f) Shell Extension Piece, (g) Packing

(c) ***Internal floating head*** In this method, one of the tube sheets is held between a floating head cover and a backing ring by the use of a flanged joint. This assembly is placed inside a fixed head attached to the shell. In some cases the floating head cover is attached to the tube sheet without any ring. This arrangement allows for a free expansion of the tubes. The tube bundle can be removed readily for cleaning the outside surface of the tubes or for making repairs.

In cases where either of the fluids passing through the exchanger undergoes a large change in temperature, a floating head is used. Figures 9.17 and 9.18 show details of two arrangements of floating heads.

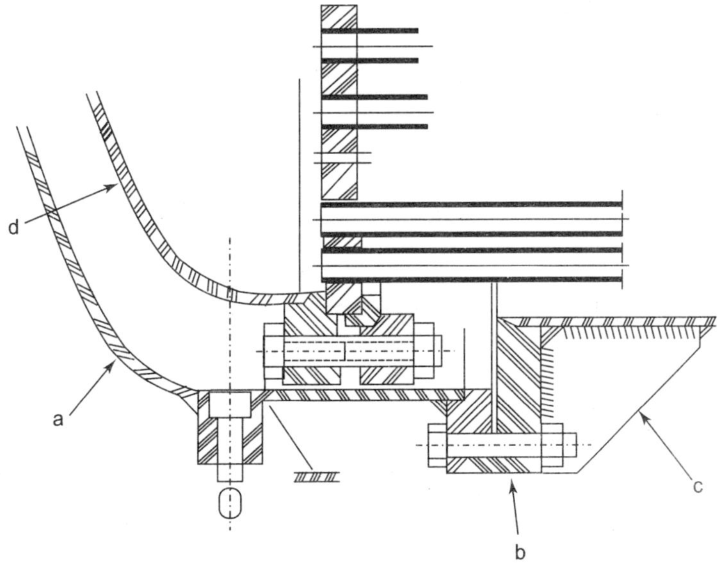

Fig. 9.17 Floating Heat Details: (a) Head, (b) Shell Flange, (c) Gusset Plate, (d) Floating Head

(d) ***U-tube arrangement*** [Fig. 9.2(d)] In this arrangement there is only one tube sheet to which both the ends of the tubes are attached. All the tubes are bent to form a U. This is a relatively inexpensive means of providing for expansion. The tubes are free to expand independently of each other. Internal cleaning of the tubes is difficult, which prohibits the use of this arrangement, when the tube-side fluid is dirty or scale forming.

(e) ***Bowed tube arrangement*** Instead of using straight tubes, a short length of the tube near the tube sheet is bent. This creates a flexibility in the tubes and allows for a limited expansion of the tubes. This method is adopted only in a few cases.

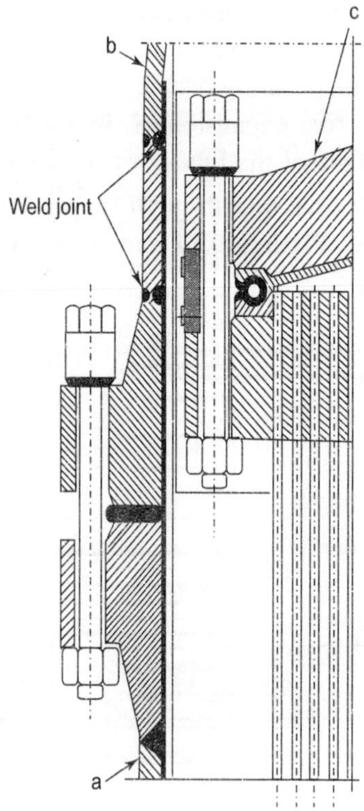

Fig. 9.18 Lined Floating Head Details: (a) Shell, (b) Head, (c) Lined Floating Head

9.3.15 Nozzles

The following nozzles are generally provided on the heat exchanger:

(a) Inlet and outlet for shell side fluid

(b) Inlet and outlet for tube side fluid

(c) Instrument connection

(d) Drain

(e) Vent

Nozzles on shells, shell covers and channels are either integral or welded. Shell nozzles have to be flushed with the inside contour of the shell. Channel nozzles may protrude inside the channel except for the drain and vent. Venting and draining nozzles may be provided in the diaphragm of an expansion bellow for horizontal heat exchanger. For a vertical heat exchanger, vent and drain can be provided through the tube sheet (Fig. 9.19).

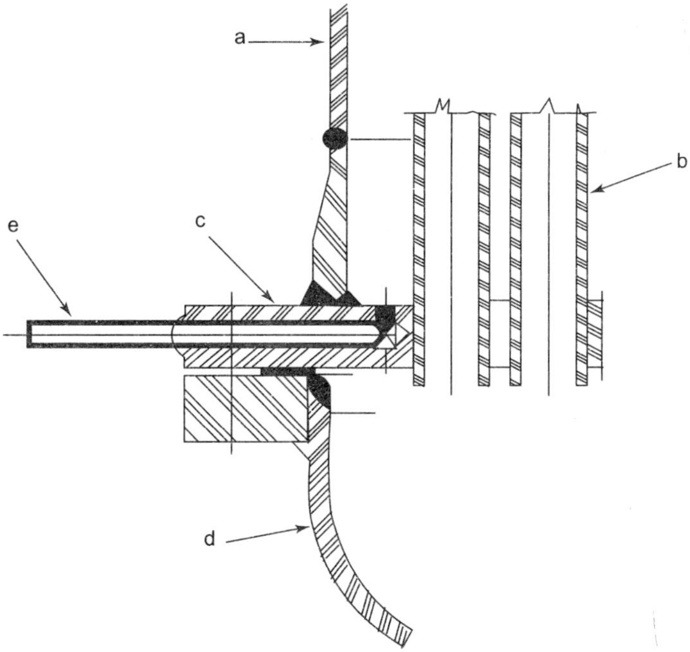

Fig. 9.19 Vertical Heat Exchangers with Drain Through Tube Sheet: (a) Shell, (b) Tube, (c) Tube Sheet, (d) Channel, (e) Drain

9.3.16 Supports

Horizontal shells are supported on two saddle supports, while vertical shells are supported on brackets. In the case of saddle supports for fixed tube sheet exchanger, one of the supports is fixed, while the other is placed on rollers, to facilitate expansion of shell (Fig. 9.20). Details regarding design of supports are given in Chapter 13 (see Table 13.2).

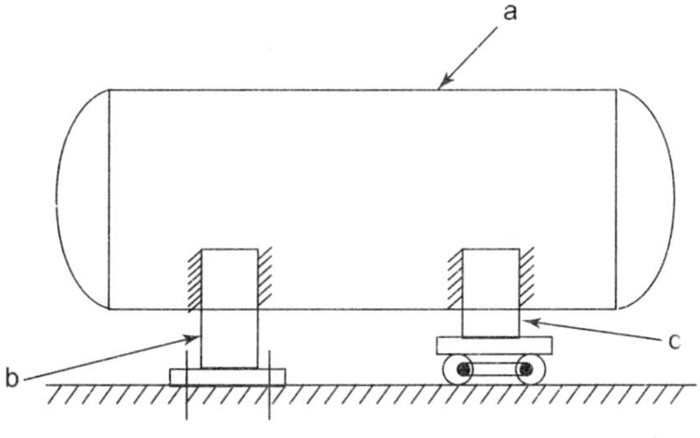

Fig. 9.20 Roller Support: (a) Shell, (b) Fixed Support, (c) Roller Support

9.4 ILLUSTRATION

The illustration presented here is for a U-tube, 2 Pass Shell and Tube heat exchanger. The attached drawing (Plate 2) is for a non-standard (non–TEMA) heat exchanger. The design pressure considered here is more than the normal 10% in excess of the operating pressure. Such a requirement may be of a typical condenser in a refrigeration system.

(a) *Shell side*

Material of construction: carbon steel (Corrosion allowance – 3 mm)

Number of shells	- 1
Number of passes	- 1
Fluid	- liquid
Working pressure	- 0.33 N/mm^2
Design pressure	- 0.50 N/mm^2
Temperature inlet	- 30°C
Temperature outlet	- 50°C

Segmental baffles (25% cut) with tie rods and spacers.

Head

Crown radius	- 400 mm
Knuckle radius	- 40 mm
Shell flange	- female facing
Gasket	- flat metal-jacketed asbestos filled
Bolts	- steel.

Nozzles – inlet and outlet – 75 mm

Vent	25 mm
Drain	25 mm
Opening for relief valve	50 mm

Permissible stress for carbon steel – 95 N/mm^2

Permissible stress for bolt material – 140.6 N/mm^2

(b) *Tube side*

Tube and tube sheet material	- stainless steel (SS 304)
Number of tubes	- 54
Outside diameter	- 18 mm
Length (maximum U)	- 12 m
Pitch (square)	- 25 mm
Fluid	- Gas

Working pressure	- 19 N/mm^2
Design pressure	- 21.5 N/mm^2
Inlet temperature	- 150°C
Outlet temperature	- 55°C
Permissible stress	- 100.6 N/mm^2

(c) ***Channel and channel cover***

Material – carbon steel (IS-2062)

Joint with tube sheet – ring facing

Gasket – steel jacketed asbestos

Nozzle – inlet and outlet dia.	- 75 mm
Permissible stress, f	- 95 N/mm^2

9.4.1 Shell Side

9.4.1.1 Shell Diameter

The exchanger is having a tube lay out on a square pitch. Therefore, from Eq. (9.3), we have,

$$a = S_T^2 = (25)^2 = 625 \text{ mm}^2$$

$$a_n = 2 \times 54 \times 625 = 67500 \text{ mm}^2$$

Area of the shell from Eq. (9.4)

$$A_s = (\pi/4)D^2 = a_n/\beta; \text{ We consider } \beta = 0.7$$

$$D_s = \sqrt{\frac{67500 \times 4}{0.7 \times \pi}}$$

$$= 350.4 \text{ mm}$$

The shell selected will be with diameter, D_s = 400 mm and will be fabricated out of plate.

9.4.1.2 Shell Thickness

The above diameter is the inside diameter. The shell thickness is then given by Eq. (5.3b);

$$t = \frac{pD_i}{(2fJ - p)}$$

The joint efficiency, J = 0.85 (Spot radiography)

$$= \frac{0.5 \times 400}{(2 \times 95 \times 0.85 - 0.5)}$$

$$= 1.24 \text{ mm}$$

As seen from Table 9.3 the minimum thickness is 6.3 mm exclusive of the corrosion allowance. Therefore, the shell of thickness, t = 8 mm will be used taking into consideration corrosion allowance.

9.4.1.3 Nozzle (Inlet and Outlet)

The thickness of nozzle, t_n

$$t_n = \frac{pD_i}{(2fJ - p)}$$ The joint efficiency, $J = 0.85$ (spot radiography)

$$= \frac{0.5 \times 75}{(2 \times 95 \times 0.85 - 0.5)}$$

$$= 0.23 \text{ mm}$$

The actual thickness with corrosion allowance would be 4 mm.

One can fabricate a nozzle from a seamless pipe, thereby the thickness may be reduced. In practice, a standard pipe having a higher next wall thickness is considered.

9.4.1.4 Head Thickness

The head used will be torispherical head. The thickness of the head is given by Eq. (5.23):

$$t_h = \frac{pR_cW}{(2fJ)} + c$$ The joint efficiency, $J = 0.85$

where,

$$W = (1/4)\left\{3 + \sqrt{R_c / R_1}\right\}$$

R_c = crown radius = 400 mm (shell I.D.)

R_1 = 6% R_c and for cold forming it is > 5 × t_h

= 5 × 8 = 40 mm

$$W = (1/4)\left\{3 + \sqrt{400/40}\right\}$$

$$= 1.54$$

Then, we have

$$t_h = \frac{pR_cW}{(2fJ)} + c$$ The joint efficiency, $J = 0.85$

$$= \frac{0.5 \times 400 \times 1.54}{2 \times 95 \times 0.85} + c$$

$= 1.9 \text{ mm} + 1.5 \text{ mm}$

$= 3.4 \text{ mm}$

The thickness of head, in practice, is kept as that of the shell, $t_h = 8$ mm.

9.4.1.5 Baffles

The baffles used will be of the transverse type with spacing = 400/5 = 80 mm.

The thickness of the baffles is 6 mm

The baffle cut is 25%.

The baffle length = 0.75 × 400 = 300 mm.

9.4.1.6 Tie Rods and Supports

Number of tie rods = 6

Diameter of the tie rod is 10 mm.

9.4.1.7 Flange Joint (Shell and Tube Sheet)

This will be made to satisfy the requirement of the flange joint between tube sheet and channel. The bolt size and the pitch circle diameter will be identical to that determined in section 9.4.2.5. Thus we have,

Number of bolts	16
Size of bolts	M48
Pitch circle diameter	525 mm.

9.4.1.8 Flange Thickness (Male and Female)

Gasket size	= 440 mm outside diameter
	= 416 mm inside diameter
Gasket dia	= (440 + 416)/2 = 428 mm
Gasket contact width (N)	= 24 mm

From Table 5.4 we have basic gasket seating width, b_0

$$b_0 = N/2 = 24/2 = 12$$

$$b = 2.5\sqrt{b_0} \quad \text{(from Fig. 5.34)}$$

$$= 8.66$$

The seating stress (Y_a) = 53.4 N/mm^2

The gasket factor (m) = 3.75

From Eq. 5.36, we have

$$W_{m1} = \pi b G Y_a$$

$$= \pi \times 8.66 \times 428 \times 53.40$$

$$= 6.2 \times 10^5 \text{ N}$$

From Eq. (5.37), we have

$$W_{m2} = \pi \times (2 \times 8.66) \times 428 \times 3.75 \times 0.5 + (\pi/4) \times (428)^2 \times 0.5$$

$$= 1.15 \times 10^5 \text{ N}$$

The flange thickness is then computed with the help of Eq. (5.42)

$$k = \frac{1}{\left(0.3 + \dfrac{1.5 W_m h_G}{HG}\right)}$$

W_m = 6.2 × 105 (higher value of W_m is considered)

h_G = (525 – 428)/2 = 48.5

$H = (\pi/4)\, G^2 p = (\pi/4)\,(428)^2\,(0.5)$
$= 71936.6$

$G = 428$

We, therefore, have $k = 0.565$

The flange thickness is then given by

$$t_f = G\sqrt{(p/k\,f)}$$

$$= 428\sqrt{(0.5/0.565 \times 95)}$$

$$= 41.3 \text{ mm}$$

Use the flange of thickness 45 mm inclusive of corrosion allowance.

9.4.2 Tube Side

9.4.2.1 Thickness of Tube

From Eq. (5.3 b), we have

$$t_f = \frac{pD_o}{(2f\,J + p)}$$

$$= \frac{(21.5 \times 18)}{\left(2 \times (100.6) + 21.5\right)} \qquad \text{(J = 1, seamless tube)}$$

$$= 1.74 \text{ mm}$$

The tube of 2 mm thickness is acceptable (IS-4503). This is checked for external pressure.

9.4.2.2 Tube Sheet

The tube sheet is held between shell flange and the channel (refer enclosed drawing plate 2). The joint on the shell side is of male and female facing type and on the channel side it is the ring facing type since the pressure on the channel side is high.

Eq. (9.5) gives tube sheet thickness.

$$t = FG\sqrt{0.25p/f}$$

$$= 1.25 \times 380 \times \sqrt{0.25 \times 21.5/100.6}$$

$$= 109.8 \text{ mm}$$

$$\text{say,} = 110 \text{ mm}$$

9.4.2.3 Channel and Channel Cover

It is proposed to make the channel and cover out of a single plate.

Eq. (9.9) gives thickness of channel plate.

$$t = G_c\sqrt{kp/f}$$

k = 0.3 as ring type gasket is used.

G_c = 380 mm

p = 21.5 N/mm^2 and f = 95 N/mm^2

Therefore, the thickness, t is:

$$t = 380\sqrt{0.3 \times 21.5/95}$$

$$= 99.02 \text{ mm}$$

Use the plate of thickness 100 mm inclusive of corrosion allowance.

9.4.2.4 Flange Joint between Tube Sheet and Channel

Gasket and bolt design

G = 380 mm; Ring gasket width = 22 mm. We then have,

$b_0 = w/8 = 22/8 = 2.75$ mm (Table 5.4) = b

$Y_a = 126.6$ N/mm^2 (Table 5.4)

m = Gasket factor = 5.5

From Eq. (5.36), we have

$$W_{m1} = \pi b G Y_a$$

$$= \pi \times 380 \times 2.75 \times 126.6$$

$$= 4.16 \times 10^5, \text{ N}$$

From Eq. (5.37) we have,

$$W_{m2} = \pi \times (2b)\ Gmp + (\pi/4)G^2 p$$

$$= \pi \times (2 \times 2.75) \times 380 \times 5.5 \times 21.5 + (\pi/4) \times (380)^2 \times 21.5$$

$$= 32.14 \times 10^5, \text{ N}$$

The higher force is considered for calculation.

Bolt area $= W_{m2}/f = 32.14 \times 10^5/140.6$

$= 22859.2$ mm^2

Number of bolts = (mean dia. of gasket in cm)/2.5

= 38/2.5

= 15.2

Therefore, we have 16 bolts.

The bolt diameter $= \sqrt{22859.2 \times 4/(\pi \times 16)}$

= 42.65 mm.

Therefore, use bolt of diameter = 48 mm, i.e., M 48 for the service.

The bolt circle diameter = 380 + 22 + 2 × 48 = 498 mm.

Therefore, we used bolt circle dia. = 525 mm

9.4.2.5 Flange Thickness

From Eq. (5.42), we have

$$k = \frac{1}{\left(0.3 + \dfrac{1.5 W_m h_G}{H G}\right)}$$

$W_m = 32.14 \times 10^5$, N (higher value of W_m is considered)

$h_G = (525 - 380)/2 = 72.5$ mm

$H = (\pi/4)\ G^2 p = (\pi/4)\ (380)^2\ (21.5)$

$= 24.38 \times 10^5$ N

$G = 380$ mm

Substituting above values we have, $k = 1.47$

Therefore the thickness, t:

$$t = G\sqrt{p/(k\,f)}$$

$$= 380 \times \sqrt{21.5/(1.47 \times 100.6)}$$

(Please note the channel cover will be in SS and not in MS. One may use cladding to reduce the cost associated with SS).

$= 144.9$

or say, $= 145$ mm.

Thus, have 145 mm thick flange.

9.4.2.6 Nozzles

Inlet and outlet nozzle size is 75 mm. Nozzle is of seamless pipe. From Eq. (5.3a):

$$t_n = \frac{p D_i}{(2 f J - p)}$$

The joint efficiency, $J = 1$

$$= \frac{21.5 \times 75}{(2 \times 95 \times 1 - 21.5)}$$

$= 9.6$ mm

Use thickness 12 mm inclusive of corrosion allowance.

Considering the size of the nozzle and pressure rating it is necessary to provide for a reinforcement pad on the channel cover refer to Section 5.8.3.2. The area required to be compensated for each nozzle is:

$$A = d \times t_h = 75 \times 100 = 7500 \text{ mm}^2 \quad (100 = \text{cover thickness})$$

Compensation will be available from the additional thickness of channel cover and nozzle. It is proposed to use a 50 mm thick pad as shown in plate II and it can be shown by calculation based on area method that it is adequate.

References and Recommended Readings

A.I.Ch.E., *Standard Testing Procedure, Heat Exchangers – Section I, Section II*, New York (1978).

Ackley, R.J., 'Heat Transfer through Glassed Steel, Chemical Engineering', **66** (8) 181 (1959).

Bailey, K.M., 'Understand Spiral Heat Exchangers, Chemical Engineering Progress', **90** (5) 59 (1994).

Bevevino, I.W., 'Tube to Tube Sheet Welding Technique', *Chemical Engineering Progess,* **70** (7) 71 (1974).

Boyer, J., and Trumpfheller, G. 'Specification Tips to Maximize Heat Transfer', *Chemical Engineering,* **100** (5) 90 (1993).

Brown, T.R., 'Use These Guidelines for Quick Preliminary Selection of Heat Exchanger Type', *Chemical Engineering,* **93** (3) 107 (1986).

Burley, J.R., 'Don't Overlook Compact Heat Exchangers', *Chemical Engineering,* **98** (8) 90 (1991).

Carlson, J.A., *Understand the Capabilities of Plate and Frame Heat Exchangers, Chemical Engineering Progress,* **88** (7) 26 (1992).

Chugh, R.D., *Methods of Tube to Tube Sheet Attachment in Heat Exchangers, Chemical Age of India,* **26** (6) 441 (1975).

Clark, D.F., *Plate Heat Exchanger Design and Recent Development, The Chemical Engineer,* **285** 275 (1974).

Copper, A., 'Recover More Heat with Plate Heat Exchangers', *The Chemical Engineer,* **285** 275 (1974).

Corsi, R, 'Specify Bayonet Heat Exchangers Properly', *Chemical Engineering Progress,* **88** (7) 32 (1992).

Foxall, D.H., and Gilbert, P.T., 'Selecting Tubes for CPI Heat Exchangers, Part I' *Chemical Engineering,* **83** (6) 99 (1976); Part II, 83 (8) 147 (1976); Part III, 83 (10) 133 (1976).

Frass, A.P., and Ozisik, M.N., *Heat Exchanger Design,* John Wiley, New York (1965).

Frenier, W.W., and Barber, S.J., 'Choose the Best Heat Exchanger Cleaning Method', *Chemical Engineering Progress* **94** (7) 38 (1998).

Greene, B., 'A Practical Guide to Shell and Tube Heat Exchangers', *Hydrocarbon Processing,* **78** (1) 79 (1999).

Hills, D.E.G., 'Graphite Heat Exchangers, I', *Chemical Engineering,* **81** (27), 80 (1974); II Chemical Engineering., **82** (2), 116 (1975).

Kern, D.Q., *Process Heat Transfer*; McGraw-Hill, New York (1950).

Kumar, Vijay, *A Systematic Approach to the Design of Exchanger Tube Sheet, Chemical Age of India,* **25** (4), 244 (1974).

Lamont, G.A., 'Consider New Plate Heat Exchanger Gasket Options', *Chemical Engineering Progress,* **89** (11) 101 (1993).

Lord, R.C., Minton, P.R., and Slusser, R.P., 'Design of Heat Exchangers', *Chemical Engineering,* **77** (2) 96 (1970).

Ludwig, R.E., *Applied Process Design for Chemical and Petrochemical Plants,* Vol. 3, Second Edition, Gulf Publishing Co., Houston, (1977).

Lunsford, K.M., *Understand the Use of Brazed Heat Exchangers, Chemical Engineering Progress,* **92** (11) 44 (1996).

Mahajan, K.K., *Design of Process Equipment* (Selected Topics), Pressure Vessel Handbook Publishing Co., USA (1979).

Mehra, D.K., *Shell and Tube Heat Exchangers, Chemical Engineering,* **90** (15) 47 (1983).

Mukherjee, R., *Don't let Baffling Baffle You, Chemical Engineering Progress,* **92** (4) 72 (1996).

Mukherjee, R., *Effectively Design Air Cooled Head Exchangers, Chemical Engineering Progress,* **93** (2) 26 (1997).

Narayan, S., *Heat Exchange: Their Applications in Process Industry, Chemical Age of India,* **20** (8) 673 (1969).

Perry, R.H., Green, D.W., and Maloney, J.O., *Perry's Chemical Engineers Handbook,* 6th Edition, International Edition McGraw-Hill Book Co., Singapore (1984).

Redman, J., 'Non-etallic Heat Exchangers', *The Chemical Engineer,* **459** 17 (1989).

Rubin, F.L., 'Practical Heat Exchangers', *Chemical Engineering Progress,* **64** (12) 44 (1968).

———., 'What's the Difference between TEMA Exchangers classes', *Hydrocarbon Processing,* **59** (6) 92 (1980).

Solan, M., 'Designing and Trouble Shooting Plate Heat Exchangers', *Chemical Engineering,* **105** (5) 78 (1998).

Someah, K., 'On line Tube Cleaning: The Basics', *Chemical Engineering Progress,* **88** (7) 39 (1992).

Trom, L., 'Consider Plate and Spiral Heat Exchanger', *Hydrocarbon Processing,* **69** (10) 75 (1990).

Vardarajulu, R., 'Have a Look at Shell and Tube Heat', *Chemical Age of India,* **27** (5), 479 (1976).

Westernan, V., 'Bellows for Heat Exchangers', *Processing,* **21** (5), 47 (1975).

Whitcraft, P.K., 'Utilizing Heat Exchanger Tubing Specifications', *Hydrocarbon Processing* **78** (1) 87 (1999).

Wilkins, C., 'Spiral Heat Exchanger Gets Around Blockages', *Process Engineer,* **58** (3) 89 (1977).

Yokell, S., 'Double Tube Sheet Heat Exchanger Design', *Chemical Engineering Deskbook,* (87) (1977).

10

Chapter

Evaporators and Crystallisers

10.1 EVAPORATORS

10.1.1 Introduction

Evaporators are used in the chemical industry for the concentration of liquids. The operation is performed normally by the use of low pressure, dry and saturated steam. For special purposes organic vapours or fuel gases may be adapted. In most cases, water, which is the solvent, is evaporated. With the vapourisation of a part of the water, the useful product is a concentrated solution. If the solution contains dissolved solids, the resulting strong liquor may become saturated, so that crystals are formed.

In a majority of the evaporators, the heat transfer is through tubular surfaces. The body of the evaporator is in the form of a cylindrical shell, in which the tubes are placed. In order to create an effective heat transfer, the liquid must flow across the heat transfer surface. As a result of the generation of the vapour, a natural circulation is created. In evaporating viscous liquids or in cases where natural circulation is inadequate, forced circulation is created by the use of a pump or by a propeller. With forced circulation, it is possible to control the velocity of flow across the heating surface. As the solution boils, a more or less stable foam is formed on the surface. The vapours generated have, therefore, liquid droplets in suspension. Sufficient vapour space (known as vapour release chamber) must be available for releasing the water vapour from the liquor. For separation of liquid droplets, various devices, such as mesh pads, baffles and impringement or centrifugal separators, are adopted.

10.2 TYPES OF EVAPORATORS

Evaporators can be classified according to the heating medium, heating surface, its disposition and the method of circulation. Solar heat, gases, organic vapours and steam are used as a heating media.

The most common heating medium, namely, steam can be used in jackets attached to vessels, as in the case of kettles or pans and in the horizontal shell and tube type of heat exchangers. Standard evaporators usually consist of vertical tubes placed in a cylindrical vessel, known as calendria, which have a provision for vapour space, either as an extension of the calendria or as a separate vessel. Evaporators can also be made with either horizontal or inclined tubes. Pipe coils, oblong or helical, submerged in the solution contained in a vessel are also used as evaporators. Table 10.1 gives a list of different types of evaporator along with the feed condition, retention times, etc. Details of construction of some evaporators are given in the following sections.

10.2.1 Jacketed Pans and Agitated Film Type Evaporators

The simplest method of concentrating liquors is by use of jacketed pans, where steam is used in the jacket to drive off excess moisture. This is particularly useful when small quantities are to be handled. The process of evaporation can be substantially improved if, instead of a jacketed pan, a jacketed vertical cylinder with an agitator is used. This is known as agitated film type evaporator. This consists of a vertical cylindrical jacketed evaporating section. A flat-blade type vertical turbine agitator located centrally, with a clearance of about 1 mm or less from the wall. The agitator is driven by a driving system at the top. The feed enters at the top of the evaporating section and flows down the walls. While it moves down it is vapourised, the concentrated liquid being discharged at the bottom of the evaporator. As the vapour rises, it is thrown against radially mounted fins which remove the entrained liquid droplets. As may be seen from Table 10.1, these evaporators are most versatile and are able to handle viscous, heat sensitive feeds. They are expensive and need considerable maintenance.

10.2.2 Standard Vertical Short Tube Evaporators (Calendria Type)

These evaporators (Fig. 10.1) comprise of a large vertical cylindrical vessel or drum, ranging from 1 metre to 6 metres in diameter, with a short vertical tube bundle, consisting of 25 mm to 75 mm outside diameter tubes, 750 mm to 2000 mm long, located at the bottom portion of the vessel. The liquor to be evaporated is inside the tubes and the steam is outside. The tube bundle is held between two tube sheets, placed horizontally and bolted or welded to shell flanges. A central downtake promotes circulation. The area of the central

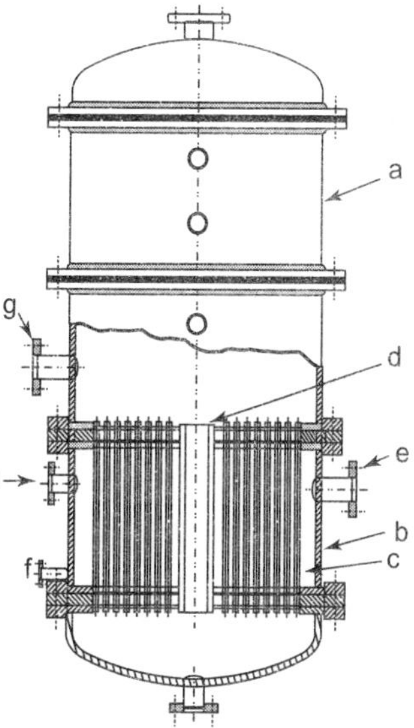

Fig. 10.1 Standard Short Tube Vertical Evaporator: (a) Drum, (b) Calendria, (c) Tubes, (d) Downtake, (e) Steam Inlet, (f) Condensate Outlet, (g) Feed (h) Vent

Table 10.1 Evaporator Types and Applications

Evaporator Type	*Feed Condition*							
	Very Viscous (above 2000 cP)	*Medium Viscosity (100–1000 cP)*	*Low Viscosity to Water (max. 100 cP)*	*Foaming*	*Sealing or Fouling*	*Crystal Producing*	*Solid in Suspension*	*Suitable for Heat Sensitive Products*
Calendia[2] (short vertical tube)		A	A	A	A	A	A	No
Forced circulation		A	A	A	A	A		Yes
Natural circulation			A	A				No[3]
Agitated film (vertical or horizontal)	A	A	A	A	A	A	A	Yes
Tubular (longtube)			A					Yes
Falling film Rising film			A	A				Yes

A = Applicable to conditions noted.

Notes:

1. Viscosities are at operating temperatures.
2. Special disengagement arrangement required for foamy liquids.
3. May be used in special cases.

downtake varies from about 40% to 100% of the total cross-sectional area of the surrounding tubes. Multiple downtakes promote an even circulation, which reduces the possibility of localized scale formation. The layout of the tubes is shown in Fig. 10.2. A baffle system is used within the steam chest, so that there is a uniform distribution of steam. A propeller can be installed in the dished or conical bottom to increase the rate of circulation. Since scaling occurs inside the tubes, it is possible to use the standard evaporator for more vigorous services than the horizontal tube evaporator. The bottom of the cylindrical vessel may be a dished head or a conical head. Considerable vapour space is provided above the tube bundle.

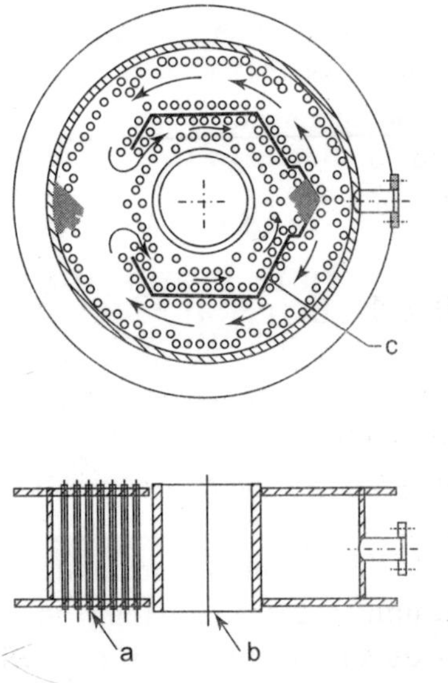

Fig. 10.2 Tube Layout of Short Tube Vertical Evaporator: (a) Tube, (b) Downtake, (c) Baffle

The main components of this evaporator are, therefore, (1) tube bundle with the surrounding shell known as calendria, (2) evaporator body or vapour release chamber and (3) two heads, one at the bottom and one at the top. The calendria is designed as a shell with an internal steam pressure, with the tube sheets of a thickness based on Eq. 9.5. It is held between the vapour chamber and the bottom head by means of flanged joints or is welded to the vapour chamber. For large diameters, a spherical vapour release chamber is more economical. The vapour chamber may be made up to cylindrical sections welded to form the necessary height or it may be made up of flanged corrugated sections, bolted together. A head is also flanged and is bolted at the top. Nozzles are provided for inlet of the solution, vapour outlet, liquor outlet, steam inlet, condensate outlet, drain, vent, manhole, sight glass, level gauge, etc. Supporting legs are of conventional structural sections. Nozzles and supports are welded to the shell.

10.2.3 Basket Type Evaporators

These (Fig. 10.3) are somewhat similar to the standard vertical evaporator. However, the tube bundle in this case is suspended on brackets, and the downtake is formed by the annular space between the bundle and the shell, instead of a central downtake. These evaporators are not used for liquids with high viscosities or large rates of scaling.

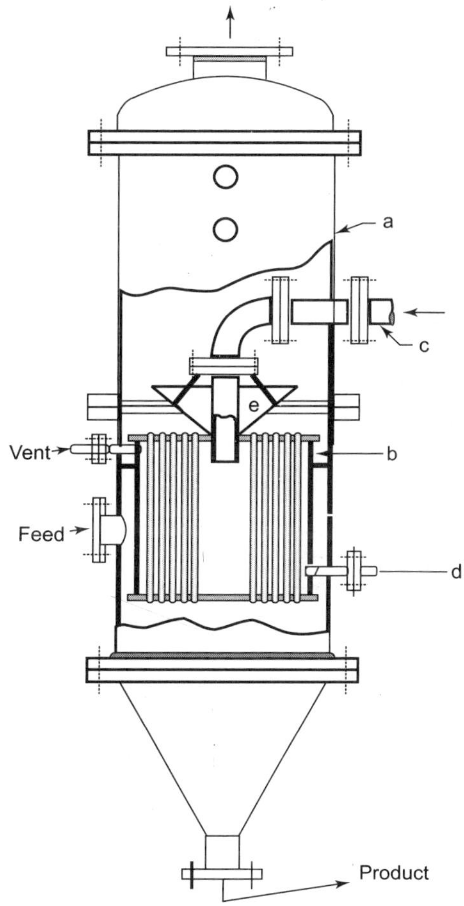

Fig. 10.3 Basket Type Short Tube Vertical Evaporator: (a) Drum, (b) Tube Bundle, (c) Steam Inlet, (d) Condensate Outlet, (e) Antisplash Conical Baffle

10.2.4 External Calendria Vertical Tube Evaporators

These consist of a short tube, single effect vertical calendria, but with a separate vapour drum placed by the side of the calendria (Fig. 10.4). The liquid level is kept low, and the foaming is, therefore, confined largely to the inside of the tubes. These tubes contain a mixture of vapour and liquid travelling vertically with high speed. The external calendria can be easily opened and cleaned.

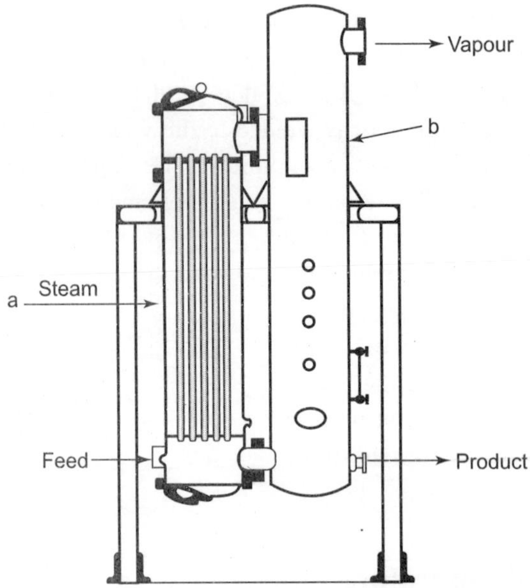

Fig. 10.4 External Calendria Vertical Short Tube Evaporator: (a) External Calendria, (b) Vapour Drum

10.2.5 Long Tube Vertical Evaporators

These evaporators (Fig. 10.5) are characterized by relatively long vertical tubes of 30 to 50 mm diameter and ranging from 3 metres to 6 metres length. The tubes are enclosed in a shell. The liquor is filled inside the tubes and steam is fed in to the shell. The solution is introduced at a fixed rate. In controlled level type evaporators the liquid level is maintained at one-third to one-half of the height of the tubes. The vapour discharges either into a vapour head directly above the tube bundle or into a separate vapour drum. The liquid is returned to the bottom of the shell through external downtakes. (There can be more than one also).

In the film-type evaporator the liquid level is not controlled. The liquor makes only a single pass through the tubes. Inclined tube evaporators are similar, except that the tubes are inclined at an angle of 45°. The liquid is circulated through an external downtake.

10.2.6 Horizontal Tube Evaporators

These evaporators (Fig. 10.6) have a vertical cylindrical or a square shell to which are attached two square steam chests. The steam is admitted inside the tubes and the liquor is outside, submerging the tube. The horizontal tubes extend between the two tube plates to which they are fastened by the packing plates or by expansion. In one method, the holes in the tube sheets are countersunk to receive conical gaskets which are fitted over the ends of the tubes and are forced in by packing plates mounted on studs from the tube plate. This allows tubes to expand freely without causing stress on the tube sheet. Tubes are of 20 to 30 mm diameter. The body of the evaporator is from

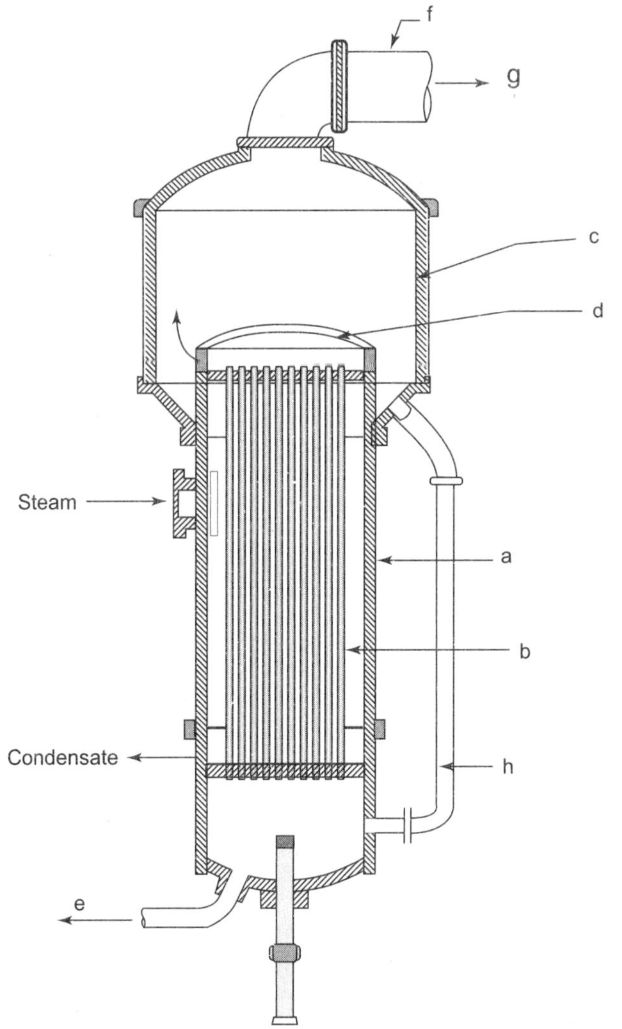

Fig. 10.5 Long Tube Vertical Evaporator: (a) Shell, (b) Tubes, (c) Drum, (d) Baffle, (e) Liquid Outlet, (f) Vapour Outlet (g) liquid Inlet, (h) External Downtake

1 metre to 4 metres diameter and 2½ metres to 4 metres height. These evaporators are unsatisfactory for fluids which form a scale or deposit salt. They are suitable for processes in which the final product is a liquor instead of a semi solid.

10.2.7 Forced Circulation Evaporators

These evaporators are similar to long tube vertical evaporators, except that a centrifugal pump is placed between the external downtake from the vapour drum and the inlet to the tube bundle. Due to

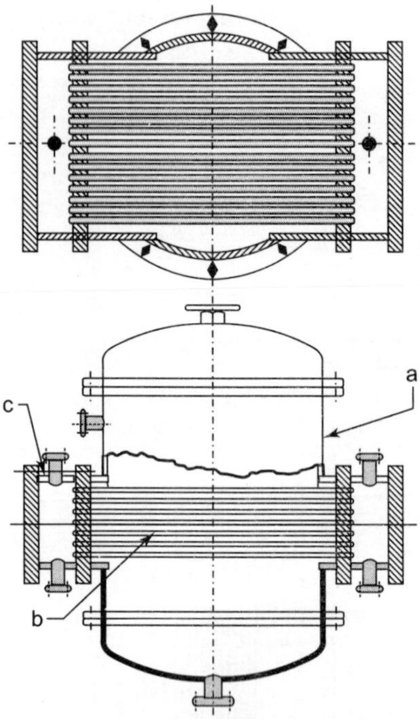

Fig. 10.6 Horizontal Tube Evaporator: (a) Drum, (b) Tubes, (c) Steam Chest

high velocities and consequently high rates of heat transfer, less heating surface is required. This is particularly applicable for a heating surface formed out of expensive materials.

The liquor which emerges from the heating elements at high velocities is directed into the vapour drum, where the vapour is separated from the liquid. The vapour drum may be either a separate drum or is located on top of the tube bundle. The liquid from the drum is fed to the pump with the help of the downtake. Forced circulation evaporators may not be as economical in operation as natural circulation evaporators, but they are necessary where the concentration problem involves a solution with poor flow characteristics.

10.2.8 Agitated Thin Film Evaporators

For certain evaporation applications, particularly those involving heat sensitive products, liquids with suspended solids, viscous materials, or chemical constituents that exhibit fouling or foaming tendencies, mechanically agitated, thin-film evaporators are often selected over the more conventional evaporators, because of their better performance and process economics.

Thin film evaporators consist of an externally heated cylindrical vessel, a rotor and a drive system (Fig. 10.7). The feed is supplied near the top of the vessel and is mechanically transported by the rotor and also by gravity so as to create a helical flow path. The rotor is so designed that it retains the feed only in the annular space in contact with the vessel wall, so that heat is transferred to the thin film of feed from the jacket. Flow of vapours can be co-current or counter-current to liquid flow. Different designs of impellers are possible, for example, spring loaded scrapper, etc.

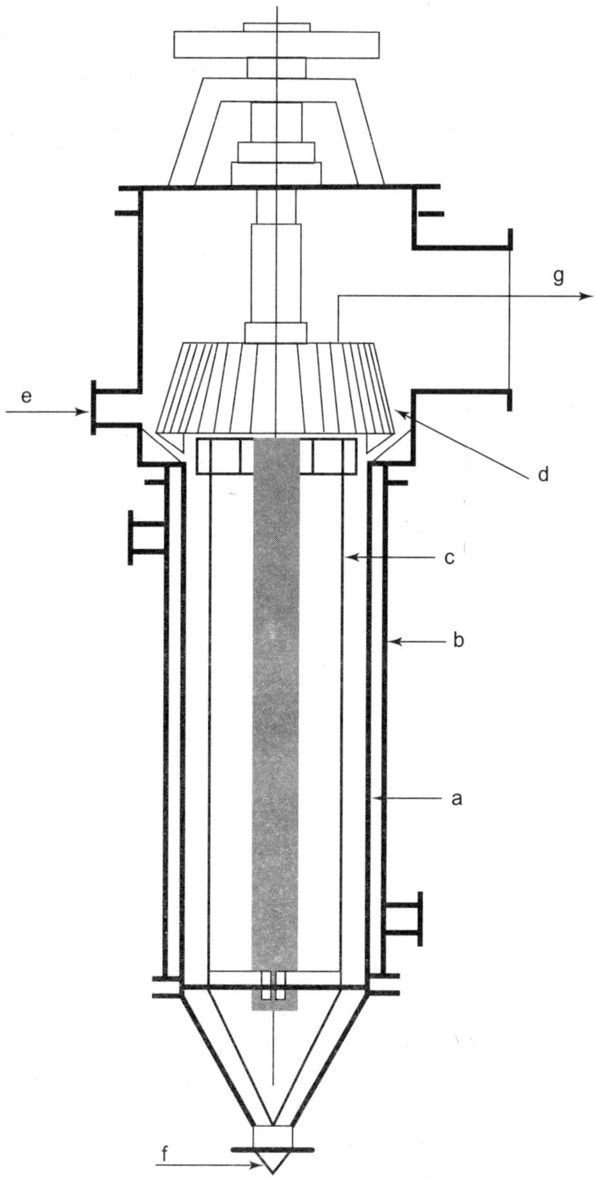

Fig. 10.7 Thin Film Evaporator: (a) Evaporation Shell, (b) Heating Jacket, (c) Rotor, (d) Separator, (e) Feed, (f) Product Outlet, (g) Vapour Outlet

10.2.9 Multiple Effect Evaporator

A number of evaporators are used in series, in order to economize the use of steam (Fig. 10.8). The vapours from each evaporator or the effect are led to the steam chest of the next, while the solution flows in the same direction, losing water as it flows. There is no theoretical limit to the number of evaporators which can be used in series. A practical limit is reached when heat transfer rates become low due to inadequate temperature difference and an economic limit is reached when the plant cost exceeds the saving of steam.

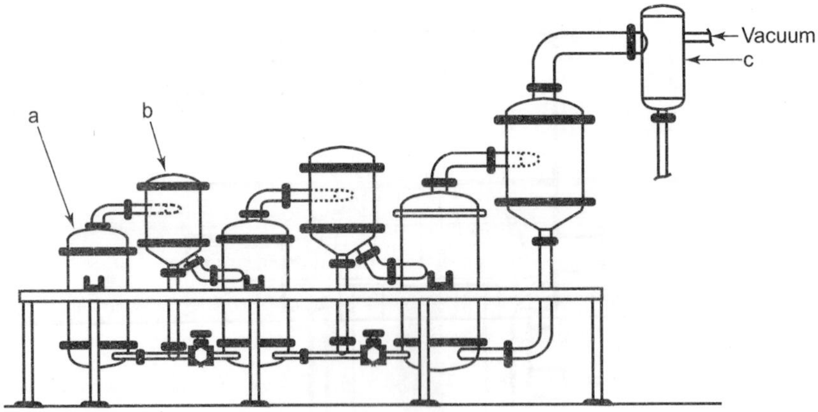

Fig. 10.8 Multiple Effect Evaporator: (a) Evaporator, (b) Separator, (c) Barometric Condenser

In order to utilize the vapours from each stage as steam in a subsequent stage, it is necessary to use lesser steam pressures in each stage. This involves passing the vapours to a surface condenser or a barometric condenser, where the vapours are finally condensed, and produce the necessary vacuum. In addition, a suitable pump is provided to remove air and non-condensables entrained by the liquor or entering through leakage.

10.3 ENTRAINMENT SEPARATORS

These are used in most evaporators to obtain dry vapours. Entrainment results wherever a vapour is generated from a liquid. When this occurs the vapour carries with it varying quantities and sizes of liquid droplets. Separators provide a means for separating a vapour from the liquid with a minimum liquid carry-over or dry vapours. Various methods are adopted for such separation.

10.3.1 Vapour Release Chamber

A large chamber is used to reduce the velocity of the vapour stream. This enables the droplets to settle out by gravitational action. The vapour release drum may either be placed just above the bundle shell or a separate unit placed adjoining to the tube bundle shell, being connected to it by a large pipe.

It may not be economically practical to make the vapour head large enough to accomplish the entire decontamination of the vapour. Further, increasing vapour space decreases entrainment of larger drops, but has no effect on the small drops.

The vapour disengagement velocity from a boiling liquid surface should not normally be more than 300 mm per second for normal solutions at atmospheric pressure and may be about 30 mm per second with crude solutions. Even allowing for sufficient vapour disengagement space, it is a common practice to provide spray traps. These traps consist merely of a series of baffles giving rapid changes of direction to the vapour stream, with a spray of clean liquid/solvent.

10.3.2 De-mister

Pads of finely woven wire (Fig. 10.9) set in the vapour release chamber at right angles to the vapour flow are known as demisters and are used for liquid entrainment separation. As the vapour and its entrained liquid pass through the pad, the liquid particles agglomerate, eventually falling back into the vapour release chamber. A highly purified dry vapour leaves the top of the pad.

Application of such pads may be difficult for vapours with suspended solids, fibres or scale forming materials, which will block the wire mesh. In such cases, washing facilities at proper intervals may be provided. The voidage is >95%.

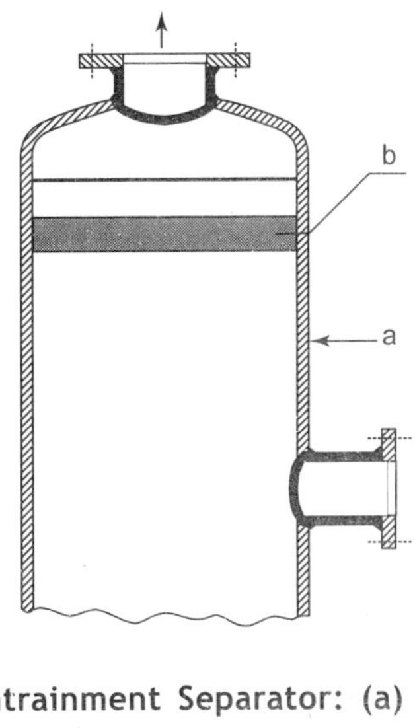

Fig. 10.9 Entrainment Separator: (a) Drum, (b) Pad

10.3.3 Vapour Release Drum Size

The size of drum provided above the tube bundle in most of the evaporators, is decided by three important considerations. They are: (a) the foaming of the liquid in the evaporator, (b) the vapour velocity, and (c) entrainment separation.

Foaming takes place above the liquid level and occupies a certain space of the vapour drum. The vapour velocity sets the minimum drum diameter. The following equation helps to determine the drum diameter, R_d.

$$R_d = \frac{(V/A)}{0.0172 \times \left(\frac{\rho_L - \rho_V}{\rho_v}\right)^{1/2}} \tag{10.1}$$

where, V — volumetric flow rate of vapour in m^3/sec

A — area of cross section of the drum in m^2

ρ_L — liquid density in kg/m^3

ρ_v — vapour density in kg/m^3.

The values or R_d may be taken as 0.5 for drums without entrainment separators. For drums having a wire mesh as entrainment separator device, R_d may be taken as 1.3.

In Eq. 10.1, if the other values are known, it is possible to get the value of the area of cross-section of the drum and hence its diameter.

The height of the drum, considered as the disengaging height which helps the process of liquid separations, is based on the drum diameter. A graph showing the relationship between the drum diameter (D) and the disengaging height (H_D) is recommended for many applications. Figure 10.10 shows the requirements of liquid level, vapour space, etc.

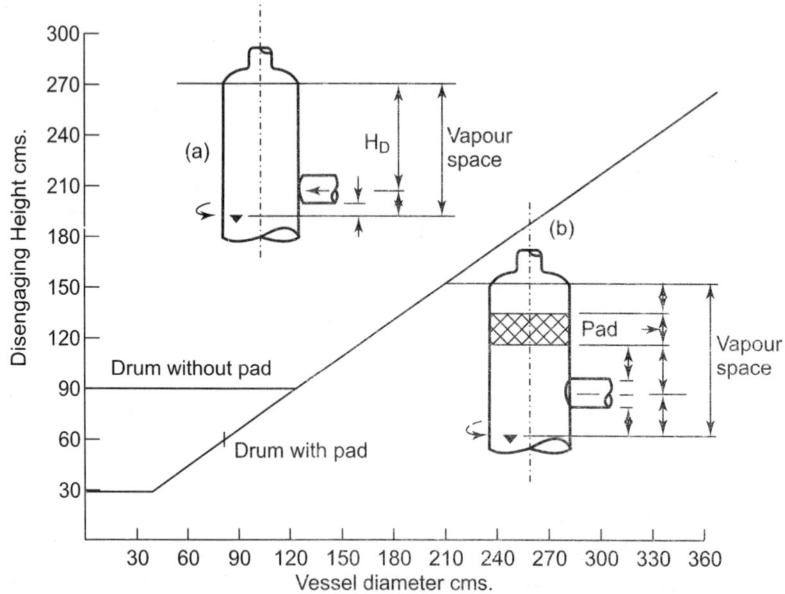

Fig. 10.10 Drum Diameter Disengaging Height Requirements: (a) Drum without Pad, (b) Drum with Pad

A thumb rule commonly employed in evaporator design of this kind is that the height of the vapour space above the calendria should not be less than one vessel diameter and the bottom space below calendria should be one-fourth of vessel diameter.

In cases where the entrainment separator forms an independent unit, the main drum can have a shorter disengaging height.

10.3.4 Centrifugal Separator

This is a separate drum in which the vapours are admitted tangentially and are made to flow in a helical path by the use of baffles. The vapours leave either from the top of the drum or through a central pipe [Fig. 10.11(A)]. A centrifugal type baffling system is shown in Fig. 10.11(B) fitted at the top of the drum. The vapours enter from the central passage and are diverted by the baffles separating the liquid in the process.

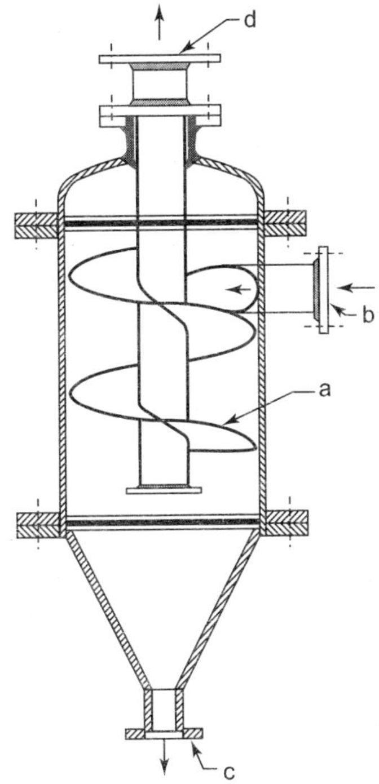

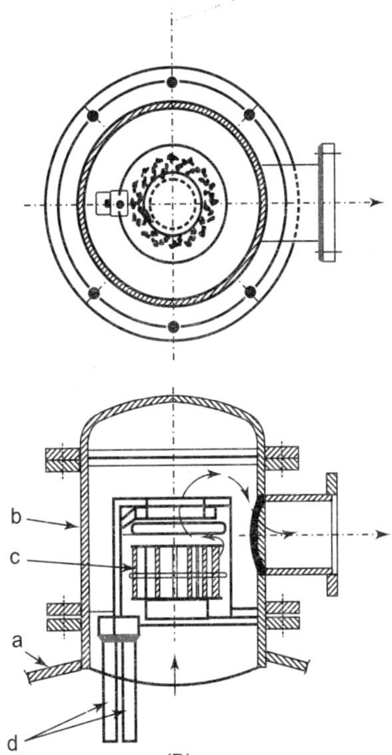

Fig. 10.11 Entrainment Separators:
(A) Helical Baffle Type;
(a) Baffles, (b) Vapour Inlet,
(c) Liquid Outlet, (d) Vapour Outlet

(B) Centrifugal Type;
(a) Drum, (b) Baffles,
(c) Vapour Outlet
(d) Liquid Outlet

10.4 MATERIALS OF CONSTRUCTION

Evaporator bodies are fabricated from mild steel, which is least expensive and is also easy to fabricate. Materials like cast iron, monel, inconel and stainless steel are used when corrosive action is to be prevented. Tubes are made of copper, nickel, aluminium or stainless steel, and tube sheet may be of cast bronze, nickel clad steel or stainless steel.

10.5 DESIGN CONSIDERATION

Evaporator drums generally operate under vacuum. These are designed for an external pressure of 0.1 N/mm^2 (see Section 5.8.1.3). The bottom head may be conical in many cases and may be designed for similar pressure rating [Eq. 5.22 and Section 5.8.2.2(b)]. The top head may be flanged or flared and dished shape or conical. The calendria which has the tubular heating surface is designed as a shell and tube heat exchanger (see Chapter 9). Since steam under pressure is usually accepted as the heating medium, the design is based on the pressure of steam. The entire evaporator body must be rigid. The conical head, the calendria and the vapour drum are connected by flanged joints or directly welded. The vapour drum may be made up of separate cylindrical pieces and joined by flanges. Large nozzles like manholes and sight glasses must be reinforced with appropriate compensating rings. Supports may be placed below the brackets welded to the vapour drum or to the calendria. External calendria is also designed as a shell and a tube heat exchanger.

10.6 CRYSTALLISERS

10.6.1 Introduction

Crystallisation is a process in which solid crystalline phases separate from liquids. The object of the process is usually the recovery of the solute from the solvent. The performance of the process is evaluated by the size, shape, structure and purity of the solute. The following equipment is used for the operation.

10.6.2 Tank Crystalliser

This is presently the simplest and perhaps the most economical unit. It is essentially a cylindrical vessel or pan which is cooled by water. Hot saturated liquor is supplied to the unit and slowly cooled. High supersaturation at the cooling surface is usually unavoidable and control is difficult. This normally results in the cooling surfaces becoming rapidly fouled with adhering crystals. The surface, therefore, might need frequent washing or scraping. Cooling can be effected either by a jacket or a coil or by the use of a separate single pass heat exchanger. An agitator is used to improve heat transfer. Jackets are preferred to coils, because the latter tend to become coated with a hard crystalline deposit and cease to function efficiently. If a cooling jacket is employed, the inner cold surfaces of the crystalliser should be as smooth as possible. This minimises crystal build-up on the cold surfaces.

10.6.3 Vacuum Crystalliser

In this unit a high vacuum is drawn in a vessel by means of a steam jet ejector or any other vacuum generator with a barometric condenser. Circulation of the liquor is effected by an agitator or a separate circulating pump.

(a) ***Draft tube type (Fig. 10.12)*** It is used frequently for vacuum cooling or chemical reaction crystallisation. The feed is supplied near the agitator and is thoroughly mixed. The agitator shaft can be made to enter from either the bottom or from the top. A draft tube surrounding the agitator helps to create circulation and prevents the short circuiting of feed. Such a unit is simple and easy to operate and, if well designed, can work for very long periods between washouts.

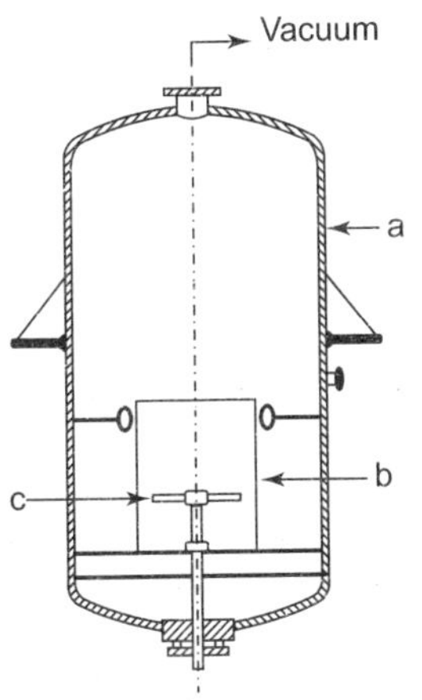

Fig. 10.12 Draft Tube Type Crystalliser: (a) Shell, (b) Draft Tube, (c) Agitator

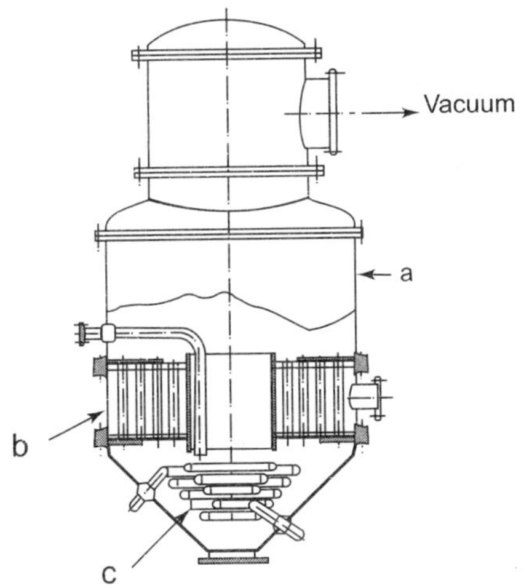

Fig. 10.13 Evaporative Type Crystalliser: (a) Drum, (b) Calendria, (c) Steam Coil

(b) ***Evaporative type (Fig. 10.13)*** The calendria type evaporator preferably with short wide tubes and a large central downtake, with the bottom in the form of a conical head can be used as a crystalliser. A barometric leg discharge can be used for vacuum production.

A unit where forced circulation of the feed is attained by a pump consists of a vessel which is large enough for vapour release and with the liquid level high enough, so as to provide a volume sufficient for retention of growing crystals (Fig. 10.14). Suction from the lower portion of the body passes through a circulation pump and through a heat exchanger back to the vessel through a tangent inlet or a vertical inlet.

This type of crystallisation equipment is probably the most widely used and available in vessel sizes from 1 m to 12 m in diameter. It is normally the least expensive equipment available, particularly when substantical amount of evaporation is required.

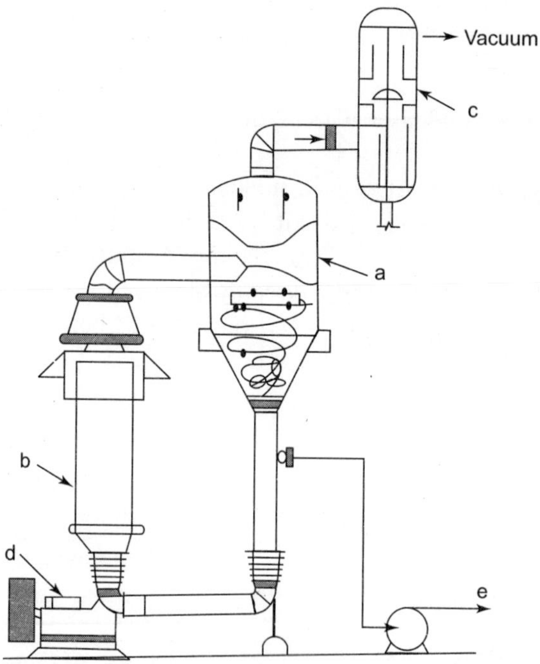

Fig. 10.14 Evaporate Type Crystalliser (Forced Circulation Type): (a) Evaporator Body, (b) Heat Exchanger, (c) Barometric Condenser, (d) Circulation Pump, (e) Product Slurry Discharge

10.6.4 Growth Type Crystalliser

This is used for the production of larger crystals with a narrow size range and for truly close control in the crystallisation process. In such a unit additional operations such as settling of the crystals in order to increase the slurry density in the crystalliser and removal of excess fines from the system are usually needed. Equipment for these functions can be added as external units to any of the crystallisers already discussed, but it is generally much more effective to combine these as an integral part of the crystalliser itself.

In the simplest unit, an agitator is provided at the bottom of the vessel with an open settling chamber at the top. Another type works under vacuum and uses draft tubes for circulation (Fig. 10.15). Agitation is either by an agitator in the vessel or by an external circulation unit like a pump. The feed liquor is introduced into the unit below the agitator in a draft tube unit, or before the pump in an external circulation unit.

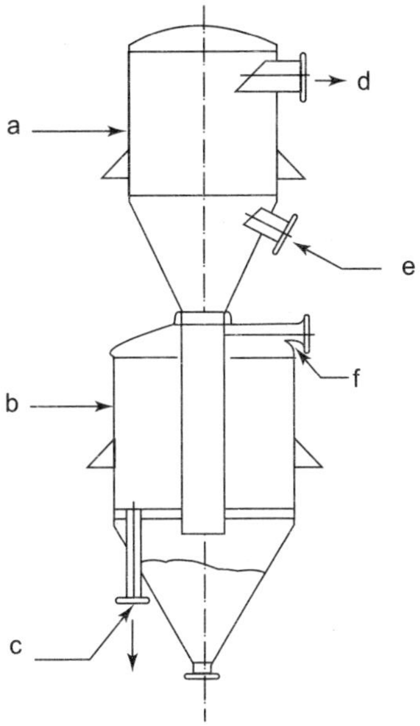

Fig. 10.15 Growth Type Crystalliser: (a) Flash Vessel, (b) Crystallising Vessel, (c) Crystal Discharge, (d) Vapour Outlet, (e) Liquid Inlet, (f) Liquid Outlet to Heater

10.6.5 Jacketed Trough Crystalliser

This (Plate 4) is probably the most widely used crystalliser. It consists of a U-shaped trough of about 3 to 6 metres long, with a jacket divided into sections, so that differential cooling may be used in the various zones. A close fitting long pitch spiral agitator keeps the crystal suspension in motion. The liquor is fed at one end, and the crystals are discharged at the other end. The solution moves counter current to the coolant in the jacket. Four troughs of 3 metre length, each may be arranged in series. If the flow of the liquor is to be more rapid than the cooling area permits, troughs are placed one over the other forming two or three decks. The incompletely crystallised liquor from the top deck is cascaded to the second deck and then to the lower deck.

The trough is formed out of metal sheets welded to an angle iron frame at the top. A jacket covering the entire trough also made of mild steel sheets is attached. The spiral agitator is supported at either end outside the trough. A stuffing box and gland attached to the cover, prevents any leakage of liquor from the opening provided for the shaft. The shaft is driven by chain or belt.

10.6.6 Double Pipe Crystalliser

This (Fig. 10.16) is somewhat similar to the trough crystalliser. It consists of a concentric double pipe, the outer pipe acting as a jacket. A long pitch agitator is rotated in the inner pipe. A set of three lengths of pipes of about 3 metres each is connected in series. The crystalliser can be formed in tiers of three or more, with a set of three lengths in each tier. The hot liquor is fed to the upper tier. The outer pipe is usually carbon steel (mild steel), while the inner pipe may be cast iron, steel or stainless steel depending upon material compatibility. The outer jacket pipe is provided with baffles, to create a spiral path for the flow of the cooling water.

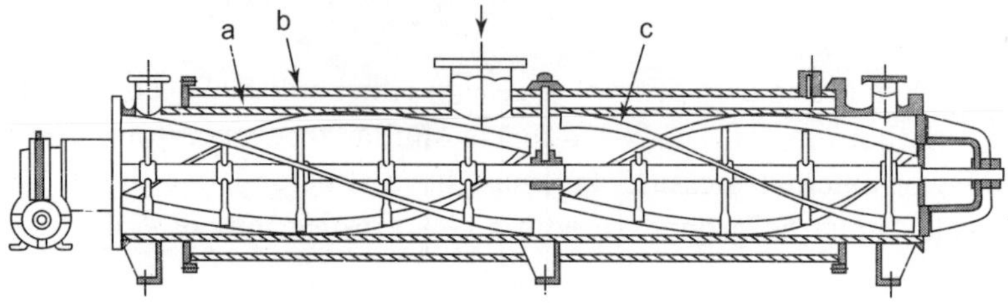

Fig. 10.16 Double Pipe Crystalliser: (a) Inner Pipe, (b) Outer Pipe, (c) Agitator

The agitators are driven by sprokets and chains at 5 to 30 rpm. The blades of the agitator scrape the bottom of the pipe. A stuffing box and gland arrangement is used to prevent leakage. The operation of the crystallizer is continuous. Liquor is fed at the top inlet and slurry is collected from the outlet continuously.

10.6.7 Coil Crystalliser

This (Fig. 10.17) consists of a horizontal cylindrical vessel, with a central shaft. An eccentric coil is mounted on the shaft, which also carries an agitator scroll, supported by diamond shaped self-cleaning arms. The cooling water enters and leaves the coil through the hollow shaft. The liquor is introduced through the charging nozzle at the top of the cylindrical vessel and discharged through the nozzle in the end plate near the bottom. The rotation of the coil is extremely slow, and varies from 1/4 to 1/7 rpm.

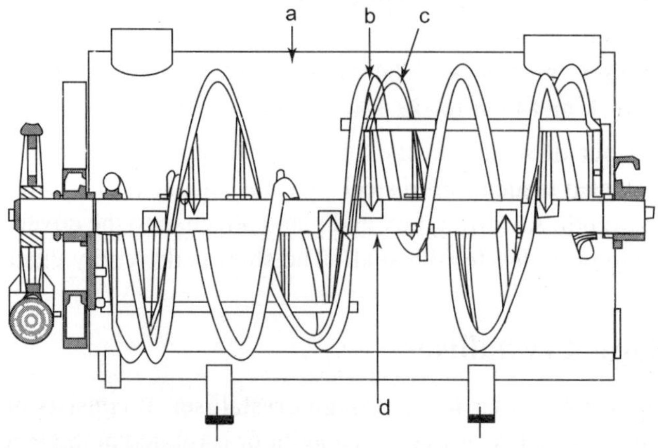

Fig. 10.17 Coil Crystalliser: (a) Shell, (b) Coil, (c) Agitator, (d) Shaft

10.6.8 Design Considerations

The tank crystalliser is essentially a cylindrical vessel or trough. Tank crystallisers are non-pressure units and can be made for minimum sheet thickness. The end closures are attached by simple flanged joints. The materials used must be such that they allow the transfer of heat and the unit must be structurally strong to resist erosion by the crystals and the action of the agitators. Fouling in the jacket is a common difficulty.

Vacuum crystalliser requires a cylindrical vessel with either a dished head at the top and a dished or conical head at the bottom. The construction is similar to evaporators. The vessel is to be designed for vacuum or external pressure and must have structural stability. In constructing this unit, all internal welds must be ground smooth and no projections or other points of possible turbulence or salt build-up must be left. Sight glasses above the boiling surface and in the vapour space ducting, are essential. Manholes on the crystallizer should be well above the liquid level. Those placed within the spray or supersaturation zone, or in the settling chamber, are focal points for salt build-up and are very difficult to open for inspection. The high points in the settling chamber must be vented to the vapour section to prevent stagnant air build-up. Stagnant air build-up results in corrosion and occasional surging of the liquor. Polished stainless steel or glass-lined mild steel are excellent materials of construction for the inner surfaces of crystallisers. Vessel design is given in Chapter 5.

The agitator must be run at a slow speed and should create only a gentle agitation. A stuffing box and gland is used to prevent leakage between the shaft and the vessel wall. Details of agitator design are given in Chapter 14.

10.7 ILLUSTRATION

Let us illustrate the mechanical design of a standard (Calendria) vertical short tube evaporator (Plate 3) with the help of the following data:

Evaporator drum	:	operating under vacuum
Amount of water evaporated	:	2500 kg/hr
Heat transfer area	:	220 m^2
Steam pressure	:	0.15 N/mm^2
Density of liquid	:	1000 kg/m^3
Material of construction (MOC):		
Evaporator	:	carbon steel IS-2062
Tubes	:	brass
Permissible stress for carbon steel	:	98 N/mm^2
Modulus of elasticity:		
Carbon steel	:	19.0×10^4 N/mm^2
Brass	:	9.5×10^4 N/mm^2

Conical head at bottom and top	:	cone angle 120°
Poisson's ratio	:	0.3

10.7.1 Calendria with Vertical Tubes

Tubes and tube lay out:
Tubes of following dimensions were selected:

Tube diameter (outside)	= 100 mm
Tube thickness	= 1.5 mm
Tube length	= 1220 mm
The effective tube length	= 1165 mm
Tube lay out : triangular pitch	= 125 mm

$$\text{Therefore, number of tubes} = \frac{\text{Heat transfer area}}{\text{Tube area}}$$

$$= \frac{220 \times 10^6}{\pi(100) \times 1165}$$

$$= 601.1$$

say, $= 605$

Area occupied by tubes, Eq. (9.2), $A = 605 \times 0.866 \times (125)^2$
$= 8.186 \times 10^6 \text{ mm}^2$
$= 8.186 \text{ m}^2$

We use proportionality factor, $\beta = 0.9$ and then from Eq. (9.4)
$A = 8.186/0.9$
$= 9.1 \text{ m}^2$

The area occupied by the central down take (down comer) is considered to be 40% of the cross sectional area of tubes,

$= 0.4 \times$ (No. of tubes) $\times$ (cross sectional flow area of a single tube)
$= 0.4 \times (605) \times (\pi/4)(97)^2$ (please note tube i.d. $= 100 - 2 \times t$
$= 100 - 2 \times 1.5 = 97$.)

$= 1.788 \times 10^6 \text{ mm}^2$
$\simeq 1.8 \text{ m}^2$.

The inside diameter, d_i, of the down take is then obtained as:

$$d_i = \sqrt{4 \times A/\pi}$$

say, $= 1.5$ m $= 1500$ mm.

The thickness is 10 mm and hence outside diameter = 1500 + 2 × 10
= 1520 mm.

The area of down take based on o.d. = $(\pi/4) \times (1520)^2$
$= 1.81$ m^2

Total area occupied by down take and tubes= area of tubes + area of down take
= 9.1 + 1.81
= 10.91 m^2
say, = 11 m^2

Therefore, diameter of tube sheet, D_T:

$$D_T = \sqrt{11 \times 4/\pi} = 3.74 \text{ m} = 3740 \text{ mm}$$

10.7.1.1 The Calendria Sheet Thickness

Steam pressure = 0.15 N/mm^2

Design pressure = 1.1 × 0.15 = 0.165 N/mm^2

From Eq. (5.3.b), we get thickness, t_s

$$t_s = \frac{p \times D}{(2 \times f \times J - p)} + c$$

$$= \frac{0.165 \times 3740}{(2 \times 98 \times 0.85 - 0.165)} + c$$

$= 3.71$ mm $+ 1.5$ mm

The corrosion allowance and rigidity to shell necessitates thickness to be 10 mm. Thus, the calendria sheet thickness is 10 mm.

10.7.1.2 The Tube Sheet Thickness

The design of the tube sheet is similar to that of a fixed tube heat exchanger (Eqs. 9.5 to 9.8). From Eq. (9.8), we have

$$k = \frac{E_s t_s (D_0 - t_s)}{E_f N t_t (d_0 - t_t)}$$

$$= \frac{19 \times 10^4 \times 10 \times (3740 - 10)}{9.5 \times 10^4 \times 605 \times 1.5(100 - 1.5)}$$

$$= 0.835$$

From Eq. (9.7) for shell side pressure, we have

$$F = \sqrt{\frac{k}{(2 + 3k)}}$$

$$= \sqrt{\frac{0.835}{(2 + 3 \times 0.835)}}$$

$$= 0.43$$

From Eq. (9.5)

$$t = FG\sqrt{0.25p / f_s}$$

The mean gasket diameter, G will be = 3740 + (25 + 25)/2 = 3765 mm

$$t = 0.43 \times 3765\sqrt{0.25 \times 0.165 / 98}$$

$$= 33.2 \text{ mm}$$

The thickness may be taken as 35 mm with corrosion allowance of 1.5 mm.

10.7.1.3 Bottom Flange of the Calendria

A flange is provided at the bottom of the calendria for fixing conical head. The selected flange has following dimensions:

Outside diameter	= 3894 mm
Pitch circle diameter	= 3825 mm
Thickness	= 40 mm
Bolt	= 20 M
Number of bolts	= 112

10.7.2 Evaporator Drum

The diameter of the drum is considered to be that of calendria, i.e., m 3740 mm.

From entrainment separation point of view we shall see adequacy of this diameter with the help of Eq. (10.1), wherein we compute factor, R_d and see whether it is acceptable or not.

$$R_d = \frac{V/A}{0.0172 \times \{(\rho_L - \rho_V)/\rho_V\}^{1/2}}$$

The volumetric flow rate, $V = 8.367\ \text{m}^3/\text{s}$

The cross sectional area of the drum = $(\pi/4)(3.74)^2$

$= 10.99$

say, $= 11\ \text{m}^2$

$$(\rho_L - \rho_V)/\rho_V = (1000 - 0.083)/0.083$$
$$= 12047.2$$

Therefore, by substituting above, we get R_d

$$R_d = 0.4$$

The drum diameter 3740 mm is adequate for R_d less than 0.5.

The column height could be 3000 mm to allow disengagement of vapour and liquid droplets.

10.7.2.1 Drum Thickness

The drum is under vacuum. The outside pressure is atmospheric (0.1 N/mm^2). Therefore, for the design purpose the drum is subjected to external pressure of 0.1 N/mm^2 as there is total vacuum inside. We shall assume the thickness and length of the shell (drum) and calculate allowable stress.

Let thickness t, be 12 mm.

From Eq. (5.14 b), we have

$$P_c = \frac{2.42 \times E \times (t/D_0)^{5/2}}{(1-\mu^2)^{3/4}\left\{(L/D_0) - 0.45(t/D_0)^{1/2}\right\}}$$

We also assume $L = 3000$ mm

The outside diameter $D_0 = 3740 + 2 \times 12 = 3764$ mm

$$P_c = \frac{2.42 \times 19 \times 10^4 \times (12/3764)^{5/2}}{(1-0.3^2)^{3/4}\left\{(3000/3764) - 0.45(12/3764)^{1/2}\right\}}$$

$= 0.37\ \text{N/mm}^2$ (critical buckling pressure)

$P_{all} = 0.37/4$

$= 0.0925\ \text{N/mm}^2$ (the factor 4 is the factor of safety).

One can follow the procedure given in IS-2825 as well.

The actual pressure is 0.1 N/mm^2 which is slightly higher than P_{all}.

Let us consider thickness to be 14 mm.

By the above procedure, we have P_{all} = 0.135 N/mm^2 which is now greater than the actual pressure 0.1 N/mm^2.

According to IS-2825 Appendix F, we get P_{all} = 0.133 N/mm^2.

Therefore, the thickness of the plate used is considered to be 16 mm (14 + 1.5) inclusive of corrosion allowance.

The compressive stress from Eq. (5.16) is then computed as:

$$f_c = \frac{pD}{2f_t}$$

$$= \frac{0.10 \times 3768}{2 \times 14}$$

$$= 13.46 \text{ N/mm}^2$$

This value is well within acceptable limit of the permissible stress value. Please note that in practice one can reduce the thickness by having stiffening rings also.

10.7.2.2 Conical Heads at Top and Bottom

The head thickness is based on external pressure 0.1 N/mm^2. Let us assume thickness 14 mm. As mentioned in section 5.8.2.2 (i), conical head with an angle of 120° is considered to be equivalent of shell with the length of shell equal to the diameter of shell (L/D_0 = 1).

Following above procedure, we get

$$P_{all} = 0.115 \text{ N/mm}^2 \text{ (from IS 2825)}$$

$$= 0.107 \text{ N/mm}^2 \text{ (using the formula for } P_c\text{)}.$$

Therefore, the thickness of above head is 16 mm inclusive of corrosion allowance.

References and Recommended Readings

American Institute of Chemical Engineers, *Equipment Testing Procedure, Evaporators*, 2nd Edition, AIChE, New York (1979).

Armstrong, A.J., 'Scrapped Surface Crystallizers', *Chemical Process Engineering,* **51** (11) (1970).

Bamforth, A.W., 'Types of Crystalliser', *The Chemical Engineer,* July/August (1974).

Couper, J.R., Roy Penny W., Fair J.R. and Walas S.M.; *Chemical Process Equipment Selection & Design,* Second Edition, Elsevier, Amsterdam (2005).

Cramer, H.W., *Chemical Engineering Practice,* Vol. VI, Butterworths, London, (1958).

Crystallization Equipment Review, *Chemical Process Engineering*, **51** (11) 67 (1970).

Jacoby, H.E., *Recent Developments in Evaporation Equipment', Chemical Engineering Progress,* **59** (1) 72 (1963).

Jorden, D.G., *Chemical Process Development, Parts I and II,* Interscience, New York, (1968).

Larson, M.A., *Guidelines for Selecting a Crystalliser, Chemical Engineering,* **85** (4), 91 (1978).

Lavis, G., *Evaporators;* 'How to Make the Right Choice', *Chemical Engineering,* **101** (4) 94 (1994).

Mullin, J.W. Crystallization 4th Edition, Butterworth (2001).

Perry, R.H., Green, D.W., and Maloney, J.O., *Perry's Chemical Engineers' Handbook*, 6th Edition International Edition, McGraw-Hill Book Co., Singapore (1984).

Richardson, J.F., Harker, J.H., and Backhurst, J.R., *Coulson & Richardson's Chemical Engineering, Vol. II,* 3rd Edition, Butterworth Heinemann, Oxford (1991).

Scheimon, A.D., 'Size Vapour-Liquid Separators', *Petroleum Refiner,* **42** (10), 165 (1963).

Sinnott R.K., *Coulson & Richardson's Chemical Engineering Series Chemical Engineering Design, Vol. 6,* 4th Edition, Butterworth Heinemann, Oxford (2005).

11

Chapter

Distillation and Absorption Towers/Columns

11.1 INTRODUCTION

Distillation (fractionation) and Absorption are the two most widely used mass transfer unit operations in the chemical industry. While distillation is performed at higher than ambient temperatures involving a wide range of pressures (nearly total vacuum to super atmospheric pressures), the absorption systems may involve cryogenic conditions and super atmospheric pressure. Regeneration/stripping operations are carried out at higher temperatures and near atmospheric (some times sub-atmospheric) pressures.

Both operations are associated with a set of auxiliary equipments such as reboiler/condenser/feed preheater or cooler, etc., in case of distillation (and stripper) and interstage coolers, feed coolers, etc., in the case of absorbers. Though apparently different operations are performed, the mass transfer contacting devices are the same (such as tray columns or packed tower) in both cases and may, therefore, be considered together from the mechanical design point of view.

11.2 BASIC FEATURES OF TOWERS/COLUMNS

Distillation and the absorption column, also known as a tower, is essentially a tall vertical cylindrical shell, with a number of nozzles. The internals of the column consist of a series of plates or trays or a variety of packings. Where the height of the column is considerable, a ladder and a platform are attached for inspection and maintenance. It can be a self-supported or a supporting structure is provided. Figure 11.1 exhibits self-supported column.

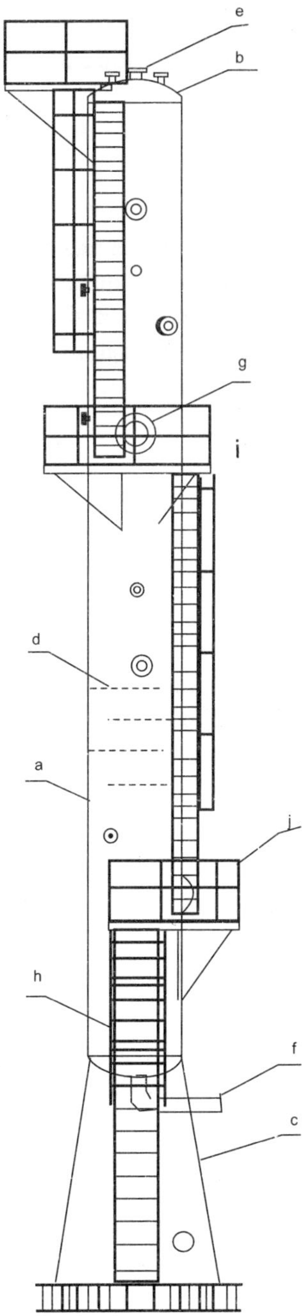

Fig. 11.1 A Self-supported Fractionating Column: (a) Shell, (b) Head, (c) Skirt Support, (d) Tray, (e) Vapour Outlet, (f) Bottom Outlet, (g) Manhole, (h) Ladder, (i) Platform, (j) Railing

11.2.1 Type

Broadly speaking, the towers/columns may be divided into two types, based on the contacting device used for mass transfer to take place, namely the plate or tray columns and packed columns.

11.2.1.1 Plate/Tray Column

The tray column consists of a plate or tray, which holds the liquid and gas/vapour which is dispersed through the pool of liquid. Various types of trays are in practice such as the bubble cap trays (oldest), sieve trays and valve trays. Here, the system is in counter-current mode of operation. Further, the liquid is in a continuous phase and gas is in a dispersed phase. The pressure drop being higher, the tray columns are not recommended for vacuum operation.

11.2.1.2 Packed Column

Here, the column (tower) is filled with proprietary packings. There are two types of packings, namely, random packings and structured packings.

As the name suggests, random packings are dumped in a column filled with water to settle them in random way. On the other hand, structured packings are introduced in a column in an organized way of stacking, packing blocks over one another, as per the recommendation of the supplier. The packings provide intimate contact between the gas/vapour and liquid facilitating mass transfer. Here, the gas/vapour is a continuous phase and the liquid is a disperse phase. A packed tower is operated in counter-current mode for the physical separation or in the case of a reversible reaction. For the system having an irreversible reaction, one can operate the column in the concurrent (co-current) mode also.

A packed tower being a low-pressure drop device, it is always recommended for distillation under vacuum, particularly with structured packings.

Both devices are also operated for liquid-liquid systems involving extraction.

11.2.2 The Tower/Column Diameter and Height

The hydrodynamic considerations will decide the column diameter, in a counter-current mode of operation, flooding will decide the same. On the other hand, flow regime or pressure drop will decide the diameter of a co-current operated column.

The height of a column is mainly governed by mass transfer considerations. Difficult separation results in taller column. Once the theoretical stages or equilibrium stages are known, the total number of plates can be found out by knowing the plate/tray efficiency. The plate spacing which depends on the type of the plate will then decide the tower height. In the case of a packed tower, the height of packing could be obtained by multiplying the number of theoretical stages by HETP (height equivalent to theoretical plate). Total height of a column is the sum of space required for mass transfer device and tower internals, such as packing support plate, hold down plate, liquid distributor, gas/vapour distributor, liquid redistributor and demister, etc.

11.3 PROCESS ENGINEERING DATA

The process design of a distillation column or absorber results in the data as depicted in a typical process data sheet Table 11.1.

Table 11.1 Process Data Sheet

	Units
Section of Column	
Tray Numbers	
Type of Trays/Packing	
Number of Trays/Depth of Packing	
Tray Spacing	
Number of Passes	
Column Diameter	
Maximum Allowable Δ P/Section	
Overload	%
Minimum Load	%
Flooding (% at Design Load)	%
System Factor (Foaming Tendency)	
Vapour Load (Design)	
To/From Section	
Fluid	
Flow Rate	
Density (Actual)	
Pressure	
Temperature	
Molecular Weight	
Compressibility Factor	
Liquid Load (Design)	
To/From Section	
Fluid	
Flow Rate	
Density (Actual)	
Viscosity (Actual)	
Surface Tension (Actual)	
MECHANICAL DATA	
Maximum Operating pressure	
Temperature	
Deck, Downcomer, Packing Material	
Valve, Cap Material	
Corrosion allowance, mm	
Deck Thickness	
Drain Holes and Location	

11.4 TOWER/COLUMN INTERNALS

The two types of columns, have different kinds of internals, as stated earlier, the details of which are as under.

11.4.1 Plate Type Columns (Table 11.2)

In these columns, the mass transfer between phases takes place in stages on a series of plates or trays or in the vicinity of either.

The minimum tray spacing recommended is 300 mm. For columns of 1500 mm internal diameter and above, spacings of 450 mm or 600 mm are preferred for easier tray access.

Depending on the method of contact and the flow of vapour and liquid. The contacting equipment may be classified as follows:

Table 11.2 Typical Tray Specification

	Tray Specifications		*Tray Specifications*
1.	Vessel i.d.	16.	Tray Spacing
2.	Item Number	17.	Downcomer Clearance
3.	Design Temp. °C	18.	Sump & Draw off Requirement
4.	I.D. Vessel Manhole	19.	No. Caps Per Tray
5.	Number & Type Trays	20.	Cap & Riser Material
6.	Tray Material	21.	Cap & Riser Assy Type
7.	Tray Corrosion Allow. (Total)	22.	Size of Caps
8.	Downcomer Area-Top	23.	Cap Spacing
9.	Downcomer Area-Bot	24.	Cap Skirt Clearance
10.	Outlet Weir Periphery	25.	Cap Slot Spec.
11.	Outlet of Weir Height	26.	Top of Slot above Tray
12.	Outlet of Weir Adjustment	27.	Type of Slot in Cap
13.	Outlet of Weir Notch	28.	I.O. & Height of Risers
14.	Inlet of Weir Height	29.	Size of Tray Support Bar
15.	Tray Manaway Height	30.	Tray Support-material

11.4.1.1 Plates without Downcomers

These are generally flat circular plates of a diameter slightly smaller than the internal diameter of the column, located with suitable spacing from the bottom to the top of the column. They are provided with a large number of holes or slots through which the liquid and the vapour pass. The Turbogrid (Fig. 11.2) is a flat plate with long rectangular slots. The grid can also be formed by closely spaced parallel bars located in one circular plane, instead of a single flat plate. The Dual flow tray is a flat plate with circular holes. The Ripple tray is corrugated instead of being flat. The Kittel tray has slots,

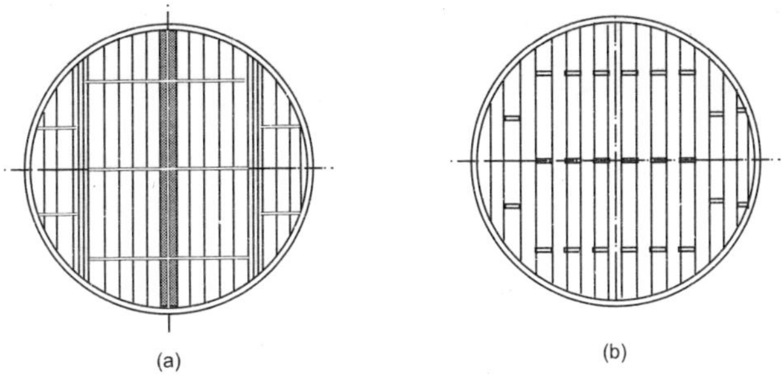

Fig. 11.2 Turbogrid Plate or Tray: (a) Top View, (b) Bottom View

which cause the vapour to move the liquid towards the centre of one tray and towards the wall of the next. Invariably, plates are assembled with slots of one plate at right angles to those on the adjacent plate, so as to prevent the liquid falling from one slot through the corresponding slot on the plate below. Holes on consecutive plates are also staggered.

11.4.1.2 Plates with Downcomers

These are circular flat plates or trays with provision of passages known as downcomers for the downward flow of the liquid. The downcomer area for each plate will be generally limited to 10% of the total area of the plate. Various arrangements of downcomers, weirs, and contacting devices are adopted for handling liquids and vapours.

(a) *Downcomers and weirs*

The function of the downcomers is to provide a passage for the downward flow of the liquid from a top tray to a lower tray. An inlet weir helps to distribute the liquid as it enters the tray from the downcomer, and prevents direct impingement of the liquid on the contacting devices. For small liquid, flows circular or pipe type downcomers (Fig. 11.3) are commonly used. A projection of a corresponding shape, provided on the downcomer acts as an exit weir. The downcomer can be formed as a segmental shape (Fig. 11.4), where a raised portion on the edge acts as a weir. A downcomer can also be formed by creating a passage between the column wall and a vertical plate extending across the chord of the tray. The weir in such a case comprises a flat plate extending across the chord, and is known as a chord type weir (Fig. 11.5). The flow of liquid in these cases is across the tray and is known as the single pass cross flow. A reverse flow is created by a weir system shown in Fig. 11.6(a). It is recommended when the liquid flow becomes small in comparison to vapour flow. A radial flow is created by downcomers placed at the centre and around the circumference of alternate trays [Fig. 11.6(b)]. For multipass arrangements the system of downcomers and weirs is as shown in Fig. 11.7. The downcomers in these cases are formed of two parallel plates extending across the diameter of the column. Such downcomers are generally tapered at the bottom to facilitate maintenance of the liquid seal. Reverse flow, radial flow and multipass arrangements help in obtaining a satisfactory liquid flow and distribution particularly in large columns. In all the cases, the weirs should be of sufficient height.

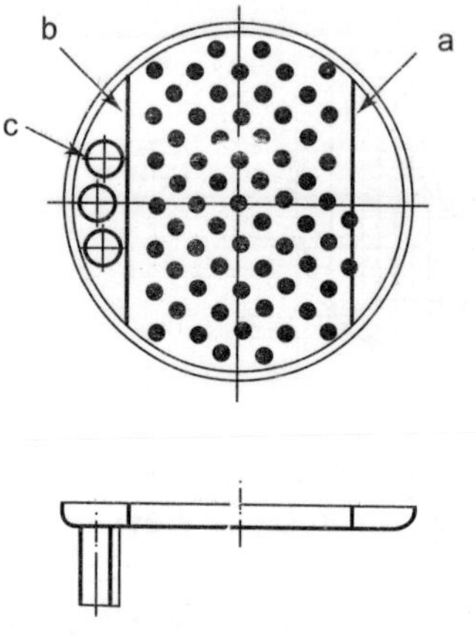

Fig. 11.3 Segmental Downcomer and Weir: (a) Inlet Weir, (b) Outlet Weir

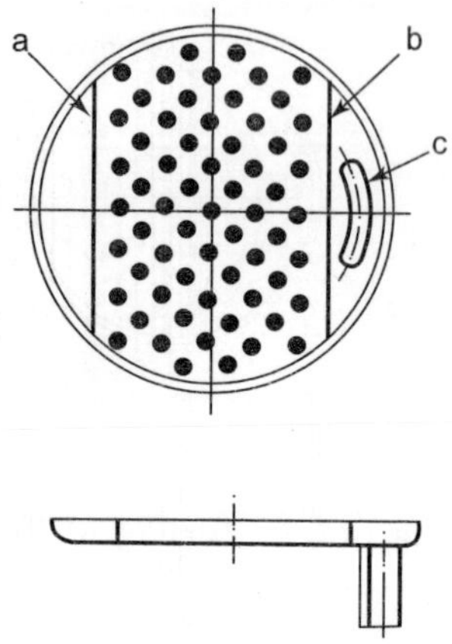

Fig. 11.4 Circular or Pipe Type Downcomer and Weir: (a) Inlet Weir, (b) Outlet Weir, (c) Down Pipe

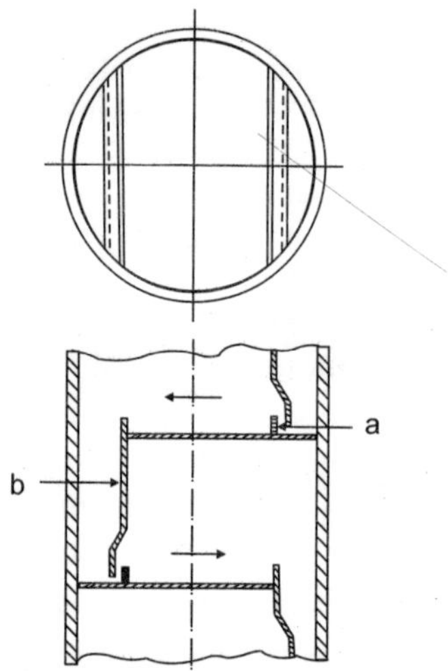

Fig. 11.5 Chord Type Downcomer and Weir (Across Flow): (a) Inlet Weir, (b) Outlet Weir with Downcomer

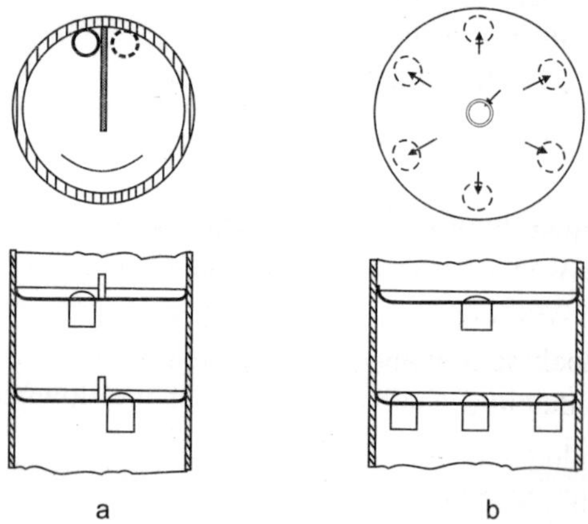

Fig. 11.6 Single Pass Flow: (a) Reverse Flow, (b) Radial Flow

Multiple downcomer trays are also designed with a row of narrow rectangular troughs spaced across the tray to act as downcomers. The bottom portion of the trough is not fully open, but is provided with slots. The position of the downcomers is staggered on consecutive trays (Fig. 11.8). The length of the weir is generally 60 to 75% of the diameter for a cross flow tray and 50 to 60% of the diameter for a multipass.

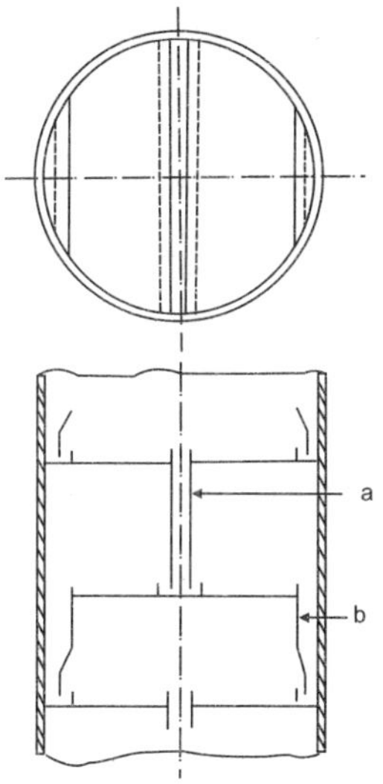

Fig. 11.7 Multipass Flow: (a) Central Downcomer, (b) Side Downcomer

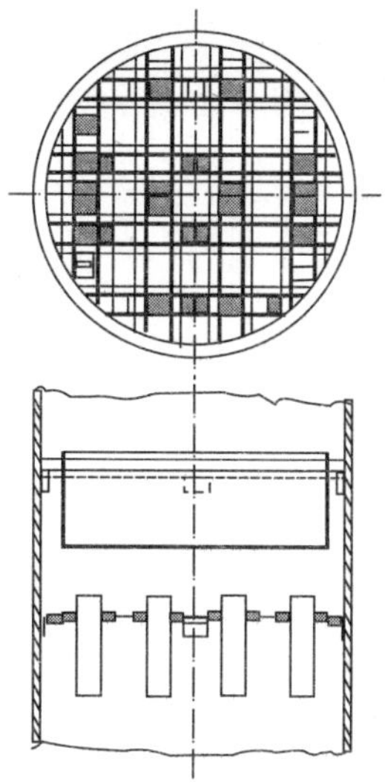

Fig. 11.8 Multiple Downcomer Tray

(b) *Contacting devices*

(i) *Bubble caps* The area of the plate or tray, between the inlet and exit weirs is provided with a large number of holes over which cylindrical risers or chimneys are placed. Caps in the form of inverted cups are located over the risers. The vapour rises up through the holes and the risers and subsequently through a number of rectangular, triangular or trapezoidal slots provided at the rim of the caps [Fig. 11.9(a) and (b)]. Table 11.3 gives the size of caps used. In special cases the rectangular risers or chimneys with tunnel shaped caps are used [Fig. 11.9(c)]. Uniflux tray is made up of interlocking S-shaped sections which form a series of transverse tunnel caps with vapour outlets on one side only.

Round bell caps of diameter 75 to 150 mm are most common while tunnel caps are ordinarily 80 to 150 mm wide and 300 mm long.

Table 11.3 Buble Cap Size

Diameter of Columns (metres)	*Cap Size (round) (mm)*
0.8	50
0.8-1.5	75
1.2-4.5	100
3-6	150

The riser diameter is normally 60 to 70% of the cap diameter. For columns of 1 meter or more in diameter, the total riser area commonly used is 10 to 20% of tower cross sectional area. The caps

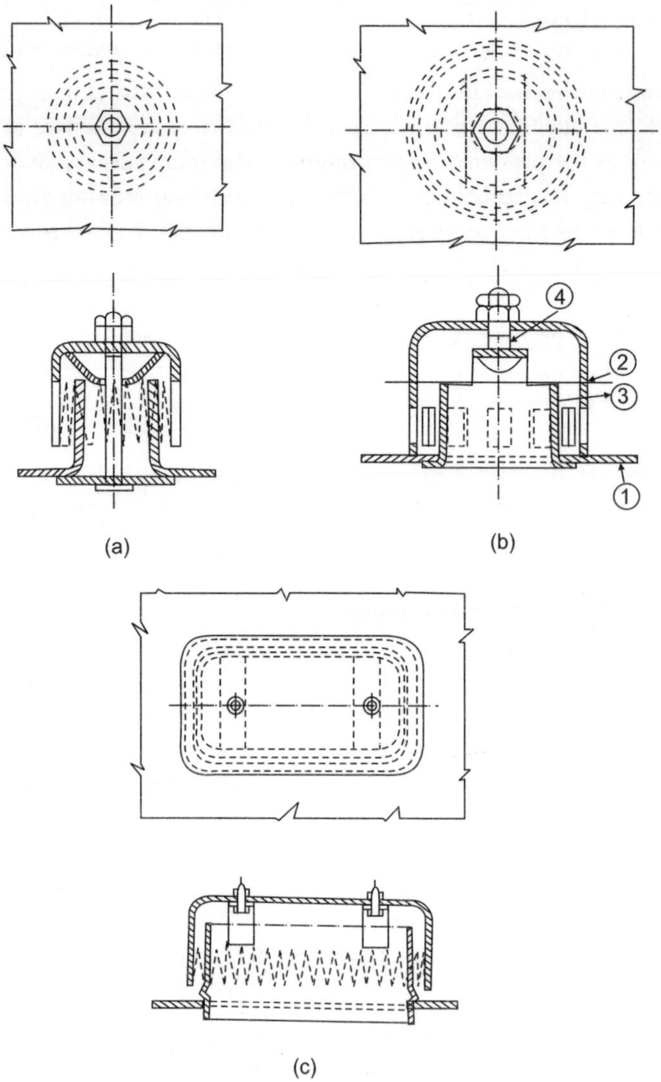

Fig. 11.9 Bubble Caps: (a) Circular Cap with Riser (Slots on Edge), (b) Circular Cap with Riser; (1) Tray, (2) Cap, (3) Riser, (4) Bolt, (c) Rectangular Cap with Riser

are slotted around the lower periphery and can be anchored to the plate with teeth touching the plate or suspended to permit a 'Skirt Clearance' between the plate and cap. In common practice, a skirt clearance of 12 to 40 mm is recommended to prevent plugging of the slots due to the residue build-up.

(ii) *Valves* The bubble caps can be replaced by poppet type valves, which are lifted by the rising vapours and act as variable orifices. These permit effective mixing of liquid and vapour over a wide range of loading. These valves are circular with domed or flat caps. Rectangular caps and caps with downward facing cones cannot rotate as the circular ones might do under certain process conditions.

Circular liftable caps are restricted in their lift by suitable constructional features such as special legs or spiral webs. Some caps are designed by a double lift system, where at low vapour loads the inner light cap lifts and as the pressure drop through the cap increases and overcomes the weight and friction of the heavier cap. It also lifts and creates an additional area for flow of the vapour. Trays with such caps are known as flexi trays [Figs. 11.10(a) and (b)]. Ballast trays have valves with flat caps or discs [Fig. 11.10(c)].

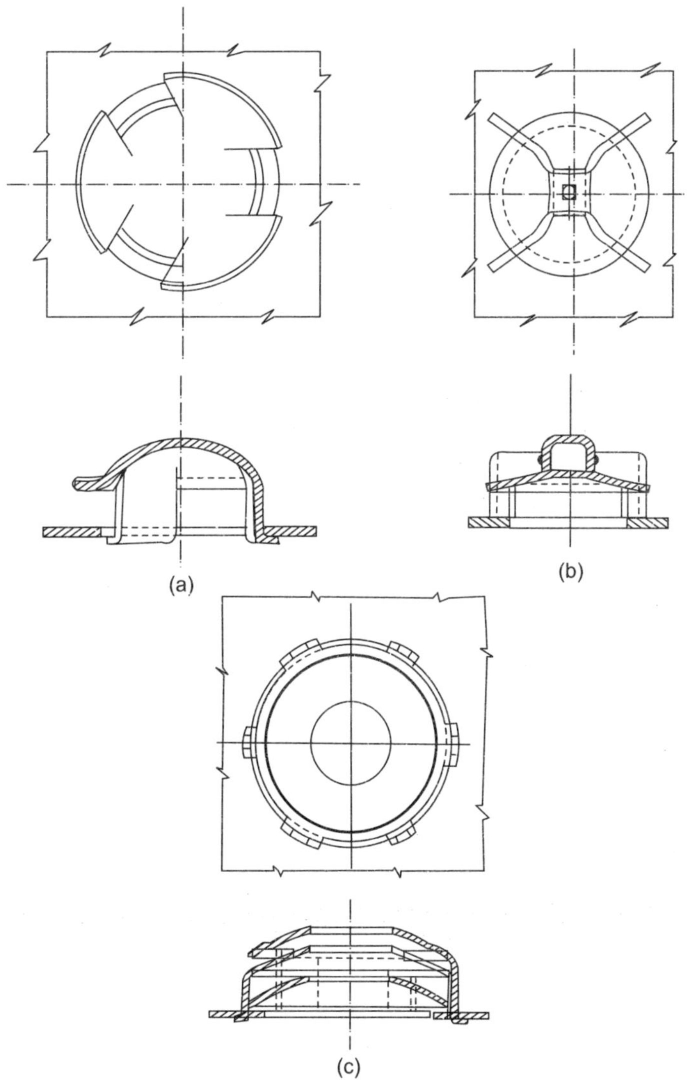

Fig. 11.10 Valves: (a) & (b) Flexi-valve Tray, (c) Glitsch, Ballast Tray

(iii) *Sieve plates* There is no separate contacting device in these types of plates. Instead, a large number of holes distributed uniformly over the plate act as passage for vapours, but at the same time the liquid is prevented from draining through the holes. In some cases, holes are punched with protruding lips. This is known as a jet tray. In another design, in addition to small holes a few angular slots are made in various positions. In a modified design of the sieve tray, a layer of expanded metal is attached to the tray. The vapour flowing upwards through the perforated tray is uniformly, deflected by the expanded metal thus, imparting a horizontal component to the vapour flow.

(c) *Distribution baffles*

With a uniform cap spacing it is not possible to locate all of the caps at the end of the rows close to the column wall. Due to excessive end clearance, the liquid will tend to short circuit such rows. If the end space exceeds the spacing by 25 mm, the gaps are closed by redistribution baffles. The clearance between these baffles and the caps should be equal to the cap spacing.

(d) *Tray drainage*

Tray drain holes are required to permit drainage of the liquid, hold-up of the column in case of a shut down. These holes must be sufficiently large, so that they may not get clogged by sediments or polymers. The current practice is to make holes of 1 to 1.5 cm diameter along the weir.

11.4.1.3 Feed Systems

Feed nozzles are sized for a flow velocity of 1 metre per second. The jet of liquid should break-up immediately on entering the column. Liquid feed to the column is above the top tray and at intermediate positions. A pipe line located above the top tray, and just opposite the inlet weir, feeds the top tray. A baffle is provided opposite the feed pipe [Fig. 11.11(a) and (b)] to create a uniform distribution. Various methods of intermediate feed arrangements are shown in Fig. 11.12. Liquid feed can be introduce at A, B, C or F. Location A may be satisfactory. However, in the case of high pressure system, where vapour disengagement out of the downcomer is extremely critical, or in cases where the temperature of the feed inlet would cause flashing of the downcomer liquid, this arrangement may limit the column capacity. Location B is generally not desirable, because of the unnecessary turbulence introduced in the downcomer area. Location C is the most convenient location. It is provided with a baffle plate and a guard plate. Location F is preferable for high velocity feeds. In case there is a large change of liquid feed rate, or the downcomer is not far below flooding, or there is any foaming tendency or flashing conditions, the feed entrance should be through a sparge pipe (Fig. 11.13). Arrangements for reflux entrance are shown in Fig. 11.14.

The vapour feed is generally located below the bottom tray as shown in Fig. 11.15. The vapour is introduced parallel to the bottom downcomer at a spacing of about 500 to 600 mm. A vapour inlet nozzle causing impingement of the vapour stream is provided for a mixed vapour/liquid feed a sparge pipe across the column centre line is provided as shown in Fig. 11.16.

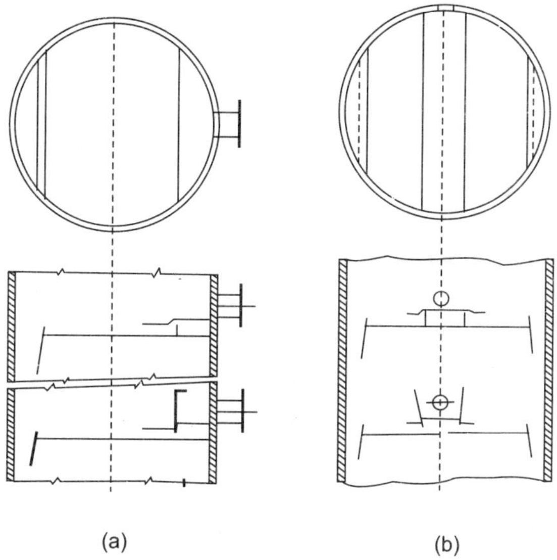

Fig. 11.11 Top Tray Feed Inlet with Weir or Baffle: (a) Single Pass, (b) Multipass

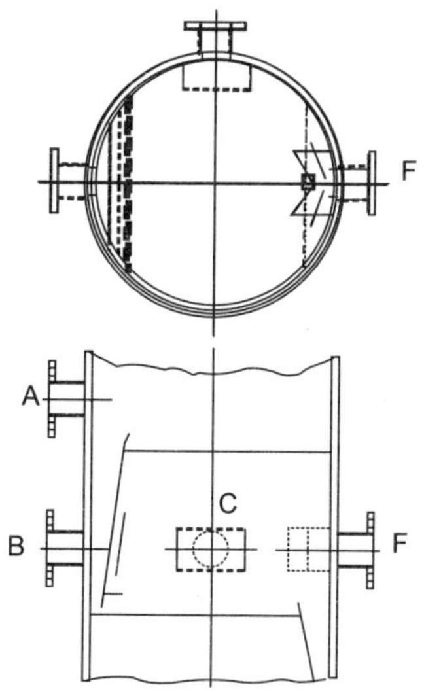

Fig. 11.12 Intermediate Feeds: A, B, C and F-alternate Liquid Feeds

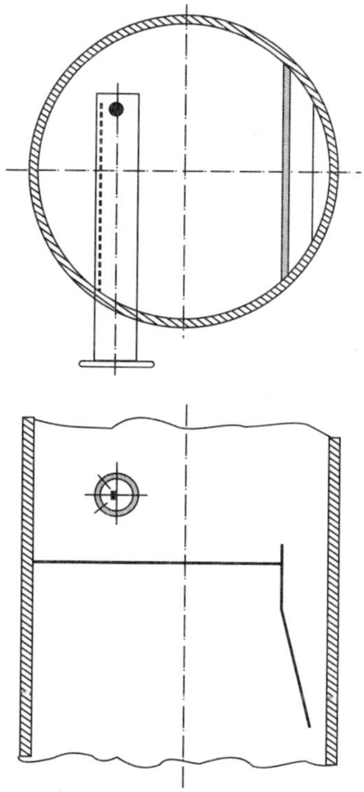

Fig. 11.13 Sparge Pipe

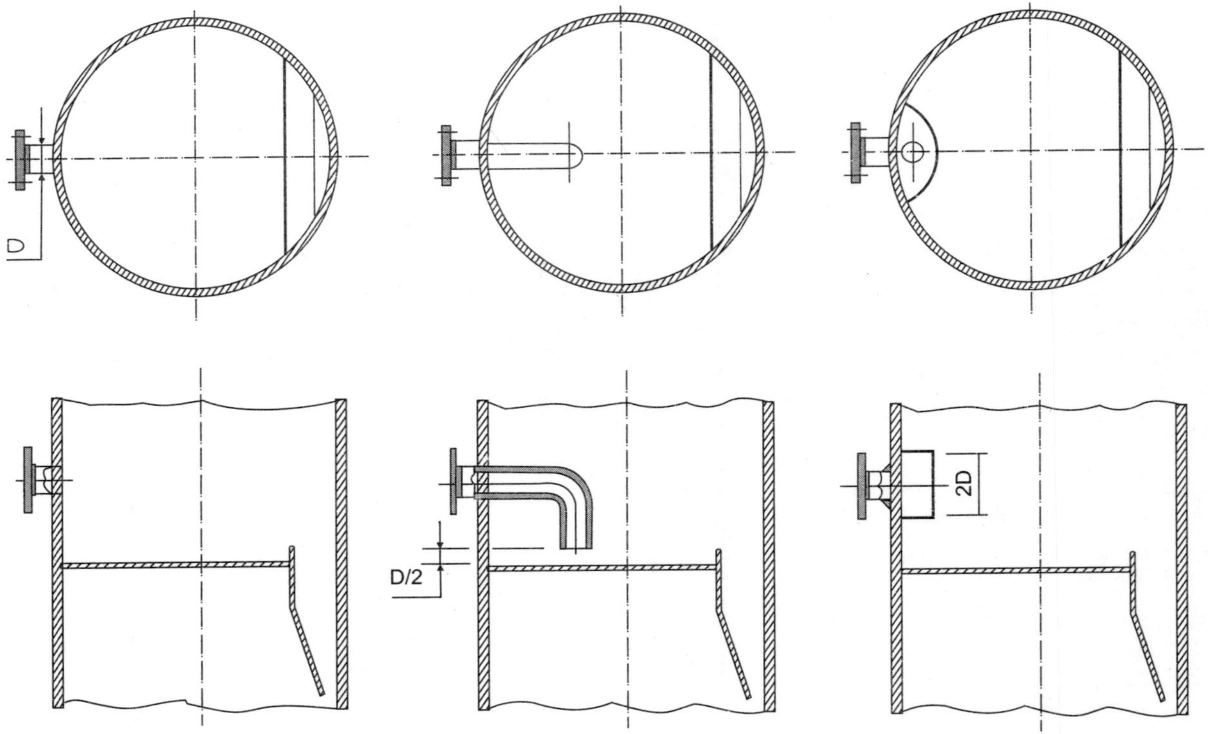

Fig. 11.14 Alternative Arrangements of Reflux Entrance

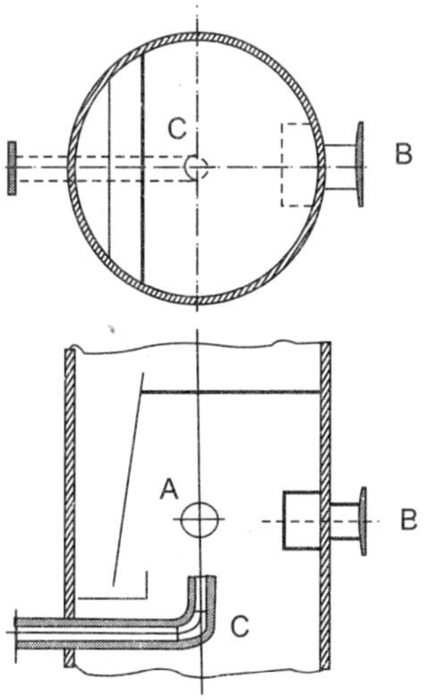

Fig. 11.15 Vapour Feeds: A, B, C-alternate Vapour Feeds

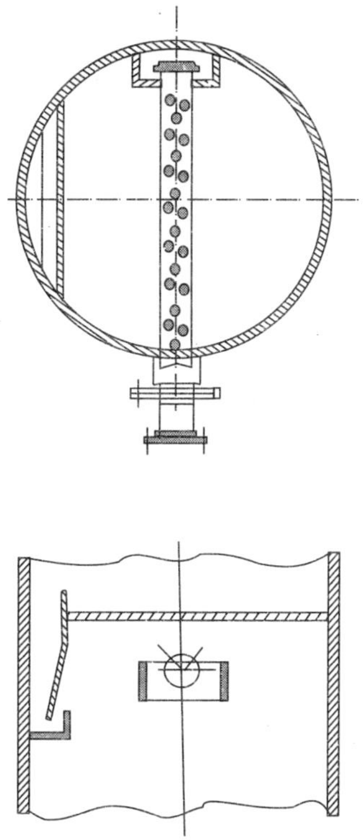

Fig. 11.16 Sparge Pipe Feed

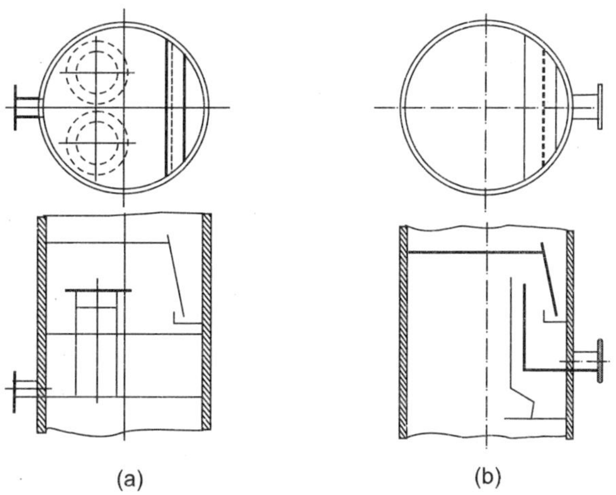

Fig. 11.17 Draw Off: (a) Chimney Tray, (b) Draw Pan

11.4.1.4 Draw Offs

The drawing of the liquid from the bottom of the column is either by the use of a chimney tray or by a draw pan as indicated in Fig. 11.17(a) and (b). The chimney tray has no downcomer and is flush with a draw nozzle. Chimneys are normally of a size of approximately 15% of the column area. These should be so located, so as to prevent directing the vapour against the adjoining downcomer.

11.4.1.5 Manholes and Handholes

Trays should be inspected for corrosion, fouling and damage, a few weeks after the start-up and at six monthly intervals thereafter. Trays should be removable both from above and below. Manholes are necessary at intervals in the column, so that each manhole serves 15-20 trays. The manhole size is usually 450 mm internal diameter or 300 × 400 mm opening. In packed columns manholes or handholes will enable each distributor and support plate to be dismantled and removed. Packing is very tedious to remove. The best way is to make the column in flanged sections, which can be taken off and emptied. However, flanged sections are extremely expensive and an alternative is to provide a large (at least 250 mm) handhole at the bottom of each bed. Each manhole must have a platform for access.

11.4.2 Packed Columns (Fig. 11.18), (Table 11.4)

Instead of using plates or trays, these columns are filled with a suitable packing material. The liquid flows downwards over the surface of the packings in the form of thin film, while the gas/vapour rises. The liquid is introduced at the top of the packings by means of a distributing plate and the vapour is introduced beneath a grid which supports the packings. Intimate contact between gas/vapour and liquid phase is provided by the packings, enabling it to carry out an efficient mass transfer operation. During this mass transfer, it is desirable that the pressure drop be kept as minimum as possible.

11.4.2.1 Type of Packings

In a packed tower we have two types of packings depending upon the way they are held in a tower. Random packings are packed randomly while structured packings are arranged in packing blocks, one over the other.

Given above are the broad classes of packings. Depending upon the vendor there could be some changes in shape. Majority of the above packings are fabricated in metal, ceramic and plastics. Some packings can also be fabricated in glass and graphite. Some of them exhibited in Fig. 11.19.

Packed columns are not suitable for dealing with liquids containing or producing large concentration of solids, as the packings will block up and/or distribute liquids unevenly. In order to overcome problems associated with 'dirty' fluids such as heavy residue (AGO/VGO), in the refinery a new type of packings known as 'grid trays' has been introduced. Packed columns are particularly useful in the field of vacuum distillation. Random packings are most suitable for corrosive service.

Table 11.4 Packed Tower Internals Data Sheet

Tower Internals		
<u>Distributor</u> ______	<u>Support Plate</u> ______	
Tower Inside Dia ______	Tower Inside Dia ______	
Material ______	Material ______	
Corrosion Allowance ______	Corrosion Allowance ______	
Liquid Flow Normal ______	Design Load (Liquid) ______	$N/m^2/hr$
Max. ______ Min. ______	Design Load (Vapour) ______	$N/m^2/hr$
Liquid Density ______	Design Temp. ______	
Design Temp. ______	Min. Dia. or Access ______	
MIn. Dia. of Access ______	Packing Size and Type ______	
Special Requests ______	Packing Depth ______	
______	Special Requents ______	
______	______	
<u>Hold-Down Plate</u> ______	<u>Bed Limiter</u> ______	
Tower Inside Dia ______	Tower Inside Dia ______	
Material ______	Material ______	
Corrosion Allowance ______	Corrosion Allowance ______	
Design Temp. ______	Design Temp. ______	
Min. Dia. of Access ______	Min. Dia. of Access ______	
Packing Size & Type ______	Packing Size & Type ______	
Special Requests ______	Special Requests ______	
______	______	
<u>Corrector Plate</u> ______	<u>Redistributor</u> ______	
Tower Inside Dia ______	Tower Inside Dia ______	
Material ______	Material ______	
Corrosion Allowance ______	Corrosion Allowance ______	
Design Temp ______	Liquid Flow Normal	
Liquid Flow Normal ______	Max. ______ Min. ______	
Max. ______ Min. ______	Liquid Density ______	
Min. Dia. of Access ______	Support Ledge ______ Wide x ______ THK	
Support Ledge ______ Wide x ______ THK	Design Temp. ______	
Max.% Leakage ______	Gas Flow ______	
No. of Sumps ______	Min. Dia of Access ______	
Gas Flow ______	Special Results ______	
Special Results ______	______	

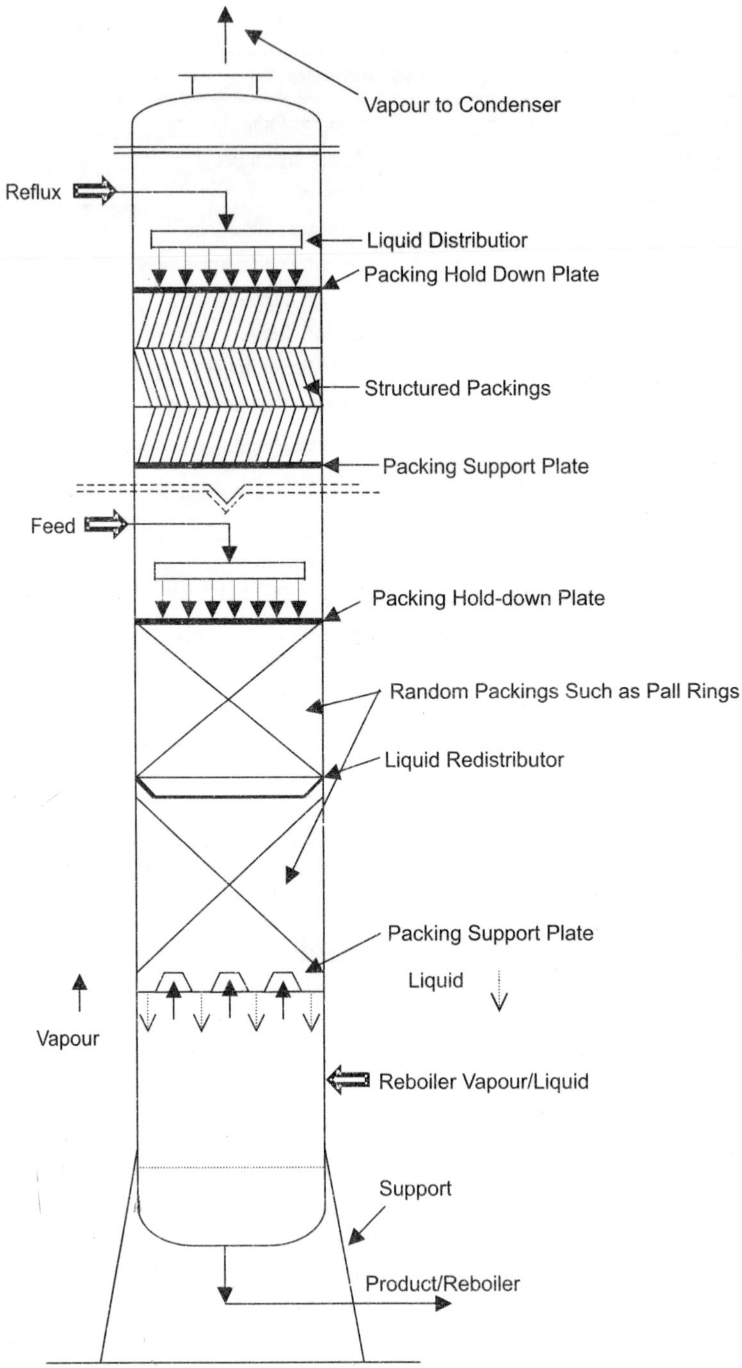

Fig. 11.18 Packed Distillation Column: Schemetic Diagram

Packings

Random Packings	Structured (regular) packings
Raschig Rings	Sulzer Mellapak
Intallox Saddles	Kerapack
Pall Rings	BX and CY wire gauze packings
Cascade Mini Rings, etc.	Grid trays, Static Mixers, etc.

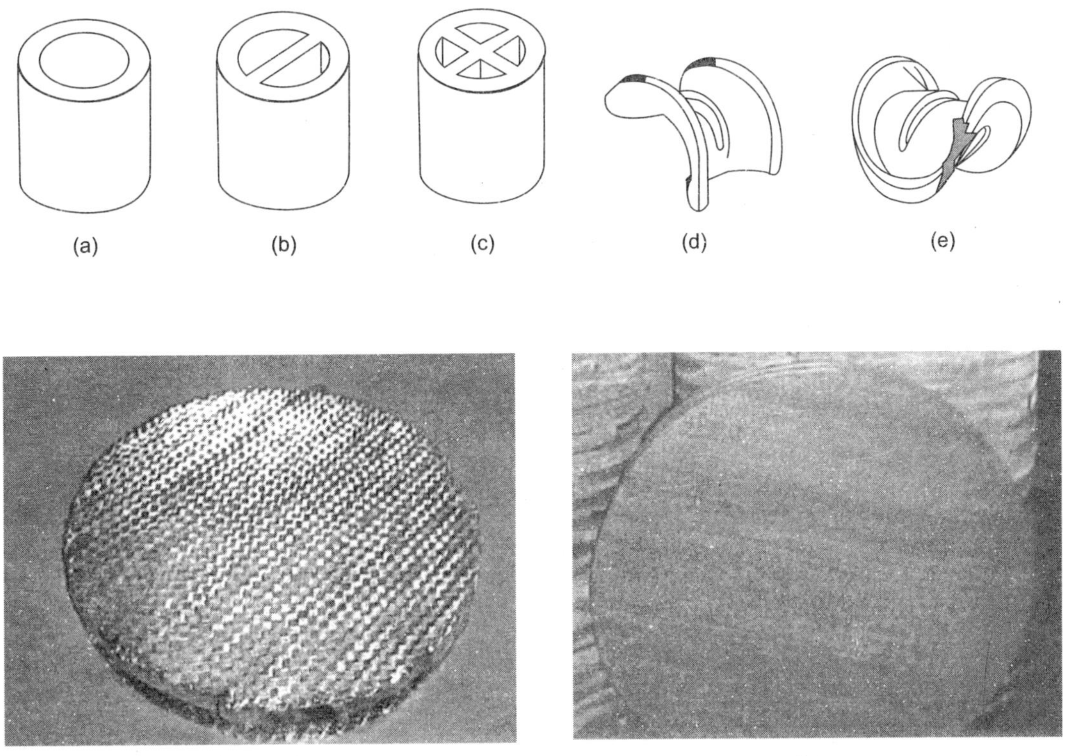

Fig. 11.19 Packings: (a) Raschig Ring, (b) Lessig Ring, (c) Cross Partition Ring, (d) Intolax Saddle, (e) Berl Saddle (f) Everpack (Sheet Metal) Structured Packing, (g) Everpack (Wire Mesh) Structured Packing

11.4.2.2 Liquid Distributors

Various methods are adopted for distribution of the liquid, to effect a proper wetting of the packing. Piping arrangements, flat perforated plates, notched troughs or weir type distributors (Fig. 11.20) are commonly used. For column diameters of less than 600 mm a single pipe nozzle or a shower with multiple pipe nozzles may be satisfactory. In the case of a single pipe nozzle, a flat plate should be provided at the centre to allow the liquid to splash on it and thereby distribute it. The distance between the nozzle tip and the splash plate should be kept between 150 and 250 mm. In the case of a shower, the spray should cover 50% of the cross section. For larger diameters, cross pipes or a spider pipe with multiple arms (Fig. 11.21) or a pipe bent to a circular form should be provided with nozzles, such that the liquid can be distributed evenly over the cross section.

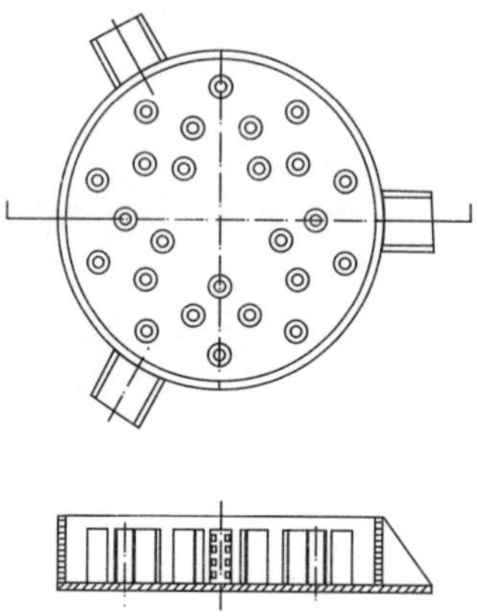

Fig. 11.20 Weir Type Liquid Distributor

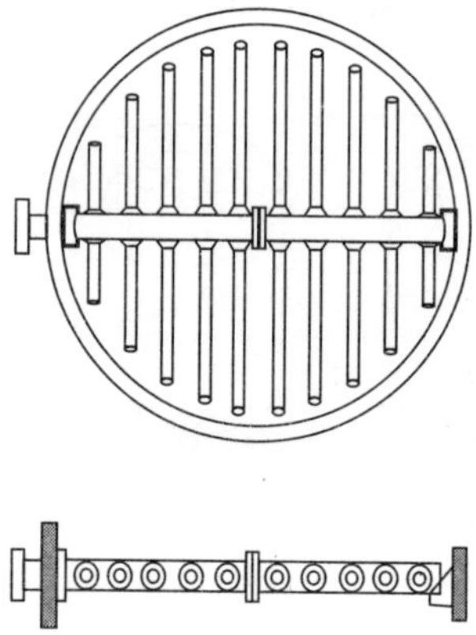

Fig. 11.21 Spider Type Distributor

11.4.2.3 Liquid Redistribution and Wall Wipers

Despite its efficient distribution on the top plate, the liquid has a tendency to drift towards the wall as it moves downwards. After a certain vertical height of liquid travel, liquid concentration falls considerably in the inner cross section. It is, therefore, necessary to direct the liquid towards the centre after it has travelled a certain distance. This is done by dividing the packing height to approximately 3 column diameters for Raschig rings and 5 to 10 column diameters for saddle packings. Between two packing sections, redistributors and wall wipers are provided to bring the liquid back to the centre from the wall (Fig. 11.22).

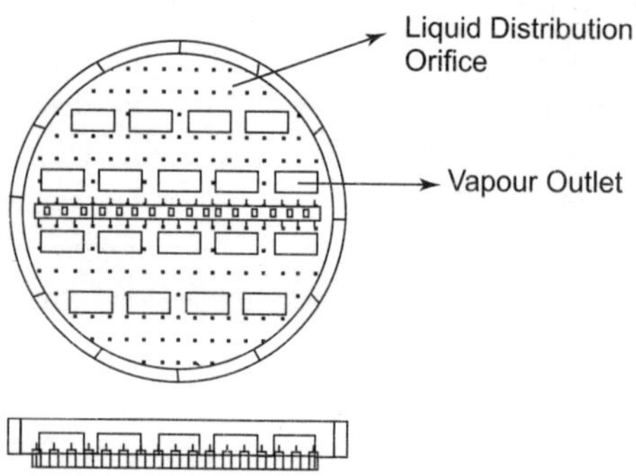

Fig. 11.22 Liquid Redistributor

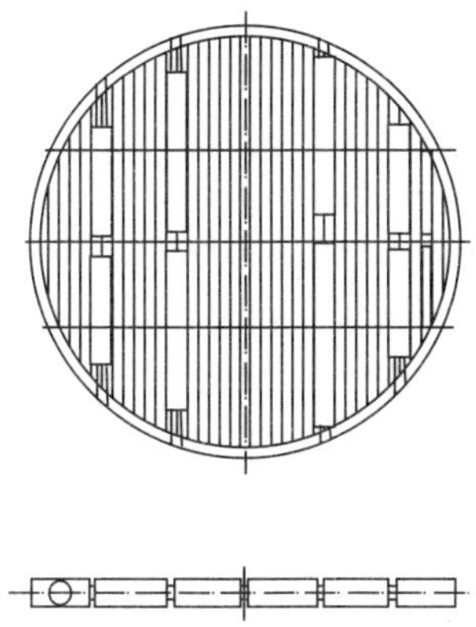

Fig. 11.23 Grid Bar Support Plate

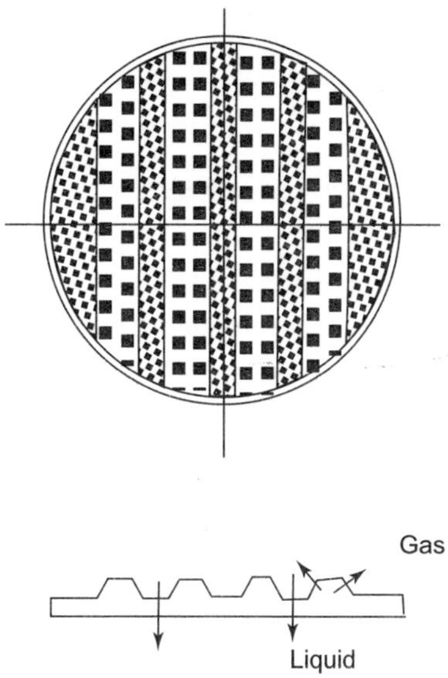

Fig. 11.24 Gas-injection Type Support Plate

11.4.2.4 Support Plates

These plates are located at the bottom of the column and support the packings. Conventional type of support plates are perforated plates or screens and grid bar assembly (Fig. 11.23). In the former, the free area is limited to 20 to 30% and in the latter to 30 to 40%. In order to achieve a better performance a minimum free area of 50% is required. In such cases, a gas-injection-weir-type (Fig. 11.24) support plate is used. These plates have several advantages over the conventional types.

11.4.2.5 Hold-Down Plates

A hold-down plate (Fig. 11.25) on top of a packed section serves an important purpose by resting directly on top of the packing and restraining the bed under conditions of high gas rates or fluctuating gas flows, when, without such restraint, ceramic or carbon packings may break and fall back through the bed, thus, reducing its capacity. The free area of the plate is decided as being as large a value of the area as possible and a value small enough to prevent the packing passing through the plate.

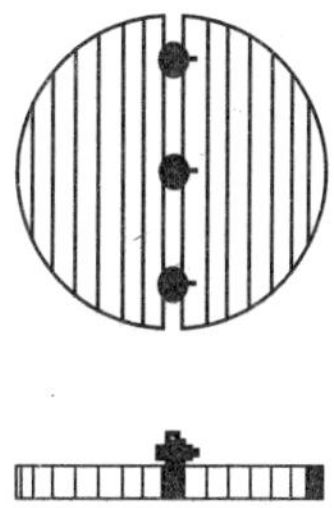

Fig. 11.25 Hold-down Plate

11.5 STRESSES IN COLUMN SHELL

The column is designed as a tall vertical vessel, either as a self-supporting structure or as a stabilised structure with guys ropes, it is appropriate to design it as a self-supporting structure. In this case the stresses in the shell of the column may be assessed as follows:

(a) Circumferential and axial stresses due to internal pressure or vacuum inside the vessel.

(b) Compressive stresses resulting from dead loads such as the weight of the vessel and its contents, the weight of insulation, and attached equipment.

(c) Stresses resulting from bending moment caused by wind loads acting on the vessel and its attachments.

(d) Stresses due to eccentricity as a result of irregular load distribution.

(e) Stresses due to seismic (earthquake) forces.

The above stresses may be combined to establish the final controlling stress. The stresses due to pressure act both in the circumferential and axial direction of the shell. The stresses due to other loads are all in the axial direction. It is possible to combine all the above stresses on the basis of shear energy criterion (distortion energy criterion, Eq. 5.11). However, the stresses in the axial direction are not the same over the entire height of the shell, and the resultant stresses in the shell are, therefore, likely to increase towards the bottom of the shell.

While analysing the stresses at various heights of the shell, calculations are made starting from the top of the column. The selection and thickness of the top head is based on the internal pressure or vacuum. The minimum shell thickness in the upper portion of the column is usually determined by the circumferential stress resulting from internal pressure or vacuum. The shell plate thickness at the top of the column can be taking into consideration the corrosion allowance. The thickness for lower portions of the column is determined by taking into consideration all the other stresses. It is assumed that the stresses due to wind load and earthquake load will not occur simultaneously. The maximum value of either is, therefore, accepted for evaluation of the combined stresses. At a height X from the top of the shell, the combined stresses in the axial direction are (excluding stresses due to eccentric loading):

(a) Internal pressure and upwind side

$$f_{t\,\max} = (f_{wx} \text{ or } f_{sx}) + f_{ap} - f_{dx} \tag{11.1}$$

(b) External pressure and upwind side

$$f_{r\,\max} = (f_{wx} \text{ or } f_{sx}) - f_{ap} - f_{dx} \tag{11.2}$$

(c) Internal pressure and downwind side

$$f_{c\,\max} = (f_{wx} \text{ or } f_{sx}) + f_{dx} - f_{ap} \tag{11.3}$$

(d) External pressure and downwind side

$$f_{c\,\max} = (f_{wx} \text{ or } f_{sx}) + f_{dx} + f_{ap} \tag{11.4}$$

where, $f_{t\,\max}$ and $f_{c\,\max}$ — maximum tensile and compressive stresses

f_{wx} — stress due to wind load

f_{sx} — stress due to seismic load

f_{ap} — axial stress from internal or external pressure (uniform over the entire height)

f_{dx} — stress due to dead loads.

11.6 DETERMINATION OF SHELL THICKNESS (AT DIFFERENT HEIGHTS)

The procedure for determination of the combined axial stresses at any distance X is outlined above. The usual criterion of shell thickness is the maximum tensile stress, which should not exceed the allowable tensile stress in the material of the shell. The maximum thickness is at the top of the shell end and is determined only on the basis of the circumferential stress.

$$t_s = \frac{pD_i}{2fJ - p} + c \tag{11.5}$$

where, t_s — shell thickness

p — design pressure

D_i — internal diameter of shell

f — allowable stress

J — joint efficient

c — corrosion allowance.

This thickness may be satisfactory up to a certain distance from the top of the shell. In case of a vacuum or low pressure the thickness t_s will be much less. A minimum value for thickness may be assumed in such cases according to the material of construction. In order to evaluate the distance X, it is necessary to determine the combined maximum stress at this distance. This stress should not exceed the allowable stress. The individual stresses at the distance X in the axial direction are given in the following.

11.6.1 Axial Stress Due to Pressure

$$f_{ap} = \frac{pD_i}{4(t_s - c)} \tag{11.6}$$

This is the same throughout the column height.

11.6.2 Stresses Due to Dead Loads

(a) Compressive stress due to weight of shell up to a distance X

$$f_{dx} = \frac{\text{weight of shell}}{\text{cross-section of shell}}$$

$$= \frac{(\pi/4)(D_o^2 - D_i^2)\rho_s(X)}{\left(\pi/4(D_o^2 - D_i^2)\right)}$$

$$= \frac{\text{weight of shell per unit height } (X)}{\pi D_m (t_s - c)} \tag{11.7}$$

where, D_o and D_i are internal and external diameters of shell

ρ_s — density of shell material

D_m — mean diameter of shell.

(b) Compressive stress due to weight of insulation at height 'X'

$$f_{\text{di(ns)}} = \frac{\pi D_{\text{ins}}\, t_{\text{ins}}\, \rho_{\text{ins}}}{\pi D_m (t_s - c)}$$

$$= \frac{\text{weight of insulation per unit height } (X)}{\pi D_m (t_s - c)} \tag{11.8}$$

where, D_{ins} t_{ins} and ρ_{ins} are the diameter, thickness and density of insulation respectively.

D_m — mean diameter of shell

D_{ins} = D_m for large diameter of columns.

(c) Compressive stress due to liquid in the column up to a height X

$$f_{d(\text{liq})} = \frac{\Sigma \text{liquid weight per unit height } (X)}{\pi D_m (t_s - c)} \tag{11.9}$$

(d) Compressive stress due to attachments, such as internals, top head, platforms and ladders up to a height of X.

$$f_{d(\text{att})} = \frac{\Sigma \text{weight of attachments per unit height } (X)}{\pi D_m (t_s - c)} \tag{11.10}$$

Total compressive dead weight stress

$$f_{dx} = f_{ds} + f_{d(\text{ins})} + f_{d(\text{liq})} + f_{d(\text{att})} \tag{11.11}$$

11.6.3 Stress Due to Wind Load at a Distance X

For details see Section 13.4.1 (b)

$$f_{\text{wx}} = \frac{M_w}{Z} \tag{11.12}$$

where, Z – modulus of section for the area of shell

$$\simeq \frac{\pi}{4} D_o^2 (t_s - c)$$

M_w is the bending moment due to wind load at a distance X.

The column is considered as a uniformly loaded cantilever beam. The stress will be compressive on the downwind side and tensile on the upwind side. The wind pressure is related to wind velocity. For calculations, a maximum value of 125 kg per square metre on a flat surface may be taken as satisfactory. For round vessels, a shape factor of 0.7 may be used. The projected are of either uninsulated or insulated tower may be taken for operation of wind load.

$$M_w = \frac{\text{wind load} \times \text{distance}}{2}$$

$$= \frac{0.7 p_w D_o \times X^2}{2} \tag{11.13}$$

where, p_w — wind pressure

$$f_{wx} = \frac{1.4 p_w X^2}{\pi D_o (t_s - c)} \tag{11.14}$$

(compare Eq. 13.22)

11.6.4 Stress Due to Eccentricity of Loads, Tensile or Compressive, According to the Position of Load

$$f_e = \frac{W_e(e)}{(\pi/4) D_o^2 (t_s - c)} \tag{11.15}$$

where, W_e — summation of eccentric loads

e — eccentricity.

11.6.5 Stresses Due to Seismic Load

For details, see Section 13.4.1 (c)

$$f_{sx} = \frac{M_{sx}}{(\pi/4) D_o^2 (t_s - c)} \tag{11.16}$$

(Ref. Eqs. 13.22)

where, bending moment M_{sx} at a distance X is given by

(Ref. Eqs. 13.24)

$$M_{sx} = 4SWX^2 \left(\frac{3H - X}{H^2} \right) \tag{11.17}$$

where, S — seismic coefficient

W — total weight of column

H — total height of column.

11.6.6 Determination of Height X

To determine the value of X, all the stresses (except stresses due to eccentricity of load and seismic load) acting in the axial direction may be added and equated to the allowable tensile stress as follows:

$$\frac{1.4p_w X^2}{\pi D_o(t_s - c)} + \frac{pD_i}{4(t_s - c)} - \frac{\Sigma \text{dead weights}(X)}{\pi D_m(t_s - c)} = f_{t\max} \tag{11.18}$$

If the joint efficiency J is taken into account

$$\frac{1.4p_w X^2}{\pi D_o(t_s - c)} - \frac{\Sigma \text{dead weights}(X)}{\pi D_m(t_s - c)} + \frac{pD_i}{4(t_s - c)} - f_{(\text{all})}J = 0 \tag{11.19}$$

In the above equation, ($f_{t\max}$) is substituted by allowable tensile stress $f_{(\text{all})}$. The possibility of the wind load and the earthquake load operating simultaneously is remote. Therefore, the stresses due to wind load and earthquake load are computed separately and the most adverse loading condition is used to calculate the maximum axial stress.

The above equation can be stated in the form of a quadratic equation

$$aX^2 + bX + c = 0$$

from which $$X = \frac{-b \pm \sqrt{b^2 - 4ac}}{2a} \tag{11.20}$$

After determining the value of X, it is described to adjust the plate thickness, t_s, for the top course so that the height of portion X will be a multiple of the plate width used. Usually, the plate thickness originally selected is satisfactory up to a considerable height. Plates below the distance X must have an increased thickness. Calculations of axial stress with the increase in thickness according to Eqs. (11.7) to (11.20) are repeated, and subsequent height ranges to correspond with increased thickness determined. The procedure is repeated till the entire height of the column in covered. Different sections of the column shall be formed out of a single plate with one longitudinal joint only. Longitudinal joints of adjacent sections shall be staggered. Weld seams shall be, to the extent possible, clear of brackets, supports, etc., so that seams are readily visible for examination.

The stressed condition of the shell indicated so far, is on the assumption that the column is in operation, i.e., it is working with all attachments, insulation, etc. and under requisite pressure and temperature conditions. However, during certain periods the column may be shut down. When stresses due to pressure conditions are absent. It is also necessary to consider the stresses during erection of the column, when there is no insulation load or internal and external attachments, liquid load, etc. The pressure conditions are also absent. Various cases of stressed conditions should be investigated individually to determine the controlling condition. If the column is designed for high pressure service, the limiting condition will usually exist when the shell is under pressure and under a wind load. Columns designed for low pressure service may have maximum stress conditions when the erected empty column is exposed to a high wind load.

11.7 ELASTIC STABILITY UNDER COMPRESSIVE STRESSES

Determination of shell thickness t_s is based on the permissible tensile stress requirements. The compressive stress produced on downwind side must be checked to ensure elastic stability. Failure due to Euler's buckling involves bending of the shell as a strut, which is generally not significant in vertical vessels. Wrinkling of the shell must, however, be checked and avoided. The safe compressive stress, which can be imposed without failure by wrinkling is given by

$$f_c = \frac{1}{12}\frac{E}{\sqrt{3(1-\mu^2)}} \times \frac{(t_s - c)}{D_o/2} \tag{11.21}$$

If, $\mu = 0.3$;

$$f_c = 0.101E\left(\frac{t_s - c}{D_o}\right) \tag{11.22}$$

where, μ — Poisson's ratio

E — modulus of elasticity

11.8 ALLOWABLE DEFLECTION

Column deflection becomes an important factor in the design when the ratio of height to diameter is relatively large. Determination of the static deflection of column is essential for accurate vibration analysis. It is a general practice to allow a maximum deflection in mm of (5 × overall height of column in metres) for vertical columns. A vibration analysis shall be made on columns with 30 metres overall height and above and with height to diameter ratio ranging from 18 to 30. The total deflection at the top of the column is calculated as an algebraic sum by the following equation:

$$\delta = \sum_{i=1}^{n}\left\{\left[\frac{Wl^3}{8EI} + \frac{(\Sigma W)l^3}{3EI} + \frac{(\Sigma M)l^2}{2EI}\right] + \sum_{i}^{l}\left[\frac{Wl^2}{6EI} + \frac{(\Sigma W)l^2}{2EI} + \frac{(\Sigma M)l}{EI}\right]_i (l_{i+l})\right\} \tag{11.23}$$

where, δ — deflection

n — number of sections from top in which the column is suitably divided

W — wind load on the section

ΣW — concentrated load on the section at the end (due to wind load on sections above)

ΣM — moment at the end of the section (due to moment of the sections above)

I — moment of inertia of the cross section

E — modulus of elasticity

l — length of the column section under consideration.

To calculate the total deflection, it is necessary to calculate the deflection of each section of the column having uniform thickness and constant value of E and I. If the wall thickness changes within the section to be calculated, then that section should be divided into two or more sections in accordance with the change in wall thickness. The topmost section of the column has no concentrated load or moment, therefore, ΣW and ΣM are equal to zero, for calculation of the deflection of this section. If the deflection estimated by the above equation is greater than the permissible deflection for the column height, the shell thickness will have to be increased. The new deflection will be recalculated with the revised shell thickness until it falls within the specification.

11.9 DESIGN AND CONSTRUCTION FEATURES OF COLUMN INTERNALS

11.9.1 Plate and Tray Construction

Several factors control the design and construction of plates and trays. These factors are:

(1) Load on the tray due to dead weight, liquid weight and impact due to downcoming liquid
(2) Expansion due to a rise in temperature
(3) Ease of installation and fabrication
(4) Ease of access and maintenance
(5) Method of support
(6) Material of construction
(7) Safety.

Plates or trays can be constructed either as one piece trays or as sectional trays. A one piece tray may be made of cast iron, cast as one piece including the risers and downcomers. It may also be formed conveniently from a thin sheet of ductile material such as copper or steel, with a thickness of 2 mm to 6 mm depending on the diameter and the material. A one piece sieve tray is essentially a flat plate, which is drilled or punched for providing the holes. Small holes are drilled, while large holes can be punched. Holes are usually of 4 to 6 mm diameter located on an equilateral triangular spacing on centres varying from 2.5 to 5 times the hole diameter. In special cases, holes can be up to 20 mm diameter. In the case of punched holes, in copper alloy or carbon steel plates the minimum hole diameter which can be punched is equal to the tray thickness, for stainless steel plates, the hole diameter must be 1½ to 2 times the plate thickness. In the case of bubble cap and valve trays, the tray is formed by cutting a circular blank from a thin sheet of ductile material and then rolling and bending the edge of the blank to form the vertical rim of the tray. Holes are drilled in the tray to suit the size of the bubble cap risers or valves. The spacing is usually on equilateral triangular positions. The sectional tray is built from sections in the form of floor plates cut from sheets, which are laid on the supporting beams and a peripheral ring. A clearance is provided between adjacent sections and clamping devices are used for fixing.

The cast iron tray is able to withstand compressive forces created due to thermal expansion within reasonable limits. The diameters of such trays are also limited to small sizes. The one piece wrought tray made of ductile material is relatively thin and has, therefore, a limited ability to absorb forces due to thermal expansion. The provision of a packing seal between the edge of the tray and the column wall (Fig. 11.26) will help to relieve these and would, therefore, prevent distortion of the tray floor.

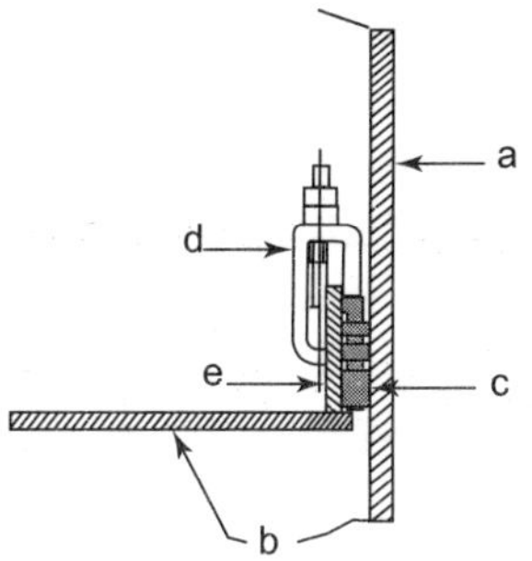

Fig. 11.26 Tray Packing Arrangements: (a) Shell, (b) Tray, (c) Packing, (d) Retaining Clamp, (e) Packing Collar

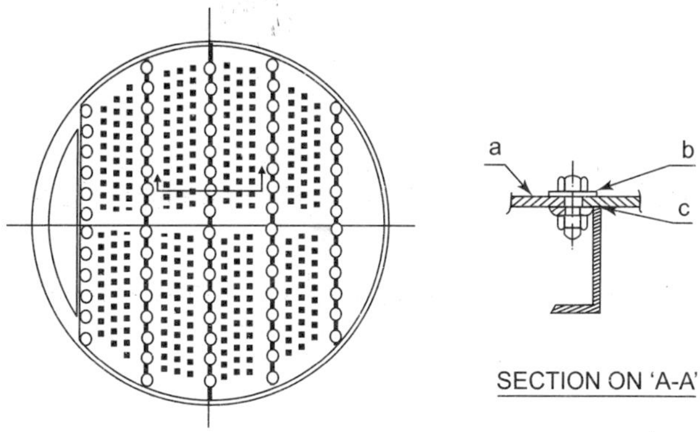

Fig. 11.27 Sectional Tray: (a) Tray, (b) Frictional Washer, (c) Packing

One of the main advantages of the sectional tray is its ability to cope with thermal expansion. The individual sections of the tray are placed on the supporting structure, with an asbestos jointing material inserted between the section and the support member. Each section is finally held by frictional clamping devices (Fig. 11.27). The sections, therefore, move slightly under the influence of the thermal forces and any tendency to distortion is prevented. However, such trays might leak if the joints between sections are not satisfactorily made.

11.9.2 Loading Conditions for Plates and Trays

Plates and trays have to be maintained flat in order to provide a uniform seal of the liquid on their surfaces. Under the action of the various loads, plates and trays are likely to deflect excessively, unless they are made sufficiently thick or and adequate supporting system is provided. The loads which cause deflection are:

(1) Weight of the tray with the contacting devices and downcomers.

(2) Weight of the liquid.

(3) Impact load of the downcoming liquid, which is given by,

$$\text{load} = \frac{wv}{g} \tag{11.24}$$

where, w — wt. of the liquid per second

v — velocity per second

g — gravitation constant.

(4) Expansion due to a rise in temperature (if prevented). It is usual to provide for free expansion. Load due to this may, therefore, be ignored.

(5) Weight of maintenance personnel and tools.

During working conditions the loads (1) to (3) need to be considered for assessment of deflection. In general, a deflection of 3 mm is permissible. In special cases this may be limited to 2 mm.

During cleaning and assembly or inspection operations, loads (1) to (5) should be considered. In this case load due to item (5) is taken as a maximum of 150 kg concentrated load at any point on the tray. The design is based not on the permissible deflection but on the permissible stresses. Certain guidelines for tray loadings are as follows:

(a) Fractionating trays shall be designed for a uniform live load of 60 N/m^2 or the weight of the liquid at maximum height of weir setting whichever is greater.

(b) Maximum deviation from the horizontal at normal tray loading shall not exceed 1/900th of the span.

(c) Areas under downcomers shall be designed for a uniform load of 3200 N/m^2 or weight of liquid for one-half the height of downcomer whichever is greater.

(d) Pans shall be designed for a uniform live load of 800 N/m^2 or the weight of liquid at the maximum operating level on the pan, whichever is greater.

(e) Baffles shall be designed for a live load of 800 N/m^2 on the projected horizontal area.

(f) Trays, baffles and pans, together with their support members shall be designed for a concentrated load of 1500 N at any point on the installed assembly independent of other design live loads and based on allowable stress at normal temperature.

11.9.3 Deflection and Stresses

Deflection and stress determination for trays will depend on the methods used for the supporting structure. Usually three methods are adopted.

(a) The tray is supported on a peripheral ring. This method is adopted only for small diameter columns (Fig. 11.28).

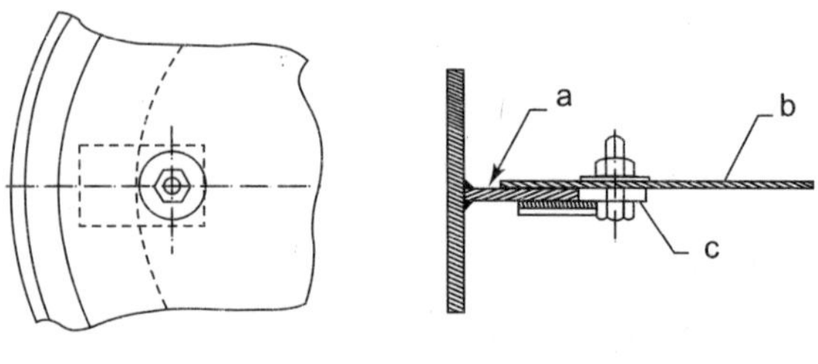

Fig. 11.28 Peripheral Ring Support for Tray: (a) Ring, (b) Tray, (c) Clamp

(b) The tray is supported on a truss made up of angle, channel or trapezoidal section members rolled or fabricated from sheets. The ends of the members of purlins of the truss, are supported by the column wall, through clamps, which are welded to the wall. Alternately, these may be supported on a support ring, welded to the wall (Fig. 11.29). For large trays, the tray is supported on a number of parallel purlins. All the purlins are made to rest centrally on a major beam specially fabricated in the form of an I-section with variable depth (Fig. 11.30). In addition, these purlins are supported and clamped at their ends to a peripheral ring. For sectional trays this arrangement is most suitable.

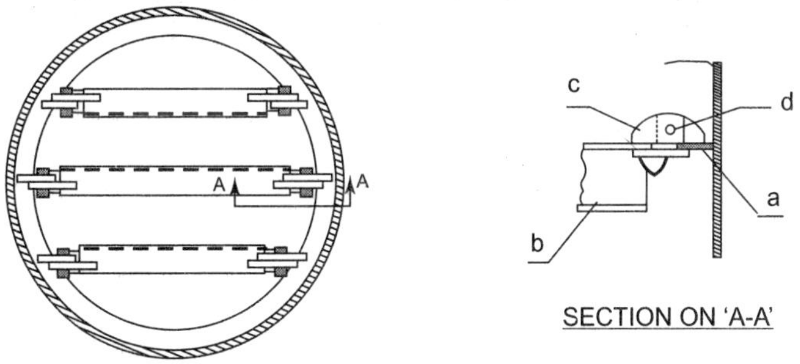

Fig. 11.29 Truss Support for Tray: (a) Ring, (b) Structural Section, (c) Clamp, (d) Fixing Bolt

(c) Vertical supports in the form of round or hexagonal bars, extend between trays and evenly spaced across the area of the tray (Fig. 11.31). This arrangement is suitable for one piece wrought trays.

The deflection and stresses in the three methods indicated can be calculated as follows:

(a) *Tray supported on a peripheral ring*

This may be considered as a circular flat plate, fixed at the circumference and subjected to a uniform load over its surface. The deflection is given by

$$\delta = \frac{3}{16}\frac{m^2 - 1}{Em^2 t^3} \cdot P_L R^4 \tag{11.25}$$

If Poisson's ratio for metal is taken as 1/3 i.e., $m = 3$

then, $$\delta = \frac{1}{6}\frac{P_L R^4}{Et^3} \tag{11.26}$$

where, E — modulus of elasticity

t — thickness of the plate

P_L — $\dfrac{\text{loads (1) to (3) causing deflection (Section 11.9.2)}}{\text{area of the plate}}$

R — radius of the plate.

The above equations may have to be modified in view of the actual construction for each type of tray.

(1) Fixing of the tray at the edge may not be complete, it may only be partial.

(2) The load may not be as uniformly distributed, as is assumed.

(3) The tray is perforated and not solid, as in the case of dual flow, turbogrid or sieve tray. In the case of bubble cap and valve tray, the holes are reinforced by risers.

The value of the constant in Eq. 11.26 may be taken as 1/2 instead of 1/6.

During cleaning and assembly operations a uniformly distributed load of tray and downcomers will produce a stress in the tray which is given by

$$f_1 = \frac{3}{4}\left(\frac{pR^2}{t^2}\right) \tag{11.27}$$

where, p — uniformly distributed load per unit area
R — radius of tray
t — thickness of tray

While due to be concentrated load of maintenance personnel and tools the stress produced in the tray will be given by

$$f_2 = \frac{3w}{2\pi t_2}\left(1.33 \log\frac{R}{x} - 1\right) \tag{11.28}$$

where, w — concentrated load at the centre of tray
t — thickness of tray
R — radius of tray
x — any intermediate radius.

(In Eqs. 11.27 and 11.28, the value of the Poisson's ratio is taken as 0.33).

The load in Eq. 11.28 is taken at the centre of tray for maximum stress condition. At the centre of the tray ($x = 0$), the stress f_2 will be infinitely large. However, in actual practice the load will be operating on a small area. The stress f_2 will be then finite. The total stress will be sum $f_1 + f_2$.

It may be seen from the above calculations of deflection and stress, that a tray supported merely on a peripheral ring has to be carefully handled with minimum loadings during actual operation and maintenance. Its application is strictly limited to small diameter columns.

(b) *Tray supported on a truss*

The size of each purlin or beam of the truss will be determined by the span and the load shared by the purlin. The central purlin will have the maximum span which is almost equal to the internal diameter of the column. Intermediate purlins will have lesser spans. The load on the tray will be shared by the purlins in proportion to the area of the tray supported by it. The deflection of each purlin has to be limited. The deflection is given by (Appendix-B).

$$\delta = \frac{5Wl^3}{384EI} \tag{11.29}$$

where, W — total load carried by the purlin, including its own weight, during working

l — span or purlin

E — modulus of elasticity

I — moment of inertia of the cross-section of purlin.

The stress is the purlin is given by

$$f = \frac{Wl}{8} \times \frac{1}{Z_1} \tag{11.30}$$

for uniformly distribute load of (1) indicated in (Section 11.9.2) and

$$f = \frac{Wl}{4} \times \frac{1}{Z_2} \tag{11.31}$$

for concentrated load of (5) indicated in (Section 11.9.2)

where, Z — modulus of section of the purlin.

(i) *Support rings* The recommended sizes of support rings for different column diameters are given in Table 11.5

Table 11.5 Support Ring Size

Column Diameter (metres)	*Approximate Ring Size*	
	Carbon Steel (mm)	*Alloy Steel (mm)*
1.5	35 × 6	35 × 6
1.5 – 2.5	50 × 6	50 × 6
2.5 – 3.0	65 × 6	65 × 6
3.0 – 3.5	65 × 8	65 × 6
3.5 – 4.5	75 × 8	75 × 6
4.5 – 6.0	90 × 8	90 × 6

(ii) *Purlins or beams* These may be fabricated in the form of angles, channels or trapezoidal sections from sheets, of approximately 2 to 6 mm thickness depending on the material of construction and the load. The following conditions may be observed as far as possible.

(1) Support beam axis should preferably be normal to the direction of liquid flow on the tray.

(2) Beam depth should not exceed 20% of the vertical distance between trays.

(3) Tray areas blanked by supporting members should be in the range of 5 to 10% of the total tray area.

(4) A minimum overlap of 25 mm should be provided between outside diameter of trays, pans, etc., and support rings.

(5) Holding down bolts should be of 10 mm diameter.

(6) Bolt spacing around trays should not exceed 100 mm and that on the downcomers should not exceed 75 mm.

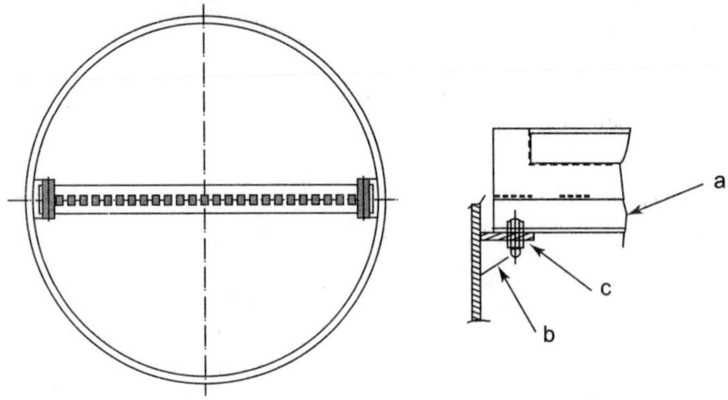

Fig. 11.30 Major Beam Support: (a) Beam, (b) Foot Rest, (c) Bolt

(c) *Vertical supports*

The load on the vertical supports is the total weight of all trays including the liquid and the total impact load. The load is shared by a large number of vertical bars. These bars are supported by the bottom head. A suitable arrangement of brackets, welded to the bottom head acts as a supporting structure (Fig. 11.31). By providing a sufficient number of vertical supports, it is possible to limit deflection of the tray to the desired range.

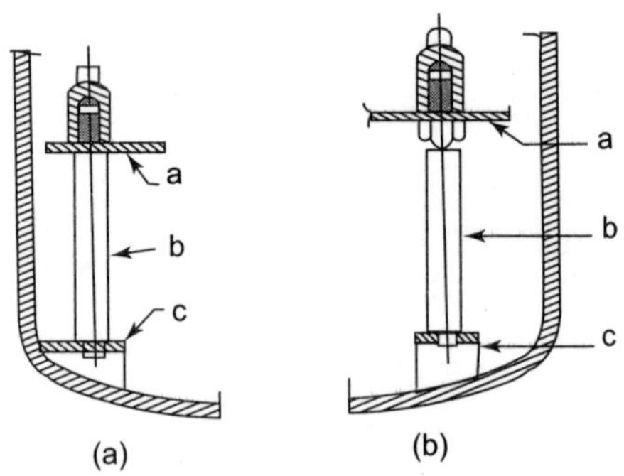

Fig. 11.31 Vertical Supports: (A) Fixed Support; (a) Tray, (b) Support Bar, (c) Bracket, (B) Adjustable Support

Support bars of carbon steel are either round or hexagonal. The recommended sizes are given in Table 11.6. Sizes for hexagonal bars are to be taken across the flats.

Table 11.6 Required Minimum Size of Support Bars (mm)

Length (cm)	*Load in N Per Support*					
	4000	*6000*	*8000*	*12000*	*16000*	*20000*
30, 37.5, 45, 60, 75, 90	22	22	22	22	22	22
105	25	25	25	25	28	30
120	28	28	28	28	32	35

11.9.4 Support Plates in Packed Columns

The load to be carried by the support plates at the bottom of the column are

(1) Weight of the packing

(2) Weight of the liquid

(3) Force due to pressure surges

(4) Weight of the intermediate supports or redistributors.

If there are no intermediate supports, the entire load will be taken by the bottom support. This is not desirable. In general, the packing height per support plate should not exceed 8 to 10 metres, depending on the type of packing. The support plate may be designed as a tray resting on a peripheral ring (Eqs. 11.27 and 11.28), since the diameters of packed columns are usually small.

11.10 SUPPORT FOR COLUMNS

Columns of 600 mm internal diameter and above are usually mounted on mild steel skirts. Smaller columns are more often mounted on legs or from the steel structure. The skirt height (usually about 2500 mm) is decided by NPSH requirement of the bottom product pump. Except for mild steel columns, the top 150 mm of skirt should be made from the same material as the column to avoid contamination of material in contact with the process. Details of support design are given in Chapter 13 (see Section 13.4.1).

11.11 ILLUSTRATION

Following worked example will assist in comprehending mechanical design procedure followed in the case of a tray column, in this case a sieve tray column. For a packed tower similar procedure is to be followed. Various loads considered are for illustration purpose.

Fractionating Column Data

(a) *Shell*

Diameter (i.d)	: 3000 mm
Working pressure	: 1.6 N/mm^2
Design pressure	: 1.76 N/mm^2
Working temperature	: 180 °C
Design temperature	: 200 °C
Base chamber height	: 2740 mm
Top chamber height	: 1050 mm
Feed chamber height	
Material—carbon steel IS-2002 Gr2A	
Permissible tensile stress	: 95 N/mm^2
Insulation thickness	: 100 mm
Density of insulation	: 120 kg/m^3
Corrosion allowance	: 3 mm

(b) *Heads – elliptical, welded to shell*

(ratio of major to minor axis – 2:1)

Material – carbon steel IS-2002 Gr2A	
Permissible tensile stress	: 95 N/mm^2
Weight of head	: 2670 kg

(c) *Support – skirt*

Height	: 4900 mm
Material – carbon steel IS-2062	

(d) *Nozzles*

	Number	Size (mm)	Location
Feed	3	100	Plate No. 34, 38 and 42
Vapour outlet	1	450	Top Head
Reflux	1	200	Plate No. 50
Liquid side-draw	1	75	Base chamber
Vapour inlet	1	500	Base chamber
Level	2	50	Base chamber
Gauge	2	75	Base chamber
Thermocouples	22	25	
Manhole	5	450	Plate No. 10, 20, 30, 42 & 50

Liquid sample	4	25	Plate No. 7, 13, 23 & 41
Reboiler feed	1	450	Bottom head
Pressure gauge	1	25	Base chamber

(e) ***Trays – sieve trays***

Number	50	
Spacing	675 mm	
Holes (diameter)	5 mm	
Number of holes	21,100 (tray No. 1 to 7)	
	24,850 (tray No. 8 to 34)	
	29,400 (tray No. 34 to 50)	
Thickness	2 mm	
Downcomer		
Centre – rectangular	size 300 × 2620 mm	
Side – chord type	size 300 × 1700 mm	
Clearance from tray surface	50 mm	
Weir		
Height above tray	25 mm	
Effective length		
(i) Centre to side	distribution	2620 mm
	overflow	1700 mm
(ii) Side to center	distributing	1700 mm
	overflow	2620 mm

Material of construction for trays downcomers and weir is Stainless steel (SS 304)

(f) ***Supports for tray*** .

Purlins – channels and angles
Live load ~ (liquid + liquid downcomer impact) – 210 kg/m^2
Material – carbon steel
Permissible tensile stress – 127.5 N/mm^2

(g)	Weight of attachments: pipes ladder, platform, etc.	140 kg/m
	Weight of liquid and tray, etc.	92 kg/m^2
	Weight of column (approx.)	200,000 kg
(h)	Wind pressure	1300 N/m^2

11.11.1 Column Shell Thickness

The minimum thickness will be at the top. From Eq. (11.5), we have

$$t_s = \frac{pD}{2fJ - p} + c$$

The entire column will be radio-graphed and, therefore, $J = 1$ (100%)

$$= \frac{1.76 \times 3000}{(2 \times 95 \times 1 - 1.76)} + c$$

$$= 28.04 + c$$

say, $t_s = 32$ mm inclusive of corrosion allowance (3 mm)

[Please note that for all calculations below to illustrate the procedure we consider $(t_s - c) = 28$ mm].

11.11.2 Head

The ellipsoidal head welded to the shell (with ratio of major to minor axis = 2) will have the same thickness as of the shell, after forming. The plate taken for forming the head should be of higher thickness.

The thickness is 28 mm exclusive of corrosion allowance. The estimated weight is 2670 kg.

11.11.3 Shell Thickness at Different Heights

Let us assume height 'H' metres from the top. We shall assume the thickness 28 mm prevails in this height. With this assumption we proceed to calculate the height as outlined below. If this height is more than the actual height of the column, the assumed thickness is acceptable. If the computed height is less than the actual height, the thickness 28 mm will prevail up to that height and beyond that it will be more than 28 mm. Repeat the procedure further.

Remember, the bottom of the tower thickness may be the same as that at the top or more, but never less.

We now calculate stresses at a distance H from the top.

11.11.3.1 Axial Stress due to Pressure

This is same throughout the column and the axial length has no meaning. From Eq. (11.6)

$$f_{ap} = \frac{pD_i}{4(t_s - c)}$$

$$= 1.76 \times 3000/(4 \times 28)$$

$$= 47.1 \text{ N/mm}^2$$

Stresses due to dead loads

(A) *Compressive stress due to weight of the shell up to height, H, is given by Eq. (11.7)*

$$fdx = \frac{\text{weight of shell}}{\text{cross section of shell}}$$

$$= \frac{(\pi/4)(D_0^2 - D_i^2) \times H \times \rho_{\text{shell}}}{(\pi/4)(D_0^2 - D_i^2)}$$

We have $D_0 = 3000 + 2 \times 28 = 3056$ mm

$D_i = 3000$ mm

and $\rho_{\text{shell}} = 7850\ \text{kg/m}^3$

Therefore, $f_{dx} = 7850 \times \text{H kg/m}^2$

$= 7850 \times 9.807 \times \text{H} \times 10^{-6}\ \text{N/mm}^2$

$= 0.077 \times \text{H N/mm}^2$

(B) *Compressive stress due to weight of the insulation*

The insulation thickness is 100 mm. Therefore, the outside diameter with insulation, $D_{\text{ins}} = 3056 + 2 \times 100 = 3256$ mm. From Eq. (11.8), we have

$$f_{\text{dins}} = \frac{\pi \times D_{\text{ins}} \times t_{\text{ins}} \times H \times \rho_{\text{ins}}}{\pi \times D_m (t_s - c)}$$

D_m = Mean diameter of the column shell = (3000 + 3056)/2

= 3028 mm

$\rho_{\text{ins}} = 120\ \text{kg/m}^3$

$$f_{\text{dins}} = \frac{3256 \times 100 \times H \times 10^3 \times 120 \times 10^{-9} \times 9.807}{3028 \times 28}$$

$= 4.52 \times 10^{-3} \times \text{H N/mm}^2$

(The factor 9.807 is to convert force in kg into N).

(C) *Compressive stress due to liquid, tray, etc.*

The weight of liquid on tray and weight of tray are given in terms of the cross section of the column. As a first approximation we assume the entire height of the column is full of trays, despite the fact that the chamber height 1050 mm does not contain any tray. The tray spacing is 675 mm.

The weight in terms of height is then= $(H/0.675) \times (\pi/4)\,(3)^2 \times 92$

$$= 963.4\ H;\ \text{kg}$$

Therefore, the stress, f_{dliq} $= 963.4H/(\pi \times D_m\,(t_s - c))$

$$= \frac{963.4H \times 9.807}{\pi \times 3028 \times 28}$$

$$= 0.035\ H\ \text{N/mm}^2$$

(D) *Compressive stress due to attachments such as: top head, platform, ladder, etc.*

This stress, f_d, is given by Eq. (11.10):

$$f_d = \frac{(2670 + 140H) \times 9.807}{\pi \times 3028 \times 28}$$

$$= 0.098 + 0.0052\ H\ \text{N/mm}^2$$

The total compressive load is the sum of all stresses A to D

$$f_{dH} = 0.077\ H + 0.00452\ H + 0.035\ H + 0.098 + 0.0052\ H$$

$$= 0.098 + 0.122\ H\ \text{N/mm}^2$$

(E) *Stress due to wind load*

The column is in open space. We, therefore, calculate stress due to wind load which is more common than that by seismic activity. We do not consider wind load and seismic load occurring simultaneously. Strictly speaking, calculations should be performed for both cases independently and then the thickness computed. Here, we will illustrate the procedure by considering wind load only. This stress is calculated based on an uninsulated column.

From Eq. (11.14), we then have

$$f_{wH} = \frac{1.4(1300)H^2}{\pi \times 3056 \times 28}$$

$$= 6.77 \times 10^{-3}\ H^2\ \text{N/mm}^2$$

In order to determine height H let us add all stresses on upward side (Eq. (11.18)) and equate to permissible stress.

The pressure load and wind load are tensile in nature, while weight loads are compressive. Therefore, from Eq.(11.1), we have

$$6.77 \times 10^{-3}\ H^2 + 47.1 - 0.098 - 0.122\ H = 95$$

(Please note the allowable stress is 95 N/mm^2 and the joint efficiency, J is = 1). Therefore, we get

$$6.77 \times 10^{-3}\, H^2 - 0.122\, H - 48 = 0$$

This quadratic equation takes on a simplified form without sacrificing engineering accuracy as

$$H^2 - 18\, H - 7090 = 0$$

By taking a positive value of H from the resulting solution, we have

$$H = 93.7 \text{ m}$$

The height acceptable is 93.7 m while the actual height is 44 m. Therefore, the column of uniform shell thickness 32 mm inclusive of corrosion allowance is acceptable.

On the downwind side, the compressive stress as per Eq. (11.3) for the height 44 m, we have

$$f_c = 6.77 \times 10^{-3}\, (44)^2 - 47.1 + 0.098 + 0.122\, (44)$$

$$= -28.53 \text{ N/mm}^2$$

It means the stress is tensile and not compressive.

If the column is considered without an internal pressure operating, then the compressive stress will be

$$f_c = 6.77 \times 10^{-3}\, (44)^2 + 0.098 + 0.122\, (44)$$

$$= 18.57 \text{ N/mm}^2$$

This is well within permissible stress for elastic stability (Eq. (11.22)).

$$f_c = 0.101 \times 190 \times 10^3\, (28/3056) = 175.8 \text{ N/mm}^2.$$

11.11.4 Nozzles

The column is provided with several nozzles. All nozzles above 50 mm diameter have to be checked for reinforcement. Calculations of reinforcements are made as per area for area method [see Section 5.8.3.2 (a)]. Similarly, all nozzles above 50 mm diameter are to be provided with standard pipe flanges.

11.11.5 Trays

A column of this type (see illustration) used in the industry, is made up of a shell, fabricated from sheets by welding. It has no flanged joints. This naturally requires a special design of trays, which can be easily dismantled and removed. Such trays are the sectional trays. These trays are made from stainless steel, with a thickness of 2 mm and with holes of 5 mm diameter. Downtakes and weirs are bolted at the appropriate positions. The downcomers on the even number of trays are located at the ends. The shape of the downcomer is segmental with a size 300 mm × 1700 mm at the top and mm × 1700 mm at the lower end. The downcomers on the odd number trays are at the centre. The shape of the downcomer is rectangular with a size 300 mm × 2620 mm at the top and 200 mm × 2620 mm at the lower end.

11.11.6 Support for Trays

The supporting system consists of six purlins symmetrically placed underneath each tray. The ends of purlins are attached to two sectors (support rings) which are fixed to the column wall. The length of each purlin is as shown in Fig. 11.32 (AA, BB, CC). The load carried by each purlin is in proportion to the area of the tray supported by purlin.

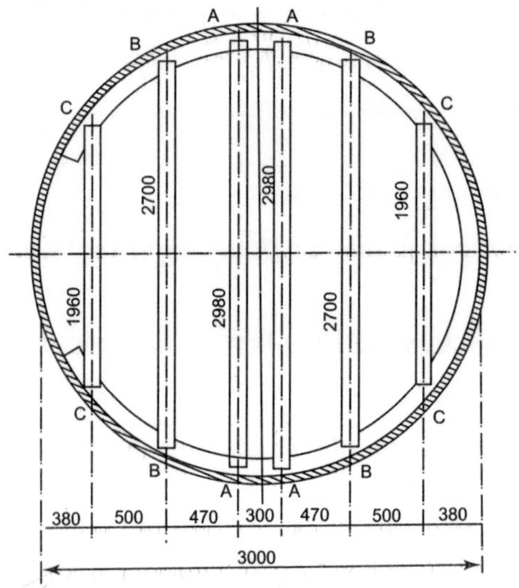

Fig. 11.32 Tray Supporting Structure

The width is taken as half the distance to next purlin on either side. The deflection, δ, is limited to 3 mm.

Purlin AA

We have,

$$\text{Length} = 2980 \text{ mm}$$

$$\text{Width} = (470 + 300)/2 = 385 \text{ mm}$$

$$\text{Live load} = 210, \text{kg/m}^2 = 210 \times 2.980 \times 0.385$$

$$= 240.93 \text{ kg}$$

$$\text{say,} = 241 \text{ kg}$$

Following are the estimated weights in kg

Tray weight	= 16
Downcomer weight	= 25
Purlin weight	= 50
Total weight (live + dead)	= 332 (= 332 × 9.807 = 3256 N)

This load is uniformly distributed. In order to find out suitable channel section, we need to find the moment of inertia (*I*) and modulus (*Z*) of the section.

Moment of inertia

From Eq. (11.29), we have

$$\delta = \frac{5 \times W \times l^3}{384 \times E \times I}$$

$$\delta = \text{deflection} = 3 \text{ mm}$$

We then have,

$$I = \frac{5 \times W \times l^3}{384 \times E \times \delta} = \frac{5 \times 3256 \times (2980)^3}{384 \times 2 \times 10^5 \times 3}$$

$$I = 186.99 \text{ cm}^4$$

say, $I = 187 \text{ cm}^4$

During the assembly and cleaning operations there are two loads namely, (i) uniformly distributed load of 91 kg (16 + 25 + 50) and (ii) concentrated load of 150 kg.

Modulus of section (Z)

From Eq. (11.30), we have for the uniformly distributed load

$$Z_1 = \frac{W \times l}{8 \times f}$$

$$= \frac{(91 \times 9.807) \times 2980}{8 \times 127.5}$$

$$= 2607.3 \text{ mm}^3 = 2.61 \text{ cm}^3$$

For a concentrated load of 150 kg, we use Eq. (11.31) to compute Z_1

$$Z_2 = \frac{W \times l}{4 \times f}$$

$$= \frac{(150 \times 9.807) \times 2980}{4 \times 127.5}$$

$$= 8595.5 \text{ mm}^3 = 8.6 \text{ cm}^3$$

The total modulus = 2.61 + 8.6 = 11.21 cm^3

A suitable channel section which satisfies the moment of inertia and modulus of section is ISJC 175 (175 × 60 × 6.9 × 3.6 all in mm). Considering corrosion allowance, use the channel of one size higher say, ISMC 225.

Purlin BB

Length = 2700 mm

Width = (470 + 500)/2 = 485 mm

Live load (liquid only, no impact load) is estimated = 60 kg/m^2

The liquid load = 60 × 2.7 × 0.485 = 78.57 kg

say, = 80 kg

The tray load will be distributed in proportion to the area:

The tray weight = 16 × (2.7 × 0.485)/(2.98 × 0.385) = 18.26 kg

say, = 19 kg

There will not be any downcomer load in this area.

The estimated purlin weight = 45 kg.

Total load = 80 + 19 + 45 = 144 kg

During assembly and cleaning operation, (i) uniformly distributed load is 64 kg (19 + 45) and (ii) concentrated load is 150 kg.

Following an identical procedure as outlined above, we get the moment of inertia I = 60.32 cm^4 and modulus = 9.45 cm^3 (Z_1 = 1.66 cm^3 and Z_2 = 7.79 cm^3).

Use channel ISJC 150. Considering corrosion allowance take one size higher say ISMC 200.

Purlin CC

Length = 1960 mm

Width = (380 + 500)/2 = 440 mm

Live load (liquid and impact load) is estimated = 210 kg/m^2

The liquid load = 210 × 1.96 × 0.440 = 181.1 kg

say, = 182 kg

The tray load will be distributed in proportion to the area:

The tray weight	= 15 kg
The downcomer load	= 25 kg
The estimated purlin weight	= 33 kg
Total load	= 182 + 15 + 25 + 33 = 255 kg

During assembly and cleaning operation, (i) uniformly distributed load is 73 kg (15 + 25 + 33) and (ii) concentrated load is 150 kg.

Following identical procedure as outlined above, we get moment of inertia I = 40.86 cm^4 and modulus = 7.02 cm^3 (Z_1 = 1.37 cm^3 and Z^2 = 5.65 cm^3). Use channel ISJC 125. Considering corrosion allowance use the channel one size higher say ISMC 200.

References and Recommended Readings

American Institute of Chemical Engineers, *Equipment Testing Procedure: Plate Distillation Column* (1962).

Arni, V.R.S., 'New Breakthrough in Distillation Tray Design', *Chemical Age of India,* **21** (10) 877 (1970).

Atkins, G.T., 'How to Specify Trays', *Petroleum Refiner,* **39** (8) 86 (1960).

Barnicki, S.D., and Davis, J.F., Designing Sieve Tray Columns, Part I, *Chemical Engineering,* **96** (10) 141 (1989); Part II *Chemical Engineering,* **96** (11) 202 (1989).

Bowman, J.D., 'Use column scanning for predictive maintenance', *Chemical Engineering Progress,* **87** (2) 25 (1991).

Bravo, J.L., 'Effectively Fight Fouling of Packing', *Chemical Engineering Progress,* **89** (4) 72 (1993).

Chen, G.K., 'Packed Column Internals', *Chemical Engineering,* **91** (5) 40 (1984).

Clay, H.A., Clark, J.W., and Munro, B.L., 'Which Packing for Which Job', *Chemical Engineering Progress,* **62** (1) 51 (1966).

Coker, A.K., 'Understand the Basics of Packed Column Design', *Chemical Engineering Progress,* **87** (11) 93 (1991).

Dwivedi, M.C. 'Mechanical Stability of Contact Tray', *Chemical Age of India,* **29** (5), 333 (1978).

Fair, L.R., 'Selecting Fractionating Column Internals', *Chemical Engineering Progress,* **72** (14) 107 (1965).

Fleming, B., Martin, G.R. and Hartman, E.L., 'Pay Attention to Reflux/Feed Entry Design', Chemical Engineering Progress, **92** (1) 56 (1996).

Harrison, M.E., and France, J.J. 'Trouble Shooting Distillation Columns Part I', *Chemical Engineering,* **96** (3) 116 (1989). Part II *Chemical Engineering,* **96** (4) 121 (1989), Part III, *Chemical Engineering,* **96** (5) 126 (1989).

Haselden, G.G., 'Scope for Improving Fractionation Equipment', *The Chemical Engineer,* **299/300** 439 (1975).

Interess, E., 'Practical Limitations on Tray Design', *Chemical Engineering,* **78** (26), 167 (1971).

Jamison, R.H., 'Distillation Internal Design Techniques', *Chemical Engineering Progress,* **65** (3) 46 (1969).

Jordon, D.C., 'Chemical Process Development, Parts I and II', Interscience, New York (1968).

Kister, H.Z., 'Design and Layout for Sieve and Valve Trays', *Chemical Engineering,* **87** (15) 119 (1980).

———, *Distillation Design,* McGraw-Hill Inc. New York (1992).

———, *Distillation Troubleshooting,* John Wiley and Sons Inc. Publication, New Jersey (2006).

Leva, M., *Tower Packings and Packed Tower Design,* 2nd Edition., U.S. Stoneware Co., Akron, Ohio, (1953).

Ludwig, E.E., *Applied Process Design for Chemical and Petrochemical Plants,* 3rd Edition, Vol. 2, Gulf Publishing Company, Houston (1997).

Mahajani, V.V., 'Figuring Packed Tower Diameter', *Chemical Engineering,* **89** (19) 132 (1982).

McClain, R.W., 'How to Specify Bubble Cap Trays', *Petroleum Refiner,* **39** (8) 92 (1960).

Olujic, Z., Stoter, F., and de Graauw, J., 'Gas distribution in large-diameter packed columns', *Gas Separation and Purification,* **5** 59 (1991).

Patton, B.A., and Pritchard, B.L. (Jr.), 'How to Specify Sieve Trays', *Petroleum Refiner,* **39** (8) 95 (1960).

Perry, D., Nutter, D.E., and Hale, A., 'Liquid Distribution for Optimum Packing Performance', *Chemical Engineering Progress,* **86** (1) 30 (1990).

Perry, R.H., Green, D.W., and Maloney, J.O., *Perry's Chemical Engineers Handbook,* 6th Edition, International Edition, McGraw-Hill, Singapore (1984).

Proctor, J.F., 'Sieve and Bubble Cap Plates', *Chemical Engineering Progress,* **59** (3) 47 (1963).

Rader, R.G., Mutsakis, M., Grosz-Roell, F., and Maugweiler, W., 'Better Absorption? Try a Static Mixer', *Chemical Engineering,* **96** (7) 137 (1989).

Robin, B.J., 'Mechanical Problems in Bubble and Grid Tray Design', *British Chemical Engineering,* **3** 840 (1958).

Robin, B.J., 'The Ballast Tray: A New Distillation Device', *British Chemical Engineering,* **4** 351 (1959).

Rukovena, F., and Koshy, T.D., 'Packed Distillation Tower Hydraulic Design Method and Mechanical Considerations', *Industrial and Engineering Chemistry Research,* **32** 2400 (1993).

Sattler, F.J., 'Nondestructive Testing Methods Can Aid Plant Operation', *Chemical Engineering,* **97** (10) 177 (1990).

Sewell, A., 'Practical Aspects of Distillation Column', *The Chemical Engineer,* **299/300** 44 (1975).

Shah, G.C., 'Effectively Trouble Shoot Structured Packing Distillation Systems', *Chemical Engineering,* **87** (4) 49 (1991).

Sloley, A.W., and Fleming, B., 'Successfully Downsize Trayed Columns', *Chemical Engineering Progress,* **90** (3) 39 (1994).

Smith, B.D., *Design of Equilibrium Stage Processes,* McGraw-Hill New York, (1963).

Sloley, A.W., 'Don't Get Drawn into Distillation Difficulties', *Chemical Engineering Progress,* **94** (6) 63 (1998).

Strigle, R.F., *Random Packings and Packed Towers: Design and Application,* Gulf Publishing Co., Houston (1987).

Tang, S.S., 'Shortcut Methods of Calculating Tower Deflection', *Hydrocarbon Processing,* **47** (11) 230 (1968).

Thrizt, G.C., 'How to Specify Valve Trays', *Petroleum Refiner,* **39** (8) 93 (1960).

Van Winkle, M., *Distillation,* McGraw-Hill, New York (1967).

The following web sites may be helpful

www.kochglitsch.com

www.sulzerchemtech.com

www.evergreenindia.com

12
Chapter

Auxiliary Process Vessels

12.1 INTRODUCTION

Auxiliary process vessels are essential for fulfilling the functioning of some of the major process equipments, for example, a wrongly selected vapour liquid separator in the evaporator (see Para 10.3) may result in loss of a valuable product or contamination of steam in case of multiple effect evaporator system. Steam contamination may require condensate treatment before it is sent to boiler as boiler feed water.

Some of the auxiliary vessels, most often used are reflux drums, liquid-liquid separators and vapour-liquid separators.

12.2 REFLUX DRUM

A reflux drum is an integral part of the distillation/fractionation columns. Generally, the reflux drum is placed in a horizontal position. Design pressure of the drum will be practically the same as the column design pressure. Fractionating columns are operated either under positive pressure, near atmospheric pressure or under reduced pressure (vacuum), depending upon the system. The design temperature would be the same as that of the condenser design temperature on the vapour side. The material of construction will be compatible with the process fluid. Diameter of the drum is based on a linear velocity of about 0.15 m/s and minimum 20% of the drum diameter is provided as vapour space. Process design of the reflux drum has been extensively discussed in the texts indicated under Reading References.

If the system operates under reduced pressure and vacuum is pulled off reflux drum, sufficient care must be taken to avoid a liquid entrainment in the vacuum system. This can be achieved by installing a demister pad as discussed under entrainment separators.

In case, there is possibility of an immiscible phase (say water as the heavier phase), a special draw off arrangement in the form of a boot is provided as shown in Fig. 12.1. The drum design is similar to the design of a pressure vessel design.

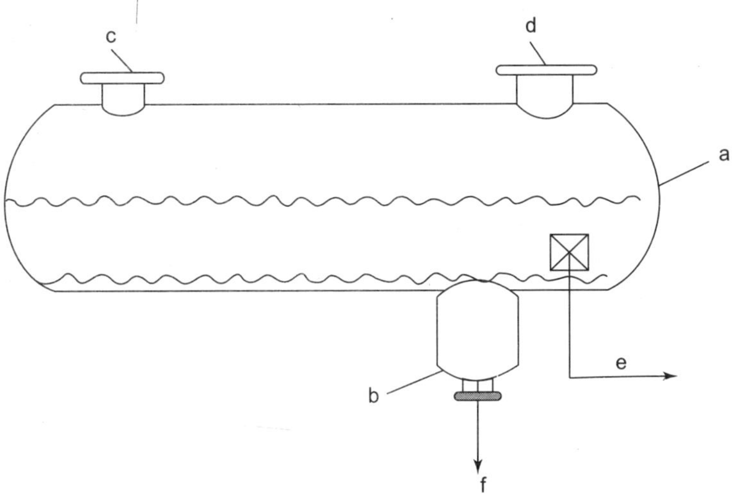

Fig. 12.1 Reflex Drum: (a) Drum, (b) Boot, (c) Liquid Inlet, (d) Non-Condensable Outlet, (e) Drum Outlet, (f) Draw-off (Heavy Phase, if Present)

12.3 COMPRESSOR KNOCK-OUT DRUM

The function of a knock-out drum is to provide enough residence time to the compressed gas/ vapour such that the liquid is separated. Residence time for gas may be about ten minutes. The exit gas moves through the demister pads to knock-out any entrained liquid.

The ratio between the length and diameter varies between 3 and 5 depending upon pressure. The design pressure depends on compressed gas pressure and so would the design temperature. The knockout drum essentially serves as an entrainment separator. In order to facilitate drainage of liquid, the drum may be installed in a slightly inclined position.

The compressed gas enters at one end and exits at other end through demister pads to knock off residual entrained liquid. The mechanical design of the drum and supports would be similar to a pressure vessel design. At times space may demand the knockout drum to be in a vertical position. In that case nozzle orientation needs change. A typical schematic diagram of knockout drum is shown in Fig. 12.2.

12.4 LIQUID-LIQUID SEPARATORS

Liquid-Liquid separators are generally horizontal vessels. For very low flow rates vertical vessels are used. The design pressure and temperature are based on fluid pressure and temperature. Process

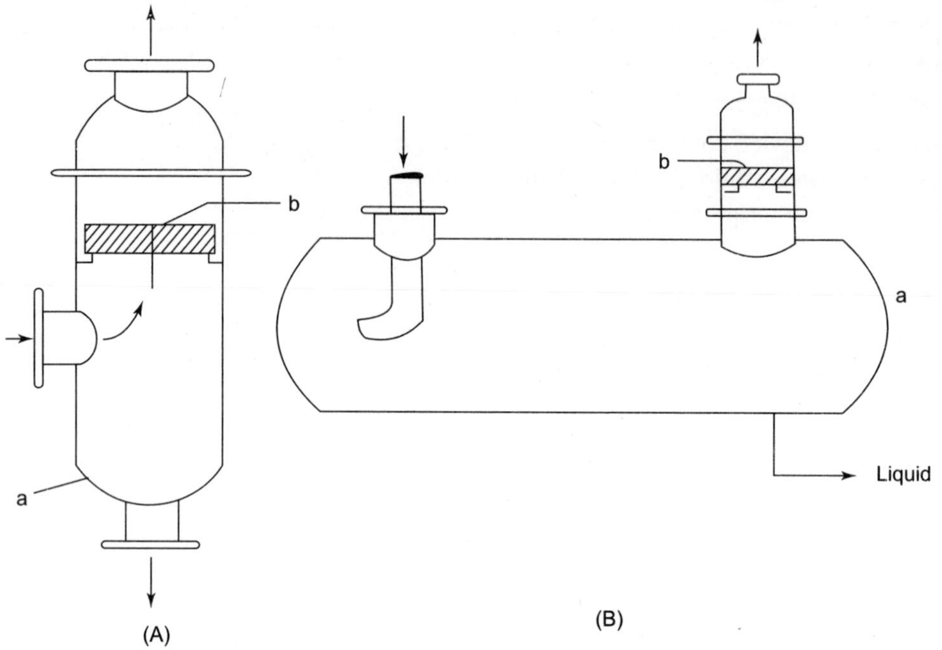

Fig. 12.2 Knockout Drums: (A) Vertical, (B) Horizontal, (a) Drum, (b) Demister Pad

fluid compatibility would decide material of construction. The feed nozzle size must be such that it would avoid 'jet' effect making separation rather difficult. A baffle placed in front of inlet feed nozzle is generally adequate to avoid the jetting effect (Fig. 12.3).

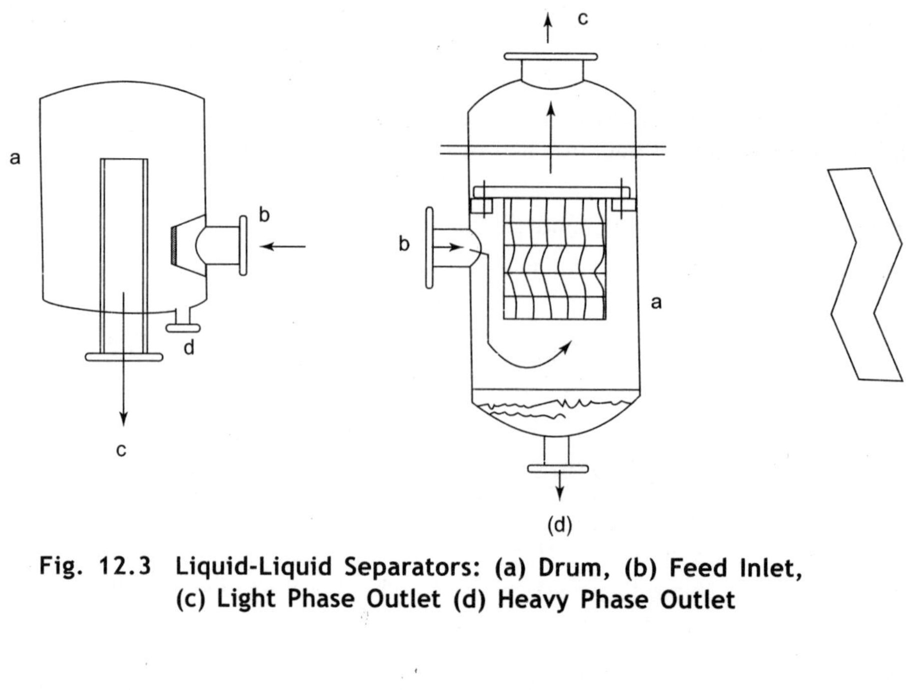

Fig. 12.3 Liquid-Liquid Separators: (a) Drum, (b) Feed Inlet, (c) Light Phase Outlet (d) Heavy Phase Outlet

If dispersion is stable (which is the case most of the time), a coalescence device is used in the separator to enhance phase separation. A stable dispersion is passed through a coalescence device, resulting in a larger diameter of droplets which can be settled is usual way.

The coalescence device could be plate type of knitted wire mesh type (wire of metal, fibre glass or plastics). The plates would also be packed depending upon difficulty of separation. It is possible to design knitted wire mesh type devices to reduce contaminant level as low as 5 ppm.

12.5 VAPOUR/GAS LIQUID SEPARATORS

Gas-liquid operations many times result in a liquid carry over in the form of fine droplets ranging from 0.01 μm to 1000 μm. For instance, the sprays may result in liquid droplets 500-1000 μm while steam coming out of vapourizer drum may carry liquid water droplets in the range typically (0.1 μm to 25 μm). Such liquid carry over in vapour is very common in evaporators, distillation columns, gas absorbers, refinery vacuum tower, knockout drums. For many process operations it is essential that stream is free of liquid (or even solid) and particulate matter. For instance if suction of a compressor contains liquid droplets, it may pose severe maintenance problems. Similarly, if steam fed to a reformer carries liquid droplets with it, the temperature of superheated steam may not reach the desired temperature. For the heat supplied will be utilized to vaporize water rather than carrying out sensible heat exchange.

A variety of gas/vapour-liquid separators are usually available. Liquid droplet size and carryover of particular matter are major factors deciding their selection. Gravity separators, cyclone separators and impingement type of separators are among the most commonly used (see Section 10.3).

12.5.1 Gravity Separators

In the gravity separators, gas or vapour enters at one end of a horizontal drum. The gas velocity is now reduced considerably due to the larger diameter of drum compared to feed pipe diameter. The liquid droplets, hence, settle and the liquid is drained through a boot. The exit gas may be passed through demister pad mounted at the other end of the drum.

12.5.2 Cyclone Separators

It is a cylindrical vessel where a gas/vapour stream enters tangentially and swirls downward and then passes through a central pipe. Due to centrifugation, liquid droplets are collected at the bottom and the accumulated liquid can then be drained. The inlet velocity should be in the range 30–50 m/s. Very high velocities may result in the gas carrying liquid due to creeping up phenomenon.

12.5.3 Impingement Separators (Mist Eliminators)

The most commonly used gas/vapour liquid separators in chemical process industry are based on droplet coalescence or capture by internal impaction, direct interception or brownian diffusion and coalescence.

In its simplest form, it consists of a baffle placed in a vapour/gas flow path perpendicular to direction of flow. The liquid droplets hit the baffle, and are drained in the form of liquid. They can hit even a pipe and then be separated while changing direction of flow.

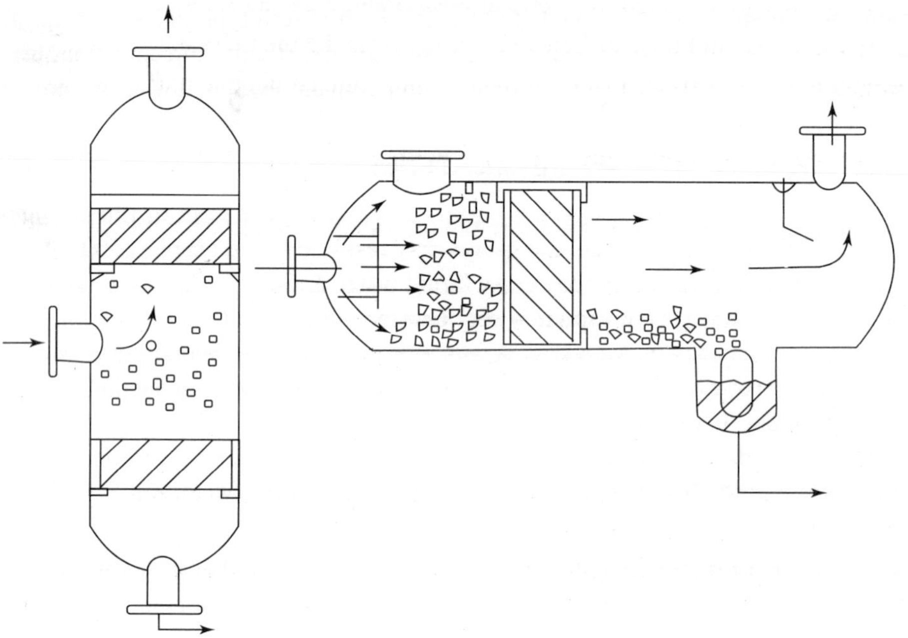

Fig. 12.4 Mist Eliminator

Some typical arrangements are shown in Figs 12.4 and 12.5. For large flow rates, vane shaped mist eliminators could be used. It consists of a series of baffles or vanes between which gas/vapour laden with liquid droplets travel. The change in direction results in impaction also. The space between two baffles could be 5 to 50 mm or more depending on allowable pressure drop and liquid carry-over. To narrow a passage may cause flooding thereby carrying liquid into exit gas stream.

All the above types are suitable for liquid droplets of 25 μm to 50 μm.

12.6 WIRE MESH MIST ELIMINATORS (ENTRAINMENT SEPARATORS)

The most popular mist eliminators are wire mesh mist eliminators with wire mesh thickness ranging from 0.1 to 0.3 mm and a void fraction of 0.95 to 0.99. The thickness of knitted wire pads could vary between 100 and 300 mm depending upon application. The material of construction may be stainless steel, monel, nickel, inconel, titanium, copper, aluminium, polypropylene, teflon depending upon compatibility with process fluid. In other words, the material could be anything that can be drawn or extruded.

Wire mesh mist eliminators can remove droplets > 3 μm. They have excellent turn down ratio of 30 to 110% for design capacity.

While installing wire mesh mist eliminators, the following points should be considered:

(1) Generally they are installed in a vertical position with gas up-flow.

(2) Gas or vapour should be distributed uniformly over the mist eliminator pad.

(3) When there are limitations inside a process equipment an internal baffle element can be used prior to wire mesh mist eliminator to ensure uniform flow.

(4) In order to remove solids adhering to the bottom face, it is desirable to spray from bottom side.

Wire mesh eliminators are not suitable for removing submicron mist as in the case of sulfuric acid plant. For such service, fibre bed mist eliminators are ideal. The fibre bed mist eliminator consists of densely packed fibre bed with steel support. Fibre diameter and bed voidage are selected in such a way as to provide mechanical stability of the element. The fibres (generally glass fibres) are such that they will not absorb liquid. For densely packed mist eliminator, pressure drop is higher than that in wire mesh type mist eliminator. Some design features are presented in Chapter 10 (Para 10.3).

Everon Vane Mist Eliminator [Courtsey: Evergreen Technologies Pvt Ltd. (Mumbai)]

Evermesh Mist Eliminator

Fig. 12.5

12.7 HYDROCYCLONE

Hydrocyclone is used for separation of two liquids as well as solids (very small quantity) from liquids. The common application of this device is in offshore oil drilling to separate oil from sea water, recovery of filter aid, recovery of catalyst, removal of silt from water, removal of oil and grease in effluent treatment plant and so on.

The separation is based on the effect of centrifugal force created in a cyclone body. The particles of dispersed phase (solid or liquid) are separated from the continuous phase by density difference.

The particle size is important. The hydrocyclone has no rotating part. The centrifugal force is created by entering fluid. The velocity of entering fluid and size of the hydrocyclone are important for successful operation. For removal of very fine particles (dispersed phase solid or liquid) smaller hydrocyclones are preferred. Considering volume of operation one can have clusters of small hydrocyclones mounted on a common big holding tank. These devices are practically maintenance free, except corrosion/erosion.

These are proprietary items and hence the proven track record of the vendor plays critical role in its implementation and successful operation.

Figure 12.6 exhibits a hydrocyclone.

The material of construction can be any metal/plastic/FRP compatible with process fluids.

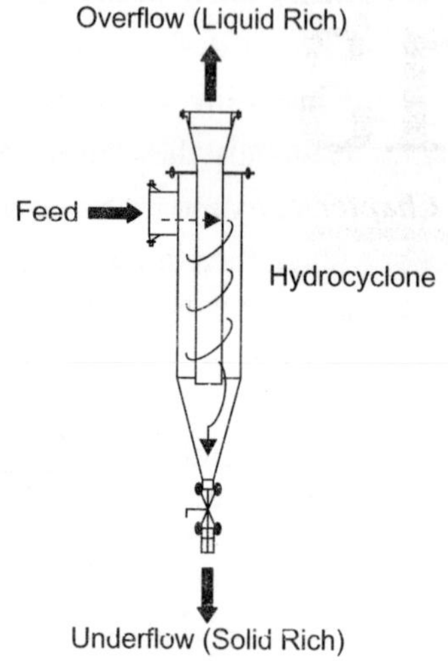

(*Courtesy*, Mamko Design and Engineering Pvt. Ltd, Mumbai -63)

Fig 12.6 Hydrocyclone

References and Recommended Readings

Carpenter, C.L., and Othemer, D.E., 'Entrainment removal by a wire mesh separator', *A.I.Ch E Journal,* **1** (4) 549 (1955).

Couper, J.R., Roy Penny, W., Fair, J.R., and Walas, S.M., *Chemical Process Equipment Selection & Design,* Second Edition, Elsevier, Amsterdam (2005).

Duros, D.R., and Kennedy E.D. 'Acid Mist Control', Chemical Engineering Progress, **74** (9) 70 (1978).

Fabrian, P., Cusack, R., Hennessey, P., and Neuman, M., Part I: *The basics Demistifying the selection of mist Eliminators, Chemical Engineering,* **100** (11) 148 (1993). Part. II: *The applications, Chemical Engineering,* **100** (12) 106 (1993).

Landis, D.M., 'Centrifugal Coalescers', *Chemical Engineering Progress,* **61** (10) 58 (1965).

Sigales, B., 'How to design settling drums', *Chemical Engineering,* **82** (13) 141 (1975), and correction in *Chemical Engineering* **82** (20) 5 (1975).

Sigales, B., 'More on how to design reflux drums', *Chemical Engineering,* **82** (20) 87 (1975).

York, O.H., 'Entrainment Separation', Encyclopedia of Chemical Processing and Design, Ed McKetta, J.J., Marcel Dekker Inc. New York and Basel, **19** 168 (1983).

Zenz, F.A., 'Cyclones', Encyclopedia of Chemical Processing and Design, Ed McKetta, J.J., Marcel Dekker Inc. New York and Basel, **14** 82 (1982).

The following web site would help

www.evergreenindia.com

www.mamko.com

13

Chapter

Supports for Vessels

13.1 INTRODUCTION

Cylindrical and other types of vessels have to be supported by different methods. Vertical vessels are supported by brackets, columns, skirts or stool supports, while horizontal vessels are supported by saddles. The choice of the type of support depends upon the height and diameter of the vessel, available floor space, convenience of location, operating temperature and the material of construction.

The attachment of supports to the vessel walls causes additional stresses in the wall which the vessel should withstand in addition to the stresses created due to the operating pressure. Calculations of bending and other stresses in the shell due to supports are very complex. It is advisable to assess such stresses only when the effect of supports is significant in comparison with internal pressure. It is also necessary to ensure that the attachments of the support to the vessel, which are usually by fillet welds, should be able to transfer the load safely from vessel to its support and that the support should be strong enough to withstand the load of the vessel.

13.2 BRACKET OR LUG SUPPORTS

These (Fig. 13.1) can be easily fabricated from plates and attached to the vessel wall with a minimum welding length. They are made to rest on short columns or on beams of a structure depending on the elevation required. They can be easily levelled. Due to the eccentricity of these supports and the resulting bending moment, compressive, tensile and shear stresses are induced in the vessel wall. These stresses must be combined with circumferential and longitudinal stresses produced in the vessel wall due to operating pressure. The shear stresses being of a smaller magnitude can be ignored.

Bracket supports are more suitable for vessels having thick walls, as these are capable of absorbing the bending stresses due to eccentricity of loads. In vessels with thin walls, it is necessary either to reinforce certain areas of the vessel wall, where the bracket is attached, or to use many brackets. In case a pad is used, it is welded to the shell. The bracket is then welded to the pad. The thickness of the pad is usually made equal to or greater than the thickness of the shell with a minimum thickness of 6 mm. It is usual to provide 2 brackets for vessels up to 0.6 metre diameter, 4 brackets for vessels up to 3 metre diameter, 6 brackets for vessels up to 5 metre diameter and 8 brackets for vessels above 5 metre diameter.

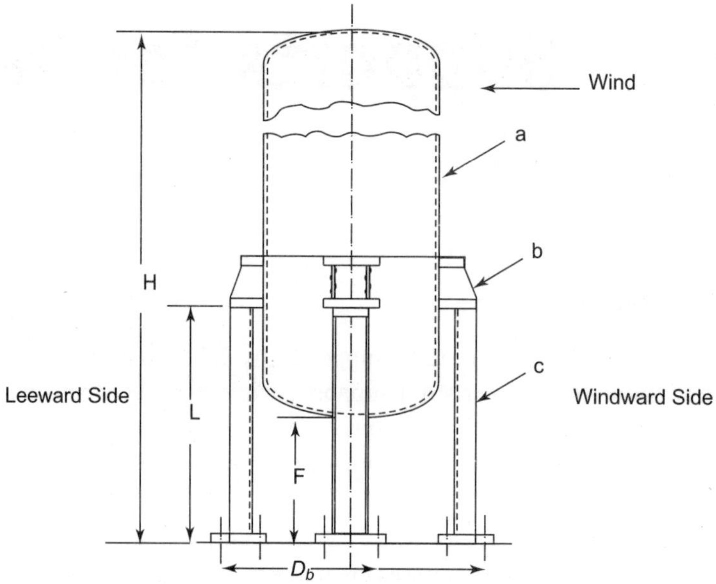

Fig. 13.1 Vessel Supported on Brackets: (a) Shell, (b) Bracket, (c) Leg

The main loads on the bracket supports are the dead weight of the vessel with its contents and the wind load. The wind load tends to overturn the vessel, particularly when it is empty. The maximum compressive stresses in the supports occur on the leeward side when the vessel is full, since dead load and wind load have a similar effect. The maximum tensile stresses are set up to the wind side when the vessel is empty, since dead load and wind load have opposing effects. Therefore, the stresses on the leeward side are the determining factor for design of supports. The maximum total compressive load in the most remote support is (Fig. 13.1).

$$P = \frac{4P_w(H - F)}{nD_b} + \frac{\Sigma W}{n} \tag{13.1}$$

where, P — total force due to wind load acting on the vessel [Eq. 13.20(a)]

H — height of vessel above the foundation

F — vessel clearance from foundation to vessel bottom

D_b – diameter of the bolt circle

ΣW – maximum weight of vessel with attachments and contents

n – number of brackets.

The wind load is neglected if the vessel is indoors or if the height of the vessel is limited. The load P would then be only due to the weight of the vessel and its contents,

$$P = \left(\frac{\Sigma W}{n}\right)$$

Three types of brackets are generally used.

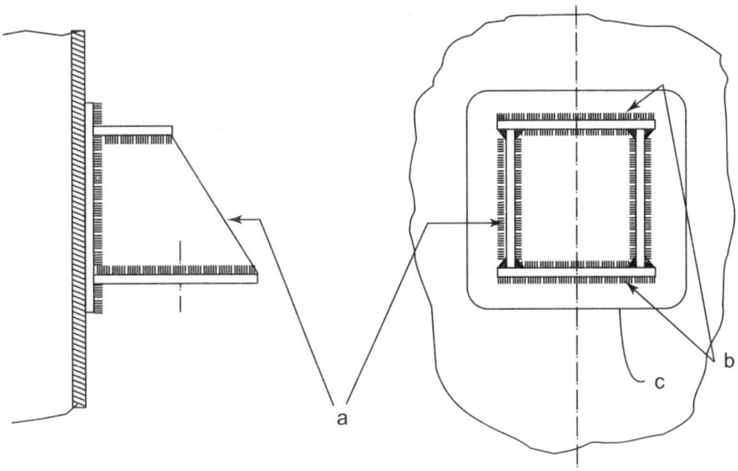

Fig. 13.2 Welding of Bracket of Shell: (a) Gusset Plate, (b) Stiffeners, (c) Pad Plate

(a) One type consists of two vertical gusset plates, with two additional horizontal plate stiffeners (Fig. 13.2). All the plates are welded to the pressure vessel shell. It is a good engineering practice to provide a pad plate (Fig. 13.2) for distributing the stress uniformly on the shell and provide stiffening at the same time. Many times for a thin shell we provide a stiffening ring or continuous pad on the full circumference of the shell. For large diameter vessels (dia. > 3000 mm) more than 4 lug supports say, 6 or 8 are provided. Both the top and bottom stiffener plates are continuously welded to the shell, as the maximum compressive and tensile stress occurs in these two plates, respectively. The gusset plates may be welded intermittently. All the welds are under a shearing load.

The compressive load P induces a reaction in the pressure vessel wall around the bracket. The bending moment created is given by

$$M = \frac{\beta^2 t^2 PbR^2}{12(1-\mu^2)Ah} \tag{13.2}$$

where,

$$\beta = \left(\frac{3(1-\mu^2)}{R^2t^2}\right)^{1/4}$$

t — shell thickness

R — radius of shell

P — compressive load (Eq. 13.1)

b — distance from the centre line of the shell plate to the centre line of the column

A — width of the compression plate (Fig. 13.3)

h — gusset height

μ — Poisson's ratio.

The axial stress in the vessel wall is, therefore, given by

$$f_a = \frac{6M_0}{t^2} = \frac{\beta^2 PbR^2}{2(1-\mu^2)Ah} \tag{13.3}$$

This is in addition to the axial stress produced by the internal pressure. A reinforcing pad is welded to the shell to take care of these additional stresses if they are of a large magnitude.

(b) Instead of using stiffeners as shown in Fig. 13.2, it is possible to strengthen the shell by the use of circumferential stiffening rings, at the top and bottom of the bracket. The size of the stiffening ring can be obtained as follows.

The longitudinal moment on shell caused by eccentricity may be given by

$$M = Pb \tag{13.4}$$

The radial force acting on the shell

$$F = \frac{M}{h} = \frac{Pb}{h} \tag{13.5}$$

The maximum bending moment on the shell stiffening ring at the bracket is

$$M_r = K\frac{Pb}{h}R \tag{13.6}$$

and the modulus of section for the ring is

$$Z_r = \frac{M_r}{f} \tag{13.7}$$

from which the size of the ring can be calculated. These rings are welded to the shell. In the above equations, the value of the coefficient K is obtained from Table 13.1 and f is the permissible tensile stress for the stiffener material.

Table 13.1 Value of K in Eq. (13.6)

No. of Brackets	Value of K
2	0.318
3	0.189
4	0.137
6	0.089
8	0.066

(c) In this case, only the vertical gusset plates are welded to the shell. The bottom plate is not welded. The method of calculating moments and stresses for this type of construction is complex and is, therefore, not considered here.

13.2.1 Thickness of the Base Plate

The plate is fixed on the edges with the load P considered as distributed over about half the area of the plate.

$$\text{Average pressure on the plate } p_{av} = \frac{P}{aB} \tag{13.8}$$

Maximum stress in a rectangular plate subjected to a pressure p_{av} and fixed at the edges is given by

$$f = 0.5 p_{av} \frac{B^2}{T_1^2}\left(\frac{a^4}{B^4 + a^4}\right) \tag{13.9}$$

where, a, B, and T_1 are as shown in Fig. 13.3.

The thickness T_1 computed by the above 'quick' formula is on conservative side and within engineering accuracy.

Since, in this case, the load is only distributed on the surface of contact between the base plate and the supporting beam, the actual stress may be taken as 40% more.

Therefore,

$$f = 0.7 p_{av} \frac{B^2}{T_1^2}\left(\frac{a^4}{B^4 + a^4}\right) \tag{13.10}$$

The thickness T_1 should be so chosen that the stress (f) should be within permissible limits.

13.2.2 Thickness of the Web (Gusset) Plates

There are two web plates for each bracket. The bending moment for each plate = $\dfrac{P}{2} \times C$

Section modulus of the plate $= \dfrac{1}{6} T_2 h^2$

Stress at the edge $f = \dfrac{3PC}{T_2 h^2}$ (13.11)

The edge is at an angle θ from the horizontal.

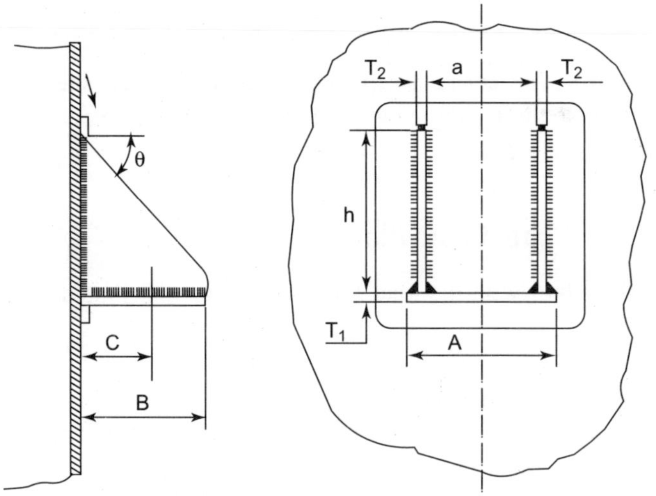

Fig. 13.3 Bracket or Lug Details

Therefore, the maximum compressive stress parallel to the edge of the web plate

$$= \frac{3PC}{T_2 h^2} \times \frac{1}{\cos\theta} \tag{13.12}$$

where, C, T_2, n and θ are as shown in Fig. 13.3. θ is generally 45° and T_2 should be so chosen that the stress shall not exceed the permissible value.

Figure 13.4 and Table 13.2 give some of the major dimensions of the brackets used in practice.

13.2.3 Column Supports for Brackets

Columns are attached under the brackets. Hence, the load on the column is not concentric with the weight of the vessel and its contents. These loading conditions act to produce an axial compression, as well as bending due to eccentricity.

The nature of stresses induced in struts, acting as column is indicated under Section 3.9. Column supports in steel are designed for slenderness ratios up to 120. Compressive stresses can be determined either on the basis of struts under direct compression or on the basis of the Rankine-Gordon formula, depending on the slenderness ratio.

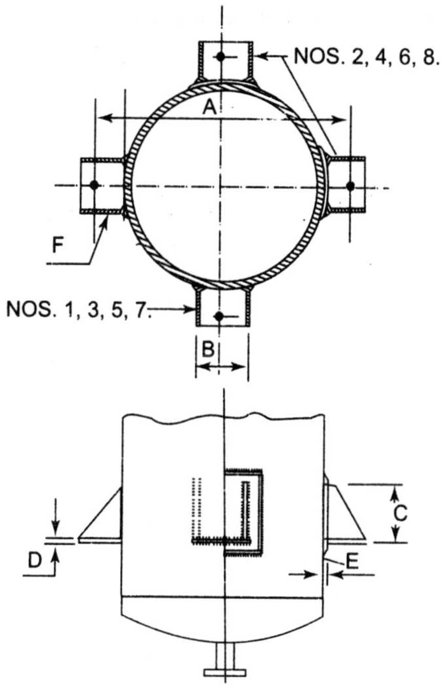

Fig. 13.4 Size of Brackets

Table 13.2 Brackets for Vertical Vessels

Centre to Centre of Bolt-holes in Lugs (Dimension A)										
Vessel dia(m)	0.75	0.90	1.05	1.20	1.35	1.50	1.80	2.10	2.40	2.70
Dim A(m)	0.90	1.05	1.20	1.35	1.50	1.65	1.95	2.35	2.65	2.95

Support No.	*Dimensions*				
	B(mm)	*C(mm)*	*D(mm)*	*E(mm)*	*F(mm)*
2	150	150	10	10	8
3	150	230	10	10	8
4	150	230	10	10	8
5	200	250	14	10	10
6	200	250	14	10	10
7	200	250	14	12	10
8	200	250	14	14	10

If the load is acting eccentric on a short column, the maximum combined bending and direct stress is given by

$$f = \frac{\Sigma W}{A \times n} + \frac{\Sigma W \times e}{n \times Z} \tag{13.13}$$

where, W — load on column

A — area of cross-section

e — eccentricity

Z — modulus of section of the cross section

n — number of columns.

If the column is long, the stress

$$f = \frac{\Sigma W}{nA}\left[1 + a\left(\frac{l_e}{r}\right)^2\right] + \frac{\Sigma W \times e}{nZ} \tag{13.14}$$

where, l_e — effective length of the column

a — constant

r — radius of gyration

$$l_e = \frac{1}{2} \text{ (for fixed ends of the column)}$$

In the case of vessels which are outside the building, column supports will also be subjected to wind load. The stress induced due to wind load will be given by

$$f_w = \left(\frac{p_w}{n} \times \frac{1}{2}\right)\frac{l}{Z} \tag{13.15}$$

where, p_w — total force acting on the vessel due to wind as per Eq. 13.20(a)

l — column height

n — number of columns.

Stress due to wind load should be combined with the stress due to dead loads.

In designing the size of an appropriate column, it is usual to choose a structural section such as a joist, a channel or a T-section of a suitable size and determine the stress as per the above equations. These stresses should be well within the range of permissible stresses. It is desirable to be conservative in the choice of the cross section of the column support and work with lower stresses than permissible.

13.3 LEG SUPPORTS

Structural sections such as angles, channels or joists (beams) can be directly welded to the pressure vessel shell to form vertical legs as shown in Fig. 13.5. Sometimes these legs can be made detachable by bolting leg plates welded to shell. The design of such a support is similar to those of column supports for brackets. The number of legs may be same as those specified for bracket supports. The legs are attached to the vessel by fillet welds. The shear stress produced in the weld by the load is given by

$$f_{ws} = \frac{\Sigma W}{0.707 \times t_w \times l_w \times n} \tag{13.16}$$

where, ΣW — load

t_w — side of weld (weld height)

l_w — length of weld

n — number of legs

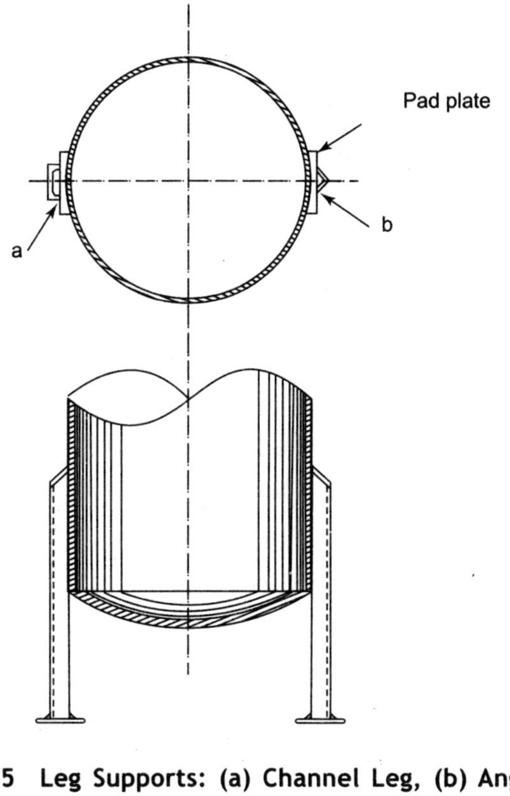

Fig. 13.5 Leg Supports: (a) Channel Leg, (b) Angle Leg

A tensile stress will also be produced in the weld due to the eccentricity of the load. This stress may be neglected.

This type of support is permitted only for small vessels. Severe local stresses are produced at the connection of the support to the vessel wall, which should be spread over a sufficient area by providing a reinforcing pad, as in the case of brackets. While determining the stresses in the supports the length may be taken as height of the leg between the lower edge of the welded portion and the base plate.

13.3.1 Base Plate for Channel Leg Support or Leg Supports

The column support is usually fabricated from a structural section such as joist (beam) or a channel. It is assumed that the load from the column is transmitted to the base plate uniformly on a rectangular area having sides equal to 0.8 times the width of the section and 0.95 times the depth of section (Fig. 13.6). It is also assumed that the base plate has a uniform bearing pressure (p_b) on its surface, which is resting on the foundation.

The leg support is welded to a pad plate which in turn is welded to a vessel. It is desirable to check that the weld length is adequate to take shear stress.

$$p_b = P \times \frac{1}{B \times C} \tag{13.17}$$

B and C are the two sides of the base plate.

If (p_b) is the allowable bearing pressure of steel plates on the foundation, it is possible to determine the dimensions B and C. The allowable pressure is based on the material used for foundations.

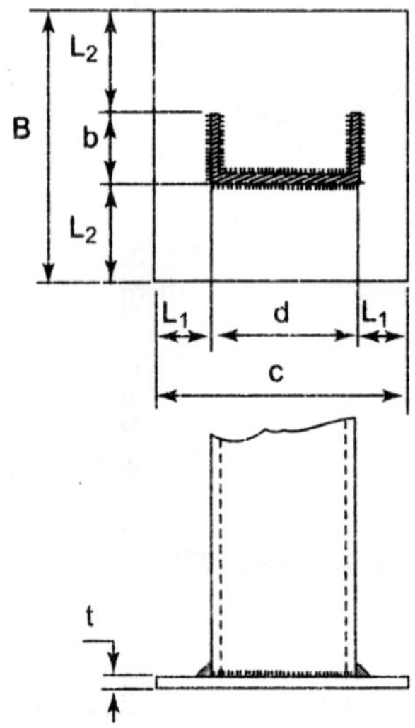

Fig. 13.6 Base Plate for Channel Leg or Leg Support

As shown in Fig. 13.6 let the projections of the plate be L_1 and L_2 and the thickness of the plate be t. Taking a unit width of the strip of base plate, the maximum bending moment in the plate is

$$M = \frac{p_b L_1^2}{2} \text{ or } \frac{p_b L_2^2}{2} \text{ whichever is greater.} \tag{13.18a}$$

Maximum stress in the plate, $f = \dfrac{6M}{t^2}$ (13.18b)

13.4 SKIRT SUPPORTS

Tall vertical vessels are usually supported by cylindrical shells or skirts (Fig. 13.7). The cross-section of the skirt is uniformly distributed at a sufficient distance from the axis. This gives a large value of the section modulus and helps to increase the resistance to bending action. The skirt is, therefore, a suitable supporting structure for tall vessels which are subjected to wind, seismic and other loads which cause a bending moment at the base of the vessel.

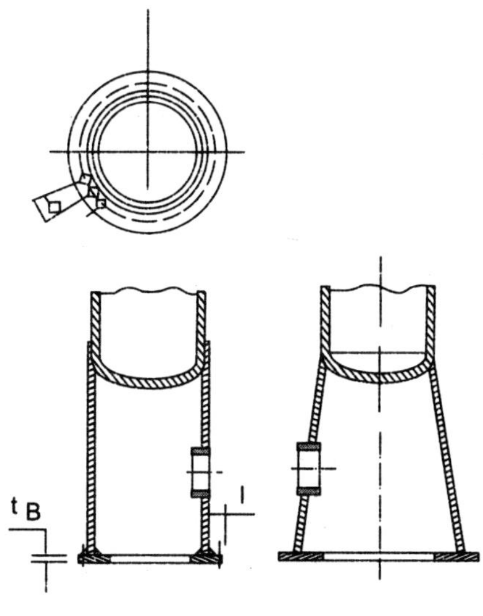

Fig. 13.7 Skirt Supports: (a) Straight Skirt, (b) Angular Skirt

The skirt is welded to the bottom dished head, either flush with the shell or to the outside of the shell. In the first case, in the absence of wind and seismic loads, the weld is in compression. In the latter method of welding the weld is in shear. A bearing plate is attached to the bottom of the skirt. This plate is made to rest on a concrete foundation and is securely anchored to the foundation by means of anchor bolts embedded in concrete to prevent overturning from the moments induced by wind or seismic loads. The bearing plate is in the form of a rolled angle or a single flat ring with or without gussets. Alternately, a bolting chair formed by two flat rings, namely a bearing plate and a compression plate, with gusset plates in between, is used for securing the skirt, to the foundation.

13.4.1 Skirt Design

The cylindrical shell of the skirt is designed for the combination of the stresses due to the vessel dead weight, wind load and seismic load. The skirt thickness is uniform and is designed to withstand the maximum values of tensile or compressive stresses.

These stresses are:

(a) *Due to dead weight*

$$f_d = \frac{\Sigma W}{\pi D_o t_{sk}} \tag{13.19}$$

where, f_d — stress

ΣW — dead weight of vessel contents and attachments

D_o — outside diameter of skirt

T_{sk} — thickness of skirt.

(b) *Due to wind load* The forces due to wind load acting on the lower and upper parts of the vessel are shown in Fig. 13.8 and are determined as follows:

$$f_{lw} = k \cdot p_1 \cdot h_1 \cdot D_o \tag{13.20a}$$

up to 20 m height

$$f_{uw} = k \cdot p_2 \cdot h_2 \cdot D_o \tag{13.20b}$$

above 20 m height

where, p_1 — wind pressure for the lower part of the vessel (400 to 1000 N/m^2)

p_2 — wind pressure for the upper part of the vessel (up to 2000 N/m^2)

The wind pressure depends upon whether the equipment is in open area or not. Further, it is also site specific.

k — coefficient depending on the shape factor (0.7 for cylindrical surface)

D_o — outside diameter of the vessel.

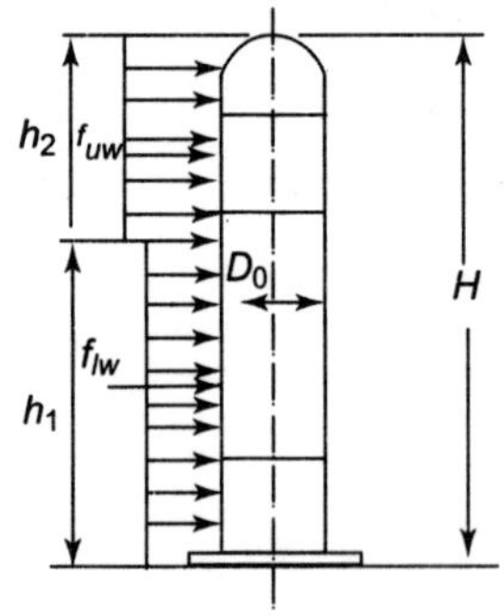

Fig. 13.8

The bending moment due to wind at the base of the vessel is determined by

$$M_w = f_{lw} \frac{H}{2} \tag{13.21a}$$

(up to $H \leq 20$m)

$$= f_{lw}\frac{h_1}{2} + f_{uw}\left(h_1 + \frac{h_2}{2}\right) \quad (13.21b)$$

$(H > 20\text{m})$

$$\text{The stress } f_{wb} = \frac{M_w}{Z} = \frac{4M_w}{\pi D_{ok}^2 t_{sk}} \quad (13.22)$$

where, Z — modulus of section of skirt cross-section

t_{sk} — skirt thickness

D_o — outside diameter of skirt.

(c) *Due to seismic load* This is a vibrational load resulting from earthquakes. The load is normal to the vessel shell and decreases linearly as shown in Fig. 13.9. The load may, therefore, be considered as acting at a distance 2/3 the height of the vessel from bottom.

$$\text{Load } F = S \cdot W \quad (13.23)$$

where, W — total weight of vessel

S — seismic coefficient.

The value of seismic coefficient, S, depends on intensity and duration of the earthquake. For a duration of more than one second, one can have $S = 0.02$ for minor damage, $S = 0.04$ for moderate damage and $S = 0.08$ for major damage (Taganov, 1968). The load is distributed as a triangular load.

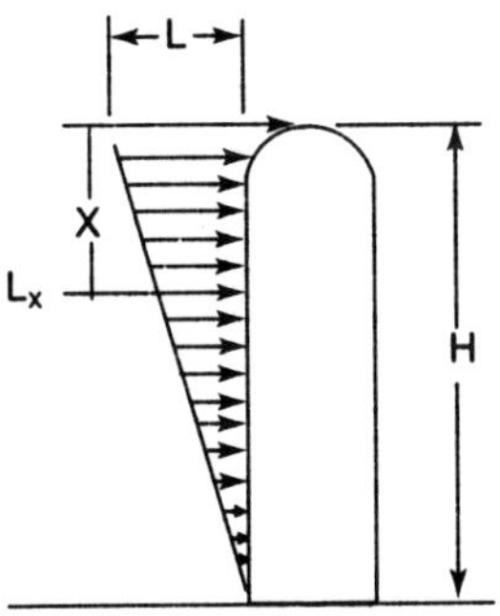

Fig. 13.9 Distribution of Seismic Load

At the plane X, we have the bending moment due to shear forces above the plane X as:

$$M_{sx} = 4SWX^2(3H - X)/H^2 \quad (13.24)$$

The corresponding bending stress is then given by

$$f_{sx} = M_{sx}/(\pi R^2 t_{sk}) \quad (13.25)$$

The maximum bending moment will be at the base, $X = H$. Thus from Eq. (13.24) by substituting H for X, we get,

$$f_{sb} = 4SWH^2 (3H - X)/(H^2 \pi R^2 t_{sk}) \tag{13.26a}$$

$$= 8SWH/(\pi R^2 t_{sk}) \tag{13.26b}$$

In other words,

$$f_{sb} = 2.6SWH/(R^2 t_{sk}) \tag{13.27}$$

The possibility of the wind load and earthquake load operating simultaneously is remote. Therefore, the stresses due to wind load and earthquake loads are computed separately and the most adverse loading condition is used to calculate the maximum resultant stress.

Max. tensile stress at the bottom of skirt

$$= (f_{wb} \text{ or } f_{sb}) - f_{db} \tag{13.28}$$

Max. compressive stress on skirt

$$= (f_{wb} \text{ or } f_{sb}) + f_{db} \tag{13.29}$$

In view of the complexity of the final equations for maximum stresses, it is usual to assume a suitable thickness t_{sk} of the skirt and check for the maximum stresses, which should not exceed the permissible values.

13.4.2 Skirt Bearing Plate

The maximum compressive stress between the bearing plate and the concrete foundation is given by the following equation:

$$f_c = \frac{\Sigma W}{A} + \frac{M_w}{Z} \tag{13.30}$$

where,

ΣW — weight of vessel, contents and attachments

A — area of contact between the bearing plate and foundation

M_w — bending moment due to wind

Z — section modulus of the area.

The maximum compressive stress (f_c) must be less than the permissible compressive stress is concrete. The thickness of the bearing plate is determined by considering it as a uniformly loaded cantilever with f_c as the uniform load. The maximum bending moment for the cantilever occurs at the junction of the skirt and bearing plate. This is given by Fig. 13.9.

$$M_{\max} = f_c bl \times \frac{l}{2} = f_c \frac{bl^2}{2} \tag{13.31}$$

where, l — difference between outer radius of the bearing plate and outer radius of skirt.

B — circumferential length.

The stress is

$$f = \frac{6M_{(\max)}}{bt_B^2}$$

$$= \frac{3f_c bl^2}{bt_B^2} = \frac{3f_c l^2}{t_B^2} \tag{13.32}$$

where, t_B — thickness of bending plate.

If the calculated thickness is less than 12 mm, a steel rolled angle may be used as a bearing plate (Fig. 13.10). If, however, the thickness is between 12 and 18 mm, a single ring bearing plate with or without gusset plates may be used (Fig. 13.11). The strengthening effect of the gusset plates will help to reduce the thickness of the bearing plate.

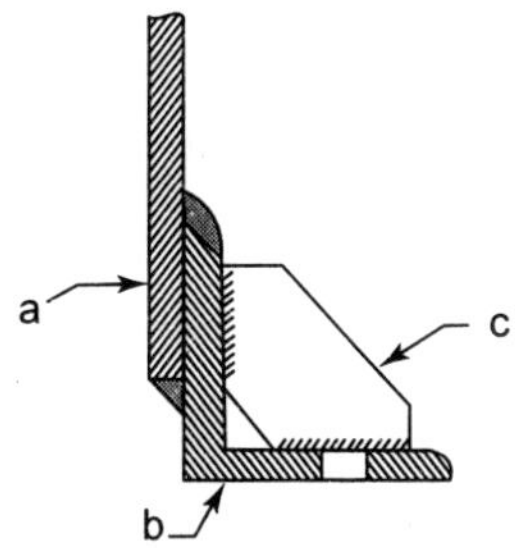

Fig. 13.10 Angle Bearing Plate: (a) Skirt, (b) Angle, (c) Gusset Plate

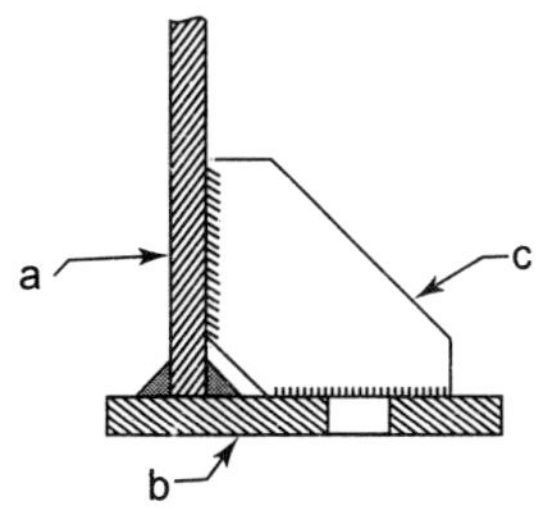

Fig. 13.11 Ring Bearing Plate: (a) Skirt, (b) Ring Plate, (c) Gusset Plate

13.4.3 Anchor Bolt Design

The minimum stress between the bearing plate and the concrete foundation will be

$$f_{c(\min)} = \frac{W_{\min}}{A} - \frac{M_w}{Z} \tag{13.33}$$

where, $W_{\min}$ — minimum weight of the empty vessel.

If the value of $f_{c(\min)}$ is positive or equals zero, it is necessary to assess the so-called coefficient of stability (Y) from the following equation:

$$Y = \frac{M_{\text{weight}}}{M_{\text{wind}}} = \frac{W_{\min} \times R}{M_{\text{wind}}} \tag{13.34}$$

where, M — moment

F — arm of force of weight of vessel

— $0.42\ D_o$.

If $Y > 1.5$, the vessel need not be anchored. In this case, the bolts are used only for fixing the vessel to the foundation. If $f_{c(\min)}$ is negative, the vessel must be anchored to the concrete foundation by means of anchor bolts to prevent overturning, owing to the moment produced by the wind or seismic load.

To assess the approximate number of anchor bolts, the following equation may be used (Fig. 13.7)

$$n = \frac{D}{600} \tag{13.35}$$

where, D — skirt diameter in mm.

This gives a minimum number of bolts necessary, but these must be a multiple of four. If the skirt is of a smaller diameter a minimum of four bolts must be provided. In general, the bolt diameter should be at least 24 mm.

The approximate value of the load on each bolt may be determined by the equation.

$$p_{\text{bolt}} = f_{c(\min)} \frac{A}{n} \tag{13.36}$$

where, p_{bolt} — load on one anchor bolt

n — number of bolts

f_c — stress determined by equation 13.33

A — area of contact between bearing plate and foundation (Figs. 13.10, 13.11 and 13.13).

The stress in the bolt

$$f_{\text{bolt}} = \frac{p_{\text{bolt}}}{A_b} \tag{13.37}$$

where, A_b — area of cross section of bolt.

The stress f_{bolt} should not exceed the permissible tensile stress in the bolt material.

13.4.4 Design of Bolting Chair

If the thickness of the bearing plate is more than 18 mm, it is advisable to use a bolting chair. This could be of the centred type (Fig. 13.12) or external type (Fig. 13.13). Centred chairs are used when the skirt diameter is more than 3 metres, and the number of bolts does not exceed 24. For large skirt diameters, and external bolting chair is suitable.

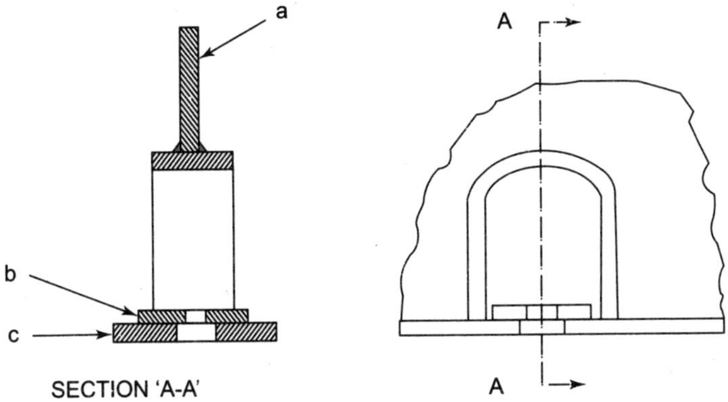

Fig. 13.12 Bolting Chair – Centred type: (a) Skirt, (b) Washer, (c) Bearing Plate

(a) *Centred chair* While calculating the bearing plate thickness for a centred chair, the plate inside the stiffeners is considered as a beam with fixed ends and with a concentrated load. The maximum bending moment in the bearing plate inside the chair will be near the bolt, and can be determined by the relation

$$M_{max} = \frac{p_{bolt} \times b}{8} \tag{13.38}$$

where, b — spacing inside chair

p — maximum bolt load.

The thickness of the bearing plate considering the reduction of section due to the bolt hole will be

$$t_{Bp} = \sqrt{\frac{6M_{max}}{(w_{Bp} - d_{bhd})f_{all}}} \tag{13.39}$$

where, t_{Bp} — bearing plate thickness

W_{Bp} — width of bearing plate

d_{bhd} — holt hole diameter in bearing plate

f_{all} — allowable tensile stress in bolt.

The thickness of bearing plate between the chair can be checked by use of Eq. (13.32).

(b) *External chair* The size and pitch of bolts may be calculated as indicated earlier. The dimensions of the chair may be determined approximately by the relationship given in Fig. 13.13.

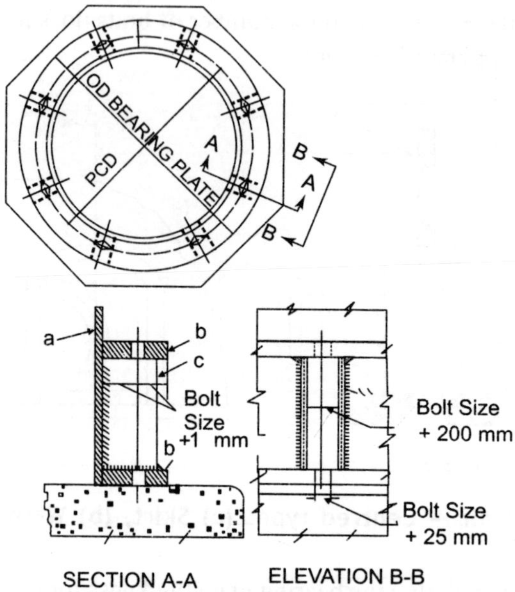

Fig. 13.13 Bolting Chair—External Type: (a) Skirt, (b) Compression Ring, (c) Gusset Plate, (d) Bearing Plate

13.5 SADDLE SUPPORTS

Horizontal cylindrical vessels are supported on saddles. These are placed at two positions. If necessary, the shell of the vessel is strengthened by stiffeners, located on the shell area surrounding the saddle (Fig. 13.17). Supports in the form of rings are preferable for vessels where supports at more than two positions are unavoidable. For large thin-walled vessels or vessels under vacuum, which are liable to be weakened by heavy loads of fittings or other supported structures, it is necessary to provide ring supports. For small vessels, simple leg supports may be used.

The location of the saddle support should be such that the distance from the tangent line to the centre line of the saddle should be less than the radius of the vessel, in order to take advantage of the stiffening effect of the head. This distance A is usually 0.4 to $0.5R$ and less than $0.2L$ (Fig. 13.14). The included angle θ should not normally be less than 120°.

Horizontal vessels resting on saddle supports behave as beams. They are subjected to longitudinal bending moments and local shear stresses due to the weight of the vessel, its contents and any attachment. The resulting stresses in the shell are longitudinal stresses, tangential shear stresses and circumferential stresses.

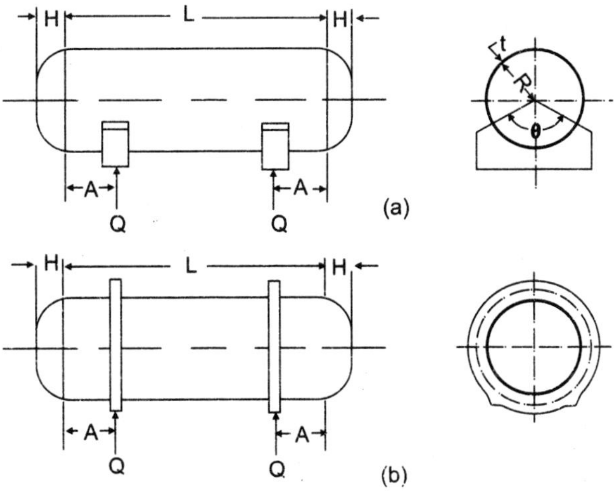

Fig. 13.14 Saddle Supports: (a) Plate Type, (b) Ring Type

13.5.1 Longitudinal Bending Moments

A cylindrical vessel with dished closures at the ends may be treated as an equivalent cylinder having a length equal to $(L + (4/3)\ H)$, where L is the length between the tangent lines and H is the depth of dished closure.

The weight of liquid and the vessel may be considered as a uniform load on the equivalent length. If two symmetrical supports are considered, each will carry a load

$$Q = \frac{w}{2}\left(L + \frac{4}{3}H\right) \tag{13.40}$$

where, w — uniformly distributed load.

In the load condition, the shell and load of heads can be taken as an overhanging beam with two supports, so that maximum bending moments are created. One maximum is at the saddle supports, the other acts in the centre of the span.

The bending moment at the supports is

$$M_1 = QA\left[1 - \frac{1 - \frac{A}{L} + \frac{(R^2 - H^2)}{2AL}}{1 + \frac{4}{3}\frac{H}{L}}\right] \tag{13.41}$$

(where, A, L, R are as shown in Fig. 13.14).

The bending moment at the centre of the span is given by

$$M_2 = \frac{QL}{4}\left[\frac{1+2\left(\frac{R^2 - H^2}{L^2}\right)}{1+\frac{4}{3}\frac{H}{L}} - \frac{4A}{L}\right] \tag{13.42}$$

13.5.2 Stress in Shell at the Saddle

The longitudinal bending stress in the shell depends on the stiffness of the shell at the cross section where the bending is maximum. If the shell does not retain its round shape under the load, a portion of the upper part of its cross section is ineffective against bending.

(a) In case the stiffness is enough to maintain a circular cross section (i.e., $A < 0.5R$), the whole cross section is effective and, therefore, the stress due to bending is given by

(i) At the topmost fibre of the cross section

$$f_1 = \frac{M_1}{k_1 \pi R^2 t} \tag{13.43}$$

(ii) At the bottommost fibre of the cross section

$$f_2 = \frac{M_1}{k_2 \pi R^2 t} \tag{13.44}$$

where, t — thickness of shell and $k_1 = k_2 = 1.0$.

(b) For $A > 0.5R$, the shell is not sufficiently stiffened by the end.

The value of the factor $k_1 = 0.107$ for $\theta = 120°$

$= 0.161$ for $\theta = 150°$

$k_2 = 0.192$ for $\theta = 120°$

$= 0.279$ for $\theta = 150°$

where, θ is as shown in Fig. 13.14.

If there is any internal pressure on the vessel, the resultant stress will be either tensile or compressive.

The tangential shear stress is generally of a small magnitude and may be ignored. The circumferential stress at the saddle due to bending may be of limited value, if the thickness of shell is greater than $0.1R$ and the unsupported length A is less than $0.5R$.

13.5.3 Stresses in the Shell at the Mid-Span

The stress at the mid-span is

$$f_3 = \frac{M_2}{\pi R^2 t} \tag{13.45}$$

which is either tensile or compressive, depending on the position of the fibre.

The resultant tensile stresses (including the axial stress due to internal pressure) should not exceed the permissible stress, and the resultant compressive stress should not exceed the permissible compressive stress or $\left(\frac{Et}{16R}\right)$, whichever is less, where E is modulus of elasticity.

13.5.4 Wear Plates and Stiffeners

The stresses in the shell band adjoining the saddle may be reduced by attaching a wear plate to the shell, directly over the saddle, somewhat larger than the surface of the saddle (Fig. 13.15). In the case of thin-walled vessels or in the case of a saddle located away from the head ($A > 0.5R$), the shell alone may not resist the circumferential bending moment. Ring stiffeners are then attached to the shell to relieve the load on the shell. These are in the form of a ring, welded either inside or outside the shell. An inside ring stiffener is more desirable, from the point of view of strength (Figs 13.16 and 13.17).

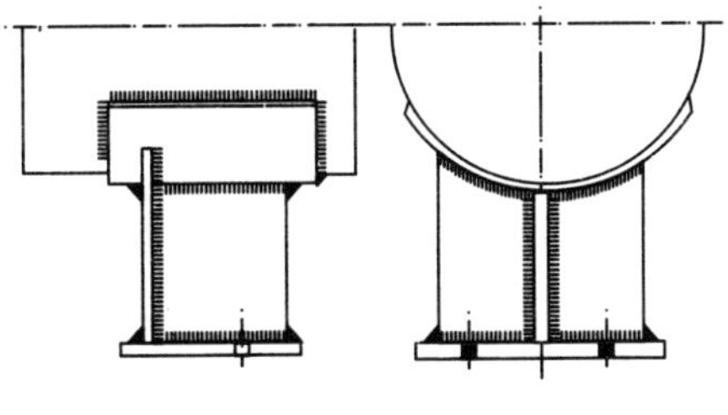

Fig. 13.15 Wear Plate for Saddle

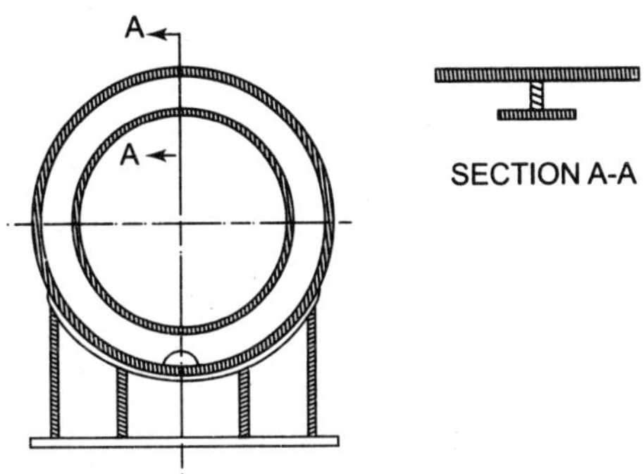

Fig. 13.16 Internal Stiffening Ring

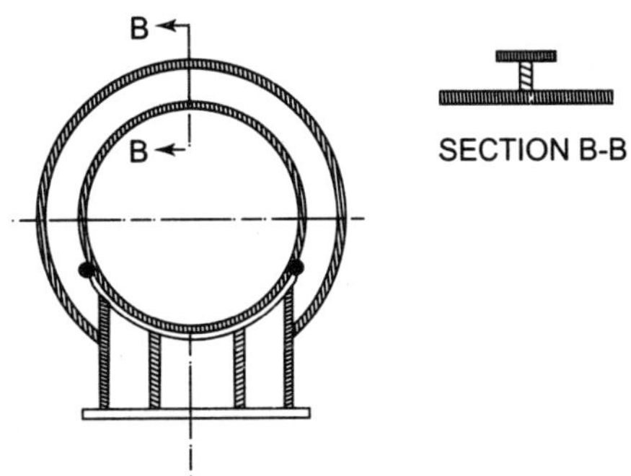

Fig. 13.17 External Stiffening Ring

13.5.5 Design of Saddles

Saddles should be strong enough to withstand the loads imposed on the vessel. Suitable sizes of saddles suggested for different diameters of vessels are indicated in Table 13.3 and Fig. 13.18.

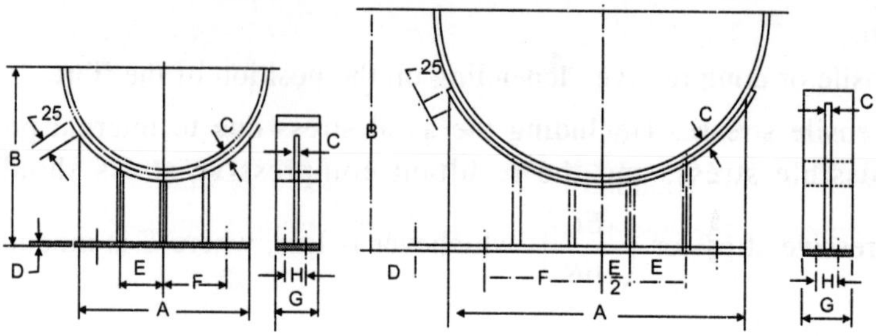

Fig. 13.18 Saddle Support Sizes: (a) Up to 120 cm Drum Diameter, (b) Over 120 cm Drum Diameter

Table 13.3 Welded Saddles for Horizontal Vessels

Drum Dia. (m)	*A(cm)*	*B(cm)*	$C_{(mm)}$	$D_{(mm)}$	*E(cm)*	*F(cm)*	*G(cm)*	*H(cm)*	*Bolt Dia. (mm)*	*Weld Size (mm)*	*Rib. Size (mm)*
0.60	55	50	10	12	19	24	15	10	19	6	70 × 10
0.90	80	65	10	12	28	34	15	10	19	6	70 × 10
1.20	110	80	12	16	37	45	20	14	22	6	90 × 12
1.50	135	95	12	16	33	58	20	14	22	6	90 × 12
1.80	160	110	14	16	40	71	20	14	22	6	90 × 12
2.10	190	125	16	20	48	81	23	15	22	8	100 × 16
2.40	215	140	16	20	53	94	23	15	25	8	100 × 16
2.70	240	155	16	20	61	106	25	18	25	8	115 × 16
3.00	265	170	16	20	66	117	25	18	25	8	115 × 16
3.30	295	185	16	20	73	129	25	18	25	8	115 × 16
3.60	320	200	16	20	80	140	25	18	25	8	115 × 16

13.6 ILLUSTRATION

The preceding discussions are illustrated below for various types of supports.

13.6.1 Bracket or Lug Support for a Vertical Vessel

We have the following data for a vessel depicted in Fig. 13.1:

Diameter of Vessel	: 1500 mm
Height of vessel	: 2000 mm
Clearance from bottom foundation	: 1000 mm
Weight of the vessel with contents	: 4000 kg
Wind pressure	: 1285 N/m^2
Number of brackets	: 4
Diameter of anchor bolt circle (Table 13.1)	: 1650 mm
Height of bracket from foundation	: 2250 mm
Permissible stresses for structural steel (IS-800)	
Tension	: 140 N/mm^2
Compression	: 123.3 N/mm^2
Bending	: 157.5 N/mm^2
Permissible bearing pressure for concrete	: 3.5 N/mm^2

13.6.1.1 Maximum Compressive Load

The force due to wind load is given in Eq. (13.20a), as the vessel height is less than 20 m.

$$f_w = k \cdot p \cdot h \cdot D_0 \qquad (k_p = 0.7 \text{ shape factor for a cylindrical vessel})$$
$$= 0.7 \times 1285 \times 2 \times 1.5$$
$$= 2698.5 \text{ N}$$

From Eq. (13.1), we have total force

$$P_{FT} = \frac{4 \times f_w \times (H - F)}{n \times D_b} + \frac{\Sigma W}{n}$$

$$= \frac{4 \times 2698.5 \times (3000 - 1000)}{4 \times 1650} + \frac{4000 \times 9.807}{4}$$

$$= 13078 \text{ N}$$

13.6.1.2 Bracket

(a) *Base Plate*

From Table 13.2 suitable base plate size is considered wherein we have,

a = 140 mm (Refer Fig. 13.3)

B = 150 mm

From Eq. (13.8) the average pressure on the plate

$$P_{av} = P_{FT}/(a \times B)$$
$$= 13078/(140 \times 150)$$
$$= 0.623 \text{ N/mm}^2$$

From Eqns. (13.9 and 13.10), we get

$$f = 0.7 \times P_{av} \times \frac{B^2}{T_1^2} \times \left[\frac{a^4}{(B^4 + a^4)}\right]$$

$$= 0.7 \times 0.623 \times \frac{(150)^2\ (140)^4}{T_1^2\ \{(150)^4 + (140)^4\}}$$

$$= 4233.4/T_1^2$$

We have the permissible bending stress, $f = 157.5$ N/mm^2

Therefore, $T = \sqrt{4233.4/157.5} = 5.2$ mm

We then use plate of thickness = 10 mm (Table 13.2, dimension *D*).

13.6.1.3 Channel Support for Bracket

It is proposed to use a channel section as support. The channel size chosen is ISMC 150

Size 150 × 75

Area of cross section (A) : 2088 mm^2

Modulus of section (Z_{yy}) : 19400 mm^3

Radius of gyration (r_{yy}) : 22.1 mm

Weight : 16.4 kg/m

Height from foundation : 2250 mm

Equivalent length of fixed ends $l_e = l/2 = 2250/2$

$= 1125$ mm

The slenderness ratio $= l_e/r_{yy} = 1125/22.1 = 50.9$, say $= 51$

The column is short column. Therefore, the stress is obtained with the help of Eq. (13.13)

$$f = \frac{13078}{2088} + \frac{13078 \times 75}{19400}$$

(The load per column was worked out in section 13.6.1.1)

Therefore, $f = 56.82$ N/mm^2.

The calculated value of '*f*' is less than permissible value (123.3 N/mm^2). Hence, the selected channel is satisfactory.

13.6.1.4 Support Base Plate for Channel Leg

It is assumed that the base plate extends 20 mm on either side of the above channel (150 × 75).

From section 13.3.1 and Fig. 13.6 we have dimensions of two sides of the base plate as:

$$\text{Side B} = 0.8 \times 75 + 2 \times 20 = 100 \text{ mm}$$

$$\text{Side C} = 0.95 \times 150 + 2 \times 20 = 182.5 \text{ mm}$$

The bearing pressure, P_b, is then obtained via Eq. (13.17).

We calculate the base plate thickness for only one leg of the lug support on leeward side which is the worst scenario, then in the normal case when the load gets distributed on each leg the plate would bear the load.

$$P_b = 13078/(100 \times 182.5) = 0.72 \text{ N/mm}^2$$

This is less than the permissible bearing pressure for the concrete (3.5 N/mm^2).

Let us compute now, thickness of the plate.

From Eq. (13.18a) we have

$$M = P_b L^2/2 = 0.72 \times (20)^2/2 = 144 \text{ N}$$

From Eq. (13.18b), we have

$$f = 6 \times M/t^2$$

The allowable stress value is $f = 157.5$ N/mm^2

Substituting M and f in above, we get

$$t = \sqrt{6 \times 144/157.5}$$

$$= 2.34 \text{ mm}$$

It is usual to select the plate of thickness 10 to 12 mm for rigidity.

13.6.2 Skirt Support for a Cylindrical Vertical Vessel

We have following data:

Diameter of the a vessel	: 3000 mm
Height of the vessel	: 37500 mm
Weight of the vessel with attachments	: 200,000 kg
Diameter of skirt	: 3000 mm
Height of skirt	: 4800 mm
Wind pressure	: 1285 N/mm^2

13.6.2.1 Skirt

(A) *Stress due to dead weight*

From Eq. (13.19)

$$f_d = \frac{\Sigma W}{\pi \times D_o \times t_{sk}}$$

$$= \frac{200{,}000 \times 9.807}{\pi \times 3000 \times t_{sk}} = 208.1/t_{sk}, \text{ N/mm}^2$$

We shall compute stress due to wind load and due to seismic load and consider the higher value for stress calculation and hence thickness.

(B) *Stress due to wind load*

Total height = 37500 + 4800 = 42300 mm

The bending moment up to the base of the vessel [Eq. (13.21a)]

$$M_w = f_{lw} \times H/2$$

where, f_{lw} = 0.7 × 1285 × 42.3 × 3 (Since wind load is the same through out the height, we do not divide height in two sections, namely, up to 20 m and above).

$$f_{lw} = 114146.6 \text{ N}$$

Therefore,

$$M_w = 114146.6 \times 42.3/2$$
$$= 2.41 \times 10^6 \text{ N m}$$
$$= 2.41 \times 10^9 \text{ N mm}$$

The stress f_{wb} can be computed via Eq. (13.22)

$$f_{wb} = M_w/Z = 4M_w/(\pi \times D_o^2 \times t_{sk})$$
$$= 4 \times 2.41 \times 10^9/(\pi \times (3000)^2 \times t_{sk})$$
$$= 340.95/t_{sk} \text{ N/mm}^2$$

Seismic and wind load cannot occur simultaneously. Therefore, we illustrate design with wind load only.

Maximum tensile stress = permissible stress.

$$\frac{340.95}{t_{sk}} - \frac{208.1}{t_{sk}} = 140$$

$$t_{sk} = 0.95 \text{ mm}$$

Maximum compressive stress $= (340.95 + 208.1)/t_{sk}$

$= (1/3)$ Yield point

$= (1/3) \times 200$

$t_{sk} = 549.05/66.7$

$= 8.23$ mm

Considering rigidity, use a plate of thickness = 20 mm.

13.6.2.2 Skirt Bearing Plate

The bolt circle diameter is assumed to be = skirt diameter + 325 mm

= 3000 + 325

= 3325 mm

For the design illustrated here, we consider bolt circle diameter is the effective plate diameter. In reality, plate OD = bolt circle diameter + 100 mm.

The compressive stress between bearing plate and concrete foundation is given in Eq. (13.30) and Eqs. (13.20 to 13.22).

$$f_c = \frac{\Sigma W}{A} + \frac{M_w}{Z}$$

We have $M_w = 2.41 \times 10^9$ N mm (refer 13.6.2.1B)

Z = section modulus of the area and is obtained via Appendix A

$$Z = \frac{\pi}{32} \frac{(D_o^4 - D_i^4)}{D_o}$$

$D_o = 3325$ mm, $D_i = 3000$ mm

$Z = 1.22 \times 10^9$ mm^3

$$A = (\pi/4)(D_o^2 - D_i^2)$$

$= 1.61 \times 10^6$ mm^2

Substituting all values we get f_c:

$$f_c = 2 \times 10^5 \times 9.807/(1.61 \times 10^6) + 2.41 \times 10^9/(1.22 \times 10^9)$$

$= 3.2$ N/mm^2

This value is less than the permissible value of 3.5 N/mm^2 used for the concrete used for foundation purpose.

Usually a composite section of the reinforced concrete is used for foundation.

The thickness of the bearing plate is now obtained via Eq. (13.32)

$$t_B^2 = 3 f_c l^2 / f_B$$

where, I = outer radius of bearing plate – outer radius of skirt

$= (1/2)\ (3325 - 3000) = 162.5$ mm

We have $f_B = 157.5\ \text{N/mm}^2$

$$t_B = \sqrt{3 \times 3.2 \times (162.5)^2 / 157.5}$$

$$= 40.11 \text{ mm}$$

The thickness of the bearing plate is say = 44 mm

The above thickness being more than 18 mm, a bolting chair is to be used [Section (13.4.4)].

13.6.2.3 Anchor Bolts

Following the discussions in Section 13.4.3, we have the minimum stress between the bearing plate and the concrete foundation give by Eq. (13.23)

We assume that the weight of the empty vessel 1.75×10^5 kg for this calculation.

$$f_c = 1.75 \times 10^5 \times 9.807/(1.61 \times 10^6) - 2.41 \times 10^9/(1.35 \times 10^9)$$

$$= 1.07 - 1.79$$

$$= -\ 0.72 \text{ N/mm}^2$$

Since f_c is negative, the vessel must be anchored to the concrete foundation by means of anchor bolts.

Assuming that the number of bolts $n = 24$, the load on each bolt is given in Eq. (13.36).

$$P_{\text{bolt}} = f_{c\min} \times A/n$$

$$= (0.91/24) \times 1.61 \times 10^6$$

$$= 61046 \text{ N}$$

The stress in bolt is given by

$$f_{\text{bolt}} = P_{\text{bolt}}/A_{\text{bolt}}$$

The bolt of nominal diameter 36 mm (M36) is considered.

From Eq. (5.41) we get $A_{\text{bolt}} = 913\ \text{mm}^2$.

$f_{\text{bolt}} = 61046/913 = 66.86\ \text{N/mm}^2$ which is within acceptable limit of $150\ \text{N/mm}^2$.

Therefore, the bolt size and number of bolts are adequate.

13.6.3 Saddle Support for the Horizontal Vessel

Let us illustrate the design procedure for the Saddle Support with the help of the following data:

Material of construction	: low carbon steel
Vessel diameter	: 1230 mm
Length of shell	: 8000 mm
Torispherical head	
Crown radius	: 1250 mm
Knuckle radius	: 6% of diameter
Total depth of	: 257 mm
Working pressure	: 0.5 N/mm^2
Shell thickness	: 10 mm
Head thickness	: 12 mm
Corrosion allowance	: 1.5 mm
Permissible stress	: 95 N/mm^2
Load of vessel and contents	: 12000 kg
Distance of saddle centre line from Shell-end	: 320 mm

13.6.3.1 Longitudinal Bending Moments

The load carried by each support, $Q = 12000 \times 9.807/2$

$$Q = 58842 \text{ N}$$

The longitudinal bending moment, M_1, is obtained from Eq. (13.41)

$$M_1 = Q \times A\,[1 - \alpha/\beta]$$

$$\alpha = 1 - (A/L) + (R^2 - H^2)/2AL$$

and

$$\beta = 1 + (4/3)\,H/L$$

By substituting the appropriate values, we get

$$\alpha = 1 - (320/8000) + [(615)^2 - (257)^2]/(2 \times 320 \times 8000)$$
$$= 1.021$$

$$\beta = 1 + (4/3)257/8000$$
$$= 1.043$$

We then have,

$$M_1 = 58842 \times 320\,[1 - 1.021/1.043]$$

$$= 3.97 \times 10^5 \text{ N mm}$$

The bending moment at the center of the span is given by Eq. (13.42).

$$M_2 = (QL/4)[\psi/\xi - 4A/L]$$

where, $\psi = 1 + 2(R^2 - H^2)/L^2$

and $\xi = 1 + (4/3)H/L$

By substituting the appropriate values we get,

$$\psi = 1 + 2[(615)^2 - (257)^2]/(8000)^2$$
$$= 1.01$$

and
$$\xi = 1 + (4/3)257/8000$$
$$= 1.043$$

We then have,

$$M_2 = (58842 \times 8000)/4 \times [1.01/1.043 - 4 \times 320/8000]$$
$$= 9.52 \times 10^7 \text{ N mm}$$

13.6.3.2 Stresses in Shell at the Saddle

From Eq. (13.43) for $\theta = 120°$, $k_1 = 0.107$ and $t = 8.5$ mm (t = shell thickness – corrosion allowance) the stress at the top fibre of the cross section:

$$f_1 = \frac{M_1}{(k_1 \times \pi \times R^2 \times t)}$$

$$= \frac{3.97 \times 10^5}{0.107 \times \pi \times (615)^2 \times (8.5)}$$

$$= 0.37 \text{ N/mm}^2$$

The stress at the bottom line of the fibre is calculated with the help of Eq. (13.44) for $\theta = 120°$ and $k_2 = 0.192$ as:

$$f_2 = \frac{3.97 \times 10^5}{0.192 \times \pi \times (615)^2 \times (8.5)}$$

$$= 0.20 \text{ N/mm}^2$$

Both stresses are well within allowable value.

13.6.3.3 Stresses in the Shell at Midspan

From Eq. (13.45), we have

$$f_3 = \frac{M_2}{(\pi \times R^2 \times t)}$$

$$= \frac{9.52 \times 10^7}{\pi \times (615)^2 \times (8.5)}$$

$$= 9.42 \text{ N/mm}^2$$

From Eq. (5.2) the axial stress in the shell due to the internal pressure is:

$$f_a = \frac{pD}{4t}$$

The design pressure p = 1.1 × internal pressure = 1.1 × 0.5

$$= 0.55 \text{ N/mm}^2$$

$$f_a = \frac{0.55 \times 1230}{4 \times 8.5}$$

$$= 19.9 \text{ N/mm}^2$$

The combined stresses $(f_a + f_1)$, $(f_a - f_2)$ and $(f_a + f_3)$ are all within acceptable limit. Therefore, the design is acceptable.

References and Recommended Readings

Brownell, L.E., and Young, E.H., *Process Equipment Design*, John Wiley, New York (1968).

Chaudhari, D., 'A Simple Way of Sizing Anchor Bolts and Base Ring', Chemical Age of India, **27** (4), 391 (1976).

Mahajan, K.K., 'Tall Stack Design Simplified', *Hydrocarbon Processing*, 54 (9), 217 (1975).

Markovitz, R.E., 'Process and Project Data Pertinent to Vessel Design', *Chemical Engineering*, **84** (21) 123 (1977).

Steel Designer's Manual, 4th Edition, English Language Book Society and Crosby Lockwood Staples, London (1972).

Taganov, N.I., *Fundamentals of Process Equipment Design*, Academic Books, Bombay (1968).

14

Chapter

Mixers

14.1 INTRODUCTION

Mixing is a very important unit operation in any Chemical Process Industry. For instance, all operations involving blending, homogenization, emulsion preparation, extraction, dissolution, crystallization, liquid phase reactions, etc., need mixing in one form or the other. Broadly speaking, two types of commercial mixing devices are available, namely,

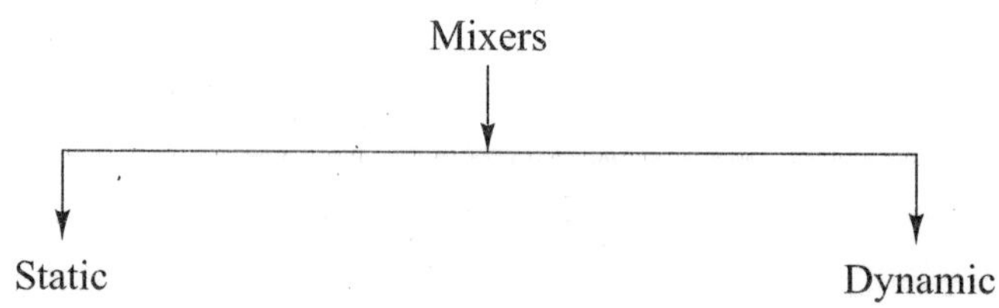

Static mixers are becoming popular among processes, and hence among plant engineers, for they are very simple to install, need minimum maintenance (practically a zero maintenance device). The accidental damage by release of fluids is reduced considerably as the static mixers, being a pipeline contactor, have considerable lower inventory of fluids than that in dynamic mixers. In this pipeline device, a mixing element is placed inside a pipe, having length typically 0.5 to 1.0 m. There are no moving parts, but mixing take place across the element as both fluids pass through it. The static mixer device is also used as a chemical reactor to carry out reactions between two fluids. The static mixers are now available for conducting highly exothermic reactions involving gas-liquid or liquid-liquid operations. The solid suspension is generally not processed through a static mixer. However, application of a static mixer to handle nano particles slurry is a distinct possibility in the near future. Where possible, static mixers are replacing dynamic mixers.

The dynamic mixer consists of the basic equipments, which may be a tank, a reaction vessel or a crystallizer having an agitation (mixer) system. Selection of an efficient agitation system will depend on the nature of the liquid, operating conditions and the intensity of circulation and shear. A variety of agitation systems are available, each one having a useful operating range. The limits of usefulness of each system are not clearly defined. The factors to be considered are: (1) type of agitator, (2) circulation pattern, (3) location of the agitator in the basic equipment, (4) shape and size of tank, (5) diameter and width of agitator, (6) method of baffling, (7) power required for agitation, (8) shaft overhang and (9) type of stuffing box or seal, bearings, drive system, etc. The detailed design of a dynamic mixer (agitator) system is much more complex than that of a static mixer system from the mechanical engineering point of view. Some of the above factors are considered below in detail.

14.2 TYPES OF AGITATORS

Mechanical agitators can be divided into seven basic groups namely, paddles, turbines, propellers, helical screws, cones, radial flow propellers and high speed discs. Mixing by agitators takes place due to a momentum transfer. High velocity streams, produced by the impeller, entrain the slower mixing or stagnant liquid areas from all parts of the vessel and a uniform mixing occurs. As the viscosity of the liquid increases, frictional drag forces retard the high velocity streams and confine them to the immediate vicinity of the impeller. Thus stagnant areas develop and uniform mixing is not achieved.

Agitators having a small blade area which rotate at high speeds, for instance, propellers, flat or curved blade turbines, are used to mix liquids having low viscosities. Agitators having a large blade area which rotate at slow speeds, such as anchors, gates and helical screws are more effective for mixing highly viscous liquids.

Agitators are mounted either vertically at the top of the tank known as top-entering or on the side of the tank known as side-entering. Top-entering agitators can also be mounted angularly or can be off-centred. Top-entering agitators are used for the larger and higher power per unit volume applications. Side-entering propeller agitators are very economical and effective. However, these cannot be used with fluids having either corrosive or erosive properties.

14.2.1 Anchor Agitator

Figure 14.1 represents some anchor agitators. The blades of these agitators normally extend close to the tank wall. They are simply pushers and cause the mass to rotate in a laminar swirling motion with practically no radial flow along the paddle blades or any axial flow. The circulation is poor and the mixing action is inefficient. The speed of rotation is very slow, and the tip speed is generally between 80 to 150 metres per minute. Highly viscous liquids and pastes are agitated by multiple blade paddles. Such anchors are made to rotate between intermeshing stationary fingers. A counter rotating pair of multiple paddles creates a high localized shear. The width of the blades is 1/8 to 1/10th of the anchor diameter. The advantage of this type of an agitator is to improve the heat transfer in highly viscous systems such as paint formulations, polymerization reactions. The clearance between blades and vessel is generally between 20 and 100 mm. In the case of a crystallizers it could be even 5 mm. Spring loaded wipers are used for surface cleaning.

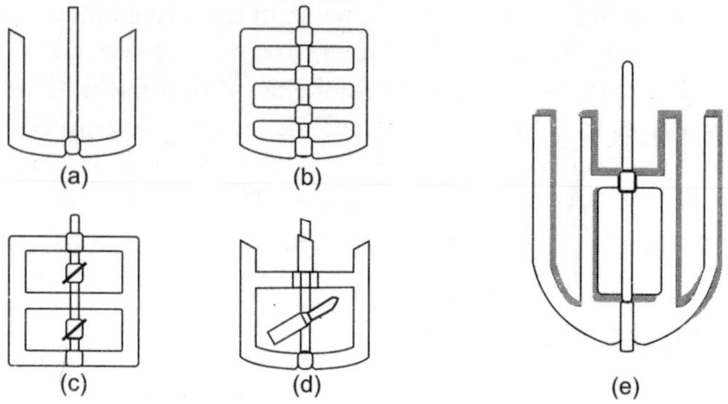

Fig. 14.1 Anchor Agitators: (a) Anchor, (b) Gate, (c) Gate with Pitched Cross Arms, (d) Anchor with Pitched Cross Arms, (e) Combined Anchor and Gate

14.2.2 Turbine Agitator

These have a variety of shapes such as radial, pitched and back-sloped (Fig. 14.2). They are capable of creating a vigorous mixing action, due to centrifugal and rotational motion, generated by them. The back-sloped turbine agitator offers maximum power economy. A stator ring surrounding this agitator gives an efficient mixing action. A radial bladed turbine gives a higher discharge velocity, but requires more power. The more commonly used type is the pitched blade turbine [Fig. 14.2(b)] which gives both radial and axial flow. The axial flow in this type of agitator is the lesser of the two flows and varies with the number of blades, the pitch angle and the blade height. The approximate peripheral speed for the turbine agitator is 200 to 300 metres per minute. All the three types of turbine agitators can be constructed either with blades directly attached to a hub or blades attached to a centre disc [Fig. 14.2 (d)]. Generally, the diameter of the agitator is kept between 0.25 and 0.6 of the tank diameter depending on the circulation required. For the turbine exhibited in Fig. 14.2d, the blade length is a fourth of the agitator diameter. Central disc is 0.6 to 0.7 of the agitator diameter. The blade angle of a pitched blade turbine agitator may vary between 30° and 60°. In general, the higher the viscosity, the greater must be the blade angle with the vertical to keep the power requirements to a minimum. The number of turbine agitators to be used is determined by the following equation:

$$\text{No. of agitators} = \frac{\text{maximum liquid height} \times \text{average sp.gr.}}{\text{tank diameter}}$$

The distance between the agitators should be 1 to 2 agitator diameters.

It is common to locate the agitator, at a height not less than 1/2 agitator diameter length from the bottom and the agitator should be submerged with liquid to a depth equal to twice the agitator diameter at low speeds and four times at high speeds. If the depth of liquid is above the agitator is more than twice the agitator diameter, it may be advisable to use two agitators.

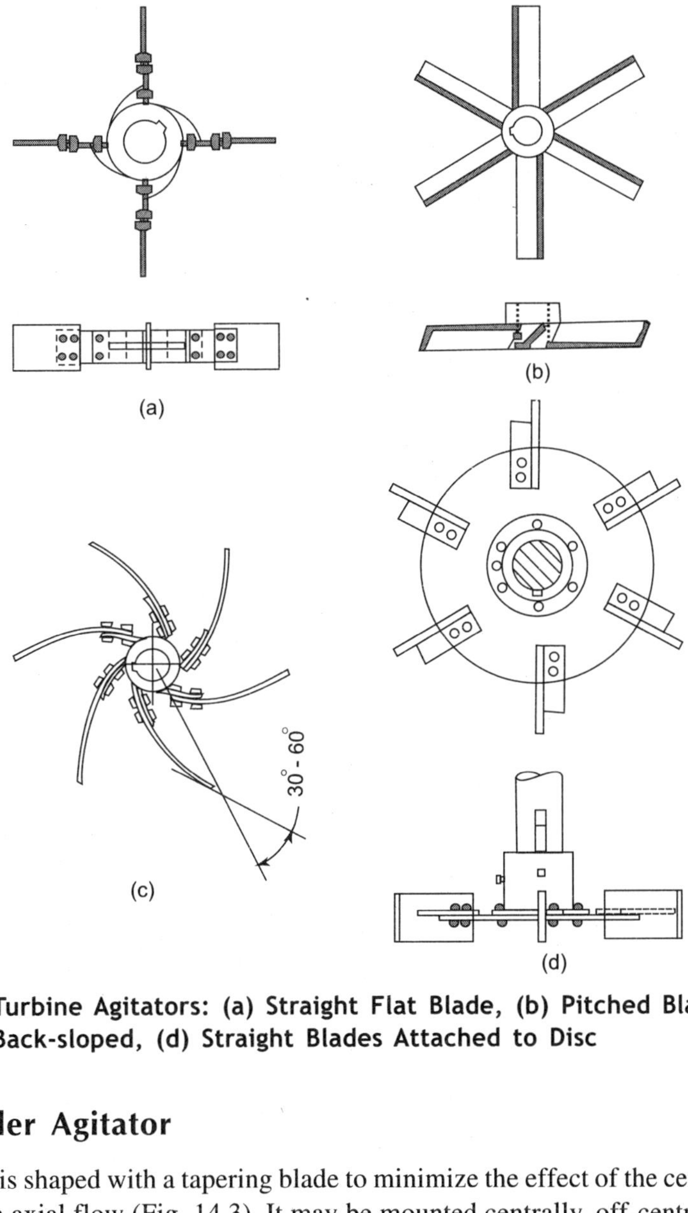

Fig. 14.2 Turbine Agitators: (a) Straight Flat Blade, (b) Pitched Blade, (c) Curved Back-sloped, (d) Straight Blades Attached to Disc

14.2.3 Propeller Agitator

A propeller agitator is shaped with a tapering blade to minimize the effect of the centrifugal force and produce a maximum axial flow (Fig. 14.3). It may be mounted centrally, off-centre or at an angle to the tank. It is a simple and portable device. The diameter of this agitator is usually between 15 and 30% of the tank diameter and its peripheral speed is generally between 300 and 500 metres per minute. It can, therefore, be directly coupled to a standard electric motor. It is considered as one of the most economical unit for the simpler mixing jobs, particularly in small tanks.

14.2.4 Helical Screw Agitator

It is in the shape of a screw, as shown in Fig. 14.1. It is an effective device for mixing high viscosity liquids. The screw drives the liquid from the vessel bottom to the top surface of the liquid. Alternatively, screws may be operated in the reverse direction to drive the liquid to the bottom of the vessel.

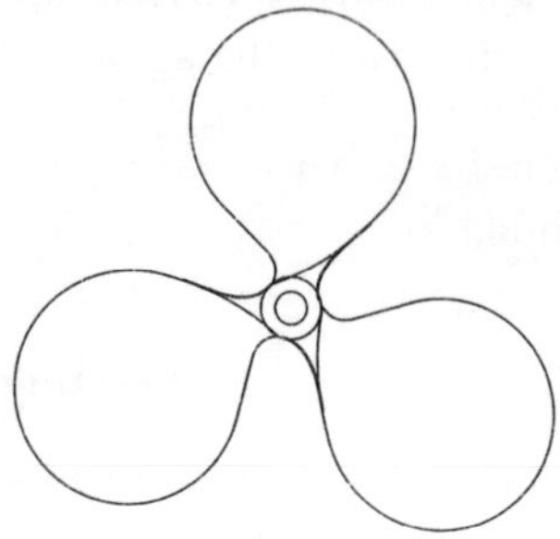

Fig. 14.3 Propeller Agitator

Baffles in a screw agitated vessel help to create turbulence. These should be set away from the vessel wall, thus, allowing the turbulence created to travel around the baffle and entrain the otherwise slowly moving liquid in contact with the vessel wall. Sometimes the screw is placed off-centre, in which case, baffles may be avoided. However, this arrangement requires more power.

14.2.5 Cone Type Agitator

This agitator is built with shallow vertical vanes on the inside periphery of the cone (Fig. 14.5). Flow is generated between the peripheries of the narrow and wide throats of the cone. It is specially used for handling fibrous and dense slurries. Its speed is similar to that of the turbine.

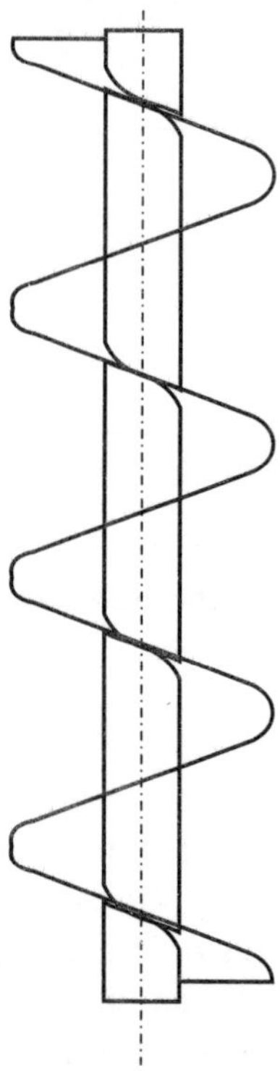

Fig. 14.4 Helical Screw Agitator

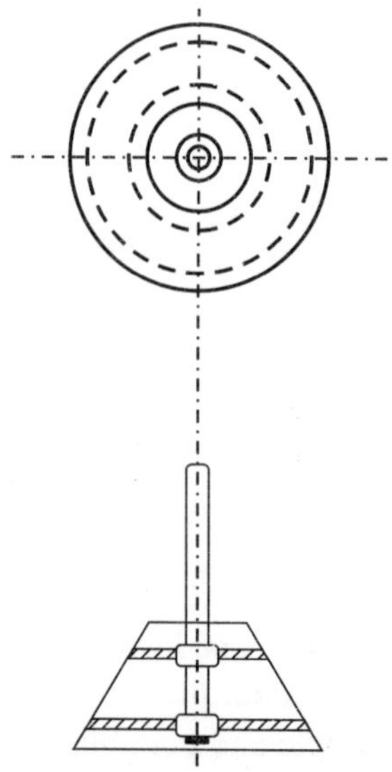

Fig. 14.5 Cone Type Agitator

14.2.6 Radial Propeller Agitator

It has blades fixed at the end of a rotating arm (Fig. 14.6). The blades are pitched to the direction of rotation. By varying the height, width, and particularly the angle of pitch, turbulence or shear can be controlled.

14.2.7 High Speed Disc Agitator

This has a disc with corrugations (Fig. 14.7). The speed of the agitator provides sufficient centrifugal force through surface friction to generate a flow. It produces high shear and helps in the disintegration of low density fibrous solids. Also used in the paint industry for dispersion of fine powder in a viscous medium.

14.2.8 Double Motion Agitator

Figure 14.8 shows a mixer screw with double motion located in a conical vessel. The screw rotates about its own axis at a fairly high speed and also rotates slowly around the vertical vessel axis. In drying application the vessel wall and the mixer both are heated.

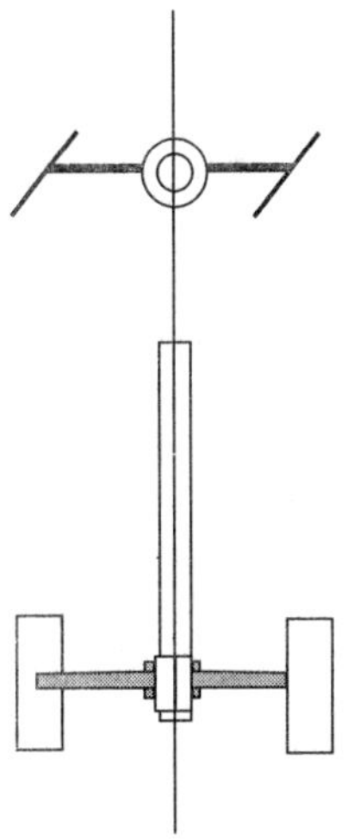

Fig. 14.6 Radial Propeller Agitator

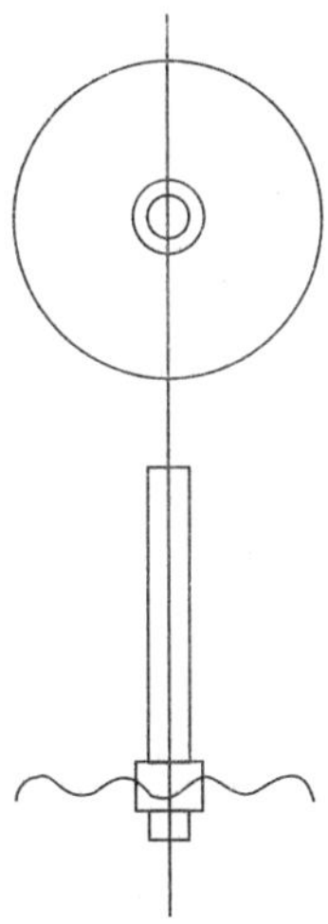

Fig. 14.7 High Speed Disc Agitator

Recently, more efficient impellers (lower N_p – power number) have been introduced by various manufacturers.

14.3 BAFFLES

Baffles are essential for an efficient mixing action. If a simple swirling motion is required no baffling is necessary. The common practice is to use baffles attached to tank walls (Fig. 14.10). Their size and disposition are as follows:

1. Four baffles should be mounted vertically on the tank wall, projecting radially from the wall and located 90° apart.
2. Baffle width should be a tenth to a twelfth of the tank diameter.
3. Baffle height should be at least two impeller diameters and approximately centred on the agitator.
4. For baffles set out from the wall, the width may be 1/12th of the tank diameter with an offset a gap of approximately 1/5th of the baffle width.
5. With coils in the tank, the baffles should be placed inside the coil.

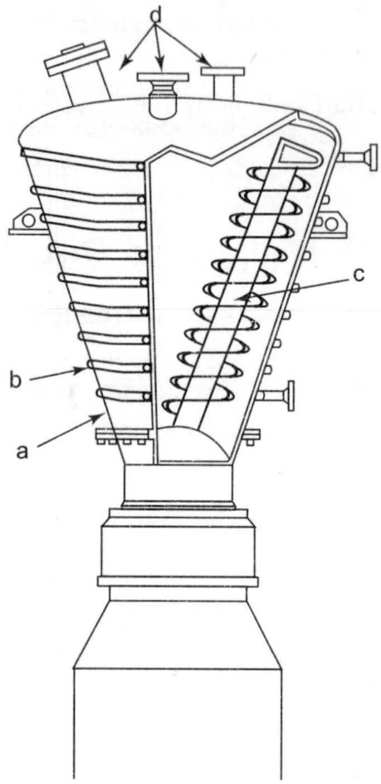

Fig. 14.8 Screw Agitator (During the Agitation with Two Rotating Movements): (a) Conical Vessel, (b) Half Coil Jacket, (c) Screw Agitator, (d) Nozzle

For large diameter tanks (6 metres and above) the number of baffles may be increased to six and they may be set out from the wall to reduce the distance from the agitator to the baffles. For high viscosity liquids (viscosity in the region of 200 poise), angled baffles are preferable. For viscosities above 600 poise, the baffles may be eliminated completely, as the viscous drag becomes sufficient to control swirl and assure vertical circulation.

Baffles are provided to avoid the vortex formation at gas-liquid interface and to provide a high degree of mixing in the tank. The increased level of turbulence induced due to the presence of baffles also results in a higher heat transfer coefficient at the wall. Further, this turbulence also aids uniform suspension of solid particles (reactants or catalyst) in the reactor.

A baffled tank has the following characteristics:

(a) top to bottom turnover with complete intermixing of all field streams throughout the entire vessel.

(b) the absence of a vortex through which air is drawn down into the field.

(c) the absence of a swirling flow pattern as contrasted to top-to-bottom turnover.

In most mixing processes a baffled flow pattern is desirable. However when it is necessary to draw down powders from the surface unbaffled flow pattern is preferable.

A chart indicating the selection of agitators for different services is given in Table 14.1.

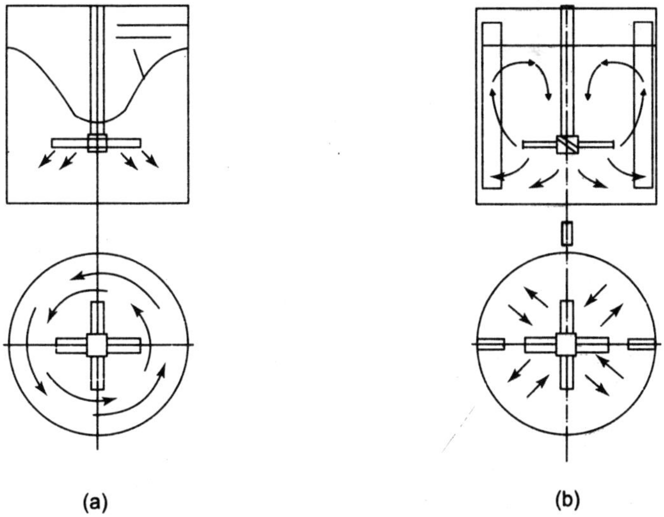

Fig. 14.9 Shows the Flow Pattern Generated by a Flat Turbine: (a) Agitator Without Baffles, (b) with Baffles

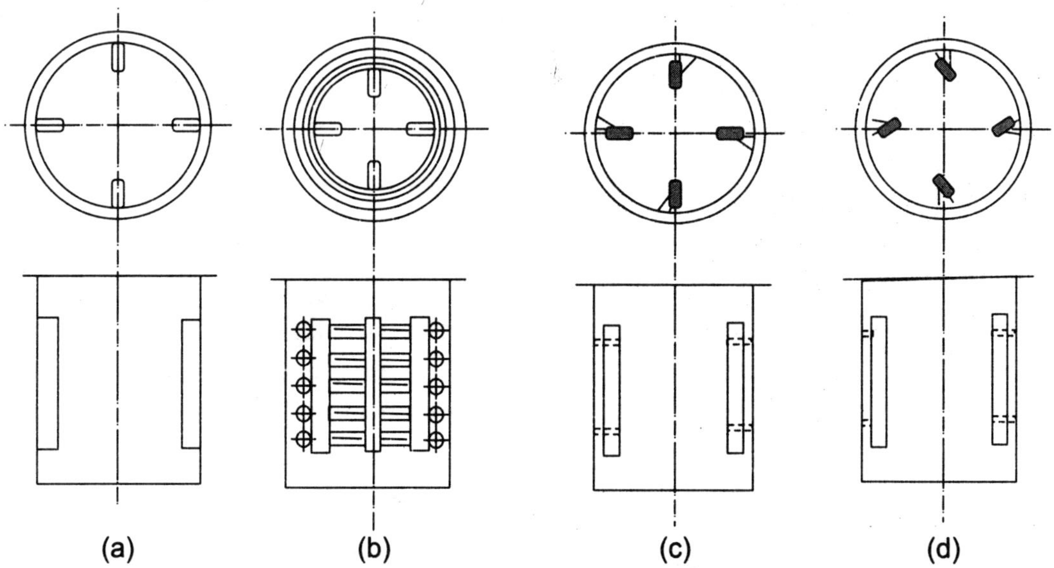

Fig. 14.10 Baffle Arrangements: (a) Normal Baffle, (b) Baffle with Coil, (c) Outset Baffle, (d) Angled Baffle

Table 14.1 Selection Chart (Shape Relationships for Turbine Designs)

Service (1)	*Mixing Device (2)*	*Criteria (3)*	*Tank Diameter to Impeller Dia., Ratio (4)*	*Tank Height to Dia. Ratio (5)*	*Impellers and Position (6)*
Blending	Turbine or propeller	1. Volume circulation	3:1 to 6:1	Unlimited	Single or multiple
Dispersion (immiscible systems)	Turbine	1. Drop size control 2. Re-circulation	3.0:1 to 3.5:1	1:1 to 1.2 IN Staged mixers	At/or below central line of liquid charge
Reactions in solutions (miscible systems)	Turbine, propeller	1. Intensity 2. Volume circulation	2.5:1 to 3.5:1	1:1 to 3:1	Single or multiple
Dissolution	Pitched turbine, propeller, paddle	1. Shear 2. Volume circulation	1.6:1 to 3.2:1	1:2 to 2:1	At/or below central line of liquid charge
Solids suspension	Pitched blade Turbine or propeller	1. Circulation 2. Velocity	2.0:1 to 3.5:1	1:1 to 1:2	Depending on particle size 1. Imp. diameter off bottom 2. On bottom
Gas applications (dispersion)	Pitched blade turbine	1. Controlled shear 2. Circulation 3. High velocity	2.5:1 to 4.0:1	4:1 to 1:1	1. Multiple-lowest impeller one diameter off bottom 2. Self-induce just below liquid level
High viscosity applications	Turbine	1.5:1 to 2.5:1	1.4:1 to 2:1	Single or multiple	
Heat transfer	Turbine propeller Anchor	1. Volume circulation 2. High velocity across transfer surface	Related to other services	Depends on other services being performed	Single or multiple impeller opposite transfer surface when using coils
Crystallization or precipitation	Turbine propeller paddle Anchor	1. Circulation 2. Low velocity 3. Shear control	2.0:1 to 3.2:1	2:1 to 1:1	Single at/or below centre line of liquid charge

14.4 POWER REQUIREMENTS FOR AGITATION

14.4.1 Power Absorbed by the Agitator

Power required to operate an agitator depends on several factors such as the properties of the liquid, height of the liquid, agitator type and size, the tank or vessel size and speed of agitation. No single equation is, therefore, capable of determining the power requirement of different agitators operating under various conditions. A convenient method accepted in many cases is the use of what are known as power curves. An individual power curve is only true for a particular geometrical configuration, but is independent of the vessel size. Numerous power curves are available, representing a wide variety of geometries.

Figure 14.11 shows the power curve for a standard tank configuration, which consists of a flat bottom cylindrical tank, with a 6-flat blade turbine agitator and filled with liquid up to a height equal to the tank diameter. The tank is also provided with four baffles. As may be seen, the curve is a log-log graph of the dimensionless power function and dimensionless power function ϕ and dimensionless Reynolds number N_{Re}, where

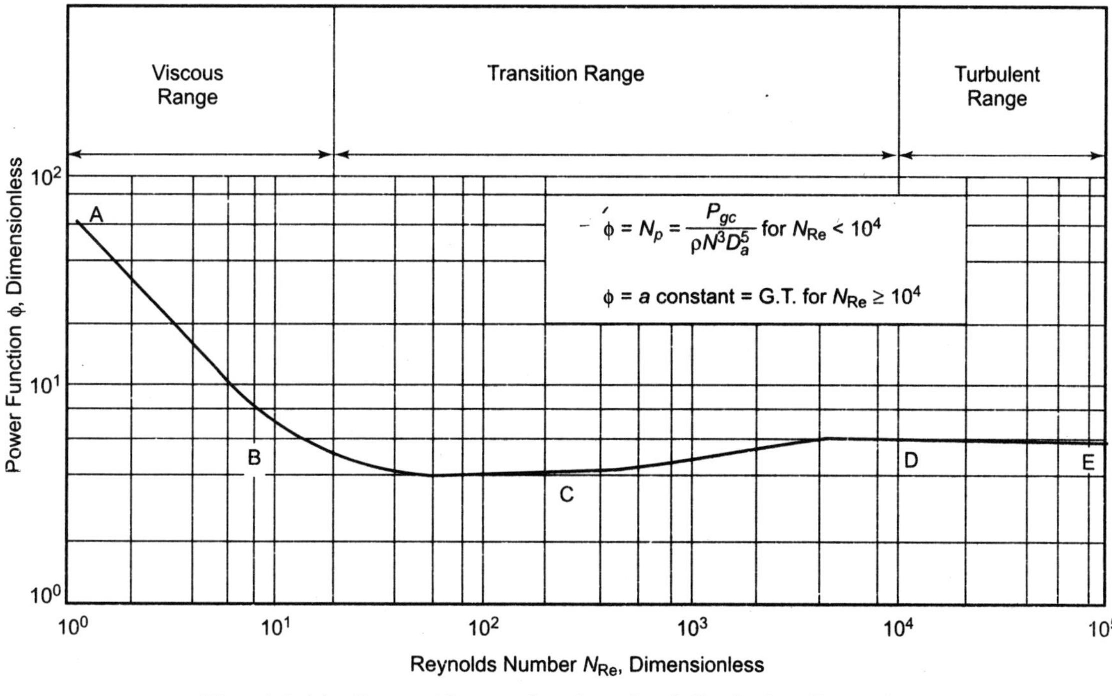

Fig. 14.11 Power Curve for Standard Tank Configuration

$$\phi = N_p = \frac{P}{\rho N^3 D_a^5} \equiv \frac{\text{imposed force}}{\text{inertial force}} \tag{14.1}$$

and

$$N_{\text{Re}} = \frac{\rho N D_a^2}{\mu} \equiv \frac{\text{inertial force}}{\text{viscous force}} \tag{14.2}$$

where, N_p — power number
P — power
ρ — density of liquid
N — speed of agitator
D_a — agitator diameter
μ — viscosity (use consistent units)

In an unbaffled vessel with similar tank configuration, the mixing action is different and the relation between function ϕ and Reynolds number N_{Re} is a shown in Fig. 14.12, where,

$$\phi = N_p = \frac{p}{\rho N^3 D_a^5} \quad \text{for} \quad N_{Re} < 300 \tag{14.3}$$

and $$\phi = N_p = \frac{(\alpha - \log_{10} N_{Re})}{N_{Fr}\beta} \quad \text{for} \quad N_{Re} > 300 \tag{14.4}$$

where, N_{Fr} is known as the Froude number and is given by

$$N_{Fr} = \frac{N^2 D_a}{g} \equiv \frac{\text{inertial force}}{\text{gravity force}} \tag{14.5}$$

Values of α and β are given in Table 14.2.

Table 14.2 6 Blades, Flat Blade Turbine Agitator

Diameter (D_a) cm	Da/D	α	β
10	0.3	1.0	40.0
15	0.33	1.0	40.0

The above curves are for a particular tank configuration, and may be used to calculate power for various agitator speeds, liquid viscosities and densities. Similar curves are available for other configurations.

14.4.2 Power Absorbed by Vessel Fittings

Fittings such as dip pipes, thermowells as also coils will cause a power loss. This loss depends on the position of the fittings. Fittings close to the shaft make no appreciable difference to the power consumption. Fittings close to the tank wall absorb more power.

14.4.3 Transmission and Gland Losses

The power loss in glands varies from 0.5 hp for smaller agitator shafts, and up to 5 hp for larger shafts. No reliable method for calculating this loss can be given. As a very rough approximation, however, the gland loss may be taken as 5% of the agitator power consumption. The gland loss depends upon the pressure inside and the speed. For helical gear, the efficiency of power transmission is above 95% and for worm gear it is 75 to 95% depending on the ratio. For 'V' belt pulley drives it is more than 95%.

14.4.4 Horse Power of Driving Motor

The required power output of electric motor is obtained by the addition of the above power consumption factors. Due allowance must be made for the possible occurrence of extra heavy loads during start-up periods or during some stage of the process. Some empirical allowances may be made. Losses in the motor at various loads can be determined from output efficiency curves. The horsepower, hp, is obtained by $h_p = W/745.7$.

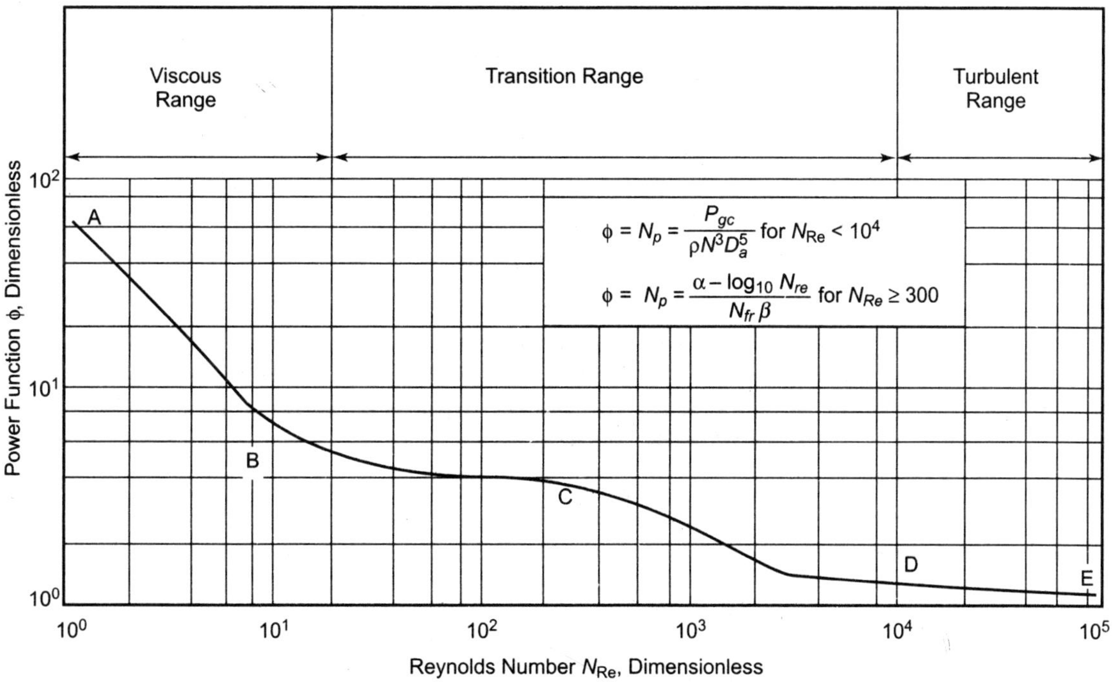

Fig. 14.12 Power Curve for Unbaffled Vessel

14.5 DESIGN OF AGITATION SYSTEM COMPONENTS

In order to rotate the agitator at the required speed, it is attached to a shaft. The driving system usually consists of shafts, coupling, bearings, gearing, pulleys and belts. The power is supplied by an electric motor, a hydraulic motor or a steam turbine. In the case of toxic, flammable or volatile materials or liquids under pressure, special provision has to be made to prevent leakage, between the shaft and the portion of the vessel surrounding the shaft. To allow free rotation of the shaft, a clearance must be maintained between these adjacent parts. A stuffing box or a seal is used for preventing leakage through this clearance. Figures 14.13 and 14.18 indicate the arrangement and components of drive system, along with an agitator as provided for a pressure vessel. Details of some of the components and the design features are considered below:

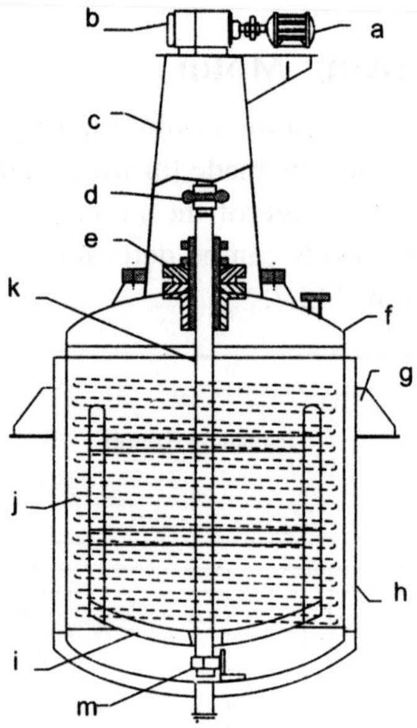

Fig. 14.13 Typical Drive System: (a) Motor, (b) Gear Box, (c) Support Bracket (Lantern Stool), (d) Coupling, (e) Stuffing Box or Mechanical Seal, (f) Vessel, (g) Bracket, (h) Jacket, (i) Anchor Agitator, (j) Coil, (k) Shaft, (m) Steady Bearing

14.5.1 Shaft Design

The shaft is attached to the agitator and may be located in horizontal, vertical or angular position. It is preferable to support the shaft by the use of bearings outside the vessel or equipment. In special cases steady bearings may be used, which may be located inside the equipment, such as bearings placed at the bottom of a vessel. A stabilizer ring just below the agitator may be used to reduce vibrations and deflection of the shaft.

The shaft design is based on criteria indicated in detail in Section 5.2.

14.5.1.1 Design Based on Torque and Bending Moment

The normal power required for the agitator and frictional losses are indicated under 'Power Requirement for Agitation' (Section 14.4). From this, it is possible to determine the continuous average rated torque on the agitator shaft.

$$T_c = \frac{p}{2\pi \times N} \tag{14.6}$$

where, N — rps,

Torque, T_c — N-m,

P — watts.

During a start up, the shaft has to withstand much higher torque. During running, apart from the torque, various forces acting on the shaft have to be taken into account. These are transient unbalanced hydraulic forces due to turbulence in the liquid or asymmetrical construction of the agitator and baffles, acting laterally on the shaft in a cyclic manner. Centrifugal forces are also present while the shaft is rotating and the agitator is out of balance. Further, there is the possibility of the agitator being jammed (or locked) when the material is added to the vessels by tipping bags or other containers. The worst cumulative conditions are assumed to be equivalent to those in which the agitator blade is jammed at a point 75% of its length from the shaft.

During starting, the shaft must be capable of resisting 1½ times the continuous average torque (T_c) at low speeds and 2½ times at high speeds. The shaft is, therefore, designed for either of these maximum values. The maximum shear stress developed is given by

$$f_s = \frac{(1\tfrac{1}{2} \text{ or } 2\tfrac{1}{2}) \times T_c}{Z_p} = \frac{T_m}{Z_p} \tag{14.7}$$

where, T_m — max. torque

f_s — shear stress

Z_p — polar modulus of section of the shaft cross-section.

If the permissible stress value of f_s is known, the shaft diameter can be determined. Another criterion of design is based on the fluctuating loads during running. Consider the agitator blade jammed at 75% of its length for *a* short period. In this case it is necessary to work out the equivalent bending moment which is the cumulative resultant of the bending moment and maximum torque (refer Eq. 5.2).

$$M_e = \frac{1}{2}\left[M + \sqrt{M^2 + (T_m)^2}\right] \tag{14.8}$$

where, T_m — (1.5 or 2.5) T_c

M — bending moment.

The bending moment M is determined as follows:

The torque T_m is resisted by a force F_m acting at a radius of $0.75R_b$ from the axis of the agitator shaft.

$$F_m = \frac{T_m}{0.75 R_b} \tag{14.9}$$

where, R_b— radius of blade.

The maximum bending moment M occurs at a point near the bearing, from which the shaft overhangs

$$M = F_m \times l \tag{14.10}$$

where, l — shaft length between agitator and bearing.

The stress due to equivalent bending moment is given by

$$f = \frac{M_e}{Z} \tag{14.11}$$

where, Z — modulus of section of the shaft cross section (Appendix A).

The stress f should not exceed the yield stress of the material or 0.2% proof stress. It is assumed that the force F_m acts as a transient force and, hence, the design is based on yield stress. From Eqs. 14.7 and 14.11 the shaft diameter may be selected. The higher value is usually chosen.

14.5.1.2 Design Based on Critical Speed

It is difficult to assess the unbalanced hydraulic forces or the unbalanced forces due to asymmetric construction of the agitator. The latter forces can be eliminated to a considerable extent, if the agitator is statically and dynamically balanced. This is usually effected by fixing certain counter balance weights in the opposite direction to the unbalanced forces. As the shaft rotates, centrifugal forces are created. If the speed of the shaft varies, it is not possible to balance these dynamic forces completely. Only a partial balance may be possible.

It is necessary to control the deflection of the shaft by adequate supports. If the shaft is overhung, it may be necessary to increase the shaft diameter for the cantilever portion to reduce deflection. Higher deflection increases eccentricity and, therefore, centrifugal forces. These forces change direction as the shaft rotates and cause vibrations. The speed at which the shaft vibrates violently is known as critical speed. It is recommended that the range of speed between 70% and 130% of the critical speed should be avoided. The shaft diameter should be so chosen that its normal working speed range does not fall within the critical speed range. Low speed agitators such as paddles and turbins normally operate between 50% and 65% of their critical speed. If dynamic balancing is done, one can operate even at 70% of critical speed (N_c). High speed impellers, such as propellers and discs, normally operate above the critical speed. Determination of critical speeds for shafts is done as follows:

Figure 14.14 shows a horizontal shaft, supported at two bearings A and B. The loads or forces w_1, w_2, are acting perpendicular to the shaft. The deflection due to each load acting independently is given by

$$\delta = \frac{Wl^3}{3EI} \tag{14.12}$$

The load w_2 is weight of the second impeller (agitator)

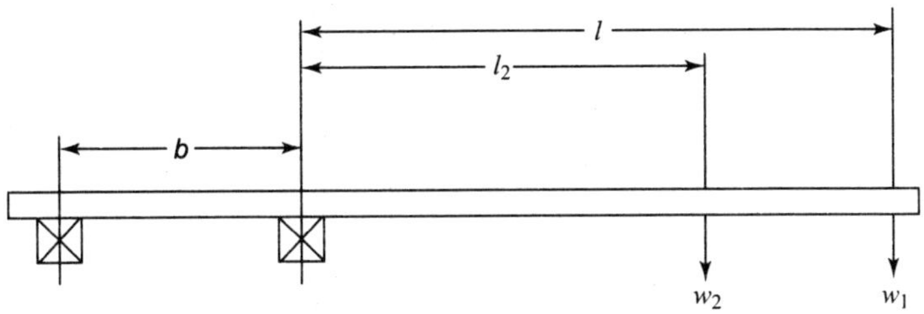

Fig. 14.14 Shaft Supported in Bearings

Similarly, the maximum deflection due to shaft weight is given by

$$\delta_s = \frac{wl^4}{8EI} \tag{14.13}$$

where, l — appropriate length

W — concentrated load

w — uniformly distributed load per unit length

E — modulus of elasticity

I — moment of inertia of the cross section of shaft.

The critical speed can be determined as follows:

$$N_c = \frac{946}{\left(\delta_1 + \delta_2 + \dfrac{\delta_s}{1.25}\right)^{1/2}} \tag{14.14}$$

where, δ_1, δ_2 and δ_s are the deflections in mm due to each load and N_c expressed in rpm.

As an approximation, we can use the above formula for a vertical shaft also. The critical speed, N_c, given in Eq. 14.14 is for the static condition only. For dynamic load prevailing in the agitator a complex procedure is involved [Fasano et al. (1995)]. Various vendors have their own computer programs. However, for a quick computation, we can assume that the force generated by the maximum torque acts at the bottom most impeller position and is perpendicular to the shaft. We then calculate the deflection, δ, due to this force and impeller with hub acting at the end of shaft via Eq. 14.12. Further, Eq. 14.13 assuming the shaft to be a cantilever with uniformly distributed load calculates the deflection due to shaft weight.

The shaft always has a bearing support. The critical speed, N_c, is then derated by a factor β given by [Uhl and Gray (1967 b)].

$$\beta = \{1/(1 + \alpha\,(I_l/I_b))\}^{0.5}; \quad \text{where, } \alpha = b/l,\ \text{I} = \text{moment of inertia}$$

For a quick estimation, β can be deemed as unity.

This procedure is well illustrated in the example given below. The recent trend is to use a hollow shaft for which similar procedure is followed with an appropriate moment of inertia (Appendix A).

14.5.2 Agitator

The main parts of an agitator are the hub and the blades. The hub is attached to the shaft by the use of keys or bolts. Agitators may either be cast as one piece or in two pieces split and bolted at the hub. Cast agitators have two basic advantages, namely, uniformity of material and a hard surface. Agitators are also fabricated from metal sheets by welding. Such agitators are liable to corrode, particularly at the welded areas, if precautions are not taken. It is desirable to restrict welding to areas near the hub, where the flow velocity is minimum. Agitators of bolted construction, with blades bolted at the hub (Fig. 14.2) are also fabricated. These are liable to corrosion and failure because of exposed threads.

Wooden agitators are useful only for relatively low speeds. They are difficult to balance and tend not to remain in balance due to working and absorption of moisture. Simple paddles or gates can be made satisfactorily. Wood is used successfully in many aqueous processes where no suitable metals are available or where the size of the equipment is large and the cost of other materials would be excessive. The lead lined (bonded) shafts used earlier are now being phased out due to health and environment concerns. Erosion or creep can cause rapid failure. Enamelled cast iron or enamelled mild steel agitators can be made in a variety of forms, but the outer contours must be smoothly rounded and the sections must be uniform as far as possible. Agitators can also be covered with rubber which offers high resistance to corrosion.

Enamel, rubber or lead lined agitators are likely to be troublesome, where high intensity agitation is required with suspended solids. In such cases, it is economical to use special alloys for impellers and shafts.

As indicated in Section 14.5.1.1, the load on the blade is assumed to act as 75% of the agitator radius. This will create a bending moment, which will be maximum at the point where the blade is attached to the hub.

$$\text{Max. BM} = F(0.75R_b - R_h) \tag{14.15}$$

where, F — force on each blade

R_b — radius of blade

R_h — radius of hub.

If the blade is flat, the stress in the blade will be given by

$$f = \frac{\text{Max. BM}}{Z} = \frac{F(0.75 R_b - R_h)}{\dfrac{b_t \times b_w^2}{6}} \tag{14.16}$$

where, b_w — blade width

b_t — blade thickness.

Since the blades are subject to high frequency oscillation, due to turbulence in the liquid, they are liable to fatigue. The permissible values of the stress should never exceed the endurance limit for the blade material.

The hub is fixed to the shaft by a key, or setscrew, which transmits the shaft torque to the impeller. The hub is subjected to bending moments due to the forces on the blades and to shear forces due to torque. It is usual to assume the outside hub diameter as twice the shaft diameter and check the shear stress due to torque. The detailed design of key and hub is indicated under Sections 5.3 and 5.4.

14.5.3 Couplings

For connecting the agitator shaft to the drive shaft, three types of couplings are generally used. These are flange coupling, both rigid and flexible, split muff coupling and sleeve coupling. They must be capable of transmitting torque, sustain the end thrust due to the weight of the agitator and should withstand the bending moments imposed upon by the unbalanced hydraulic and centrifugal forces. For small agitators with shafts up to 40 mm diameter and 1.25 m length, a sleeve coupling may be used. The details of construction and design are indicated in Section. 4.4.

14.5.4 Bearings

It is common to use bearings as far as possible outside the equipment [Fig. 14.18(h)] where agitation is carried out. However, intermediate bearings and bearings at the free end of the shaft may also be used. In deep tanks with long agitator shafts, excessive vibrations can be reduced by using an auxiliary or steady bearing at the bottom of the shaft [Fig. 14.11(m)]. These should be mounted on a support fastened to the bottom or side of the tank. The mounting should allow the shaft to be adjusted for alignment of the agitator shaft. Servicing of a steady bearing involves the inconvenience of shutting down the equipment. These bearings also create problems of alignment, lubrication, corrosion and maintenance. Whenever possible, the use of such bearings should be avoided. If it is absolutely necessary to use these types of bearings, replacement sleeves for shaft and bearing surfaces are advantageous. Bearings outside the vessel are nearly always of the ball or roller type. However, plain bearings with busing may be satisfactory for lower shaft speed, although difficult to install. Bushes may be made of cast iron, phenolic material, phosphor bronze, rubber, etc. Grease lubrication is normal and protection against entry of the condensate, dust or fumes from the vessel should be provided. Details of bearing design are given under Section 4.5.

Bearings are available in 5 mm increments of shaft diameter, i.e., bearings are available for shaft diameter 25, 30, 35, ..., 60, 65, 70 mm and so on. For shafts larger than 150 mm the vendor may be consulted.

14.5.5 Stabilisers

As mentioned earlier, the critical speed range for the rotation of shaft should be avoided. However, with variable speed agitators, it may be necessary to rotate the shaft within this range. The unbalanced forces may, therefore, cause the shaft to vibrate vigorously. To dampen such vibrations, a stabiliser is sometimes used. This is essentially in the form of a ring (Fig. 14.15), welded to the lowest agitator at the bottom end of the cantilever shaft. When the transient unbalanced hydraulic force pushes the agitator away from the axis of rotation, the movement creates a counter balancing drag force on the stabiliser. This drag force provides the damping and reduces the vibrations and stresses.

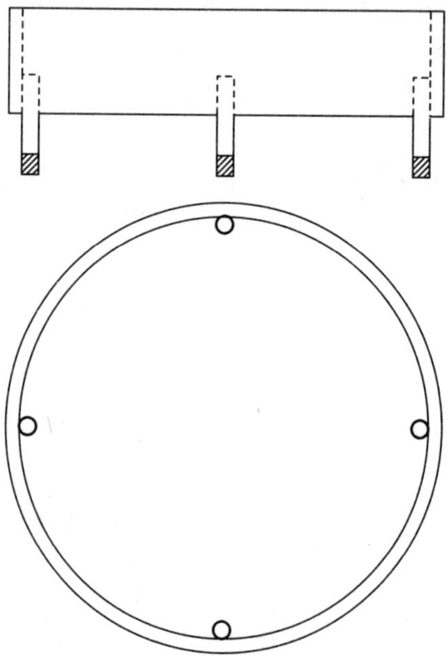

Fig. 14.15 Stabiliser Ring

14.5.6 Shaft Seals

In the case of toxic or flammable materials, and during the evaporation of volatile materials, or when the process pressure is different from the atmospheric pressure, the agitator shaft must be sealed to isolate the contents of the equipment from the atmospheric surroundings and prevent an outflow of the fluid from the equipment or an inflow of foreign material into the equipment. The most common methods for sealing are stuffing boxes and mechanical seals. Details regarding both the methods are indicated in Sections 5.12, 5.13 and 5.14.

A stuffing box may be located partially within the vessel (Figs. 14.16, 14.17 and 14.18) or outside the vessel. In either case, it is fixed to a pad welded to the top head of the reaction vessel. Figure 14.16 shows a bush in the base of the stuffing box. This may be necessary when shafts are likely to be deflected or where a high standard of performance is expected of the gland. Cast iron, plastic impregnated asbestos or cotton packings are very often satisfactory where the vessel and the agitator are lined with lead, enamel or rubber. For severe duties, as may be seen in the figure, it is possible to provide a shoulder on the shaft with a wearing sleeve of resistant material up to the shoulder and through the gland. These sleeves may be of hastelloy, high silicon iron, phosphor bronze, etc. The gland may be constructed out of steel with a sleeve of soft material. Figures 14.17 and 14.18 show a method of lubricating the packing, through a lantern ring located in the centre of the stuffing box. It is supplied with a lubricating fluid. A heavy duty stuffing box is shown in Fig. 14.19. It is provided with a cooled jacket, supplied with a circulating liquid. A few springs are also placed in the centre of the stuffing box. These springs exert a pressure on the two rings placed on either side. The pressure is transferred to the packings, which helps in creating a uniform pressure within the stuffing box, thus, reducing the leakage from the gland. Mechanical seals have already been discussed earlier in Chapter 5.

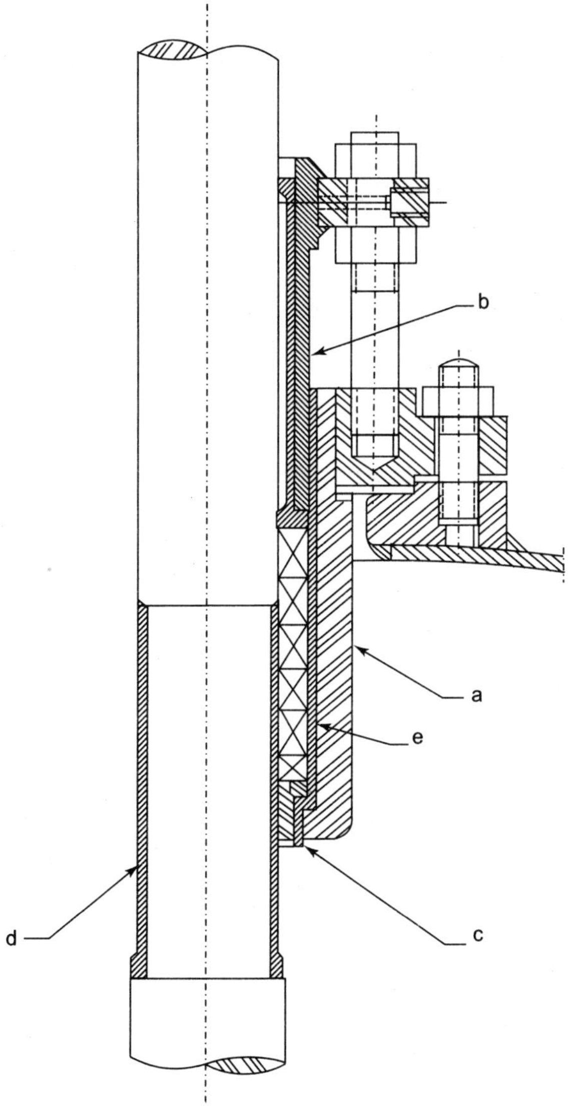

Fig. 14.16 Stuffing Box and Gland: (a) Stuffing Box, (b) Gland with Sleeve, (c) Bush, (d) Shaft Lining, (e) Packing

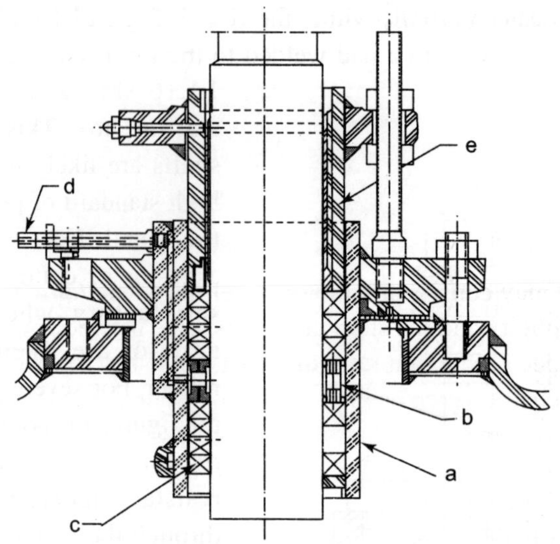

Fig. 14.17 Lubricated Stuffing Box and Gland: (a) Stuffing Box, (b) Lantern Ring, (c) Packing, (d) Inlet of Lubricating Fluid, (e) Gland with Sleeve

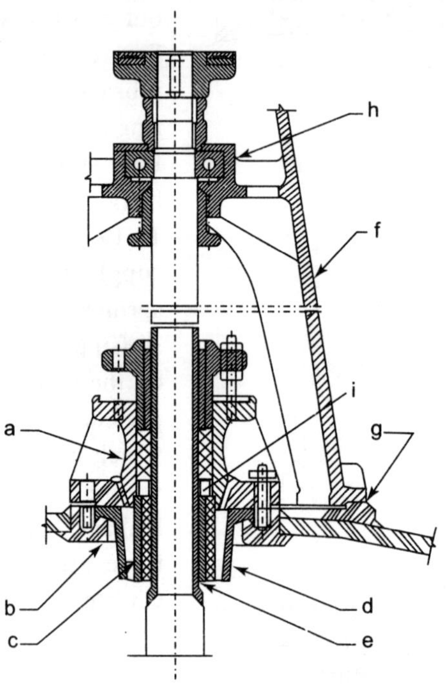

Fig. 14.18 Lubricated and Jacketed Stuffing Box: (a) Stuffing Box, (b) Mounting pad, (c) Bush, (d) Cooling Jacket, (e) Shaft Lining, (f) Cast Support for Bearing, (g) Pad for Fixing the Frame to Head, (h) Bearing with Housing, (i) Lantern Ring

14.6 DRIVE FOR AGITATORS

The power can be estimated as

$$P = 2\pi NT$$

$$N \equiv rps$$

$$T \equiv \text{N.m}$$

$$P \equiv w$$

Agitator drive systems may consist of a motor, a belt drive and/or a gear drive. Usually, a V-belt is preferred due to its simplicity and the short centre distance between motor pulley and shaft pulley. It is satisfactory up to a speed reduction ratio of 3 to 1 for V-belt, 5 to 1 for gear (helical and bevel type) and for worm gear it can be even be higher than 50. Spur, helical, bevel and worm gearing may be used according to convenience and shaft arrangement. Helical gear teeth have greater strength and smoother engagement. The transfer of power is also gradual. The teeth are also subjected to less shock. Helical gear trains are efficient in power transmission and are preferred to other types of gears. Worm gearing is suitable for transmission of power between perpendicular shafts. With this gearing it is possible to obtain high velocity ratios in minimum of space but usually by sacrificing efficiency. Gear boxes are expensive.

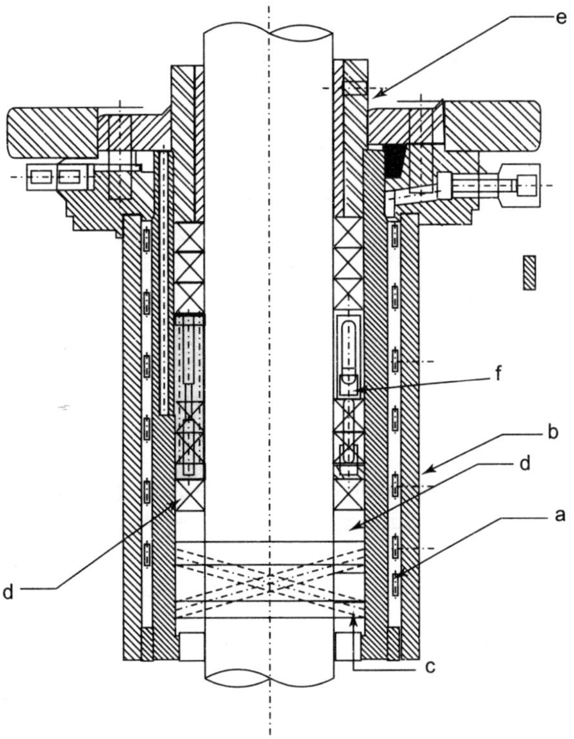

Fig. 14.19 Jacketed Stuffing Box: (a) Stuffing Box, (b) Jacket, (c) Spring, (d) Packing, (e) Gland Pusher with Sleeve (f) Lantern Ring

14.6.1 Vertical Drives

In the vertical agitator applications, the motor may be placed in a vertical position or in a horizontal position. In most applications, a reduction in speed is desired, since the agitators rotate at speeds lesser than the speed of the motor. This involves a single, double or triple reduction, depending on the required agitator speed. High speed agitators such as propellers require only one set of reduction gears. Low speed agitators such as paddles require two or three sets of reduction gears. The vertically mounted motor can be directly coupled to the input shaft on an in-line vertical reducer and the output shaft of the reducer can be directly coupled to the agitator shaft. In the case of a horizontal motor, it is necessary to provide for the transitions from the horizontal to vertical position of the agitator shaft. This can be achieved with either a worm or a bevel gearing, along with a belt drive or a gear train.

14.6.2 Side Entering Agitators

In this arrangement, the agitator shaft is horizontal. The motor may be either directly coupled to the agitator shaft (Fig. 14.20) or through a single reduction gear unit of a V-belt arrangement. The reduction is between the parallel shaft of the agitator and that of the motor. In special cases, a bevel or spiral bevel gear may be used on the agitator shaft with the input pinion mounted vertically along with the motor.

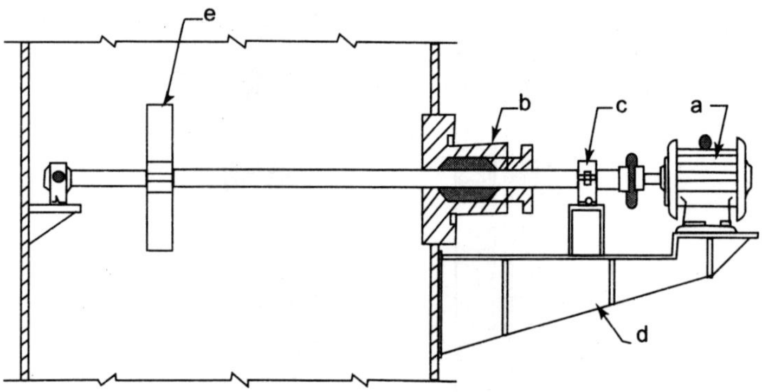

Fig. 14.20 Side Entering Agitator Shaft: (a) Motor, (b) Stuffing Box and Gland, (c) Bearing, (d) Support Bracket, (e) Agitator

In a large tank (storage tank), there could be more than one side entering agitator, locally supported or just suspended.

14.6.3 Special Drives

Instead of using an electric motor, a hydraulic drive system may be used, particularly where high torque at low speeds is required. The hydraulic drive motor can be directly mounted to a vertical shaft, eliminating gear boxes. A wide range of speeds is feasible. However, this needs additional equipment such as pumps, valves, etc.

14.6.4 Mountings

The drive system assembly must be located at the top of the vessel with preferably an independent supporting structure. In the case of an open tank, the assembly can be supported between a support bracket and a horizontal girder placed on top of the tank. Alternately, it can be supported between the brackets, over which girders can be placed.

In vessels with heads, it is necessary to consider the rigidity of the support to eliminate vibration or flexure, detrimental to stuffing box operation and packing life, and also the effect of load on the head thickness.

One method of avoiding any load on the head is to provide a beam support, the ends of the beams can be extended down the vessel shell to distribute the load over a wider area. In the case of low drive assembly load, it may be mounted on pads fixed to the head with a stool support [Figs. 14.13(c), 14.18(f) and (g) and 14.21]. In such cases, and with a full vacuum in the vessel, additional agitator drive load must be given special consideration.

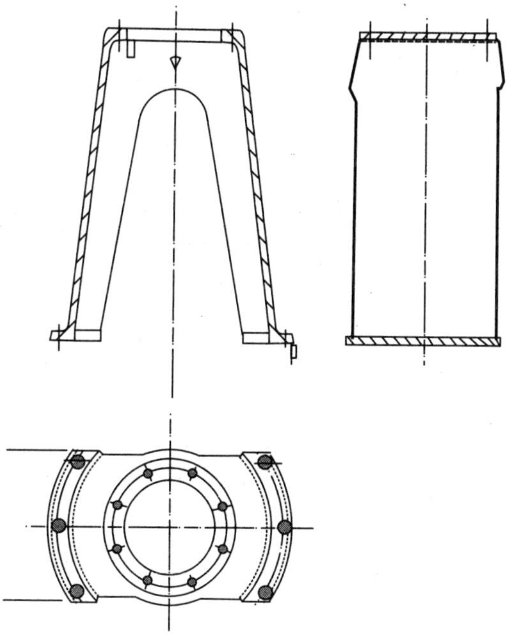

Fig. 14.21 Stool for Supporting Drive System (Welded Construction)

The effect of an eccentric load on the mounting arrangement may cause stresses in the joint. This is particularly relevant with horizontal motors. Special supports or brackets must be provided beneath the centre of the drive assembly in such cases.

AGITATOR

Process Data Sheet

VESSEL SERVICE: Vessel Item/Spec. Number

Type:

Entry: ☐ Top ☐ Bottom ☐ Side ☐ Eccentric ☐ Angled

Driver Type: Working Standby Driver Item No.

OPERATING CONDITIONS

Duty: ☐ Continuous ☐ Batch No. Starts/hour ☐ Varying Liquid L

Mixing Process: ☐ Blending and Dispersion (Gas/Liquid/Liquid)

☐ Suspend Solids ☐ Emulity

☐ Maintain Solids Suspension ☐ Heat Exchange Flow rate

☐ Dissolving ☐

Direction of Flow: ☐ Up ☐ Down ☐ Reversible

Required Process Result of Agitation:

MATERIAL CHARACTERISTICS

Component	Quantity	Viscosity	% Solids	Density	Settling Velocity	Te
Units			%			
TOTALS						

Liquid Behaviour ☐ Newtonian ☐ Non-Newtonian

Foaming Tendency: ☐ None ☐ Mild ☐ Strong

Description of Solids:

Vessel Dimensions: Location: ☐ Indoors ☐ O

Liquid Level: Normal Maximum Minimum

Baffle Requirements:

Impeller Type: Number of Impellers

Vessel Operating Pressure: Normal Max. Min RV Set Press

Shaft Sealing:

Materials of Construction:

Special Process Requirements:

Side entering agitators must be mounted such that the drive assembly has the least possib overhang. The unit has to be supported by tie-rods or brackets of sufficient strength (Fig. 14.20). is advisable to support outboard machinery with any structure anchored to the ground, since slig shifts in the position of process vessel lead to incorrect alignment between the shaft and the sea This results in excessive shaft vibration, wear and leakage of stuffing box or seal.

14.7 STATIC MIXERS

Static mixers, also known as on-line mixers are gaining popularity in the industry over the mo conventional dynamic mixers. As the name suggests they are also referred to as motionless mixer as there is no motion of the mixing elements. The mixing elements are specifically designed for t application in mind. Each element is deemed to be stirred tank reactor. The mixing is achieved fi with a high level of turbulence and then by molecular level diffusion. The extent of turbulence reflected in energy dissipation (energy loss) in the form of pressure loss, ΔP.

We then have pump power,

$$P_p = Q \times \Delta P; \quad W \tag{14.1}$$

where, Q = flow rate; m^3/s

ΔP = pressure drop; N/m^2

The power for centrifugal pump depends upon pump efficiency and motor efficiency.

Typically, the pump efficiency, η_p, in the absence of data from the vendor, can be obtained as estimate via [Smith (2005)]:

$$\eta_p = 0.86 - 0.14 \ln(Q) - 0.01 [\ln(Q)]^2 \tag{14.1}$$

$$2.8 \times 10^{-4} < Q < 0.28$$

The motor efficiency, η_m is assumed to be 0.95

$$\text{The power required} = Q \times \Delta P/(\eta_p \times \eta_m) \tag{14.1}$$

The pressure drop and the residence time distribution depend upon the elements, which ha proprietary design.

Figure 14.22 exhibits a typical static mixing concept. A typical mixing element with housing exhibited in Fig. 14.23.

The mechanical design consists of:

- Internal mixing elements, which are proprietary in nature.
- Housing design, which is nothing but a pipeline design described elsewhere consider material of construction and design pressure of fluid streams.
- The housing containing mixing elements is integrated with the existing system throug flange joint preferably.

We, thus, see that the mechanical design of static mixers is simple, compared to the dynar mixers. The maintenance is also very simple as there are no moving parts.

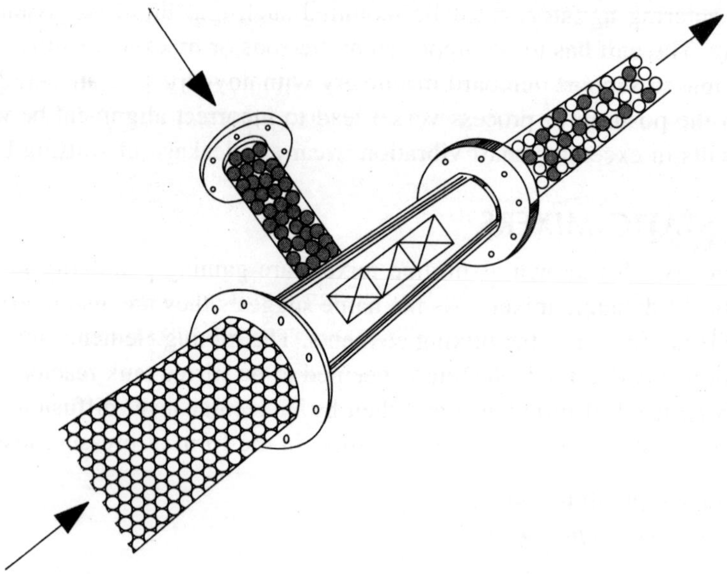

(Courtesy: Mamko Design and Engineering Pvt. Ltd. Mumbai 400063)

Fig. 14.22 Static Mixing Concept

(Courtesy: Mamko Design and Engineering Pvt. Ltd., Mumbai 400063)

Fig. 14.23 Housing with Elements

The static mixers are now used not only for mixing operation but also for liquid-liquid and gas-liquid reactions with simultaneous heat and mass transfer. They find applications in process intensification. The static mixers could be constructed in any metal as well as poly propylene, (PP) fibre reinforced plastic (FRP), poly vinyl chloride, high density polyethylene (HDPE) and Teflon(PTFE) .

14.8 JET MIXERS

Like static mixers jet mixers are very simple in design involving no rotational / moving part. They are practically maintenance free. They are very popular in tank mixing application. In jet mixing, a fast moving stream of one fluid (Motive fluid) is injected in practically stationary bulk liquid. The jet entrains the bulk liquid in its direction of flow and thereby creates mixing. The jets are very good in blending operations or homogenization of material in the tank. The jets are positioned near the bottom or they can have side entry to the tank. One needs to have a pump for motive fluid. Power requirement is higher as compared to agitator.

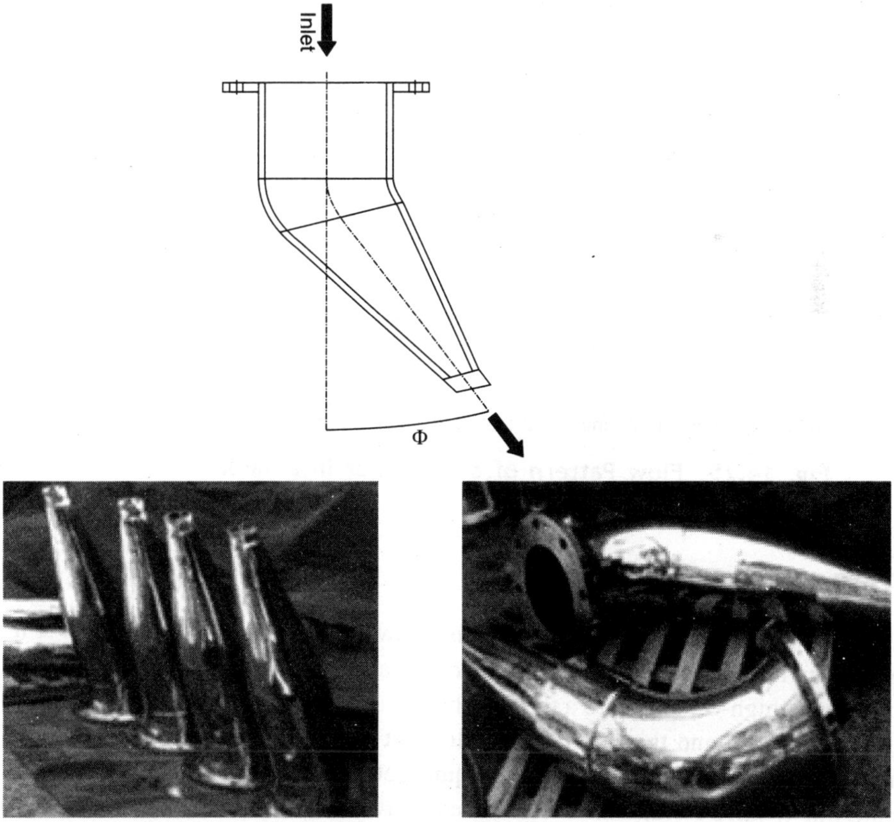

(Courtesy: Mamko Machine Fabricators Pvt. Ltd. Mumbai - 400063)

Fig. 14.24 Jet Mixer Nozzle

The jets are many times combined with eductor or static mixers to have intense mixing. The jets nixers find applications in large tanks wherein mechanical agitation is not attractive. For example, in rude oil storage tanks wherein hard sludge deposition could be practically eliminated by having jet nixers. They can be used for tank aeration application also. The jet mixer design is a specialized job vherein one has to have very good understanding of fluid dynamics, mixing time and end pplication. Fig 14.24 exhibits jet mixer nozzles and Fig 14.25 shows flow patterns in a tank.

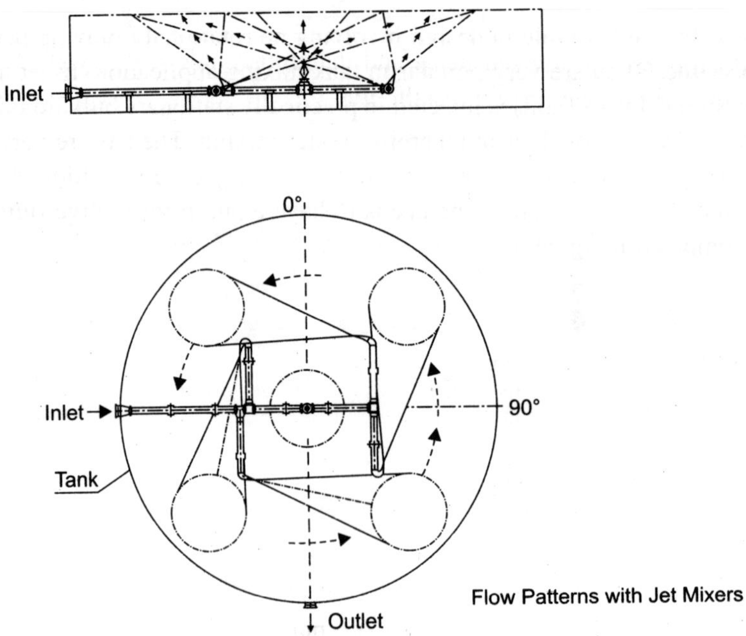

(Courtesy, Mamko Machine Fabricators Pvt .Ltd. Mumbai - 63)

Fig. 14.25 Flow Pattern of a Jet Mixer in a Tank

14.9 EDUCTOR

When an ejector is used for liquid-liquid application it is known as eductor. It is a liquid–Jet liquid ejector. In its simple form it is a venture as depicted below in a schematic diagram (Fig. 14.26). The motive liquid is pumped at high velocity into a throat. This in turn creates suction of the other liquid. In a throat both get well mixed and then they are ejected out in the bulk. That mixed stream then creates another mixing zone via eddies and then the liquid gets well mixed. When we want to mix two liquids or would like to have storage tank well mixed to avoid variation in product quality from batch to batch, it is common practice to use eductors. One can use eductor with jet mixer also to get advantage of both. A static mixing element in a throat can result in very efficient mixing.

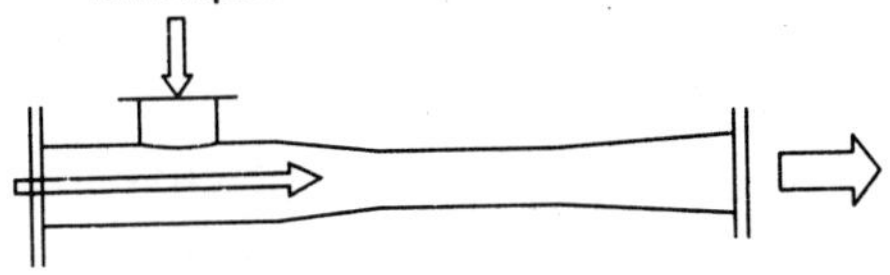

Fig. 14.26 An Eductor: A Schematic Diagram

n eductor can be used even to lift liquid from underground level up to above ground level tank. a simple device practically free from maintenance. It can be fabricated in any metal or plastic as rial of construction.

atic mixers, jet mixers and eductors are specialized items and therefore vendors with proven record are always recommended for fabrication and installation.

10 ILLUSTRATION

illustration gives a first hand experience of how to carry out the mechanical design of an agitator reaction or mixing vessel having a standard tank configuration.

/e have the following data:

The vessel diameter	: 1500 mm
Internal pressure	: 0.5 N/mm^2
Diameter of an agitator (impeller)	: 500 mm
Width of the blade	: 75 mm
Agitator blades (flat)	: 6
Thickness of blade	: 8 mm
Speed	: 200 rpm
Liquid in vessel	
Specific gravity	: 1.2
Viscosity	: 600 cP
Overhang of the agitator shaft between bottom bearing and agitator	: 1300 mm
Baffles at tank wall	: 4
Shaft material – commercial cold rolled steel (IS-2062)	
Permissible shear stress in shaft	: 55 N/mm^2
Elastic limit in tension (yield stress)	: 246 N/mm^2
Modulus of elasticity	: 1.95×10^5 N/mm^2

Permissible stresses for the key (carbon steel)

Shear : 65 N/mm^2

Crushing : 130 N/mm^2

Stuffing box (carbon steel)

Permissible stress : 95 N/mm^2

Studs and bolts (hot rolled carbon steel)

Permissible stress : 58.7 N/mm^2

14.10.1 Power

We have from Eq. (14.2) Reynolds Number as

$$N_{Re} = D_a^2 N\rho/\mu$$

$$= \frac{(500/1000)^2 \times (200/60) \times 1200}{600 \times 10^{-3}}$$

$$= 1667$$

say, $= 1670$

From Fig. 14.11, we have Power Number

$$N_p = 4.5$$

Therefore, power can be calculated with the help of Eq. (14.1):

$$P = N_p \times \rho \times N^3 \times D_a^5$$

$$= 4.5\ (1200) \times (200/60)^3 \times (500/1000)^5$$

$$= 6250 \text{ W}$$

The gland loss is 5% and the efficiency of power transmission is 95%

Therefore, motor rating = 6250/(0.95 × 0.95) = 6925.2 W

= 6925.2/745.7

= 9.28 hp

say, = 10 hp

Thus, we shall have a motor of 10 hp (or say, 7.5 kW)

14.10.2 Shaft Design

For the design of shaft we consider power = 10 hp = 7457 W. Strictly speaking, power input to shaft be considered. The design at higher rating is, therefore, very conservative.

We have from Eq. (14.6) torque T_c

$$T_c = P/2\pi N = 7457/(2\pi\ 200/60)$$
$$= 356 \text{ Nm}$$

The maximum torque during start up = 1.5 × 356

$$T_m = 534 \text{ N/m}$$
$$= 534 \times 10^3 \text{ N/mm}$$

The permissible shear stress = 55 N/mm^2

From Eq. (14.7) the polar modulus of the, Zp, of the shaft is

$$Z_p = 534 \times 10^3/55 = 9709.1 \text{ mm}^3$$

Also, $Z_p = (\pi/16)d^3$

Therefore, the shaft diameter, d = 36.7 mm; say = 40 mm

Let us now check for the stress in the shaft.

The force F_m is given by Eq. (14.9) as

$$F_m = T_m/0.75R_b = 534 \times 10^3/(0.75 \times 250)$$
$$= 2848 \text{ N}$$

The maximum bending moment M at the bearing point from where the shaft overhangs. From Eq. (14.10), we get

$$M = 2848 \times 1300 = 3702.400 \text{ Nmm}$$
$$= 3702.4 \text{ Nm}$$

The equivalent bending moment is obtained by Eq. (14.8)

$$M_e = (1/2)\left\{M + \sqrt{M^2 + T_m^2}\right\}$$
$$= (1/2)\left\{3702.4 + \sqrt{(3702.4)^2 + (534)^2}\right\}$$
$$= 3721.6 \text{ Nm}$$
$$= 3721600 \text{ N mm}$$

Therefore, we have

$$f = 3721600/\{(\pi/32)(40)^3\}$$

$$= 592.3 \text{ N/mm}^2$$

This is higher than the permissible elastic limit 246 N/mm^2.

Therefore, the use of a shaft of diameter 60 mm for which the stress, f = 175.5 N/mm^2 we get from above. This is less than the permissible value of 246 N/mm^2. Therefore, we have the shaft of diameter = 60 mm.

The critical speed consideration

We use the method discussed earlier, wherein a dynamic load on the shaft is also considered. The deflection of the shaft due to this force, F_m = 2848 N can be given by Eq. 14.12.

$$\delta_1 = Wl^3/(3EI)$$

where, $W = F_m = 2848$ N

$$l = 1300 \text{ mm}$$

$$E = 1.95 \times 10^5 \text{ N/mm}^2$$

$$I = \text{(From Appendix A)} = (\pi/64)d^4$$

$$= (\pi/64)60^4$$

$$= 6.36 \times 10^5 \text{ mm}^4$$

Substituting all in consistent units we get δ_1 = 16.82 mm.

The deflection due to the impeller and hub weight, δ_2 is computed as follows, with the help of Eq. 14.12:

The impeller and hub weight is estimated to be = 20 kg

$$\delta_2 = \frac{(20 \times 9.807) \times (1300)^3}{3 \times 19.5 \times 10^4 \times 6.36 \times 10^5}$$

$$= 1.16 \text{ mm}$$

The shaft weight per unit length,

$$w = (\pi/4)d^2\rho$$

$$= (\pi/4)\ (60/1000)^2\ 7850$$

$$= 22.2 \text{ kg/m} = 0.0222 \text{ kg/mm}$$

the help of Eq. (14.13) the deflection, δ_s due to the shaft weight is given as:

$$\delta_s = \frac{Wl^4}{8EI} = \frac{(0.0222 \times 9.807) \times (1300)^4}{8 \times 19.5 \times 10^4 \times 6.36 \times 10^5}$$

$$= 0.63 \text{ mm}$$

then have from Eq. (14.14) the critical speed, N_c, as

$$N_c = \frac{946}{[\delta_1 + \delta_2 + \delta_s/1.25]^{1/2}}$$

$$= \frac{946}{[16.82 + 1.16 + 0.63/1.25]^{1/2}}$$

$$= 220 \text{ rpm}$$

present speed is 200 rpm that is about 90% of the critical speed and, therefore, the shaft er 60 mm is not acceptable.

us consider the shaft diameter 75 mm.

then have,

$I = 15.53 \times 10^5 \text{ mm}^4$

$\delta_1 = 6.9 \text{ mm}$

$\delta_2 = 0.52 \text{ mm}$

shaft weight w = 34.7 kg/m = 0.0347 kg/mm

$\delta_s = 0.4 \text{ mm}$

stituting, we get N_c = 340 rpm

actual speed is 200 rpm which is 59% of the above critical speed.

he shaft and impeller assembly is dynamically balanced the speed can be up to 70% of the l speed. Otherwise the speed up to 50% of the critical speed is on safer side.

refore, the shaft with diameter = 75 mm is acceptable, if the system is dynamically balance.

).3 The Blade

ssential to see whether impeller blades will bear the stress or not. From Eq. (14.15) the num bending moment, BM_{max}, is given by

$BM_{max} = F_m(0.75R_b - R_h)$ where, F_m is from Eq. (14.9)

and from Eq. (14.16) the stress, f, is then obtained as

$f = BM_{max}/Z$ where, Z is the section modulus.

$Z = (1/6)b_t(b_w)^2$

The diameter of hub is assumed to be 2 times diameter of the shaft.

$$R_h = 150/2 = 75 \text{ mm}$$

$$R_b = 500/2 = 250 \text{ mm}$$

We have, $F_m = 2848$ N

$$BM_{max} = F_m (0.75 R_b - R_h)$$

$$= 2848 (0.75 \times 250 - 75)$$

$$= 320400$$

Therefore,

$$f = \frac{320400}{(1/6)(8)(75)^2}$$

$$= 42.72 \text{ N/mm}^2$$

which, is acceptable.

14.10.4 Hub and Key Design

Hub diameter of the agitator = 2 × shaft diameter

= 2 × 75 = 150 mm

Hub length = 2.5 × 75 = 187.5 mm

say, = 200 mm

Key length = 1.5 × 75 = 112.5 mm

say, = 115 mm

From Eq. (4.6) we have,

$$T_m/(d/2) = b \times l \times f_s$$

Rearranging, we have,

$$b = 534 \times 10^3/[(75/2) \times (65) \times 115] = 1.9 \text{ mm}$$

say, = 2 mm

Also, $T_m/(d/2) = (l \times t/2) \times f_c$

$$t = 534 \times 10^3/[(75/2) \times (130) \times 115/2] = 1.9$$

say, $= 2$ mm

In order to have rigidity use 6 mm × 6 mm × 115 mm key.

14.10.5 Stuffing Box and Gland

The vessel internal pressure is = 0.5 N/mm^2

Design pressure = 0.55 N/mm^2

Please refer to section 4.13.

From Eq. (4.30 and 4.31) internal diameter of the stuffing box (b) and thickness (t) is given as

$$b = d + \sqrt{d} \qquad \text{where, } d \text{ is in cm}$$

$$= 7.5 + \sqrt{7.5}$$

$$= 10.23 \text{ cm}$$

say, $= 100$ mm

The thickness,

$$t = [(p \times b)/2f] + 6; \text{ mm}$$

$$= [(0.55 \times 100)/(2 \times 95)] + 6$$

$$= 6.3 \text{ mm}$$

say, $= 8$ mm

The load on the gland is obtained via Eq. (4.32)

$$F = (\pi/4)p\,(b^2 - d^2)$$

$$= (\pi/4) \times 0.55 \times (100^2 - 75^2)$$

$$= 1889.9 \text{ N}$$

say, $= 1890$ N

We assume number of studs = 4.

The stud diameter can be obtained from Eq. (4.33)

$$F = (\pi/4)d^2 \times n \times f_t$$

$$d = [4 \times 1890/(\pi \times 4 \times 587)]^{0.5}$$

$$= 3.2 \text{ mm}$$

The minimum stud diameter is 15 mm.

The flange thickness = $1.75 \times 15 = 26.25$

say, = 30 mm

14.10.6 Coupling

The coupling used is a clamp coupling. Please refer to Section 4.4.2 and Fig. 4.4.

We assume eight number of bolts ($n = 8$). From Eq. (4.14), we have force per bolt P, as g

$$P = 2 \times T_m/(\pi \times \mu \times d \times (n/2))$$

where the coefficient of friction, μ, is assumed to be 0.35 (between cast iron and steel).

Therefore,

$$P = 2 \times 534 \times 10^3/(\pi \times 0.35 \times 75 \times (8/2))$$

$$= 3237.7 \text{ N}$$

say, $= 3238$ N

Area of bolts $= P/f$

$$= 3238/58.7$$

$$= 55.2 \text{ mm}^2$$

From Eq. (5.41) we see that M10 bolt, i.e., bolt with 10 mm nominal diameter is adequate

The overall diameter of coupling is $= 2 \times d = 2 \times 75 = 150$ mm

The above procedure illustrates how the system design can be done. Not only it understand but also it would help in quickly checking the design of the agitator system ve

References and Recommended Readings

A.I.Ch.E., *Equipment Testing Procedure, Mixing Equipment (Impeller Type),* 3rd Edition, Institute of Chemical Engineers, New York (2001).

Baker, J.R., 'Motionless Mixers Stir up New Uses', *Chemical Engineering Progress*, **87** (6) 32 (19

Casto, L., 'Practical Tips on Designing Turbine-Mixer Systems', *Chemical Engineering*, Desk (1977).

Dickey, D.S., and Hemrajani, R.R., 'Recipes for Fluid Mixing', *Chemical Engineering*, **99** (3) 83 (1

Dickey, D.S., and Hicks, R.W., 'Fundamentals of Agitation, Chemical Engineering', **83** (3) 93 (197

Fasano, J.B., Miller, J.L., and Pasley, S.A., 'Consider Mechanical design of Agitators', *Engineering Progress*, **91** (8) 60 (1995).

Fasano, J.B., Bakker, A., and Roy Peeny, W., 'Advanced Impeller', *Chemical Engineering*, **101** (8)

Gates, L.E., Henley, T.L., Fenic, J.F., 'How to Select Optimum Turbine Agitator', *Chemical Engineering*, **82** (26) 110 (1975).

Hicks, R.W., Morton, J.R., and Fenic, J.G., 'How to Design Agitators for Desired Process Response', *Chemical Engineering*, **83** (9) 102 (1976).

Holland, F.A., and Chapman, F.S., *Liquid Mixing and Processing in Stirred Tanks*, Reinhold, New York (1966).

Ludwig, E.E., Applied Process Design for Chemical and Petrochemical Plants, Vol. I, Second Edition, Gulf Publishing Co., Houston (1977).

Muller, H.K., and Nau, B.S., 'Keep Mechanical Seals in Shape and Running Smoothly', *Chemical Engineering*, **105** (13) 92 (1998).

Myers, K.J., Bakker, A., and Ryan, D., 'Avoid Agitation by Selecting Static Mixers', *Chemical Engineering Progress*, **93** (6) 28 (1997).

Netzel, J., and Volden, D., 'Suitable Seals Lower the Cost of Ownership', *Chemical Engineering*, **105** (13) 9[illegible] (1998).

Oldshue, J.Y., Fluid Mixing Technology, McGraw-Hill Publications Co., New York (1983).

Parker, N.H., 'How to Select Double Arm Mixers', *Chemical Engineering*, **72** (18) 121 (1965).

Penny, W.R., 'Recent Trends in Mixing Equipment: Report', *Chemical Engineering*, **78** (7) 86 (1971).

Ramsey, W.D., and Zoller, G.C., 'How the design of Shafts, Seals, and Impeller affects the agitator performance', *Chemical Engineering*, **83** (18) 101 (1976).

Smith, R. 'Chemical Process Design and Integration', J. Wiley and Sons (2005)

Uhl, V.W., and Gray, J.B., *Mixing Theory and Practice*, Vols. I and II, Academic Press, New York (1967).

15

Chapter

Filters

15.1 CLASSIFICATION

The function of the filtration equipment is to separate the mixture called more or less completely from the liquid. Depending on the nature of the solid-liquid slurry, the driving force required for separation may be divided into four categories, namely, gravity, vacuum, pressure and centrifugal.

(a) *Gravity* This is the simplest method of filtration. But the equipment is generally very bulky and the filtration is not complete.

(b) *Vacuum* The equipment used for generating the vacuum is either a vacuum pump or an ejector. But the filtration equipment may be bulky and relatively inexpensive. For example, we have what is commonly known as a nutsche filter. The expensive system in vacuum filtration is vacuum belt filter.

(c) *Pressure* The method gives higher outputs. The equipment may, therefore, be smaller in size. It can handle volatile liquids without much difficulty. These filters are widely used.

(d) *Centrifugal* This method utilizes maximum separating forces due to the centrifugal action. High outputs can be obtained.

It is proposed to consider the design and construction features of some of the more important types of filtration equipment. Such equipment may be either for batch production or for continuous production.

15.2 VACUUM FILTERS

15.2.1 Batch Type

There are two batch type filters: (i) Tank filter (Nutsche Filter) and (ii) Leaf filter (Moore filter).

15.2.1.1 Tank Filter

This filter is simple in construction and is used widely for small-scale filtration work. It consists of a tank or cylindrical vessel, divided into two chambers by a perforated plate. The plate can be covered with filter cloth or wire screen, placed loosely in position or sealed around the edge by means of a caulking ring or a joint ring. A stoneware porous plate can also act as a filter medium. The lower chamber is connected to a vacuum line, the vacuum being created by either a vacuum pump or a steam ejector. The size of the vessel is determined by the quantity to be separated in each batch. Where large capacities are desired, the lower chamber may be evacuated of the filtrate intermittently.

The vacuum produced in the lower chamber draws the liquor down through the filter medium leaving the solids in the upper chamber. The accumulated solid or cake may be washed. The tank or the cylindrical vessel must be designed for vacuum. If a rectangular tank is used it must be sufficiently reinforced, so that buckling is prevented. A cylindrical vessel has, obviously, a greater strength.

Agitated nutsche filter is a recent innovation in the filtration technique. It is extensively used in dyes, paints, pesticides, and fine chemical industries. Safety requirement and environmental concerns due to solvent evaporation lead to development of this type of filter, where filtration under vacuum or pressure can be carried out in a closed vessel. Even solids can be discharged straightway into a dryer, mechanically, without any manual operation.

Filter Features

It consists of a vessel with dished ends and a perforated plate in between. A filter cloth is kept over the perforated plate. The entire vessel can be kept at a desired temperature by having a limpet coil or jacket through which heat transfer medium can be passed. The vessel can be made leak proof for vacuum or pressure service.

A multipurpose agitator is the unique feature of this system. The agitator performs the following operation, by lowering it and by even direction of rotation.

(a) Slurry contents can be kept fluidized till most of the mother liquor is filtered through.

(b) When filtration is over, the cake built develops a crack causing upsets in the vacuum operation. This does not allow for the removal of the last quantity of mother liquor left. The agitator can be lowered and can be used to press the cake.

(c) Washing of the cake can be performed in the same tank by reslurring the cake. Thus, the cake can be thoroughly washed, a feature, most uncommon in other filters.

(d) After washing, filtration can be performed. The cake can then be discharged through the discharge port. This is done by slowly lowering the agitator and rotating it in such a way so as to bring all of the cake towards the discharge port.

(e) From the discharge port through a conveyor it can be fed to a dryer.

Advantages

(a) Vacuum or pressure filtration is possible.

(b) Inert gas atmosphere can be maintained.

(c) Very high solvent recovery is possible.

(d) Solvents are in a closed system. No let off of toxic solvent vapours in the atmosphere. Also, flammable solvents with or without inert gas (nitrogen) can be handled safely.

(e) Solids are also handled in a closed system. No manual operation is involved. Thus, personnel safety is of a high standard.

(f) Heat transfer surfaces can be provided to maintain the desired temperature.

(g) Considerable saving in manpower.

15.2.1.2 Leaf Filter

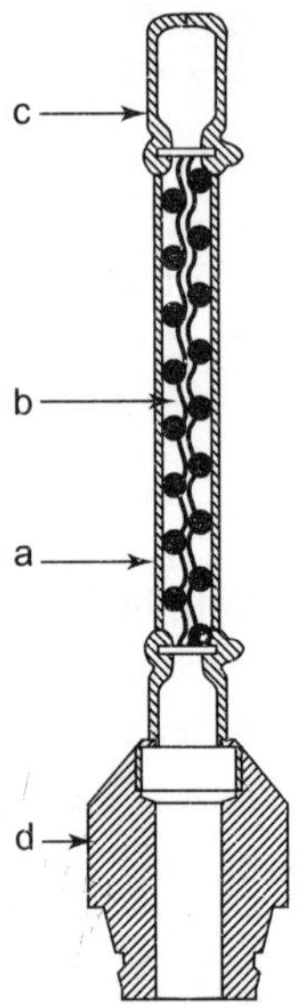

Fig. 15.1 Leaf Filter: (a) Filter Medium, (b) Screen, (c) Hoop, (d) Coupling

There are a number of ways in which a leaf can be formed. In a simpler method of fabrication, a screen or perforated plate is welded to a channel frame with a drainage connection at the top. The filter medium is welded to the screen. If a fabric material is to be used as a medium, it is made to cover the screen to form a bag which is fastened at the top. A basket is formed by assembling the leaves in parallel and held in position by two parallel beams which are placed at right angles to the axis of the leaves (Fig. 15.1).

The basket is submerged in a feed liquor tank. Vacuum is applied at the top of the leaves, through a hose connection, which is attached to each of the submerged filter leaves. After the cake is deposited the basket of leaves is lifted out of the filter tank with a crane and is placed in a wash tank. Vacuum is maintained during the transfer by means of flexible hoses. Wash liquor draws off through the cake in the same manner. The leaves are then transferred to a tank, where an air blow discharges the cakes.

The leaf should be sufficiently rigid. The channel frame and the screen or plate must be able to withstand the pressure corresponding to the vacuum applied. The cross section of the channel and the thickness of the screen will depend on the size of the frame. The channel can be formed from a sheet of metal. The leaf filter can also be used for pressure filtration with appropriate changes in the system.

15.2.2 Continuous Type

(1) Rotary drum filter

(2) Rotary disc filter

(3) Horizontal rotary filter.

15.2.2.1 Rotary Drum Filter

This filter (Fig. 15.2) is one of the types most widely used in industry, finding application both, where cake washing is required and where it is not required. Constructionally, vacuum drum filters fall into two groups, namely, those in which the whole drum is under vacuum and those in which vacuum is applied only to an outer annular series of compartments. The second category is by far the most common. The filter consists of the following parts:

(a) Drum,
(b) Filter medium,
(c) Automatic filter valve,
(d) Filter tank,
(e) Agitator,
(f) Cake discharge system,
(g) Shaft,
(h) Bearings,
(j) Drive system,
(k) Supporting framework.

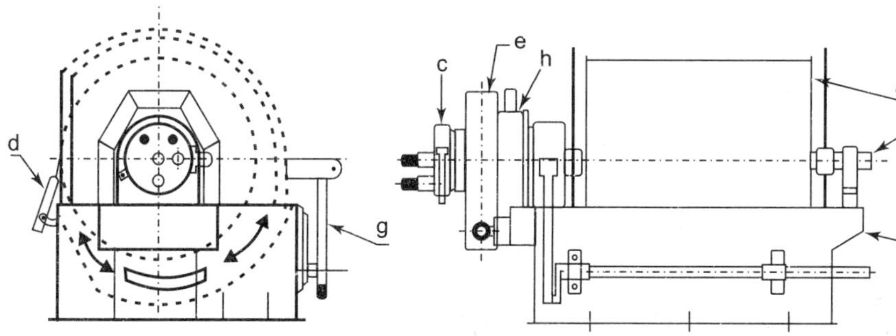

Fig. 15.2 Layout of Rotary Drum Filter: (a) Drum, (b) Shaft (Stub), (c) Filter Valve, (d) Scraper, (e) Driving Gear, (f) Slurry Tank, (g) Oscillating Agitator, (h) Bearing

(a) *Drum*

The drum (Fig. 15.3) diameters vary from about half a metre to about four metres, and the face width from about half a metre to eight metres. Drums are made of wood, cast iron or steel. The surface of the drum is corrugated or covered with corrugated sheets made of copper, stainless steel or other materials to facilitate liquid flow and a more effective utilization of cloth. The surface is divided into a number of shallow sections, usually 12 to 24, depending on the drum diameter. The sections are sealed from each other by means of longitudinal strips, and are covered with screens or grids for supporting the filter medium. The strips are made from metallic sheets, by bending a flat strip into an approximate U-shape. Each section forms an independent compartment. The screen or grid must be so shaped that adequate drainage (about 2.5 cm depth) is provided between the filter cloth and surface forming the bottom floor of the section. Each section has one or several outlets, depending upon the length of arm and amount of filtrate to be handled. The outlet pipes from each section are connected to a larger common pipe leading to the automatic filter valve. Where maximum separation of filtrate and wash liquor is necessary. The outlets are normally arranged along the trailing edge of each section. If both the cake dryness and good separation are important, a compromise can be made by using both leading and trailing edge piping. In very special cases, where it is essential to have both a sharp separation of liquor and maximum cake dryness, two automatic filter valves can be provided, one mounted on each end of the drum. All leading edge piping is connected to one valve and all trailing edge piping to the other. The ends of the drum may be open or closed, the open design being slightly cheaper. It also facilitates access for maintenance. The drum diameter can, typically, vary between 0.5 to 4 m and length from 1 to 8 m.

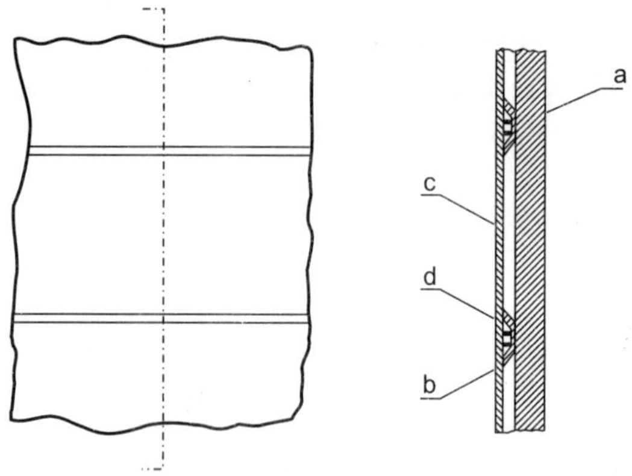

Fig. 15.3 Rotary Drum Section: (a) Drum, (b) Screen, (c) Filter Cloth, (d) Longitudinal Strip

(b) *Filter medium*

The filter medium is usually made of cotton duck, metal woven cloth or plastic materials in the form of cloth. The common method of screening the filter medium to the screen surface is by caulking the cloth into the longitudinal divisional strip. The cloth is held taut in position inside the U-shaped strip by blocks of rubber sheathed with a metallic cover. The ends of the cloth are sealed either by caulking or by the use of bands or wire. In addition, the cloth may be held taut, sometimes by a wire wound at a pitch of about 4 cm. In certain cases, the filter cloth is not fixed to the screen or grid, but is in the form of an endless belt which remains in contact with the drum for only a part of its travel.

(c) *Automatic filter valve*

An important part of the filter is the filter valve. As the drum rotates, the valve connects each section on the drum to suction line or to a wash line or to a pressure line as required by the sequence of operations. The valve consists of three parts [Fig. 15.4(A)].

(i) ***Rotating tube plate (moving sector block)*** This is made as an integral part of the rotating trunnion or shaft which supports the drum. It is essentially a ring with circular ports, to which the filtrate piping from each section is connected. The tube plate, the drum, the filtrate piping and the shaft are interconnected and, therefore, rotate at the same speed.

(ii) ***Valve head (fixed distributor block)*** It is in the form of a stationary block with an annular channel, facing the circular parts in the tube plate. The annular channel is divided into several segments by means of bridge pieces, which may either be fixed or movable. The segments of annular channel are so shaped that they form a piping manifold on the outside. The valve head has one of its surface facing the tube plate and the other side has the manifold for connecting to a suction line, a wash line and a compressed air line. Figure 15.4(B) shows an alternative arrangement of valve head.

(iii) ***Wear plate*** It is in the form of a ring which is placed between the rotating tube plate and the stationary valve head. It helps to prevent leakage of the filtrate as it flows from the tube plate to the valve head. The assembly is held tight by a stud with a spring. Which is located at the centre of the valve head. The rotation of the tube plate causes wear of the wear plate surface, which may be reduced by lubrication with the help of a grease cup. The wear plate is replaced when worn out.

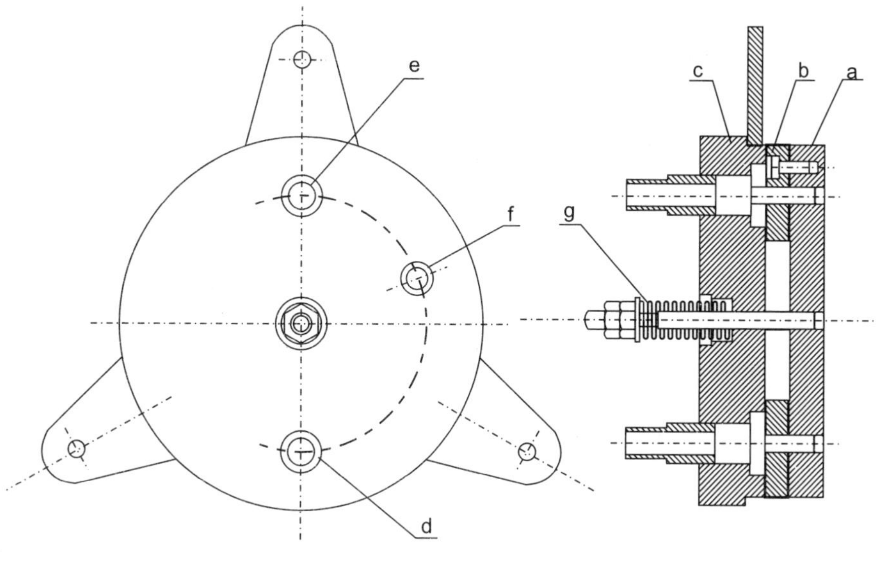

Fig. 15.4 (A) Filter Valve Components: (a) Rotating Tube Plate, (b) Wear Plate, (c) Valve Head, (d) Suction Line, (e) Wash Line, (f) Compressed Air Line, (g) Spring

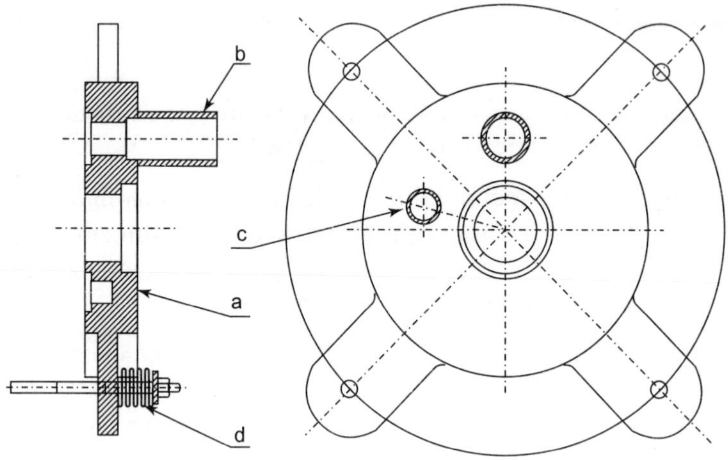

Fig. 15.4 (B): (a) Valve Head, (b) Suction Line, (c) Compressed Air Line, (d) Spring

(d) *Filter tank and (e) Agitator*

The tank is placed below the drum and covers a portion of the drum over its entire length. The tank is fabricated from sheets of metal usually about 4 mm to 6 mm thick. The most common filter tank is designed for about 40% submergence of the drum. The slurry in the tank is stirred by means of a horizontal agitator, which is placed close to the tank bottom. The agitator may be of the paddle type. A better agitation system consists of oscillating cross bars supported by arcs hung from the tank. The agitator moves back and forth at a slow rate.

(f) *Cake discharge system*

The cake formed on the filter cloth surface, is discharged, as each section of the drum moves towards the cake discharge point, while the filter valve shuts off the vacuum on that particular section. Depending on the type of cake, it is removed by one of the several available methods.

(i) Scraper discharge for easy filtering of solids.

(ii) Roll discharge for materials which are sticky and thin but will peel off.

(iii) Longitudinal wire under considerable tension for discharge of semi-sticky cakes.

(iv) String type discharge for cakes which tend to adhere to the filter cloth.

(v) Belt discharge where clean cloth is essential.

(vi) Precoat scraper where a measured amount of cake is removed for feeds low in solids and producing a very thin cake.

A scraper discharge arrangement (Fig. 15.5) consists of a narrow scraper supported by, and attached to the filter tank. It is free to oscillate about a hinged support, so that it can float and follow the contour of the cloth, which is slightly inflated by air pressure, to facilitate cake discharge. The scraper is balanced so that it rests slightly on the filter medium. The scraper may have a rubber edge or a metal edge.

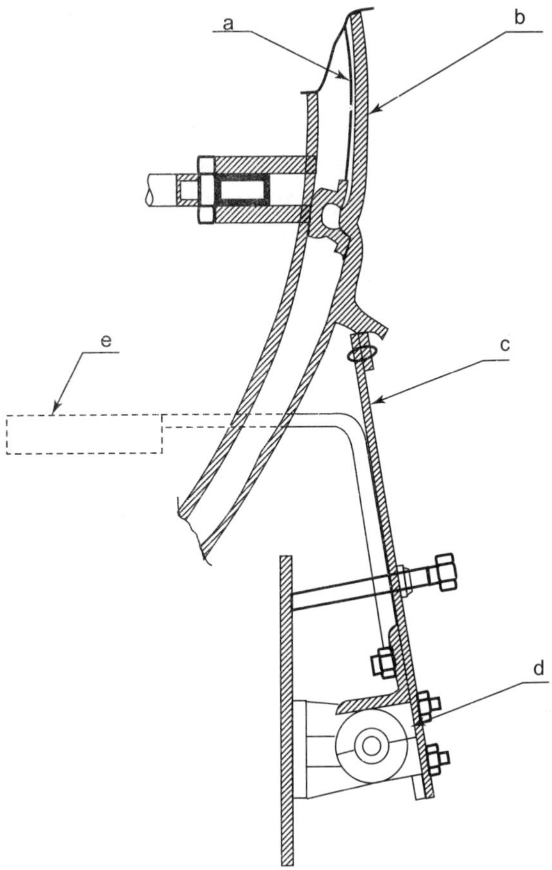

Fig. 15.5 Scraper Mechanism: (a) Filter Medium, (b) Filter Cake, (c) Scraper, (d) Scraper Support, (e) Counter Weight

(g) *Shaft*

The drum is fixed to the shaft through the end covers by providing suitable hubs at the centre. For an open end drum the shaft is fixed through the spider frame. The shaft extends on either side of the drum, and is supported by bearings located close to the drum. In most cases the shaft is not continuous, but is in the form of a trunnion fixed to the cover at either end. The drum, therefore, acts as a connecting component between the two turnnions. On one side of the drum the shaft is extended beyond the bearing for connection to the drive system, and for fixing the rotating tube plate of the filter valve. The loading of the shaft will be as shown in Fig. 15.6. It will transmit power required for the following loads:

(i) Acceleration of the drum from the stationary conditions. The rotation of the drum is extremely slow, usually less than one revolution per minute.

(ii) Friction at the bearings, wear plate and scraper knife.

(iii) Resistance of the slurry in the tank, in which the drum is partially immersed.

Each of the above can be assessed in terms of the torque required, depending on the radius at which the above loads are acting. The shaft can also be considered as a beam with loads due to the weight of the drum, driving gear and the tube plate, and supported at the bearings. From these loads, it is possible to calculate the maximum bending moment. From Eqs (4.2) and (4.3), the shaft diameter can be determined, with the values of maximum torque and bending moment. Since the shaft is generally not continuous, but in the form of a turnnion at either end, each turnnion will require an independent assessment of the stresses. The drum will also transmit power in this case.

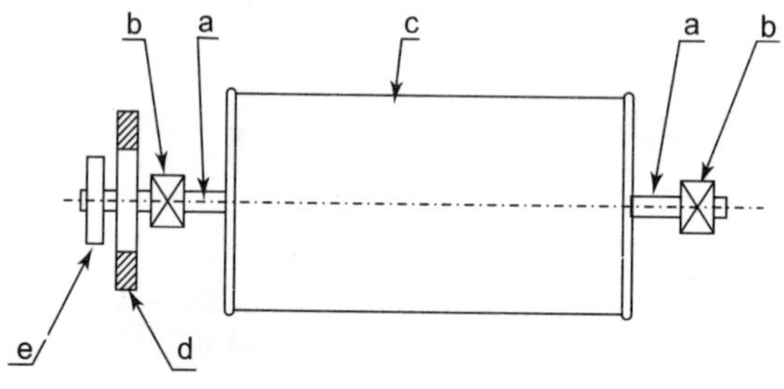

Fig. 15.6 Shaft Loading: (a) Shaft, (b) Bearing, (c) Drum, (d) Gear, (e) Valve

(h) *Bearings*

Usually, two cup lubricated journal bearings are provided for supporting the shaft. The load is primarily radial. Assuming a permissible bearing pressure of 120 kg/cm^2 the size of the bearing bush can be determined (Eq. 4.16). The features of the bearings are as shown in Fig. 4.6.

(i) *Drive system*

Considering the extremely slow speed at which the drum is to rotate, the appropriate choice of the driving gear is the worm gear (Fig. 4.15). The worm gear is mounted on the drum shaft, while the worm is fixed to a perpendicular shaft. The worm shaft is driven through a chain or V-belt drive from an electric motor. The major speed reduction is obtained from the worm and worm gear. Ball bearings are used to support the other shafts. A multistage helical gear system might be considered for a higher efficiency.

(j) *Supporting framework*

It is essential to build a framework surrounding the drum. This is made up structural sections and plates welded to form a rigid structure. The framework has to support the load of the drum, through the bearings, automatic valve, tank with agitator, discharge mechanism and washing equipment.

15.2.2.2 Rotary Disc Filter

The vacuum disc filter comprises one or more circular discs mounted on a common horizontal central shaft which is driven so that the discs can rotate (Fig. 15.7). The lower part of each disc is submerged in a tank of slurry. During passage through which, a cake is deposited on each side of the disc as a result of suction through the piping located inside the central shaft. The cake is then dried by further suction as it rotates through air and is finally discharged from the disc, generally by a reversed blast of air combined with the use of a scraper.

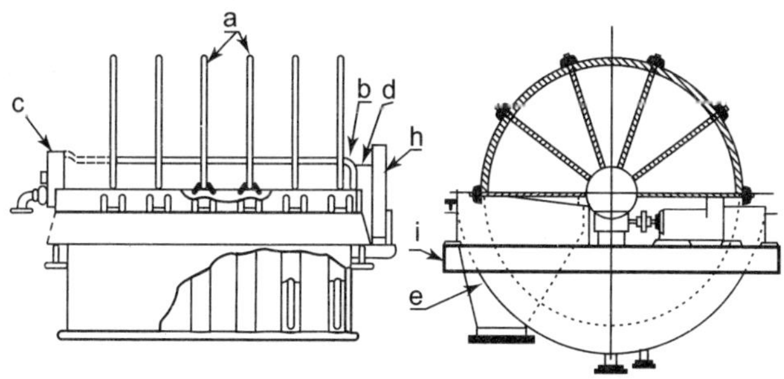

Fig. 15.7 Rotary Disc Filter: (a) Filter Disc, (b) Shaft, (c) Rotary Valve, (d) Bearing, (e) Slurry Tank, (h) Gear Drive, (i) Supporting Frame

The main components of the filter are: (a) filter disc, (b) shaft, (c) rotary valve, (d) bearings, (e) slurry tank, (f) agitator, (g) cake discharge system, (h) drive system, (i) supporting framework.

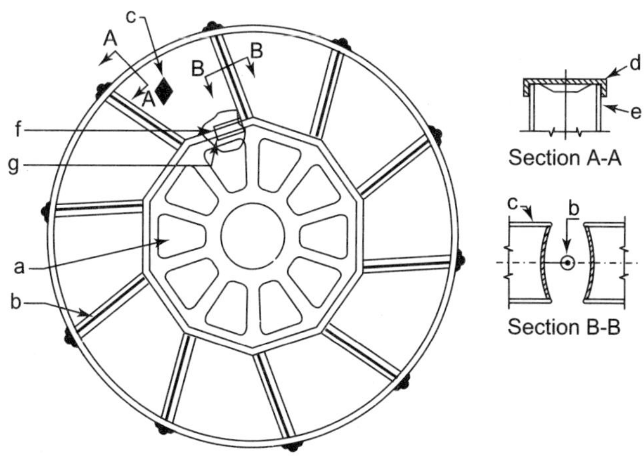

Fig. 15.8 Disc Details: (a) Channel for Filtrate, (b) The Rod, (c) Sector, (d) Sector Clamp, (e) Perforated Plate with Wire Mesh, (f) Pipe

(a) *Filter disc*

The filter disc (Fig. 15.8) is constructed from a number of segmental elements which can be detached. A segment may either be of solid construction with each face grooved or it may be hollow with each face perforated. The disc is covered with a filter cloth in the form of a bag, which slips over the sectors. The sector narrows to an apex through which a radial pipe nipple protrudes, which is plugged into the central shaft. The bag is folded over and clamped at the edges.

Table 15.1 Dimensions of Rotary Disc Vacuum Filter

Diameter of Disc m	*No. of Discs*	*Filtering Area (m^2)*
1.83	2 to 8	9 to 35
2.13	3 to 9	19 to 56
2.44	4 to 10	34 to 84
2.74	5 to 11	54 to 120
3.05	6 to 12	82 to 164
3.35	7 to 13	117 to 218

The discs are made of wood, carbon steel, stainless steel or bronze. The diameters of discs range from half a metre to 5 metres. Each disc is composed of 10 or more sectors. Multiples of up to 12 discs are assembled on one shaft, which rotates at speeds varying between 2 and 5 revolution per minutes.

(b) *Shaft*

The central shaft (Fig. 15.9) is hollow, with a large central hole and a number of channels (manifolds) concentric with the central hole but separated from it. These channels are closed at one end, while at the other, they lead into a rotary valve. In view of the hollow shape of the shaft, it is easier to construct the shaft in cast iron or by forging.

(c) *Rotary valve*

The valve (Fig. 15.10) is somewhat similar to the type used in rotary drum filters. It has a large part for filtrate outlet, and a smaller connection for the blow-back of air.

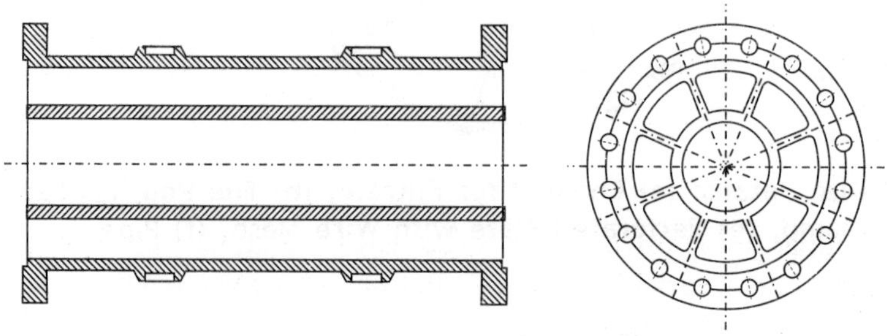

Fig. 15.9 Shafts for Disc Filter

(d) *Bearings*

The shaft is extended on either side by a flanged connection and is supported on bearings at either end. The bearings are similar to those in the drum filter.

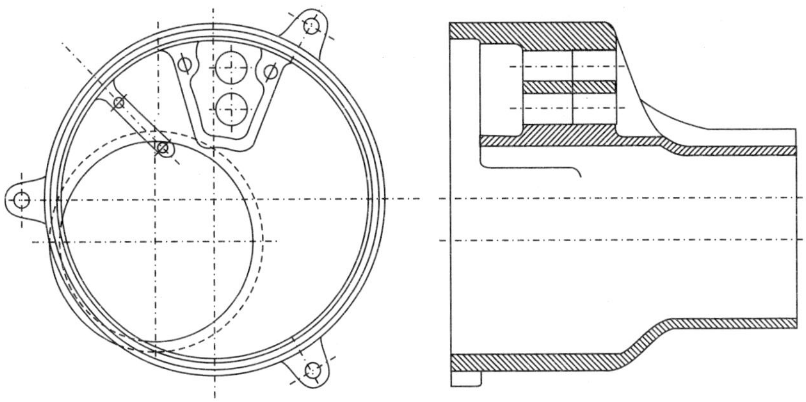

Fig. 15.10 Valve for Disc Filter

(e) *Slurry tank and (f) Agitator*

The tank is semi-circular in cross section and a part of it is provided with crenelations—one for each disc. The level of the slurry in the tank is critical, since it must always be sufficient so that a segment, after the cake discharge, is fully immersed before it is connected to the vacuum source. The exact point of rotation where suction is applied is controlled by the location of one of the bridge pieces on the rotary valve.

If the drive shaft is to be partly or wholly submerged in the slurry, a stuffing box has to be fitted. The slurry must not be so high as to foul the cake discharge knife. The surface available for cake formation is restricted to 20 to 45% of the drum's diameter.

The agitator is somewhat similar to the one used in the slurry tank of the drum filter.

(g) *Cake discharge systems*

The discharge of the cake is facilitated by a pair of radial blades touching the disc on either side. As the shaft revolves, the sectors are brought in contact with the radial blades, while an air-blow bulges out the filter cloth. The blade flattens out the bulge so that the cake falls off and drops, guided by a deflector through the spaces provided between the crenelations to a chute or conveyer.

(h) *Drive system*

This is somewhat similar to the drive system suggested for a drum filter. The speed of the disc is extremely slow and is also variable. The calculations of power required may be made from acceleration and frictional loads. The motor power is usually between ¼ hp and 10 hp depending on the filter size.

(i) *Supporting framework*

The framework is built up from structural sections and plates by welding. The thickness of sections or plates may vary between 6 mm and 12 mm.

15.2.2.3 Horizontal Rotary Filter

This filter (Plate V) is essentially a rotating horizontal circular table divided into a number of sectors. Each sector is a separate compartment. Vacuum is applied from the automatic valve concentrically located beneath the filter. The bottom of each compartment slopes towards the centre of the filter. The filter medium is supported on a punched plate or wire mesh drainage deck and caulked into dovetailed grooves at the periphery of the sectors. An elevated centre island and a circumferential vertical rim help to keep the slurry and the liquors on the filtering surface.

Feed is applied by a pipe and weir box distributor about the filter, which extends radially across the width of the surface. Dewatered and washed cake is discharged by a spiral, located just ahead of the feed. A radial dam is placed between the scroll and the point of feed to prevent slurry running back to the cake discharge zone.

There is no filter tank. The unit is mainly controlled by rotation speed with a variable speed drive and by the rate of slurry feed.

The filter may be built of steel, lead or rubber lined steel or stainless steel. The rotating table is supported on the lower side, the filter medium being located on the upper side. The drive may be either by a gear ring attached to the rim of the table, with a driving pinion meshing with it or by a sprocket and chain or a V-belt arrangement. In either case, a variable speed drive must be provided so as to permit operation in the range of 1-10 minutes per revolution.

15.3 PRESSURE FILTERS

15.3.1 Batch Pressure Filter

(1) Filter press

(2) Mechanical filter press

(3) Pressure leaf filter

15.3.1.1 Filter Press

This filter (Fig. 15.11) consists essentially of a skeleton framework made up of two end supports connected by horizontal bars. On these bars a varying number of filter chambers are assembled. These chambers are formed either by cloth covered recessed filter plates or by alternate frames and cloth covered plates. The chambers are closed and tightened by a screw or hydraulic ram which forces the plate or plates and frames together, making a gasketed joint of the filter cloth. The slurry is forced into these chambers, which are designed in such a way that the filtrate cannot leave the filter press except by passing through the filter medium. The filtered cake is held back by the cloth.

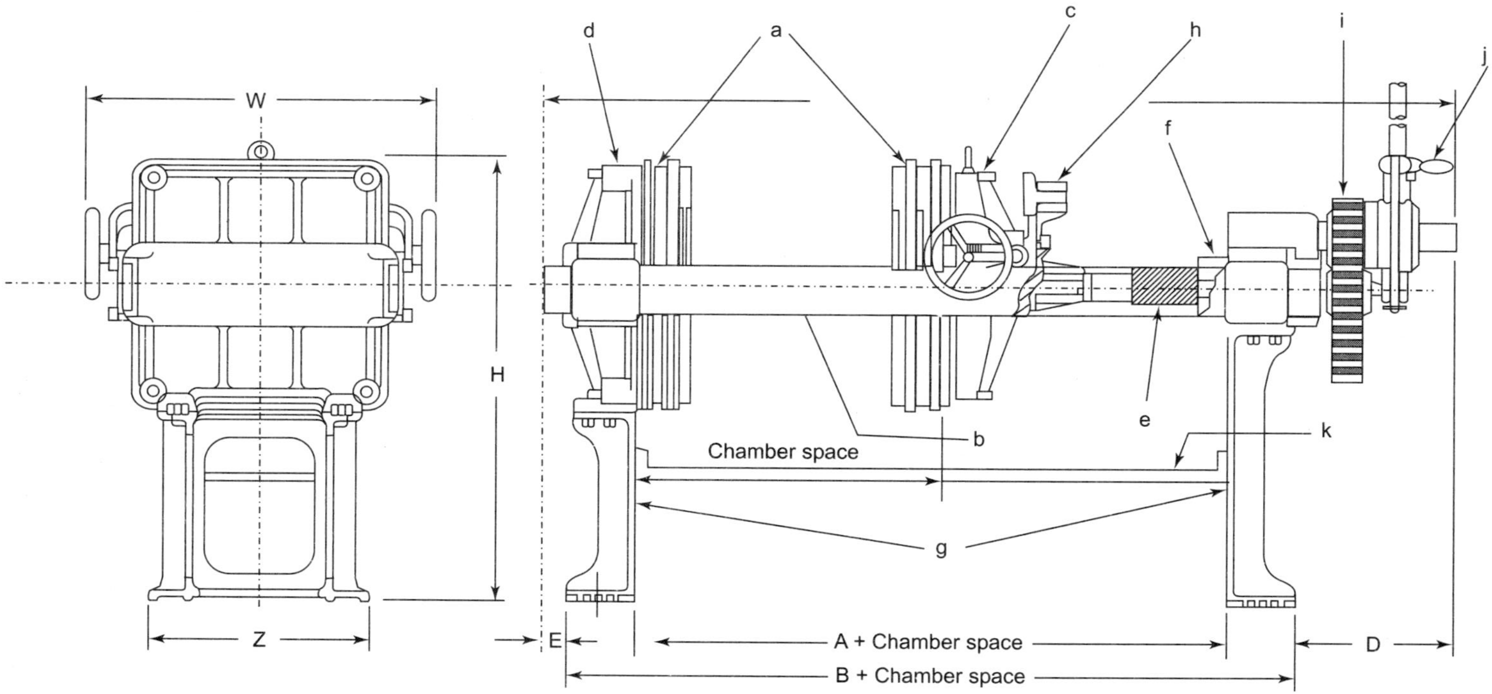

Fig. 15.11 Filter Press: (a) Plates and Frames, (b) Side Bar, (c) Movable Head, (d) Stationary Head, (e) Screw, (f) Screw Standard, (g) Supporting Legs, (h) Thrust Block, (i) Gears, (j) Wheel (k) Drip Collector Tray

The flow of the slurry through the filter press is not from chamber to chamber consecutively, but through all the chambers simultaneously, so that each chamber fills at the same rate.

Different parts of the filter press are: (a) plates and frames, (b) side bars, (c) movable head, (d) stationary head, (e) screw, (f) screw standard, (g) supporting legs.

(a) *Plates and frames*

These are square or rectangular in shape (Fig. 15.12). They are made of cast iron, cast steel, stainless steel, aluminium, lead, bronze, nickel, monel or special alloys. With non-ferrous materials plates are made of wood, hard rubber polypropylene, or (other plastic materials). Coated materials are also used.

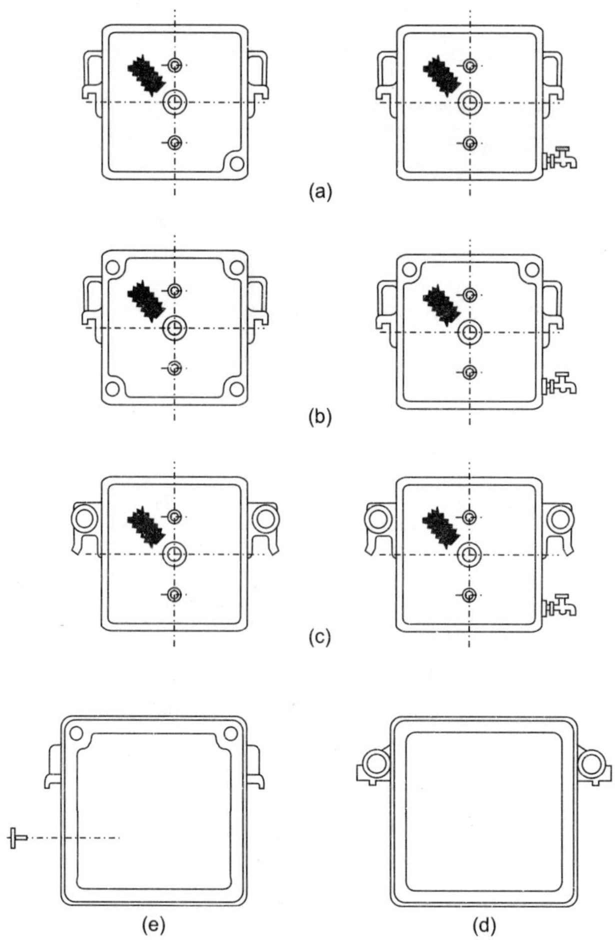

Fig. 15.12 Plates and Frames: (a) Plates with Centre Feed Corner Discharge, (b) Plates with Feed and Discharge at Corners, (c) Plates with Side Feed and Corner Discharge, (d) Frame, (e) Cross section of Side of Frame

(i) *Plate arrangements* Depending on the type of filter press, there are two general types of arrangements: (a) flush plate and frame, (b) recessed plate.

The first arrangement is made up of alternate plate and frames forming chambers. The plates generally have the drainage surfaces practically flush with the joint surfaces. The frames are hollow and provide the space for filter cake and can be made for any desired thickness of cake, the usual thickness range being 25 to 50 mm. Filter cloths are placed over each plate to cover the plate surface on both sides. The joints between plates and frames are made tight on closure by gasket joints formed by filter cloth. The feed and the discharge holes are located either in the corners of the plates and the frames or on lugs placed on the sides.

Table 15.2 Typical Dimensions of Filter Presses

	Plate Thickness		*All Dimensions in mm as per Fig. 15.11*							
Size of Plate	*Standard Cast Iron*	*Standard Wood*	*A*	*B + D + E = C*				*H*	*W*	*Z*
175	12.5	-	178	284	94	16	394	300	238	138
300	12.5	31	294	468	156	49	668	525	400	250
450	19	44	575	836	238	50	1124	875	725	375
600	25	44	725	1060	450	64	1574	1100	850	550
750	25	44	825	1228	412	72	1712	1275	1100	600
900	25	-	1090	1590	405	81	2076	1400	1250	725
900	-	56	915	1415	405	81	1901	1400	1250	725
1080	28	-	1350	2000	496	125	2508.5	1475	525	1050
1080	-	56	1175	1828	496	125	2336.6	1475	1525	1050
1200	31	62	1495	2140	512	112	2764	1575	1725	1050
1400	-	69	1450	2100	500	112	2712	1557	1950	1100

The second arrangement has only plates, the chamber being formed in the recesses in the adjacent plates. The depth of recess seldom exceeds 15 mm making a cake of 30 mm. Where necessary, frames may be inserted between recessed plates to permit building up of a thick cake. The feed is generally located in the centre of the plate. Filter cloth on the recessed surface of each side of the plate is sealed around the feed opening by two cloths sewn together at the hole or by clip nuts. The plates are not under any pressure, but a slight pressure might be exerted due to cake resistance. The plate thickness is such, so as to provide for enough rigidity. It varies according to the material. The frames are however under pressure, which acts on the inner periphery of the frame. The frames are held in position against the plate and, therefore, receive considerable support. The bending moment due to pressure acting on each side of the frame is given by

$$M = \frac{P\omega l^2}{16} \tag{15.1}$$

where, P — pumping pressure

l — length of the side of frame

ω — width of the frame.

The stress in the cross section of the side of frame is given by

$$f = \frac{M}{Z} \tag{15.2}$$

where, Z — modulus of section of the cross section of the side of frame.

If the permissible stress (f) is known, the size can be determined. The cross section is rectangular as shown in Fig. 15.12(e) for which modulus of section (Z) can be determined. Alternately, assuming a suitable cross section, the stress can be assessed. The stress should not exceed the permissible value.

(ii) *Method of filtrate discharge* Filtrate discharge may be from outlets in each plate, with or without control cocks. This is known as 'open' discharge. A closed channel in the plates through the fixed head is connected by ports to each plate. This channel may be located in the joint surface or outside it, and carries the filtrate discharge of each plate. Such plates are known as closed discharge. Test and shut-off cocks can be provided to permit examination of filtrate from any plate of a closed discharge filter press and to shut off the flow from any plate when necessary.

(iii) *Drainage surface of the plate* The truncated pyramid type drainage surface is commonly adopted to obtain maximum amount of cloth support without losing effective filtering area. Radial drainage grooves or corrugated surface metal plates are also used. For wood or rubber plates a corrugated surface is provided.

(iv) *Control of leakage* Certain materials, particularly very thin fluids, tend to leak through gasket joints made by the filter cloth between plates and frames. To overcome this difficulty, several types of plates and frames are used. Frames are cast with special grooves in the joint surfaces. The grooves are connected to openings forming a channel which discharges through the head of the filter press.

Alternatively, plates are made with cast or milled grooved into which rubber gaskets are fitted. In another arrangement, the filter cloth is held against the plate by means of metal bands screwed against a rim inside the joint surface. A flat rubber gasket is cemented to the joint surface.

(b) *Side bars*

All the plates and frames, and the movable head with the thrust block are supported on two side bars. According to the number of plates the load may be resting only on the part of the length of the bars. However, assuming that the entire span of the bars is used to support the maximum number of plates, it is possible to determine the size of the bar. The ends of the bar are fixed. The bending moment due to the total load divided between two bars is given by

$$M = \frac{Wl}{16} \tag{15.3}$$

where, W — load on each bar = half the total load (frames, cloth, plate and cake)

l — length of bar

The stress, $f = \frac{M}{Z}$ (15.4)

where, Z — modulus of section for the cross section of the bar.

The cross section of the bar is rectangular with the depth 4 to 6 times the thickness. Assuming a suitable cross section, the stress (f), which should not exceed the permissible value, can be calculated.

(c) *Movable head ((follower head)*

The movable head, Fig. 15.13, is cast to the same size as that of the plate. On one side it has a surface similar to that of the plate, which is in contact with the adjoining plate or frame, and, therefore, acts as a plate. The outer side is provided with radial ribs terminating into a centred boss. All the plates and frames are pressed together by the movable head towards the fixed head. The movable head is pressed by either a screw or a hydraulic ram, for holding the plates and frames tight. A thrust block is mounted on the head and the screw is fitted in the thrust block. Teh function of the thrust block is to facilitate quick drawback of the movable head and to distribute load uniformly. The movable head is supported on the two side bars and is provided with rollers and hand wheels on either side for easy movement.

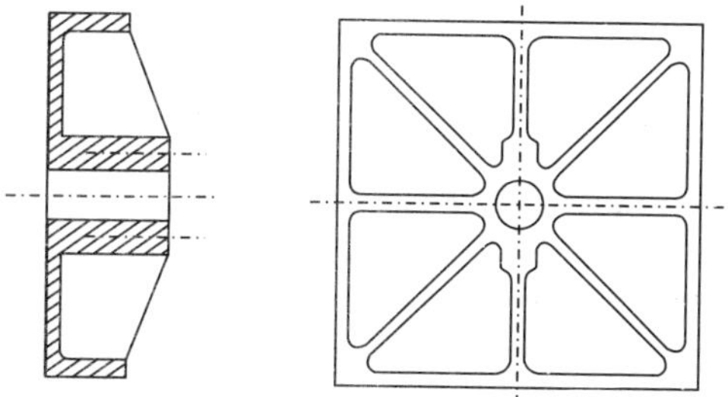

Fig. 15.13 Movable Head

The thickness of the movable head can be determined on the basis of the loading conditions. On one side of the head the screw exerts a force at the centre of the head. This force is required to keep the plates and frames tight, through the gasket joints. The force required on the joints can be calculated as (Eq. 5.36)

$$W = l \times b \times Y \tag{15.5}$$

where, W – force

l – length of the gasket joint (periphery of the plate forming the joint)

b – effective seating width of the filter cloth acting as gasket

Y – design seating stress for filter cloth material.

The above load may be considered as a uniformly distributed load, with the support in the centre of the span. The maximum bending moment will be given by

$$M = \frac{W \times L}{16} \tag{15.6}$$

where, L — length of the longer side of the plate.

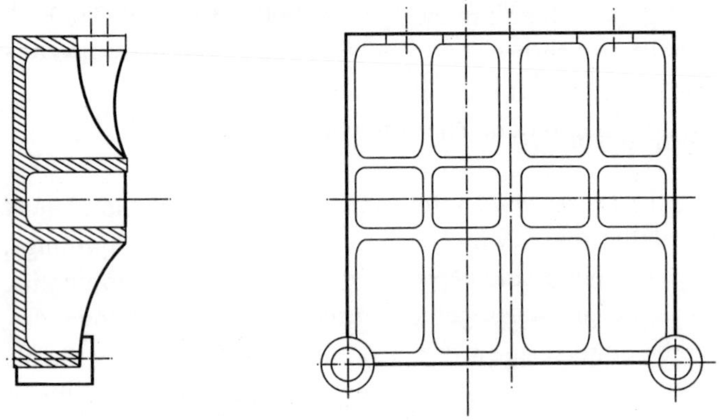

Fig. 15.14 Stationary Head

The cross section of the head will be as shown in Fig. 15.13. Assuming a suitable thickness, and knowing the size of the plate, it should be possible to determine the modulus of section (Z). The stress will be given by

$$f = \frac{M}{Z} \tag{15.7}$$

This stress should not exceed the permissible value. The average minimum thickness for grey iron castings is about 6 mm for parts up to about 45 cm in length. It gradually increases to 18 mm for large and heavy castings. Ribs are added to make the construction more rigid. While they always increase rigidity, they may fail to increase strength. The provision of ribs for the construction of the head helps to make it rigid and limits deflection. This is essential to maintain a uniform pressure on the gasket joints.

(d) *Stationary head*

The stationary head (Fig. 15.14) is cast to be the same size as the plate. On one side it has a surface similar to that of the plate and acts as a plate. On the other side, it has cross ribs. The head is supported on its edge by the legs, and is attached to it by bolts. The load on the stationary head is given by the equation 15.5. If the load is considered as being a uniformly distributed load the maximum bending moment is given by

$$M = \frac{WL}{16} \tag{15.8}$$

where, L — length of the longer side of the plate.

The cross section of the stationary head is as shown in Fig. 15.14. The thickness may be assumed and the stress may be checked on the same basis as the design of movable head. Ribs are provided for rigidity which is expected to ensure uniform pressure on the joints.

(e) *Screw* and *(f) Screw standard*

The plates and frames are pressed between the two heads by means of a screw acting on the central boss of the movable head, or on a thrust block which is interposed between the end of the screw and the movable head. The screw standard is a fixed casting block supported on the legs. The side bars are made to terminate within this block. The screw standard has a nut located in the centre which meshes with the screw. The screw is tightened directly by a wheel fixed at its end or by a gear and pinion mechanism attached to the end of the screw. A tommy bar of suitable length may give additional leverage for turning the screw. On large size filter presses, the screw is replaced by a hydraulic ram, and the associated power pack.

The screw is made of steel and has square threads. The load on the screw is mainly the tightening force W (Eq. 15.5). It is, therefore, subjected to the following stress.

(i) Compressive stresses resulting from the load W

$$f_c = \frac{4W}{\pi d_1^2} \tag{15.9}$$

where, d_1 — root diameter of screw.

If the length of the screw exceeds eight times the root diameter, the screw must be treated as a column.

(ii) Torsional shear induced in the screw by the external turning moment T (force applied to wheel) to overcome the force W and thread friction.

$$T = \frac{1}{2} W d_m \tan(\alpha + \phi) \tag{15.10}$$

where, d_m — mean diameter of thread

α — helix angel of screw

ϕ — angle of friction

$$\text{Shear stress } f_s = \frac{T}{\dfrac{\pi}{16} d_m^3} \tag{15.11}$$

The design of screw is based primarily on the permissible compressive stress f_c. The other stresses are checked for permissible values.

The nut placed in the screw standard is made of brass or bronze. The number of threads to be provided are given by

$$n = \frac{W}{\frac{\pi}{16}(d_1 - d_2)p_b} \tag{15.12}$$

where, d_2 — inside diameter of nut threads

d_1 — outside diameter of nut threads

p_b — bearing pressure.

Safe bearing pressure for steel screw and bronze nut is between 160 kg/cm^2 and 220 kg/cm^2.

(g) *Supporting legs*

The legs are usually made of cast iron with the cross section in the form of an angle or T-section. They are mainly under compression due to the weight of the fixed head on one side and the screw standard on other side. The rest of the weight consisting of the plates and frames, side bars, movable head, etc., is eccentric and will create a bending stress in addition to the compressive stress. It is usual to choose a suitable cross section of the leg and check for the stresses.

15.3.1.2 Mechanised Filter Press

A modification of the standard plate and frame filter is a mechanised unit, which consists of circular plates fixed in the vertical position and supported at the centre line by a horizontal shaft on each side of the filter. The sets of circular frames mounted 180° apart are attached to one of the side shafts, which can be rotated. When one set is in the operating position, the other is in cleaning or discharge position. The plates are made up as two movable parts sealed to each other by O-rings. Between them they form a closed compartment.

15.3.1.3 Pressure Leaf Filter

This consists of a series of uniformly spaced leaves mounted in a vertical or the horizontal cylindrical or rectangular pressure vessel. In general, the leaves consist of a heavy wire drainage screen mounted in a tubular frame, which acts as a support and the filtrate conduit. Filter medium may be either fine wire mesh cloth or any natural or synthetic fabric. The wire cloth is sealed between flanges on the frame by rivets, bolts, etc.

This vertical pressure vessel can be equipped with a quick opening door. On the horizontal vessel type the leaf and a manifold below it are attached to the movable head. Leaves and the movable head are removed from the vessel for discharging and cleaning by internal rollers and a monorail.

15.3.1.4 Centrifuge Discharge Filter (Fig. 15.15)

In a centrifuge discharge filter, the internal stack consists of circular plates mounted on a tubular shaft driven by an external motor. Slurry to be filtered is supplied from the top of the vessel and filters through the media into the tubular shaft. The clean filtrate then flows out of the vessel while the cake remains on the horizontal plates. When the cake is ready for discharge, the shaft is rotated by the motor. The cake is released which drops to the bottom of the filter.

15.3.1.5 Helical Filter (Fig. 15.16)

In this filter, the feed is pumped in a vessel from the top. The filter elements are in the form of pans covered with filter media. These pans are attached to a central shaft forming a continuous helical pattern. The filtration occurs across the surface of the filter elements. The filtrate flows to the inner radius of the element and moves along the helix to lower levels and ultimately into the central shelf at the last helical turn. An outlet is provided at the bottom of the central shaft which helps to exit the filtrate out of the vessel. The cake gets deposited on the helically located filter pans. The cake is removed by back-washing in the form of slurry.

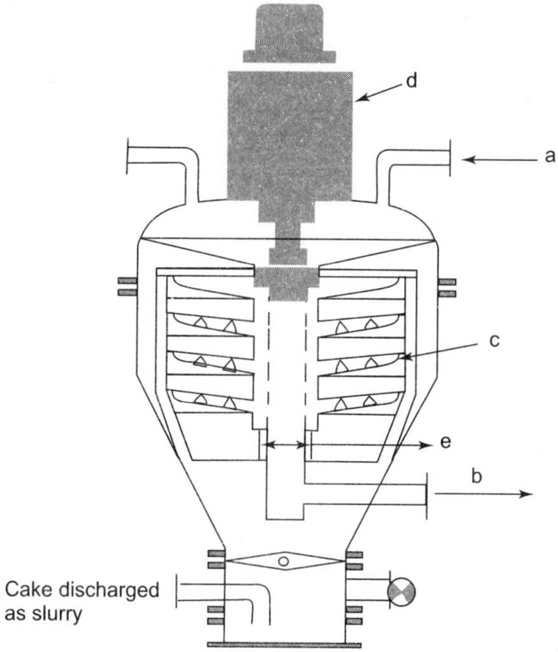

Fig. 15.15 Centrifuge Discharge Filter: (a) Slurry Inlet, (b) Filtrate Outlet, (c) Plate Stack, (d) Motor, (e) Seal

15.4 CENTRIFUGES

These machines are used for separating solids from liquids either by filtration or by sedimentation. Advantage is taken of the centrifugal force created due to rotation of a container in which the slurry is fed. There are two main types of centrifugal machines.

(i) Machines in which the solids are retained by a porous medium, while the liquid is forced through.

(ii) Machines in which the solids are separated as a cake while the liquid is drained off by overflow.

Both types can be operated either as a batch operation or as a continuous operation, with or without facilities for washing.

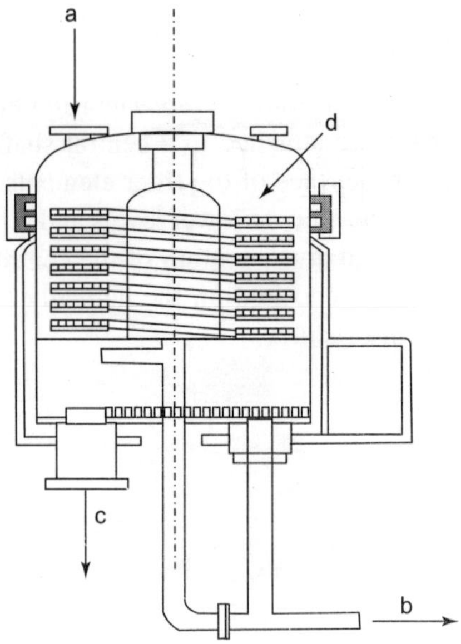

Fig. 15.16 Helical Filter: (a) Feed Inlet, (b) Filtrate Outlet, (c) Cake Outlet, (d) Helical Filter Element

15.4.1 Batch Centrifuge

15.4.1.1 Centrifuge with Basket

This is the most widely used centrifuge in the chemical and process industry. It consists of a perforated metal basket. The axis of the basket is either vertical or horizontal. In the vertical mounting the basket may be either underdriven (Plate VI) or overdriven (Fig. 15.17).

Surrounding the basket is a stationary monitor case or curb (solid shell), with an annular gutter at the bottom provided with a discharge pipe. Slurry is generally fed through a pipe which directs it against the side of the rotating basket. Occasionally, a device such as a cone moving on the vertical axis is used to promote the formation of a more even cake from the top to bottom of the basket. The solids are deposited on the basket liner, while the filtrate passes through and is collected in the curb and flows through the annular gutter to the discharge. The solids may be washed most efficiently by spraying water up and down the vertical wall. The cake is discharged through the openings in the bottom. An annular valve which forms part of the bottom is opened to allow the cake discharge. The valve is closed after the discharging action is complete. The solids drop to a chute, a conveyer or a rotating screen. The cake is dislodged from the basket walls manually by use of a wooden spade or automatically by a plough blade operated by a rack and pinion to allow its up and down movements and a hinge for sideways motion. In the case of free flowing crystals, baskets have a conical bottom which is self-discharging. An alternative cake release device is known as a scrapper, which comprises a narrow blade which moves back and forth across the face of the cake while it simultaneously advances towards the screen surface.

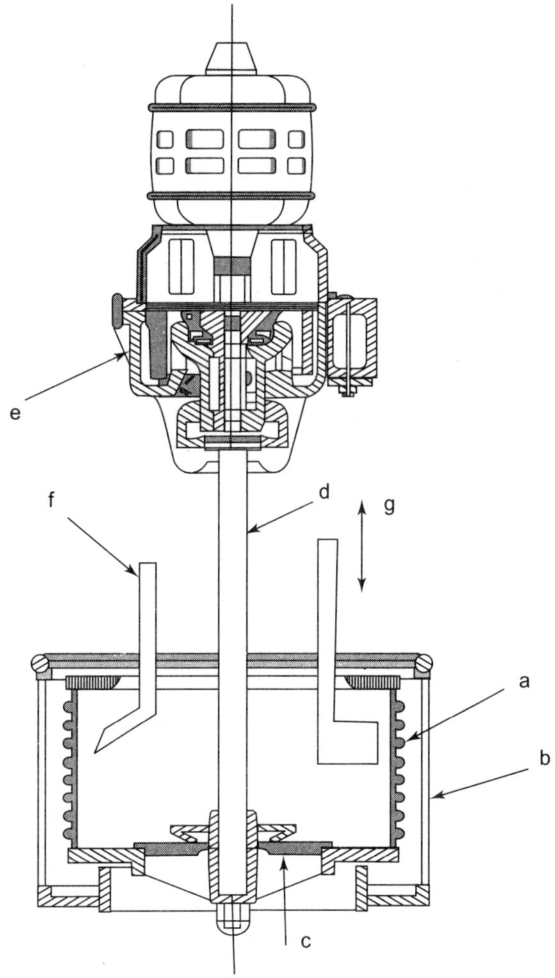

Fig. 15.17 Overdriven Basket Centrifuge: (a) Basket, (b) Curb, (c) Valve, (d) Shaft, (e) Drive System, (f) Feed, (g) Scrapper

In a horizontal axis basket centrifuge, the basket is directly mounted on the extension of the electric motor shaft and rotates at the same speed as the motor. The shaft is supported on tapered roller bearings on one side of the motor, while the basket is overhanging. The charge is fed in the basket through a horizontal spout. After washing and drying, a travelling knife slices out the cake. The movement of the knife is well controlled so that its edge stops at 1 mm from the filter cloth. The movement of the knife is controlled by a mechanism located on the end cover.

The main components of the basket centrifuge are: (a) basket, (b) mesh or screen with filter cloth, (c) plough blade, (d) curb, (e) cover, (f) shaft, (g) bearings, (h) drive system, (i) brake.

(a) *Basket*

The main parts of the centrifuge such as the basket, screen, valve, etc., are constructed of mild steel, copper and various alloys such as bronze, monel and stainless steel. In case of carbon steel, they may be lined or coated with lead, rubber, tin, etc. The basket may be either underdriven with open top or overdriven with a suspended basket. The open top machines are preferred for intermittent service and for handling of volatile solvents and fumes. The suspended basket type is used for high running speeds, bottom discharge of solids and the greatest stability with unbalanced loads.

Table 15.3 gives sizes of baskets. The speed of rotation varies between 50 and 300 rpm. Depending on size, speed can be higher also. Basket shells have perforations and may be strengthened by hoops located evenly over the depth of the baskets. The number of hoops might vary from one to five depending on the depth. These hoops may be made of materials of higher tensile strength than the basket materials. Table 15.4 gives a list of materials used for basket shells and hoops, and the allowable stress values.

The load on the basket is due to the centrifugal force of the rotating mass of the basket, hoop and material charged which may be taken as distributed over the entire surface.

(i) *Unhooped basket – The centrifugal force*

$$f = \omega^2 r \tag{15.13a}$$

$$\omega = 2\pi N \tag{15.13b}$$

$$r = D/2 \tag{15.13c}$$

$$F = 2\pi^2 N^2 D \times W \tag{15.13d}$$

where, W — mass of basket shell and material charged, kg
N — revolutions per second
D — mean diameter of basket shell, *m*
F — Newton.

Surface area of the basket (A) = πDL (15.14)

where, L — depth of basket

Table 15.3 Sizes of Baskets of Centrifuge (Typical)

Mean Diameter of Basket Shell D (mm)	*Recommended Basket Depth at Lip (mm)*
400	250
500	275
630	325
750	350
900	350/450
1000	400/500
1200	450/600/750
1250	450/750/830

Internal pressure on the basket wall due to force F (assumed to be uniformly distributed)

$$p = 2\pi^2 N^2 D \times \frac{W}{\pi DL} = 2\pi N^2 \times \frac{W}{L} \quad (15.15)$$

The circumferential stress due to internal pressure

$$f = \frac{pD}{2t}\left(\frac{P}{P-d}\right) = \frac{\pi N^2 DW}{Lt}\left(\frac{P}{P-d}\right) \quad (15.16)$$

where, t — thickness of basket

P — axial pitch of perforations

d — diameter of perforations

(ii) *Hooped basket* (basket shell and hoops made of same material). The stress in basket due to centrifugal force

$$f_b = \pi^2 N^2 D \times \frac{W_1 K}{(LtK + A)} \times \frac{P-d}{P} \quad (15.17)$$

The stress in the hoop

$$f_h = \pi N^2 D \times \frac{W_1}{(LtK + A)} \times \frac{1}{J} \quad (15.18)$$

where, D — mean diameter of basket shell

N — revolution per second

W_1 — mass of basket shell, hoop and material charged

L — depth of basket

A — cross sectional area of hoops.

K — $\frac{P_1 P - 2d^2}{P_1 P}$ (a factor allowing for reduction of stiffness due to perforations)

P_1 — circumferential pitch of perforations

t — thickness of basket

d — diameter of perforations

J — joint efficiency

The values of D and L are determined by filter requirements. Assuming a shell thickness (t), the stress (f) may be determined, which should not exceed the value given in Table 15.4. A minimum basket thickness recommended is given in Table 15.5. The built-up basket shall be adequately balanced by static or dynamic methods appropriate to the shape and size of the basket.

(b) *Mesh or screen*

This is built up from a coarse wire screen placed against the wall of the basket covered by a filter cloth of cotton, wool or metal wire fabric. It is held in position either caulking in the circumferential grooves at the ends of the basket or by means of an expandable ring.

Table 15.4 Materials of Construction for Basket Shells and Hoops of Centrifuges

Materials	*IS Specifications*	*Tensile Strength in kgf/mm²*	*Maximum Allowable Stress Value in kgf/mm²*
Mild steel	IS: 2002-1962		11
	(a) Grade 1	37 to 54	
	(b) Grade 2	42 to 50	04
Stainless steel	IS: 1570-1961	55 min	14
	(a) 04Cr19Ni9		
	(b) 07Cr19Ni9Nb70		
	(c) 07Cr19Ni9Mo2Ti20		
	(d) 05Cr18Ni11Mo3Ti20		
Copper sheets (fully annealed)	IS: 1972-1961	22.5 min	5.5
Jointless copper shells electrolytically deposited	—	25 to 28	6

Table 15.5 Minimum Thickness of Basket Shells

Basket Mean Diameter, D, mm	*Minimum Thickness, t mm*	
	Unhooped	*Hooped*
Below 650	3.2	2.5
650 up to but not including 1250	5.0	3.2
1250 and over	6.0	5.0

(c) *Plough blade*

The function of the blade is to dislodge the thick hard crystalline cake compacted against the filter cloth or mesh, due to centrifugal pressure. The action of the blade causes considerable wear and tear. The blade holder may be fixed to the curb or casing can be operated by a rack and pinion mechanism so that the blade can be moved vertically along the depth of the basket. The blade presses against the cake during the operation and may be withdrawn if required.

(d) *Main casting or curb*

This is a stationary cylindrical vessel placed outside the basket and is supported at the bottom by legs. The curb may be fabricated from mild steel or stainless steel sheets of about 6 to 10 mm thickness. It carries a hinged cover at the top and a valve at the bottom.

(e) *Cover*

It is made from metal sheets, and fixed to the curb by a gasketed flange joint. A portion of the cover (lid) is made separate and is attached to the main cover by hinges and bolts and wing nuts to facilitate frequent opening. There is a safety switch to stop the basket from spinning, when the cover is open.

(f) *Shaft*

In the case of overdriven or underdriven vertical centrifuges, the drive shaft is held vertical by bearings located at the top in the first type and at the bottom in the latter type. In the case of horizontal centrifuge, the shaft is horizontal and is supported at one end. In certain cases, the shaft extends into the basket and is supported in a bearing, located at the centre of gravity of the entire rotating mass.

The horse power required is determined from the considerations of the torque required for acceleration of the basket and the frictional torques. The nature and the position of the forces acting on the shaft will depend on whether the shaft is vertical or horizontal and on the drive system.

In the case of overdriven or underdriven vertical centrifuge, the axial forces acting on the shaft are mainly due to the weight of the basket and its contents and any coupling or pulley mounted on the shaft. The axial forces will create a tensile stress in the case of overdriven and a compressive stress, in the case of the underdriven centrifuge. Forces perpendicular to the shaft will be acting on both types due to the tensions of belt or chain drive. A resultant stress has to be determined from the shear stress duc to thc torque, tensile or compressive stress due to axial load and a bending stress. Based on Eqs 4.1, 4.2, 4.3 and 4.4, the minimum diameter of the shaft can be determined. The actual shaft provided may not be of the same diameter over its length. In the case of horizontal centrifuge there will be no axial force on the shaft. The design will, therefore, be based on torque and bending moment due to normal loads. In all cases, it is necessary to check the unbalanced forces, deflection and the critical speed of the shaft. Due to the variation of speed of the basket it may not be possible to obtain a perfect balance of the centrifugal forces. Deflection of the shaft should be limited to 0.02 mm at critical points. Vibrations can be limited if the normal speed is below 60% of the critical speed (Sections 4.2.5 and 14.5.1.2).

(g) *Bearings*

Axial forces on the shaft can be supported by the provision of thrust bearings. In addition, ball bearings are also provided for radial loads. For an overdriven shaft a self-lubricating bearing, shaped in the form of a sphere, helps to maintain the shaft in the vertical position, in spite of the vibrations created due to imbalance of the rotating mass. The choice of bearings may be made according to the loading conditions.

(h) *Drive system*

The basket may be driven either by connecting the shaft directly to the motor shaft or by the use of a flexible coupling. For variation of speed, a belt or chain drive may be used. The motor may be supported by brackets fixed to the casing or to a framework. Lately, VFD drives are used to reduce the motor power required.

(i) *Brake*

A brake in the form of a shoe with a lever is applied to the basket for reducing speed and ultimately making the basket stationary.

15.4.1.2 Pusher Type Centrifuge

This consists of a horizontal basket mounted on a shaft (Fig. 15.18) . Feed enters through a feed pipe, which is made to enter the basket. The slurry from the feed pipe is accelerated gradually to the basket speed. A cake is deposited on the screen of the basket. To remove the cake, a hydraulically operated pusher moves axially, shearing the cake along the perforated basket, through a short distance. The pushers stroke is controlled by a piston, operating in a cylinder under oil pressure. Dry cake flies off the end of the basket into the portion of casing (housing), placed in front of the basket. From the casing the cake passes out through the discharge. The pusher returns to its initial position. The length and frequency of the stroke of the pusher may be adjusted and is generally set to handle slightly more than the anticipated maximum production of solids. As the cake moves along the basket, the filtrate is thrown out in the casing, surrounding the basket. This part of the casing is sealed from the front portion of the casing to prevent any leakage of the filtrate. A wash water pipe supplies the water for washing.

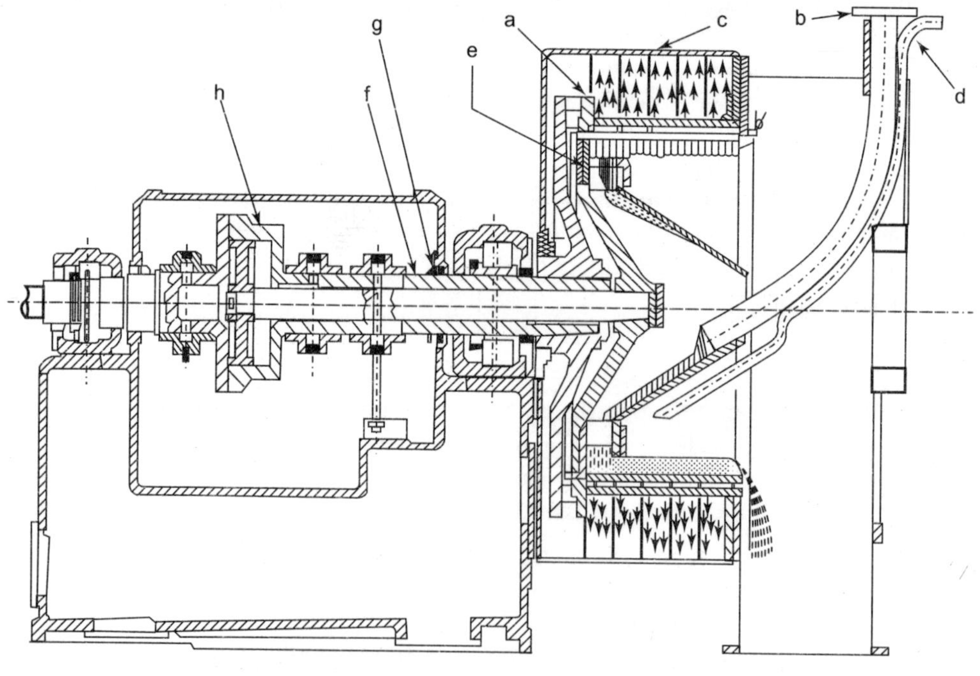

Fig. 15.18 Pusher Type Centrifuge: (a) Basket, (b) Feed Pipe, (c) Casing, (d) Wash Water Pipe, (e) Pusher, (f) Main Shaft, (g) Pusher Shaft, (h) Hydraulic Cylinder for Operation of the Pusher Shaft

The basket diameters vary from 600 to 2400 mm. The pusher shaft is solid and is located inside the hollow shaft of the basket. The hydraulic cylinder forms a part of the hollow shaft. The cylinder cover is fitted to the cylinder by a flanged joint. The operating oil is supplied from either side through an annular gap in the hollow shaft. The hollow shaft and the cylinder are to be designed for transmission of torque and axial thrust due to oil pressure acting on the cylinder. In addition, the cylinder and the hollow shaft have to withstand internal oil pressure. The stresses produced are circumferential, axial and also a shear stress due to torsion. The hollow shaft is supported between two tapered roller bearings; the rest of the shaft along with the hydraulic cylinder and other mechanisms are enclosed in a housing. The basket overhangs from one of the bearing supports, while the drive to the shaft is from a pulley placed near the other bearing. From the loading on the shaft, namely, the weight of the hydraulic mechanism, basket, pulley, pusher rod shaft weight, etc., a bending moment diagram can be drawn to assess the maximum bending moment and stresses due to bending can be determined. The design of the shaft is finally based on the permissible resultant stress.

15.4.2 Continuous Centrifuges

There are two types of continuous centrifuges.

(1) Screw conveyor type (Fig. 15.19)

(2) Spiral scraper type.

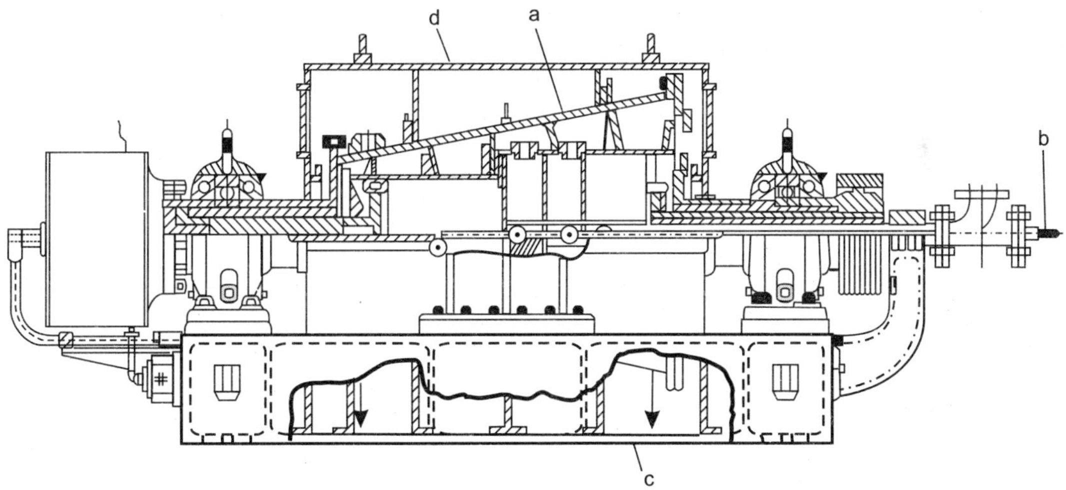

Fig. 15.19 Screw Conveyor Type Centrifuge: (a) Solid Bowl, (b) Feed Pipe, (c) Outlet, (d) Casing

15.4.2.1 Screw Conveyor Type

A solid bowl is mounted horizontally with an internal conveyor scroll. Both the bowl and the conveyor revolve in the same direction. The centrifuge is driven by a V-belt and pulley fixed to the shaft towards the feed end. Rotating of the bowl is transmitted through an epicyclic gear train, placed in a box towards the other end of the conveyor, which is made to rotate at a slower speed than the bowl.

Feed slurry enters into the chamber inside the conveyor assembly through an axially located pipe (hollow shaft). Parts around the periphery of this chamber distribute the material into the bowl. Centrifugal force holds the material against the bowl wall. The conveyor moves the solids along the bowl wall and leads to the solid discharge part. The liquid collects in the lower part of the bowl and flows towards the filtrate part at the other end.

The bowl is cylindrical in shape (partly tapered) preferably with a taper at one end. It may be fabricated from sheets of mild steel, stainless steel, copper, etc., or cast in two pieces joined together by a flanged joint. The bowl is provided with covers or heads at either end attached by flanged joints. A hollow shaft is fixed to the cover and carries a V-pulley at one end. Pedestal ball bearings are placed at either end to support the two ends of the shaft. The pedestal bearings are in turn supported by a base frame that is cast or fabricated. The outer casing surrounding the bowl is divided into compartments. It is connected to a hopper for slurry feed and discharge. The casing is fabricated from steel sheets.

15.4.3 Sedimentation Centrifuges

These are settling devices employing centrifugal force for the acceleration of the settling effect. They are employed for a wide variety of separations from relatively coarse particles of 6 mm down to submicron sizes.

15.4.3.1 Bowl Centrifuges

These centrifuges are used to remove solids from liquids and to separate immiscible liquids. There are several types of these centrifuges. The simplest is the solid bowl type, which is used mainly for liquid/liquid separations. Others are the nozzle type, for recovering large volumes of soft solids and the solids ejecting and nozzle valve types, which discharge solids intermittently.

All these types induce powerful centrifugal forces, equivalent to thousands of gravities, and offer a large amount of settling surface in a small volume. So, they are often a good alternative to the gravity settlers, filters and other means of separating solids and liquids.

A disc centrifuge consists of a bowl in which a stack of discs is located. The number of discs are usually between 30 and 200. The discs are held apart by metal spacers, the space between the discs being 0.5–1.0 mm. They make an angle of 35-45 degrees with the axis of rotation of the bowl (Fig. 15.20).

Feed material is being introduced through the axis of rotation and moves outward through a distribution chamber, and then enters the disc stack. As the less dense fluid moves upward between the discs, the solid particles and the denser fluid migrate outward. When the solids reach the disc, they move along the surface and settle at the periphery of the bowl. The arrangement of discs greatly increases the settling area available within a given volume and considerably reduces the radial distance a particle has to travel before settling down on a surface.

15.5 SELECTION OF FILTERS (See Process Data Sheet)

Classification of filters and details of construction have been discussed so far. For selection of a filter, a number of factors must be taken into account. These are, the specific resistance of the filter cake, the quantity to be filtered and concentration of solids. For the free filtering materials a rotary vacuum filter is generally found to be the most satisfactory, since it has a very high capacity for its size and does not require much manual attention. If the cake needs be washed, a rotary drum is preferred to the rotary leaf. If the specific resistance is high, a positive pressure filter like the filter press is suitable, particularly if the solid content is not so high that frequent dismantling of the press is necessary. For efficient working, a leaf filter is effective, since it facilitates preparation of a very thin cake and reduces the risk of channelling during washing. In case, very small quantities of solids are present in the liquid, an edge filter is employed. The centrifuges are frequently used for drying crystals and for separating emulsions into their constituent liquids.

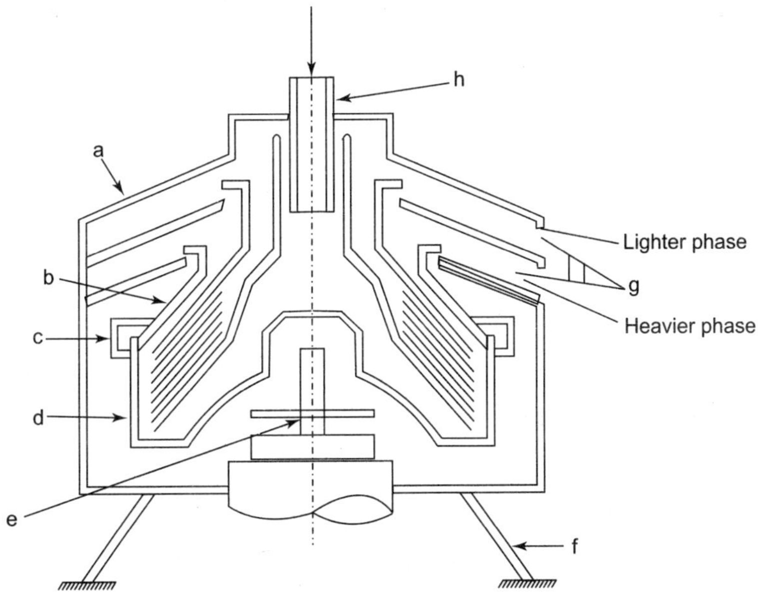

Fig. 15.20 Disc Bowl Centrifuge: (a) Casing, (b) Bowl Hood, (c) Main Lock Ring, (d) Bowl, (e) Spindle, (f) Support, (g) Liquid Outlet, (h) Feed Pipe

FILTER

Process Data Sheet

OPERATING CONDITIONS				
Type				
Operation	☐ Continuous	☐ Batch	☐ Vacuum	☐ Pressure

FEED				
	Design Capacity:		Average Capacity	
	Batch Size		Frequency	
	Temperature	Density	Viscosity	Inlet Pressure
	Filtration Rate	Filtration Area		Active Area
	Precoat Required: ☐ Yes ☐ No		Material	

FEED COMPOSITION		
	Composition	
	Concentration (Wt % in feed)	
	Screen Analysis:	
	Characteristics: ☐ Granular ☐ Flaky ☐ Crystalline ☐ Friable ☐ Abrasive ☐ Clogging	
	☐ Builds Up and Hardens ☐ Toxic ☐ Corrosive ☐ Flammable ☐ Explosive	
	Composition	
	Density	Viscosity · Vapour Pressure
	Characteristics: ☐ Toxic ☐ Corrosive ☐ Flammable ☐ Volatile	

FILTRATE		CAKE	
Filterate is: ☐ Product ☐ Waste		Cake is ☐ Product ☐ Waste	
Recycled: ☐ Yes ☐ No		Max. Allowable Liquid (Wt. %)	
Max. Allowable Solids (Wt. %)		Density	Rate
Filtrate Composition		Composition	
Filtrate Rate		Discharge ☐ Dry ☐ Wet	
Density		Sluicing Medium	Rate

Number of Washes:	Reason for Washing

FIRST WASH			SECOND WASH		
☐ Countercurrent	☐ Cocurrent		☐ Countercurrent	☐ Cocurrent	
Washing Fluid	Flow		Washing Fluid	Flow	
Temp.	Density	Visc.	Temp.	Density	Visc.
Vap. Press.	Inlet Press.		Vap. Press.	Inlet Press.	
Max. Allow Liquid in Cake (Wt%)	Wash Ratio		Max. Allow Liquid in Cake (Wt%)	Wash Ratio	
Filtrate Composition			Filtrate Composition		
Filtrate Flow	Wash Cycle Time		Filtrate Flow	Wash Cycle Time	

THIRD WASH			
☐ Countercurrent	☐ Cocurrent		Location: ☐ Indoors ☐ Outdoors
Washing Fluid	Flow		
Temp.	Density	Visc.	Materials of Construction:-
Vap. Press.	Inlet Press.		
Max. Allow Liquid in Cake (Wt%)	Wash Ratio		
Filtrate Composition			Filter Medium:-
Filtrate Flow	Wash Cycle Time		

References and Recommended Readings

Brown, G.G., *Unit Operations*, John Wiley and Sons, New York (1956).

Chamlets, J.M., Elledge, L.R., and Porter, R.E., 'Filter'. *Chemical Engineering*, **62**(6), 191 (1955).

Chen, W.U., 'Solid Liquid Separation via Filtration', *Chemical Engineering*, **104** (2) 66 (1997).

Couper, J.R., Roy Penny, W., Fair, J.R. and Walas, S.M., *Chemical Process Equipment Selection & Design*, Second Edition, Elsevier, Amsterdam (2005).

Day, R.W., 'Technique for Selecting Centrifuges', *Chemical Engineering*, **81** (10) 98 (1974).

Dickey, G.D., *Filtration*, Reinhold, New York (1961).

Fleming, D.K., and Dahlstrom, D.A., 'Enclosed Continuous Filtration Equipment', *Chemical Engineering Progress*, **59** (7) 62 (1963).

Jordon, D.G., Chemical Process Development, Interscience, New York (1968).

McKenna, J.V., 'Centrifuges Push ahead in Specialities', Chemical Engineering, **105** (10) 90 (1998).

Patnaik, T.P., 'When a Rebuilt Centrifuge is a Viable Option', Chemical Engineering, **105** (10) 92 (1998).

Perry, R.H., Green D.W., and Maloney, J.O., Perry's *Chemical Engineers' Handbook*, 6th Edition, International Edition, McGraw-Hill, Singapore (1984).

Purchas, D.B., 'The Sizing of Rotary Vacuum Filters, Part 1', *British Chemical Engineering*, 2196 (1957).

———., *Industrial Filtration of Liquids*, Leonard-Hill, London (1971).

Schweitzer, P.A., *Handbook of Separation Technique for Chemical Engineers*, 2nd Edition, McGraw-Hill Book Co., New York (1988).

Schneider, G.G., Horzela, T.I., Spiegel, P.J., Cooper, P.J., 'Selecting and Specifying Electrostatic Precipitators', *Chemical Engineering*, **82** (5) 94 (1975).

Schroeder, T., 'Selecting the Right Centrifuge', *Chemical Engineering*, **105** (10) 82 (1998).

16

Chapter

Dryers

16.1 INTRODUCTION

The function of the dryers is to remove water/solvent from wet solids or from slurries by evaporation. The process involves heat transfer and diffusion. Accordingly, the process of drying can be divided into two stages. During the first stage, in which the material surface is wet, the rate of evaporation is constant. This is known as constant rate period. In the next stage, the surface being dry, the water/solvent must force itself to the surface by diffusion, which is slower than evaporation. This, therefore, known as the falling rate period. The entire process, therefore, involves heat transfer as well as mass transfer. Heat transfer can be achieved either by direct contact between the heating medium and the material or by indirect contact in which the material and the medium are separated by a wall. Heat transfer and diffusion can be facilitated if the material is agitated or sprayed to facilitate drying. Although evaporation is the main operation in drying of the materials, in certain cases it is equally essential that the physical appearance and properties of the dried product be preserved. A critical choice of processing conditions and the method of heat transfer is necessary in such cases.

16.2 TYPES OF DRYERS

A large number of dryers have been designed for individual groups of materials, with different arrangements and methods of heat transfer, as also the heating mediums. Dryers can, therefore, be classified according to physical state of the material, solids or liquids, the method of heat transfer, i.e., direct or indirect, or according to the heating medium, i.e., air, steam, hot water, etc. On the basis of the method of heat transfer, dryers can be classified as (a) direct dryers, (b) indirect dryers. In each category they can further be divided into batch and continuous types. As a general rule, production rates of 5,000 kg per day are best handled by batch dryers and rates over 5,000 kg in a continuous dryer. However, the choice depends upon local conditions, for instance a fully automated plant can have a lower capacity.

Details of construction and design features of some of the important types of dryers are considered. Flash dryers, fluid bed dryers, spray dryers have complex design features and accessories and these are beyond the scope of this book.

16.3 BATCH TYPE DRYERS

16.3.1 Tray Dryer

The simplest type of batch dryer is the tray dryer (Fig. 16.1), which is essentially a cabinet or a large compartment with a number of trays. These trays may either be fabricated from sheets or from screens. In these dryers, steam, gas or electrically heated air is used as a drying medium. The air is passed by means of a fan over a radiator or over finned tubes and then over the trays. A portion of the air is let out at the discharge, the remainder is reheated and recirculated.

An amount of fresh air equivalent to the volume discharged is admitted at the fan. In modern designs of the dryers hot air is allowed to travel only a short distance, usually not more than that corresponding to two or three stacks of trays. The air is reheated after each passage over trays, with partial rejection of humid air and its replacement by fresh air. Secondary heating tubes are placed in the path of the air to restore its temperature and heat content.

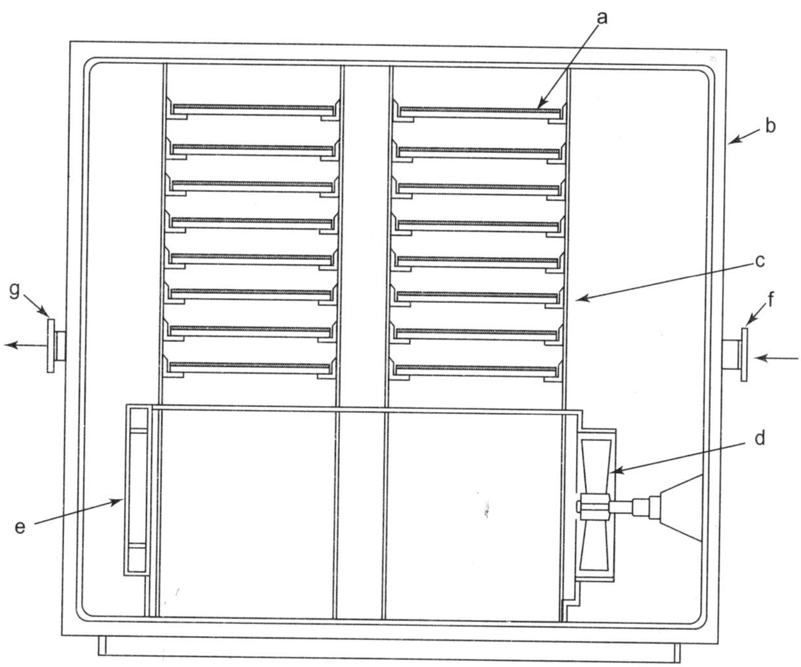

Fig. 16.1 Tray Dryer (Door Removed): (a) Tray, (b) Cabinet, (c) Shelf (trolley), (d) Fan, (e) Radiator, (f) Air Inlet, (g) Air Outlet

The cabinet or the chest, in which the shelves and trays are located, is rigidly built from a framework of structural sections, provided with panels and doors of double metallic sheet construction having heat insulation material in-between. The spacing of trays are such, so as to maintain low pressure losses and the spacing may be about 75 mm. Air velocities of the order of 7500 mm per minute are used between the trays, but where there is a danger of entrainment of the material as drying proceeds, lower velocities should be used. A high volume propeller fan is adequate. It is common to use two air passes, the middle tray acting as a horizontal partition between two passes. Uniform distribution of air is important. Stagnant air pockets should be avoided and for this reason curved deflector baffles are provided at corners. Where extremely uniform drying of material throughout the drying cycle is important, the design of the cabinet is such, so as to permit a frequent reversal in the direction of air flow.

Batch vacuum shelf or tray dryers are generally used for materials which are excessively heat sensitive. They consist of a cast iron rectangular or steel cylindrical chamber, fitted with a vacuum-tight charging door. The door is either of the quick-acting type or is provided with several wing-nuts and swing bolts.

Steam-heated shelves are connected to a steam manifold at one end and a condensate discharge at the other end. A number of trays are arranged on the shelves. The material to be dried is loaded into these trays. The chest is evacuated during each cycle before steam is supplied.

For vacuum tray dryers, the framework with the covering must be strong enough to withstand external pressure of about 1 bar. All joints must be airtight. The walls and the door can be strengthened by welding horizontal and vertical stiffners on the outside surface of the sheets.

In some cases the trays are mounted on trolleys with wheels which may be pulled out of the cabinet for loading and unloading. Several such trucks may be placed in the dryer.

The trays are usually 600 mm wide, 900 to 1800 mm long and 30 to 40 mm deep. They are made of mild steel, stainless steel, enamelled iron or other special materials, and are fabricated from sheets of 14 to 18 gauge (1 to 2 mm thick). A shelf is fabricated from steel structural sections like angles or trees, on which about 10 to 20 trays may be supported. The shelf must be sufficiently rigid to avoid deflection of the framework due to the dead load of trays and the material. It is necessary to ensure that the trays remain flat under the load of the material.

The framework consists of horizontal structural sections, acting as beams, while the vertical sections will act as columns. The deflection δ, of each horizontal section may be determined by the following equation,

$$\delta = \frac{5WL^3}{384\,EI} \tag{16.1}$$

where, W — load carried by each horizontal section. The total load of trays and material is distributed between all horizontal sections

L — length of section between supports

E — modulus of elasticity of the material of the section

I — moment of inertia of the cross-section.

The cross section of the structural member should be so chosen that the deflection will be between 1 and 2 mm. The vertical structural members should be designed as columns as per Eq. 3.16.

16.3.2 Jacketed Pan Dryer

This dryer (Fig. 16.2), consists of a flat bottom shallow cylindrical pan with a steam jacket at the bottom and on the sides. An anchor shaped scraper type of agitator is used to move the wet material over the surface heated by steam. The material is removed through an outlet at the bottom.

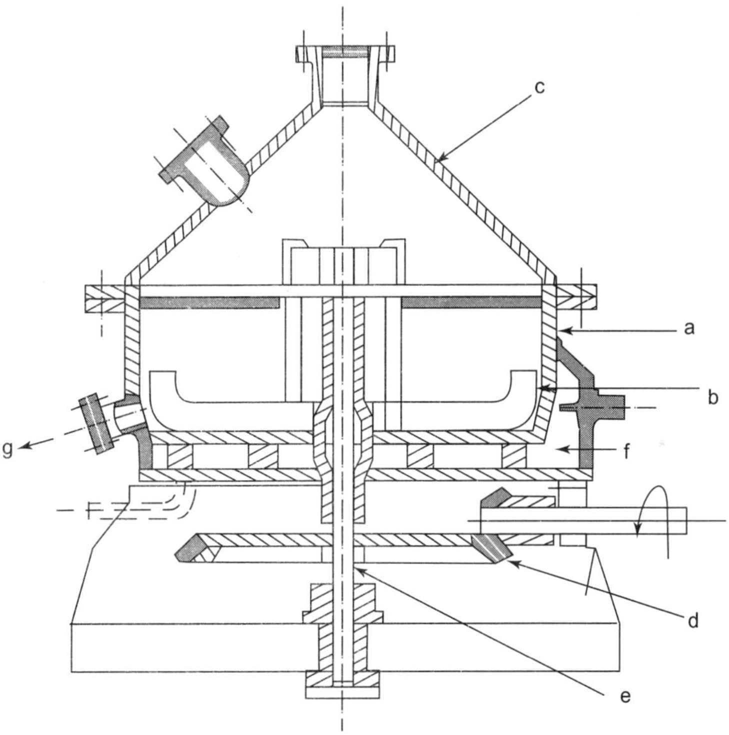

Fig. 16.2 Jacketed Pan Dryers: (a) Pan, (b) Anchor Agitator, (c) Cover, (d) Bevel Gear, (e) Shaft, (f) Steam Jacket, (g) Outlet

If the pan dryer is to work under vacuum, it is fitted with an airtight conical cover with nozzles. The vapour generated by heating is sucked by a vacuum pump.

The scraper is rotated at speeds usually between 2 and 20 rpm with the intention of merely preventing the solids from forming a hard cake on the heating surface. When drying slurries and wet pastes, clearance between the scrapers and heated surfaces must be minimal to prevent a skin of dried material building up on these surfaces. The driving system must, therefore, be designed for heavy torque conditions. The usual size of the pan is up to 3 metres in diameter, with capacities up to 5000 litres. The drive to the stirrer is either from the bottom or from the top of the pan with a bevel reduction gear, which in turn is driven by a V-belt and motor.

The pan is designed for an external pressure of about 3 to bar, depending on the pressure of steam. In addition, if a vacuum pan is to be designed, the design pressure will be increased further by 1 bar.

The design of the pan dryer is similar to that of a reaction vessel, discussed in Chapter 7. The stirrer design is also indicated in Chapter 14.

16.3.3 Agitated Vacuum Dryer (Fig. 16.3)

This is similar in principle to a pan dryer, where a jacket is used for the heating medium and an agitator is used for moving the wet material over the heated surface. The unit consists of a horizontal jacketed cylindrical shell, closed at the ends by suitable heads. The dryer is built in a variety of sizes, ranging from about 750 mm diameter by 4 m long to about 1.6 m diameter by 12 m long. A central shaft is supported on bearings outside the shell. It is sealed by stuffing boxes against leakage through the holes in the heads.

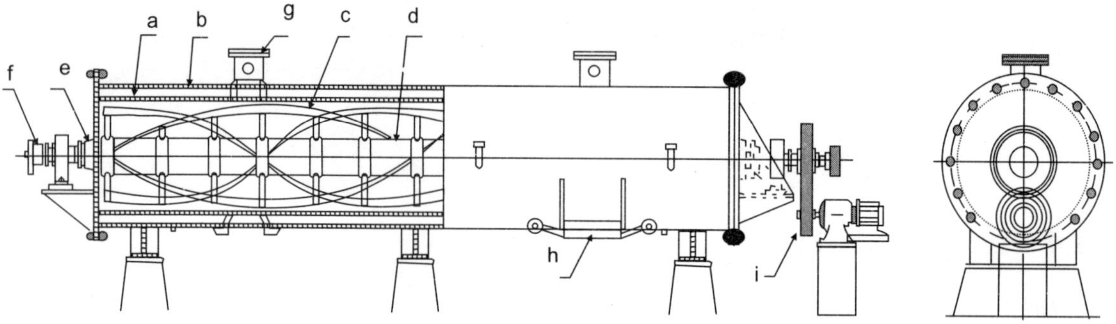

Fig. 16.3 Mechanically Agitated Dryer: (a) Shell, (b) Jacket, (c) Agitator, (d) Shaft, (e) Stuffing Box, (f) Rotary Joint, (g) Charging Nozzle, (h) Discharge Door, (i) Drive System

The shaft carries a spiral agitator or paddle with two sets of blades. One set of blades moves the material in one direction and the other set in the opposite direction. Doors are provided on the shell, at the top for charging the material and at the bottom for discharging. Hot water, steam or a suitable fluid is supplied to the jacket for heating. There is also a provision for feeding steam into the shaft of the agitator. This involves the use of a rotary joint (Fig. 16.4), which provides a joint between a rotating steam pipe and a stationary condensate piping.

The shell is designed by considering external pressure as discussed earlier and the shaft is designed as outlined in Chapter 14. One has to consider 'fatigue' also because the shaft is horizontal. This treatment is excluded from the scope of this book but one may refer to the Mechanical

Engineer's handbook. It is desirable to consider characteristics of the material before and after drying while designing the impeller.

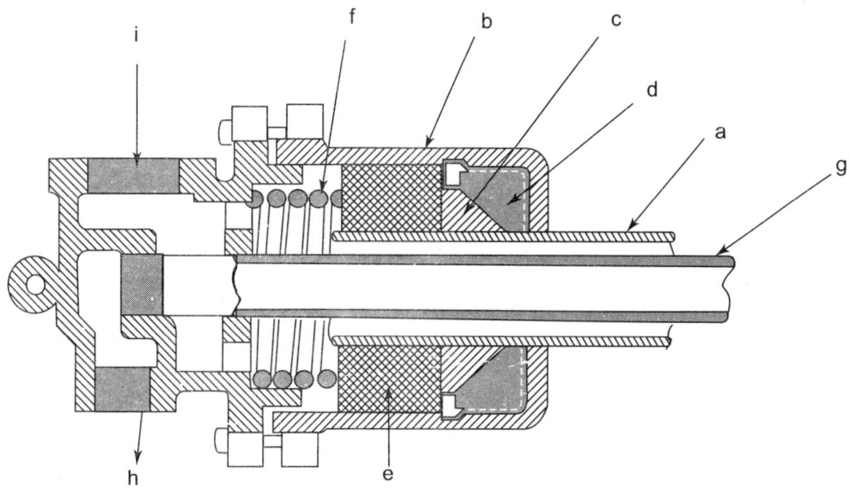

Fig. 16.4 **Rotary Joint: (a) Rotating Steam Pipe, (b) Joint Housing, (c) Shoulder, (d) Graphite Bushing, (e) Graphite Seat, (f) Seating Spring, (g) Condensate, Discharge Pipe, (h) Condensate Discharge Connection, (i) Steam Inlet**

16.3.4 Tumbler Dryer

This type of dryer (Fig. 16.5) is, to some extent, replacing the agitated rotary vacuum dryer. It has two opposing jacketed cones on a common jacketed short cylindrical base. When the cones are in a vertical position the dryer can be rapidly discharged. The unit is provided with supporting trunnions running on suitable bearings. The trurnnions are hollow. A vacuum connection is made through one of the trunnions. The connection pipe is turned upwards in the cone and is fitted with a dust filter as its end. Inlet and outlet pipes pass through the outer trurnnion for supply of a suitable heating medium to the jacket. A special design of a rotating valve is used for this purpose. One of the trunnions is driven through a reduction gear and chain drive.

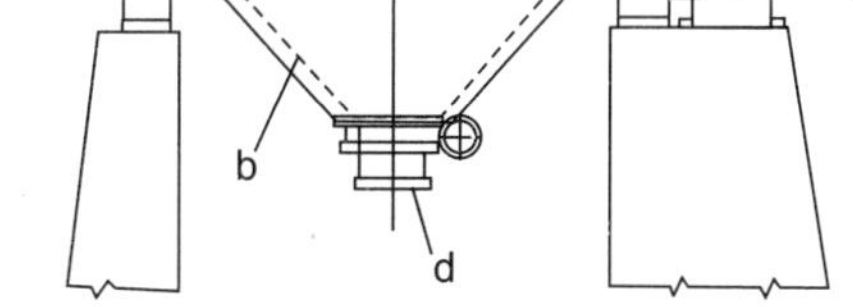

Fig. 16.5 **Tumbler Dryer: (a) Conical Shell, (b) Jacket, (c) Cover, (d) Discharge, (e) Steam Inlet, (f) Vacuum Connection, (g) Dust Filter, (h) Bearing, (i) Drive System**

Rotational speeds will range from 12 rpm for small units to 3 to 4 rpm for large commercial installations. The horse powers vary between ½ and 15. The base cylinder diameters are about 1 m to 3 m.

16.4 CONTINUOUS DRYERS

16.4.1 Band Dryer

With the trays on trucks, the dryer can be made continuous by passing the trays with the wet material continuously through a drying chamber. Trays may be placed on a conveyor instead of using trucks. In some cases the material may be placed or attached directly to a conveyor belt. The conveyor belt is made of perforated metal plate or woven wire mesh.

A section of the dryer is used for locating the heater which may be usually of steam heated finned tubes. A fan which is placed above the heater pushes the air through the heater and circulates it through the wet material. Recirculation and reheating may be automatically controlled.

Conveyor widths vary between 40 cm and 250 cm. Lengths range up to 50 m, so that the material can be retained in the chamber for a sufficient time. The entire chamber is fabricated out of the structural steel sections with steel sheets welded to it. Doors are provided at either end with the necessary ports. The chamber is properly insulated.

16.4.2 Turbo Dryer (Fig. 16.6)

It consists of a vertical rotating structure placed in a cylindrical housing which carries annular tray assemblies, vertically fixed, one above the other. The material which is loaded in trays drops down from one level to the next lower level by means of a stationary scraper. The structure rotates about a central hollow portion in which a vertical shaft carrying a number of multiblade fans are located.

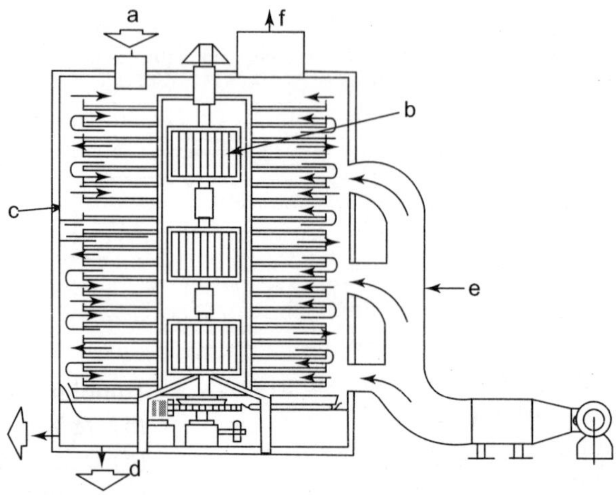

Fig. 16.6 Turbo Dryer (Schematic): (a) Feed, (b) Turbo Fan, (c) Rotating Tray, (d) Product Outlet, (e) Inlet Hot Drying Gas, (f) Exhaust Gas

The action of the fans is to draw air radially inwards over the material in the trays, and to recirculate it outwards to the housing which encloses the rotating structure. Heating elements are placed in the space between the rotating structure and the housing. The wet material enters at the top, and the dry discharge is removed from the bottom. The air enters from the bottom and leaves from a stack at the top.

16.4.3 Rotary Dryer

It consists of a long cylindrical shell mounted horizontally with a slight slope (Fig. 16.7). The shell is either rotated or may be kept stationary. If it is stationary, an agitator is made to revolve within the shell at a slow speed. The wet material is fed at the upper end, that moves gradually towards the lower end due to the rotation of the cylinder or movement of the agitator. Warm air or a stream of hot gas travels counter-current to the material. The rate of feed, the speed of rotation or agitation, the volume of the heated air or gases and their temperatures, are so regulated that the material is completely dried before it is discharged at the lower end.

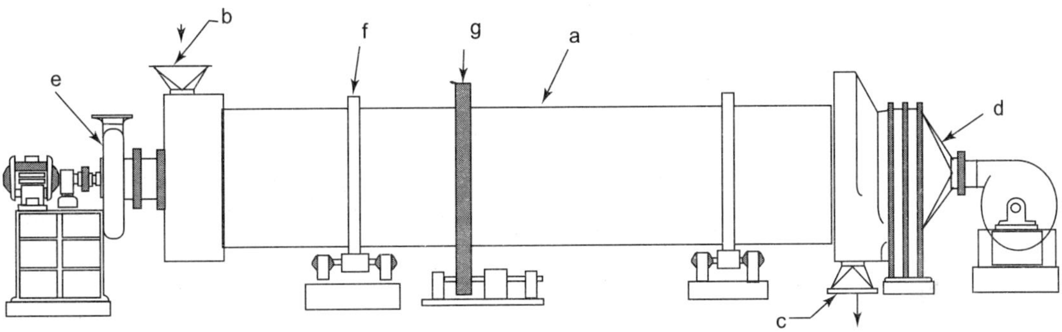

Fig. 16.7 Rotary Dryer: (a) Shell, (b) Feed Hopper, (c) Discharge Hopper, (d) Air Preheater, (e) Exhauster, (f) Tyre, (g) Gear Wheel (Girth Gear)

With the rotation of the shell, the material is carried upwards, one-fourth of the circumference and then rolls back to the lower level. The action of lifting the material is carried out more efficiently if lifting devices (Fig. 16.8) such as shelves or baffles are provided on the inner circumference of the shell along the entire length. These are of different designs, and are able to lift the material to half-way round the circumference of the shell and then drop from the topmost position, through the central part of the shell, where the air is hottest. The simplest shelves are longitudinal baffles, plain or serrated, of about 5 cm to 10 cm width on the periphery of the cylinder. Cross shelves covering the entire cross section of the cylinder give better distribution. Baffles with louvers form a cone section with the narrower end at the feed. In some cases the cylinder is divided in sections, each section being provided with radial and cross baffles.

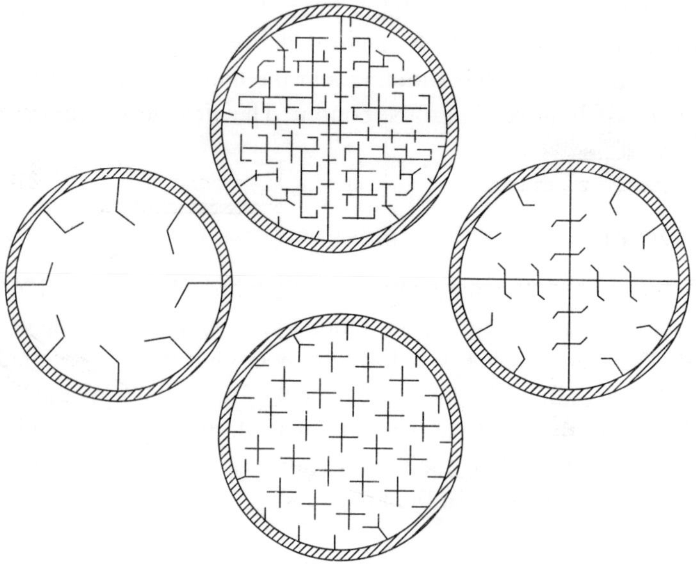

Fig. 16.8 Lifting Devices

The wet material is fed continuously through a hopper at the upper end and a vibrating feeder or screw regulates the flow of the material. The dry material which drops in a hopper at the lower end may be removed continuously by a bucket elevator or conveyor (screw or belt). Free flowing feedstock will usually be fed through an inclined chute, rotary valve, vibrating feeder, rotary table or a belt conveyor.

The retention time of the material in the dryer will be determined by several factors such as the slope of the dryer shell, its speed of rotation and the length, the arrangement of flights, etc. The retention time at a given speed of rotation is inversely proportional to the slope of the shell. The rotational speed in rpm × dryer diameter in metres lies between 75 and 105. The slope is 18 mm to 54 mm per m.

A sealing arrangement at either end has to be provided between the rotating shell and the stationary end feed box forming the material feed and the hot inlet and exhaust gas openings. These seals act to prevent the leakage of air into the cylinder or leakage of hot gases out of the dryer. These seals (Fig. 16.9) can be constructed (A) by use of a friction brake lining-attached to the rotating shell and pressed by a spring on the stationary wall of the end box, (B) by use of a labyrinth with intermeshing projections and (C) by use of cloth fastened to the end box and pressed against the rotating shell by a metal band. Whichever type of seal is used, it is important that the dryer shell should have an absolute minimum of eccentricity in its construction.

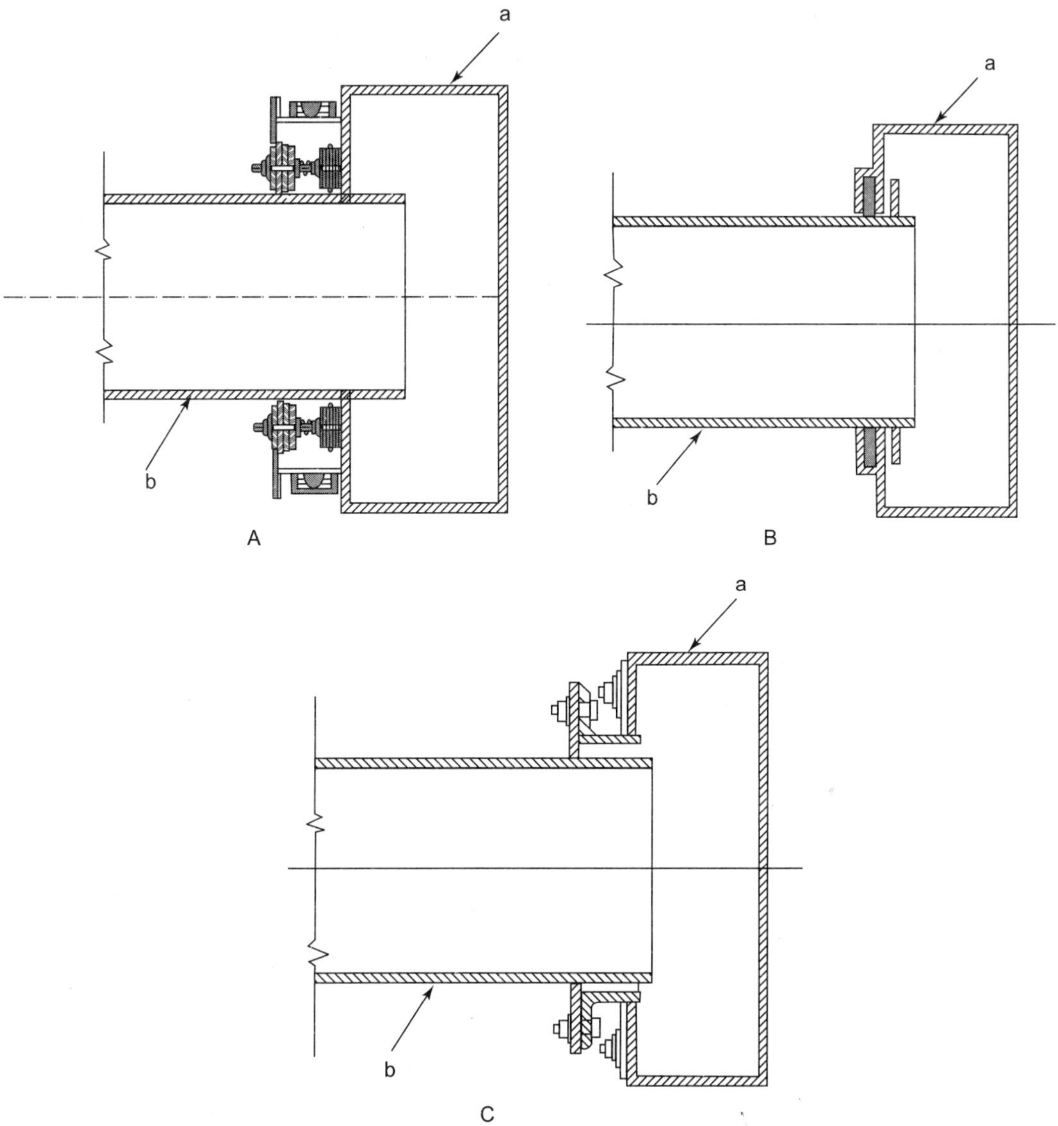

Fig. 16.9 Sealing Arrangements: (A) Friction Type Seal; (a) Feed Box, (b) Dryer Shell, (B) Labyrinth Type Seal, (C) Flexible Cloth Rubbing Seal

Heating of the material is done by passing flue gases, or superheated steam or preheated air, which enter at the lower end and leave through an exhauster at the upper end. The flow is thus counter-current to the flow of the material. When the material is sensitive to heat, the flow may be concurrent. In some rotary dryers, steam heated tubes are placed along the entire length of the dryer (Fig. 16.10). It is also referred to as steam tube dryer.

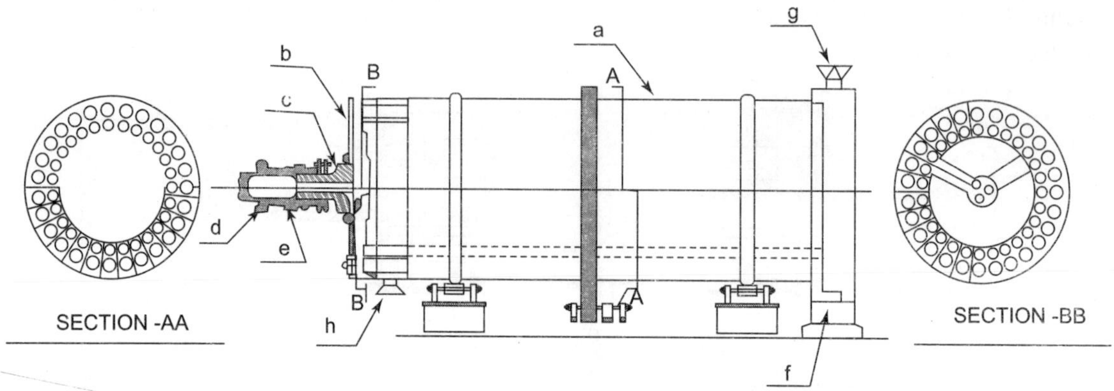

Fig. 16.10 Steam Heated Rotary Dryer: (a) Shell, (b) Steam Manifold, (c) Steam Neck, (d) Gland, (e) Steam Inlet, (f) Condensate, (g) Feed Hopper, (h) Discharge Hopper

The steam enters through a central inlet pipe and is distributed to all tubes through a header. The condensate from each tube is discharged into a common ring placed at the end of the dryer. The ring is concentric with the shell. The tubes are loosely supported in a plate to allow for free expansion.

The rotating shell diameter varies between 0.3 and 3 m, while the length of the shell may range between four and ten times the diameter. The shell is generally 6 to 8 mm thick, at lower diameter, and can be up to 40 mm thick for higher diameter, and is made as one piece. It may be fabricated from mild steel, stainless steel clad or lined. In order to support the shell, two rings, known as tyres (girth rings), are fixed to the shell at two intermediate positions. The tyres are mounted on tyre bands which are a part of the shell wall, and are held in place by retaining blocks. On relatively light dryers, the tyres may be hollow, but for heavy loads, the tyres are solid and are additionally supported by filler blocks, which are metal bands surrounding the tyre band.

Tyres are made to rest on bearing wheels, while the inclined rotating shell is kept from sliding off the bearing wheels by 'thrust wheels'. The entire assembly of bearing and thrust wheels is mounted on a bearing wheel base (Fig. 16.11). The bearing wheels are supported by means of pressure lubricated roller bearings on stationary shafts. The thrust wheels which roll against edges of the tyres are also supported by pressure-lubricated ball bearings on fixed shafts. Tyres and supporting bearing wheels should be of ample face width and diameter and should be of machined cast steel. The detailed engineering is excluded from the scope of this work.

A girth spur gear wheel is fixed to the shell near one of the tyres and is driven by a pinion. The pinion is run by an electric motor with a reduction gear. The pinion and the gear are designed to transmit frictional torques. The details of gear and shaft design are indicated in Sections 4.2 and 4.8. The total power required can be determined by taking into consideration the frictional power, power for acceleration and the loss of power in the drive system.

In double shell dryers, the hot gases first pass down a central tube coaxial with the dryer shell and return through the annular space between the tube and the shell, the material being supplied to the annulus. The material is heated by direct contact with the returning gases in the annulus and also by conduction from the hot central tube.

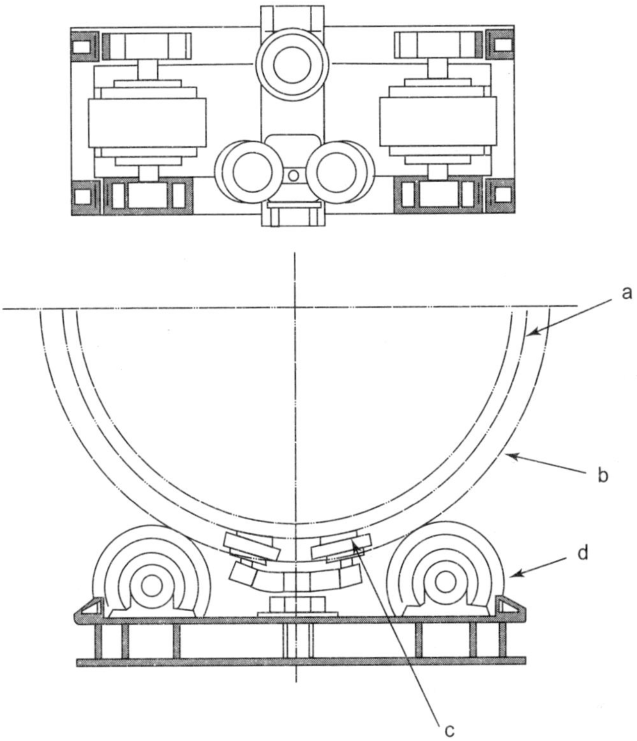

Fig. 16.11 Supporting Wheel Arrangements: (a) Dryer Shell, (b) Tyre, (c) Thrust Wheel, (d) Bearing Wheel

The rotary louvre dryer is similar to a through circulation rotary dryer. The heating medium is admitted through longitudinal channels formed by a series of radial plates fitted to the inside of the horizontal rotating shell (Fig. 16.12).

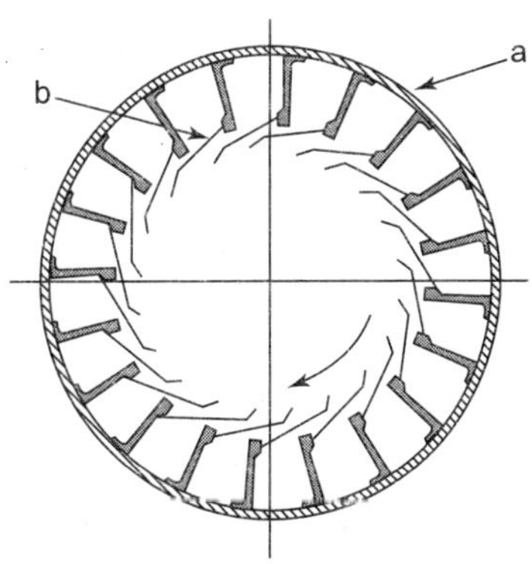

Fig. 16.12 Arrangement of Louvres: (a) Shell, (b) Louvres

The channels diminish in depth towards the discharge and are covered by tangential louvres which overlap, forming an inner drum designed to provide a passage for the heating medium to the inside of the drum. The louvres also help to prevent the charge of material dropping in the channels. The material travels to the discharge end with a spiral motion forming a compact bed. Heating medium is admitted only to those channels whose louvres are covered by the bed of the material. The heating medium passes through the bed of material before being exhausted through the vane discharger

ROTARY DRYER

Process Data Sheet

PROCESS CAPACITY	Normal Minimum	
	Normal Maximum	
	Peak	
DUTY	Duty Cycle	
	Special Duty	
Special Material Characteristics		
Specific Heat of Dry Solid		
Moisture of Inlet wt. %		
Moisture at Outlet wt. %		
Material Temperature at Inlet/Outlet		/
Temperature to which material may be heated without damage		
Process Heat Load: Minimum/Maximum		/
Drying/Heating Medium – Name		/
	- Flowrate Inlet/Outlet	/
	- Inlet Temperature Normal/Maximum	/
	- Outlet Temperature Normal/Maximum	/
	- Inlet/Outlet Pressure	/
Gas Flow Direction		
Method of Heating		
Drum Speed		
Drum Slope		
Residence Time		
Maximum % Loading (Drum Volume)		
Method of Feed		
Method / Location of Discharge		
Nominal Length		
Nominal Diameter		
Recommended Flight Form		
Location		☐ Indoors ☐ Outdoors
Hazard Classification		
SPECIAL PROCESS REQUIREMENTS		
MATERIALS OF CONSTRUCTION		

into an exhaust head situated at the discharge and of the dryer. This may be used for drying coarse solids without fines, for example catalyst tablets, pellets, extrudets, etc.

16.4.4 Film Drum Dryers

These dryers are operated either under atmospheric conditions or under vacuum. In the latter case, the dryer is enclosed in a vacuum-tight chamber. They have a single-drum or double-drum arrangements. The feedstock is supplied continuously to the effective drying surface of the drums and the dried product is removed by a scraper knife. In a single drum dryer, the drum dips in slurry contained in the feed trough. The dried material is removed by a scraper (or doctor knife). In a double drum dryer the feed is introduced into a trough above the nip formed between the two drums and of spring loaded end plates. The direction of rotation of the two drums is almost invariably downwards towards the nip between the cylinders. The clearance at this point is adjustable by changing the position of one drum relative to the other.

The material is fed by various arrangements. Pasty material is fed by feed rollers from top. Viscious material is fed at the nip between drums. Slurries are normally fed by submerging the drum partially in a trough. In some arrangements instead of submerging the drum, a provision of rollers or sprays is made in the trough. Figure 16.13 shows a trough with spray feeding for a drum.

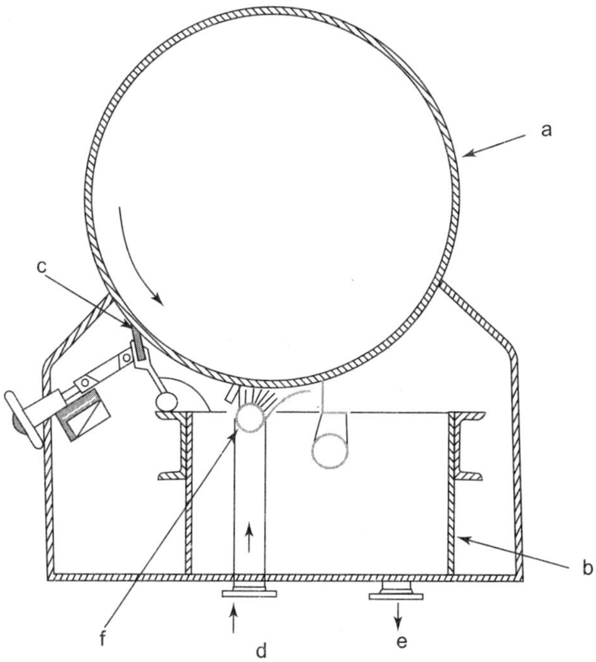

Fig. 16.13 Drum Dryer Knife: (a) Drum, (b) Feed Trough, (c) Scraper (Knife), (d) Feed Inlet, (e) Feed Outlet, (f) Spray Feeding

The drum, which may be of ½ m to 2 m diameter and 1 m to 4 m length can be fabricated from a plate or cast to the required shape. The materials used are cast iron, bronze, chromium plated steel or stainless steel. The surface of the drum should be hard and smooth. The drum is heated internally by steam and has, therefore, to withstand internal pressure. The ends of the drum are closed by flanged type of covers, with trunnions supported on bearings. The steam is supplied through a pipe which passes through the centre of the trunnion. The condenstate is discharged by means of a scoop or siphon through the second turnnion by means of a rotary joint (Fig. 16.4).

Drums are rotated at speeds in the range of 0.3 to 6 rpm. As the drum rotates, a film of about 1 to 3 mm thickness is formed, which is scraped by doctor knives. Three types are generally used, stationary single-bladed, oscillating single-bladed and multiple adjustable abutting or overlapping bladed. For materials which form a film of dried material, that is easily removed by a knife; a single knife extending the whole length of the drum, with adjusting screws at frequent intervals is satisfactory. The knife is of silicon-carbon steel with fairly small thickness. Sectional knives are usually of higher thicknesses. Each section is adjustable by hand wheels. For materials which are extremely abrasive, the knives are often tipped with abrasion resistant metal. Where corrosion is likely to occur or where metal contaminations is to be completely eliminated, the knives may be of suitable plastic material or even toughened glass. Knives are set at an appropriate angle at the point of contact. A sheet, metal or stainless steel chute plate is fitted below the knife assembly to deflect the dried product away from the knife adjustment gear.

The drum is rotated by a gear and pinion. The pinion is driven through a belt or chain or through a reduction gear box connected to a motor. A variable speed drive may be provided for regulating the speed of the drums. The design of the components is similar to that of a drum filter (see Section 15.2.2.1).

In recent years, double or twin drum dryers have replaced single drum dryers in several applications, due to their more efficient operation, wider range of products and high production rates.

A double drum dryer with a pendulum feed is shown in Fig. 16.14.

Drum dryers are considered to be extremely flexible to operate since all variables may be exchanged independently. These variable are:

(a) Steam pressure, which controls the temperature of the drum surface.

(b) Speed of rotation, which determines the time of contact between film and preheated surface.

(c) Film thickness, which is controlled by the gap between drums and speed of drums.

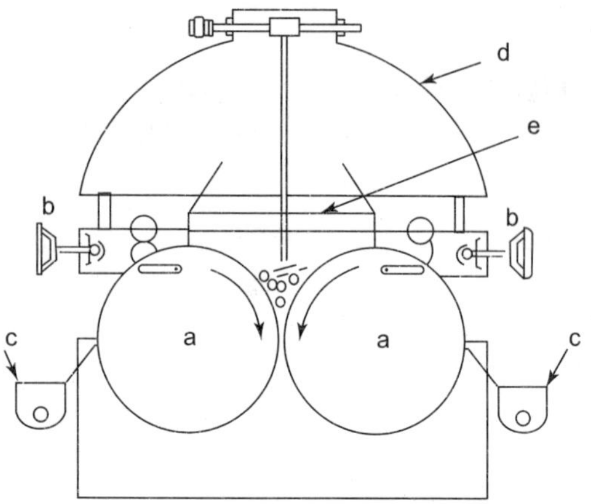

Fig. 16.14 Double Drum Driver: (a) Drum, (b) Knife, (c) Conveyor, (d) Vapour Hood, (e) Feed Pipe

(d) Condition of feed material, i.e., concentration and temperature of the feed reaching the drum.

16.4.5 Flash Dryer

The name flash dryer originates from the feature that the drying is carried out in a very short span of time, 0.5 to 3 secs. In this type of dryer systems, pneumatic conveying duct is the heart of the system, wherein drying operation is carried out. There exists a homogeneous suspension of drying material due to hydrodynamic effect caused by upward motion of hot gas (conveying medium).

In this dryer surface moisture is easily removed. Due to short residence time, heat sensitive materials can also be handled by this dryer.

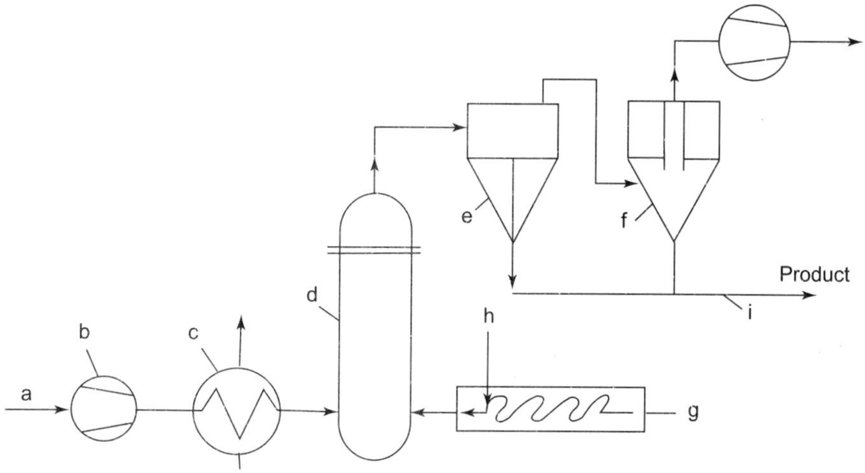

Fig. 16.15 A Typical Flash Dryer (Schematic): (a) Air Inlet, (b) Blower, (c) Air Preheater, (d) Dryer, (e) Cyclone, (f) Filter, (g) Mixer, (h) Feed Inlet, (i) Product Outlet

A finely powdered wet material is introduced with the help of a screw conveyor into the hot gas (air) stream. The material is pneumatically conveyed through the duct. During its conveyance the moisture is removed. The dried material is separated in a cyclone. The product collected at the bottom of the cyclone is transported to a silo via the screw conveyor. If the wet cake is difficult to introduce, a specially designed disperser can be used.

Fragile crystalline/particulate matter which gets easily disintegrated cannot be processed.

16.4.6 Fluid Bed Dryer

Also known as fluidized bed dryer. Fluid bed drying systems are becoming popular among process engineers for the following reasons:

(a) absence of moving parts (except blower) results in ease of maintenance.

(b) high heat transfer rates.

(c) rapid mixing means more or less isothermal operation leading to uniform drying.

In this type of a dryer hot gas is passed through the material at a velocity sufficiently high to fluidize the wet material but not high enough to cause pneumatic conveying as in flash dryer. For a fluidized bed having particle density between 1000 and 2000 kg/m^3, typical velocities are:

Particle size (avg) μm	*Velocity m/s*
100–300	0.2 to 0.4
300–800	0.4 to 0.8
800–2000	0.8 to 1.2

The bed could be operated in a backmixed or plug manner. When only surface moisture is to be removed backmixed mode of operation may work. On the other hand, in a plug flow mode of operation bound moisture of heat, sensitive material can be removed because residence time could be well controlled. Further, it is possible to use feed which is too wet to fluidize in a backmixed mode of operation. However, for plug flow mode, feed is required which can be easily fluidized.

In addition to the basic system the fluidized bed system also needs an air blower, hot air generator, feed conveyor, cyclone separator, product conveyor. The bag fitters and/or electrostatic precipitators are required depending on air discharge quality. A typical system is exhibited in Fig. 16.16.

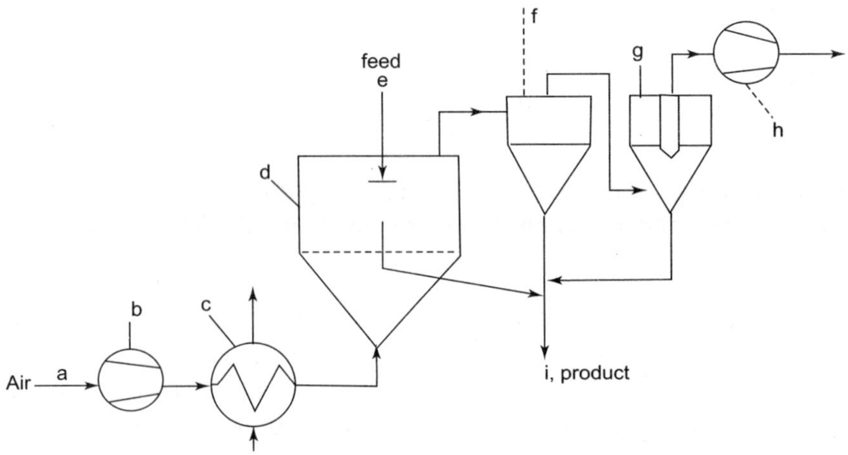

Fig. 16.16 Fluid Bed Dryer System (Schematic): (a) Air Inlet, (b) Blower, (c) Air Preheater, (d) Fluid Bed Dryer, (e) Feed Inlet, (f) Cyclone, (g) Bag Filter, (h) Exhaust, (i) Product Outlet

16.4.7 Spray Dryers

Spray dryers are most popular in the dairy industry, food industry and pharmaceutical industry. It is now finding its application in chemical industry, detergent industry as also in the dyes industry.

A pumpable fluid stream is atomized by high speed rotating atomizer or by a spray nozzle into a very hot gas stream (usually air is the hot gas). The temperature of air could be as high as 600°C. The inlet temperature range 150–250°C. The contact time is so short that despite very high air temperature and, hence, high evaporation rates, the product quality remains unaffected. Thus, heat sensitive materials may be conveniently dried in a spray dryer, e.g., milk powder from milk.

The key to a successful operation of spray dryer lies in getting very fine droplets. The rotating atomizing wheel (100–200 m/s) can produce particle size in a range of 1–500 μm. While pressurised spray nozzle can give particles in the range 10–800 μm.

The spray dryer consists of three main components:

(a) Atomizing device.

(b) Drying chamber.

(c) Product recovery system.

(a) *Atomizing device*

For effective drying in a very short contact time droplet sizes in the range 5 to 500 μm are essential depending upon the system under consideration. Pressure nozzles, pneumatic nozzles and atomizing wheel are commonly used. Pressure nozzles could produce droplets in the range 10–800 μm, pneumatic nozzles produce the same between 5 and 500 μm. Rotating atomizing wheel can yield droplets in the range 1–500 μm. Rotating atomizing wheels can handle abrasive material and different feed rates.

Drop-size can be controlled by changing revolutions. However, it needs a drying chamber of large diameter. It is also energy intensive and the atomizing wheel being of specialized design, is expensive. On the other hand, nozzles as atomizing device are simpler and more compact. There are no moving parts and, therefore, cheaper. They consume less energy also. However, nozzles are not suitable for abrasive fluid for abrasion results in erosion of nozzle tip, thereby, changing droplet dispersion and size characteristics.

(b) *Drying chamber*

Drying chamber shape is decided by the atomizing device. For instance, if atomizing wheel is used, due to broad spray the diameter/height ratio is generally greater than one. The bottom discharge portion is conical. For nozzle as atomizing device, diameter/height ratio is far less than one. Hot air can be in co-current or counter-current mode. For heat sensitive material co-current mode is preferred mode of operation.

(c) *Product recovery/Collection*

Cyclones. These are most commonly used to recover spray dried material from drying chamber exhaust gases. Depending on discharge standards and carry over of particulate matter, cyclones are connected to bag filters, electrostatic precipitator or a scrubbing system.

It should be noted that dryers and filters are considered to be proprietary equiment. Generally, the system is purchased from the specilized vendors.

References and Recommended Readings

A.I.Ch.E., Equipment Testing Procedure Dryers (Rotary Continuous Direct Heat Dryers) (1998).

Belcher, D.W., Smith, D.A., Cook, E.M., 'Design and Uses of Spray Dryers', Part I, *Chemical Engineering*, **70** (20) 83 (1963); Part II, Chemical Engineering, **70** (21) 201 (1963).

Devahastin, S., Mujumdar's *Practical Guide to Industrial Drying*, Exergex Corporation, Montreal (2000).

Foust, A.S., Principle of Unit Operations, John Wiley and Sons, New York (1960).

Keey, R.B., *Drying Principles and Practices*, Pergamon, Oxford (1972).

Mujumdar, A.S., *Handbook of Industrial Drying*, Third Edition, CRC Press, Taylor & Francis Group, New York (2006).

Nonhebel, G., and Moss, A.A.H., *Drying of Solids in the Chemical Industry*, Butterworths, London, (1971).

Papagiannes, G.J., 'Select the Right Dryer', *Chemical Engineering Progress*, **88** (12) 20 (1992).

Parker, N.H., 'Aids to Dryer Selection', *Chemical Engineering*, **70** (13) 115 (1963).

Perry, R.H., Green, D.W., and Maloney, J.O., *Perry's Chemical Engineers' Handbook*, 6th Edition, International Edition, McGraw-Hill, Singapore (1984).

Raouzeos, G., 'The Ins and Outs of Indirect Drying', *Chemical Engineering*, **110** (13) 30 (2003).

Sloan, C.E., 'Drying Systems and Equipment', *Chemical Engineering*, **74** (13), 167 (1967).

Williams-Gardner, A., *Industrial Drying*, Leonard-Hill Books, London, (1971).

17

Chapter

Process Hazards and Safety Measures in Equipment Design

17.1 INTRODUCTION

The previous chapters dealt with the selection of materials, analysis of stresses, determination of strength and the design procedure adopted in equipment design. If the design procedure is strictly adhered to, the selection of the material is appropriate, and the fabrication and final inspection are carefully carried out, it is expected that the equipment should not fail under service condition. However, failures in service conditions or environments are not entirely eliminated. Singly, or in combination, many specific causes can lead to these failures. Certain safety measure can help to prevent such failures.

17.2 HAZARDS IN PROCESS INDUSTRIES

Modern technology has been quite successful in developing tailor-made chemicals. However, this effort has also introduced some additional problems, since manufacturing and handling experience is frequently inadequate to deal with the question of hazards. The ever increasing production of flammable organics, the rush to bring new products from the laboratory into full scale production, the problem of familiarization with a stream of new technology have all extended the probabilities of hazards. Toxic and corrosive chemicals, fires, explosions, falls and faulty mechanised equipment are the major hazards encountered in the operation of plants in chemical industries. The design engineer must be aware of these hazards and must make every attempt to present a design which provides the maximum protection for the plant personnel and the minimum chance for the occurrence of accidents.

17.3 ANALYSIS OF HAZARDS

An analysis must be made of the probable sources of hazards. It is also necessary to identify and evaluate the hazard potential. Where practical, recognised hazards should be eliminated, but such hazards can be tolerated if proper procedures are used to guard against accidents. The sources of hazards can be divided into two categories, namely, material hazards and process hazards.

17.3.1 Material Hazards

These are either due to the nature and properties of the materials or the reactions the materials are likely to undergo during processing.

(1) Combustible solids, liquids or gases
(2) Explosive or blasting agents
(3) Radioactive materials
(4) Nuclear materials
(5) Oxidising materials
(6) Materials that react with water to produce a combustible gas
(7) Materials subject to spontaneous heating, spontaneous polymerisation, or explosive decomposition
(8) Corrosive materials
(9) Toxic materials

17.3.2 Process Hazards

These hazards might arise due to several factors, some of which are enumerated here.

(1) Storage, handling and physical change of materials, such as transfer of flammable liquids or their mixing in open containers, etc.
(2) Exothermic reactions, where there is a strong possibility of the reaction getting out of control.
(3) Low pressure processes operating at sub-atmospheric conditions so that air or contaminants leak into the system.
(4) High pressure processes, which may involve pressures of 200 kg/cm^2 or above. A hazard results from the potentially large expansion of fluid released to the atmosphere from elevated pressures.
(5) High temperature processes, which may involve temperatures above flash point, boiling point or auto-ignition.
(6) Dust or mist explosion hazard may arise in processes which involve handling of materials that could create dust or mist.
(7) Sparking of electric motors, switches and alarms.

(8) Electrostatic charges.

(9) Hazards arising out of equipment and/or instrument failure.

In many cases the equipment or instrument fails to function due to instrument air failure, thermocouple burn out, loss of electric-power, steam or cooling water failure, possible reversal of flow, plugging of lines or equipment, loss or pressure, etc.

17.4 SAFETY MEASURES

The very nature of chemicals handled poses process hazards in the form of fire and toxicity of operators. The Material Safety Data Sheet (MSDS) of all chemicals used in the plant including raw materials, intermediates, products by-products, catalyst, etc., deals with the toxicity. The word hazard is associated with flammability or combustibility. Flammable liquids burn more easily. On the other hand, combustible liquids do not burn that easily. According to the National Fire Protection Association (NFPA) of Washington, USA, the flammable liquids are the ones having a flash point below 100°F (approx. 40°C) and combustible liquids have a flash point above 100°F (Schmidt 2002).

Let us try to understand the commonly used terms associated with fire hazard.

Flash Point

It is the temperature at which enough vapour pressure exists to burn the chemical in air.

For fire or flame to sustain we need fuel (chemical vapours), oxidizer (oxygen from air) and ignition source (flame, static electricity, spark), to start the fire.

Ignition Temperature

The temperature at which vapour will burn even without outside ignition source leading to uncontrolled fire and explosion. The flash point is an indication of potential danger.

Flammability

When the concentration of the chemical in air is very low, the vapour will burn, but the adiabatic temperature rise is not enough to either propagate or sustain the flame. This is the lower flammability limit. On the other hand, when concentration of vapour in air is large, the vapour will exclude O_2 as a result combustion does not produce enough heat to sustain the flame. This is the upper limit of flammability. In between these two limits, the flame sustains. The range is determined experimentally by open or closed cup method. For example, the lower limit of flammability for methane is 5% (v/v) and the upper limit is 15%. Butanes are a major component of domestic LPG. Their lower limit is 2% and the upper limit is 8.5% (v/v). Narrower the band between two flammability limits; safer is the chemical to handle. Flammability limits for various chemicals can be obtained from various sources (Anil Kumar 1981). From the safety point of view, it is always desirable to work outside the flammability limits (upper and lower limits). In other words, if the vapour air mixture is having a composition within flammability limits, there is every chance of a fire (explosion) due to a flame, spark or static electricity.

Explosion

In the case of explosion there is a sudden release of energy. There are two types of explosions namely: Deflagration, where there is slow burning and Detonation, where combustion is very very rapid and the flame velocity is more than that of the velocity of sound. Please note that localized flame is not at all dangerous. It is the speed of propagation that causes damage. The detonation flame causes much more damage.

Based on Petroleum Rules of 1976 (Rego 1991) liquids can be classified as per their flash point as:

Class A: Liquids with Flash Point <23°C

Class B: Liquids with Flash Point (FP) as 23 < FP < 65°C

Class C: Liquids with Flash Point (FP) as 65 < FP < 93°C

The regulations for handling, storage and transportation of chemicals as per the Factory Act and Petroleum Rules are out of scope of this book.

In every process industry a wide variety of preventive and protective safety measures must be provided. Such measures can be of two types: (1) measures to minimize or avoid mishaps, (2) measures which will limit the extent of damage, if the situation gets out of control. Some of the measures can be considered as basic preventive and protective features. Others can be recommended as the minimum features, while some others as specific preventive features appropriate to each individual situation or equipment.

Prior to use, the properties of each material in a proposed process should be examined to determine if it is unstable at normal temperature, or if it can explode from mechanical shock, from exposure to high temperature while confined or ignited. The flammability of each chemical should also be determined. Similarly, precautions must be taken while handling toxic chemicals. Radioactive materials should not be used unless appropriate storage, decontamination, testing and disposal facilities are available.

Some of the basic preventive and protective measures are: adequate supply of water for fire protection, over pressure relief devices, segregation of reactive materials in process lines and equipment, proper and adequate grounding of electric equipment, safe location of auxiliary electric gear, drainage from spills, fire fighting water from hose nozzles or sprinkler heads, insulation of hot surfaces that heat to within 80% of auto-ignition temperature, proper building and equipment layout, limitation on glass devices in flammable or hazardous service. Certain protective features which are recommended as minimum, where hazards are limited, are: protection of supports for equipment with fire-resistant materials, water spray to equipment and structure, special 'fail-safe' or remote operated valves and instruments, quick and effective removal of blowdown or spill material, instruments and techniques to ensure that explosive mixtures do not form inside process equipment, monitoring systems which actuate protective systems such as sprinkler systems or ventilation fans, building ventilation, segregating hazardous operations and installation of barrier walls.

Specific preventive measures can be any of the above measures suggested or additional measures to suit the material or process. For instance, in case of high temperature process it is essential to provide: (1) instrumentation and/or special devices to minimise flow of flammables, (2) combustible gas monitor with the alarms, (3) combustible gas monitors that turn on the deluge system or safety shut down equipment, (4) special vent and dump system.

Fire Fighting Agents

The fire fighting agents or chemicals used, work on one of the following principles or combination of the same:

(a) reduction in oxygen availability for combustion.

(b) reduction in fuel (combustible material) supply either by totally stopping the supply or dilution by liquid or blanketing of liquid.

(c) cooling the flammable material to reduce the temperature below the combustion temperature.

Water is by far the most commonly used fire-fighting agent. Liquid water takes away heat, thereby, lowering vapour pressure of the liquid and the resultant steam formed reduces oxygen availability. However, water reacts with some chemicals such as thionyl chloride resulting in toxic harmful products. Therefore, it cannot be used in the area where there is storage of such chemicals. Water is not suitable for extinguishing electrical fire because there exists a possibility of electrolysis of water resulting in formation of more explosive H_2 and O_2. There also exists a possibility of the floor becoming conductive endangering people working around.

Foams are water based and are used for extinguishing fires of light hydrocarbons. The foam acts as blanketing agent preventing oxygen supply for combustion. The water after dripping down cools the liquid under fire as well.

The use of *vaporizing liquid* to cool the fire and blanketing the oxygen from air is discouraged nowadays because of toxicity associated with such vapours and their ozone depletion potential. For example, chloro, chloro- bromo and chloro – bromo – fluro chemicals (carbon tetrachloride, chloro bromo methane, bromo chloro difluro methane respectively) were used earlier.

Dry powders anti fire agents include sand, sodium bicarbonate and potassium bicarbonate. Though they are effective in masking oxygen supply, cannot cool liquid effectively. Therefore, after dry powder application foam is used to extinguish fire.

The *carbon dioxide gas* (CO_2) is by far the best for fire extinguisher for costly electronic devices, analytical equipments, libraries, archives, museums and electrical appliances.

For the detailed treatise on this subject, refer to book by Lees (1980).

17.5 SAFETY MEASURES IN EQUIPMENT DESIGN

So far, a large number of safety measures for preventing or controlling hazards are indicated. Some of these are relevant in equipment design problems. Items or equipment considered in this book are either pressure vessels such as reactors, heat exchangers, etc., or those involving rotary motion such as filters, agitators, etc. Some of the important safety measures which need consideration during the design stages of these items of equipment are discussed in some detail.

17.5.1 Material of Construction

Materials available for construction of equipment, their properties, problems of corrosion and protective coatings have been considered in Chapter 2. In some cases, process conditions vary a good deal, which makes the choice of the material difficult. Too often, the choice of material is based on similar or previous experience, but even small differences in the constituents of the fluid streams may make considerable difference to the material to be used. Many vessel failures, not all catastrophic, of course, are attributable to a bad choice in the material. A critical analysis of the processing conditions and material handled must be made to ensure safe operation.

17.5.2 Precautions in Design and Construction

A wide variety of precautions is necessary to ensure the safe working of an equipment and prevent failures which may result in hazards and accidents. These will vary according to the type of equipment. A well designed piece of equipment will have safety and loss prevention features built in. The following design considerations are important to prevent failures:

1. Unit reliability
2. Ease of operation
3. Unit flexibility
4. Provisions for future expansion, inspection and maintenance
5. Adequate emergency shutdown facilities
6. Equipment standardization for rapid replacement
7. Design to anticipated pressure range – with overpressure controls
8. Design to anticipated temperature range – with over temperature controls.

It is not possible to give a list of precautions for each individual unit of equipment. Some of them are indicated below.

17.5.2.1 Pressure Vessels

(1) The design and construction of pressure vessels and storage tanks should follow Indian Standards Codes and vessels should be tested at 1.3 times the design pressure.

(2) Attempt should be made to keep the vessel as simple as possible, consistent with the fulfilment of its function. It should not be overburdened with manways and inspection openings for which there is only a remote chance of usage.

(3) Careful attention should be given to any welds made on a vessel, particularly if they are thick.

(4) If the vessel is to work under pressure cycling, system changes, vibrations or similar factors which are likely to create fatigue conditions, it is essential to check the fatigue strength.

(5) Careful attention should be given to design criteria indicated under Section 5.7.2 of Chapter 5 on Pressure Vessels.

(6) Flange joints must be leak proof. This can be ensured by checking the gasket position, flange faces and bolt holes.

(7) Pressure relief devices should be provided on all pressure vessels.

17.5.2.2 Heat Transfer Equipment

Equipment such as heat exchangers, vaporisers, reaction kettles, furnaces, require some method of heating, which may be directly fired with the help of fuel, electric heating, heat transfer media such as steam, or organic fluids. For such units, special precautions should be taken, which would prevent overheating, fire and/or explosion. The following precautions in the design and construction of such equipment will help in a safe operation.

(1) Sufficient heating surface must be provided to avoid excessive rate of heat input per unit surface, so as to minimize excessive liquid film temperatures.

(2) The heat absorbed by the tubes must be removed by the circulating fluids and the liquid film coefficient inside the tubes must be sufficiently high to prevent an excessive temperature rise through the liquid film.

(3) Inspection openings must be sufficient in number and location to permit complete periodic inspection of the vaporiser.

(4) Vent valves should be provided at all high spots in the equipment.

(5) Liquid phase systems must include an expansion or surge tank.

(6) Heater tubes must be tightly secured to headers and vapour drums.

(7) In case of furnaces, combustion space should be designed to avoid flame impingement on tubes.

(8) Pressure relief devices must be designed for high temperatures.

(9) Care should be taken to allow for stresses due to thermal expansion.

(10) Appropriate choice of insulation must be made consistent with the material handled by the equipment.

17.5.2.3 Equipment Involving Rotary Motion

This type of equipment involves a mechanical drive system which should be protected by guards. Bearings should be well lubricated and cooled, if necessary, to reduce temperature.

17.5.2.4 Equipment Involving Electrical Energy

All electric installations require safety features which are controlled by Electricity Regulations. IS-5571 (1970) is a guide for selection of electrical equipment for hazardous areas. Like Petroleum Rules, we do have a classification of hazardous areas for electrical work also. The rationale behind this is simple as outlined below:

Zone 0: There is always presence of explosive vapour/gas and air mixture. In other words, there always exists a vapour within the flammability limit.

Zone 1: In normal operation explosive mixture is most likely to occur.

Zone 2: Explosive mixture will not occur in normal operation but will occur in abnormal situation and that also for a short time.

Non-hazardous: materials are outside the flammability limit.

Dust being a potential source of explosion, it should be properly considered.

IS-5572 (1970) gives classification of hazardous areas for electrical installations.

(1) The entry of vapours or gases into unpressured electrical equipment cannot be prevented even by the tightest of gaskets. To avoid an explosion from spreading the enclosure of the equipment should be strong enough to contain explosion, and its joints must be long and clearances small so that flame would not propagate to the outside. Such equipment is said to be explosion proof.

(2) Purging or pressuring enclosures with clean air or inert gas will help to prevent the entrance of hazardous dust.

(3) Static electrification can be prevented by earthing, static eliminators and effective surface treatment of equipment.

17.6 PRESSURE RELIEF DEVICES

One of the most important safety devices used in process equipment, is for the prevention of failure from over pressure. The more common causes of over pressure are external fire, closed outlets, liquid expansion, failure of reflux, cooling water, electric power and heat exchanger tubes. IS-2825 gives the conditions under which relief devices are to be used, their types, capacities and installation requirements.

The available relief devices generally fall into six broad categories:

(a) ***Safety valve:*** Actuated by static pressure upstream of the valve, this device is characterized by the rapid full opening or pop action. It is used for steam, gas of vapour service.

(b) ***Relief valve:*** Unlike the above, this device opens in proportion to the increase in pressure over the opening pressure and is used primarily for liquid service.

(c) ***Safety relief valve:*** This is suitable for use as either a safety or relief valve depending upon the application.

(d) ***Rupture disc:*** This device normally consists of a metal disc and a vacuum support when required, which is held between two special flanges.

(e) ***Relief valve rupture disc combination:*** Here the disc is either used in series with a relief valve or is used as a secondary relief device in parallel with the relief valve.

(f) ***Pressure-vacuum combination relief valve:*** This device is generally used on atmospheric-type storage vessels. It is frequently designed to relieve both pressure and vacuum.

17.6.1 Relief Valve

A relief valve should be of sufficient capacity to discharge the maximum quantity of the fluid contained in the vessel without permitting a rise in the pressure vessel of more than 10% above the set pressure, when it is discharging. Equations to determine vapour relief and liquid relief are given in IS-2825 (1968).

No matter how carefully the valve has been sized and selected, poor practice in handling and installation can result in a completely unsatisfactory performance. Fig. 17.1 shows a valve mounted on a pressure vessel without a shut-off valve, while Figure 17.2 shows the arrangement with shut-off valve. The pressure drop in the nozzle and valve together should not be more than 3% of set pressure. The long-radius elbow should be supported. In all the above cases, no incoming processing piping connection should be made between the vessel and the valve.

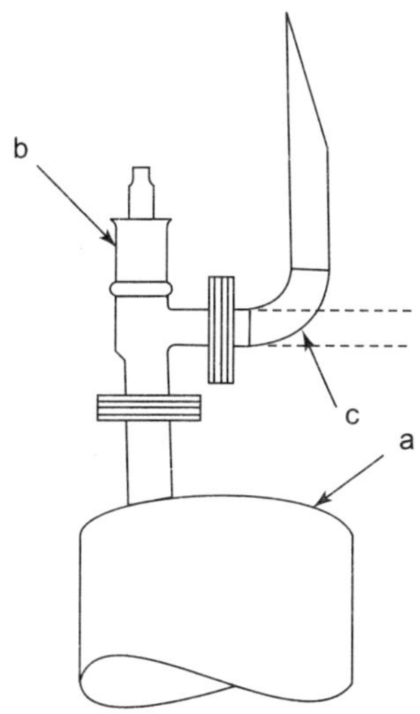

Fig. 17.1 Relief Valve Mounted on Pressure Vessel: (a) Vessel, (b) Relief Valve, (c) Elbow

Figure 17.3 shows alternate relief valve positions on a fractionating column. Location of the valve can influence the size of the relief valve as well as influence the pressure available for the relief system design. Location 'A' is usually preferred when the relief valve is tied up in a closed relief system because of maintenance convenience, a shorter length of line to the relief header, and cooler relieving temperature when air fan cooler is used. Pressure drop through the condenser must be calculated as also the relieving condition, to assure proper setting of relief valve. Location 'B' is preferred when the column is to be relieved to the atmosphere. This practice is preferred when relieving low molecular weight gases, or when dispersion is such that the vapour is out of the flammability range before it reaches an elevation low enough to be dangerous. Location 'C' is used on low pressure columns designed for the same top and bottom pressure. This is to account for the

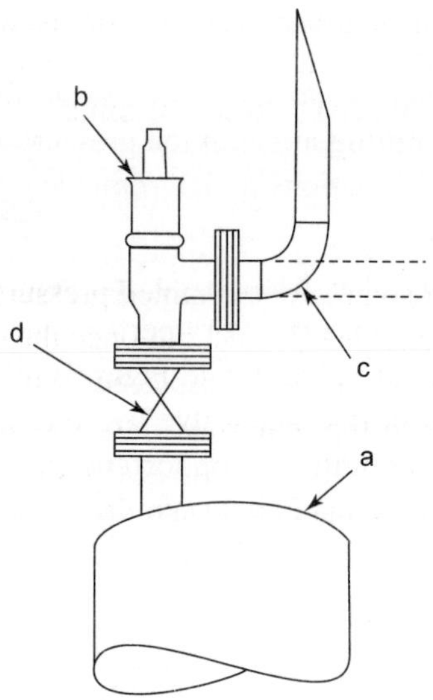

Fig. 17.2 Relief Valve with Shut-off Valve: (a) Vessel, (b) Relief Valve, (c) Elbow, (d) Shut-off Valve

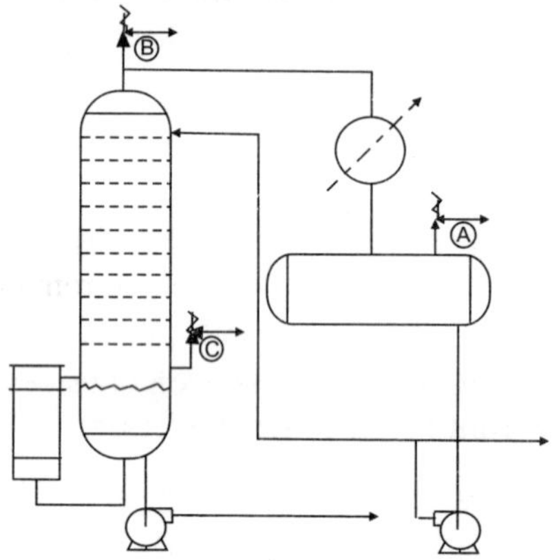

Fig. 17.3 Relief Valve Positions on Fractionating Column

pressure drop across the column. All relief valve installations should be built without pockets to avoid trapping liquids.

From a safety standpoint, any relief valve which may discharge flammable liquid should be connected to a closed system. Flammable vapour may be discharged at any point which will not endanger personnel or equipment; however, it should not condense in appreciable quantities. Regardless of whether the relief valve discharges to the atmosphere or to a closed system, safe plant design requires a means of rapidly depressurising the plant and in most instances safe disposal of the contents of the vessels and equipment either to storage or elsewhere under emergency conditions. If there is any probability of liquid discharge into the flare system, a knock-out drum should be provided adjacent to each process unit or a group of units.

17.6.2 Rupture Disc

This device is designed to burst at some predetermined pressure in a chemical equipment. This may be used as a primary or the sole relieving device, but once the disc relieves pressure, the equipment is open to atmospheric pressure until the disc is replaced. With highly toxic, poisonous or corrosive materials, it is necessary to release the materials into a vent surge or flare header system. The advantage of the rupture disc is that it will not allow leakage under normal conditions. Under normal conditions the rupture disc is sealed tight and protects the relief valve. If the maximum allowable pressure is exceeded, the disc will break and the relief valve will start relieving the pressure. As the pressure drops the valve will shut and re-close the process.

Types of Rupture Discs

Conventional or Standard Rupture Disc

This is a metallic disc fabricated out of special metal. The operating pressure is roughly less than 75% of the set pressure (burst pressure). When the disc ruptures, fragments of metal fly out with vapours and, therefore, it is never installed upstream of relief valve.

Reverse Buckling Rupture Disc

This is a more rugged rupture disc and costlier also. The convex side faces process stream. When the pressure increases and approaches the set pressure, the pressure reverses the disc and then either it is ejected out or the segments cut by a knife blade remain attached to the rim. There is no fragmentation. It is a good engineering practice not to have the same on the upstream side of the pressure relief valve unless it is warranted by corrosion considerations. The working pressure can be as high as 90% of the set pressure.

The disc is made of alloy steel and for corrosion protection even gold plated rupture discs are also available. Flat discs of graphite can also be used for corrosive service.

It is a good engineering practice to have the rupture disc from a reputed vendor, having a proven track record.

Quick Sizing of Rupture Disc

For the discharge of release vapours to the atmosphere, one can use the following expression to quickly size the rupture disc to get a rough estimate of the size. This is based on the method outlined by Saxena (2004).

Following assumptions are made

- discharge coefficient 0.62,

- back pressure is one atmosphere as the discharge is to atmosphere,

We then have diameter of the disc: D in mm as:

$$D = 0.72\sqrt{W/P}\,(T/\beta M)^{0.25}; \quad \text{mm} \tag{17.1}$$

Here, W — flow rate of released vapour, kg/hr

P — pressure of bursting, bar absolute

T — temperature, K

M — molecular weight of vapour.

And

$$\beta = k\left[\frac{2}{k+1}\right]^{\frac{k+1}{k-1}} \tag{17.2}$$

k = c_p/c_v, ratio of specific heats

Please note that, W vapour flow rate can be computed by knowing rate of heat release and latent heat of vapour.

$W = \Delta H_R/\lambda$; where, $\Delta H_R = kJ/hr$ = kJ/kg and λ

In case of an external fire, W is the rate of heat absorbed by the tank. If the containment vessel is provided which is the case, normally one must consider backpressure. Ludwig (1977) has extensively covered the subject of sizing release devices namely, relief valve, and rupture disc and vent valve. After having the disc size the nozzle size is then considered as next higher standard pipe size. For example if the disc size is 45 mm then nozzle size would be 50 mm.

NB: A typical rupture disc arrangement is shown in Fig. 17.4.

17.6.3 Vents and Flame Arresters

Vents are provided on storage tanks, heat exchangers vessel, jackets, etc. The sizes of breather vents used for tanks are given in Table 7.4. Emergency venting is required for cases of excessive internal pressure within the tank from the liquid boiling due to fire exposure. In some cases, the number and size of the venting devices per tank will be governed by pressure requirements, whereas in other cases this will be decided by vacuum requirements. If the tank contains a volatile liquid, a breather valve, which is suitable for both pressure and vacuum operation, set to open at a specified setting, is fitted to vent opening to minimize evaporation losses to the atmosphere.

Vent valves are not regarded as providing the required protection against fire and explosion to tank contents; consequently, flame arresters should be employed in conjunction with a breather valve where necessary. A flame arrester is placed at or near the outlet end of a vent pipe to protect the tank contents by preventing the propagation of flame. In many cases, normal venting requirements may be inadequate to handle the abnormal vapour efflux generated by heat from an exposure to fire.

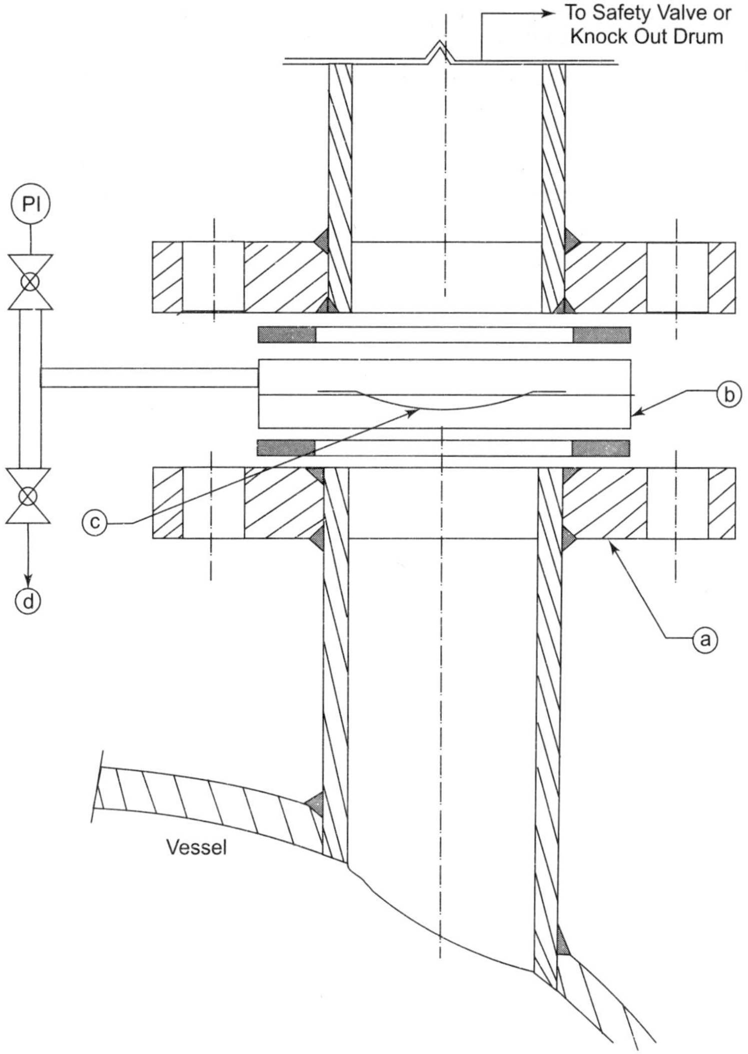

Fig. 17.4 Rupture Disc: Schematic Arrangement: (a) Vessel Nozzle, (b) Rupture Disc Holder, (c) Reverse Buckling Type Rupture Disc, (d) Drain

Consequently, some form of safety device that will relieve excessive internal pressure is necessary. These may consist of additional vents, self-closing gauge hatches, lifting type manhole covers, etc.

Emergency vents are usually set at a slightly higher setting than the pressure settings on breather vents, but not higher than the pressure in the tank can safely withstand.

References and Recommended Readings

Anil Kumar, 'Chemical Process Synthesis and Engineering Design' Tata McGraw-Hill Publishing Company Ltd. (1981).

Albraht, A.R., and Seifert, W.E., 'Accident Prevention in High Temperature Heat Transfer Fluid Systems', *Loss Prevention*, 4 (67) CEP Technical Manual (1969).

Bartknecht, W., 'Ignition capabilities of Hot Surfaces and Mechanically Generated Sparks in Flammable Gas and Dust/Air Mixtures', *Plant/Operations Progress*, **7** (2) 114 (1988).

Burklin C.R., 'Safety Standards – Codes and Practices for Plant Design, Part I – Design Consideration, *Chemical Engineering*', **79** (22), 56 (1972); Part II 'Testing Standards and Insurance Association', *Chemical Engineering*, **79** (23) 113 (1972).

Ebadat, V., 'Electrostatic Hazards in the CPI', *Chemical Engineering*, **103** (7) 141 (1996).

———, 'Reduce Electrostatic Hazards', *Chemical Engineering*, **104** (7) 86 (1997).

Englund, S.M., 'Design and Operate Plants for Inherent Safety – Part 1' *Chemical Engineering Progress*, **87** (3) 85 (1991).

Fire and Explosion Safety and Loss Prevention Guide; *Hazard Classification and Protection*, CEP Technical Manual American Institute of Chemical Engineers (1973).

First, K.E., and Huff, J.E., 'Design Charts for Two-Phase Flashing Flow in Emergency Pressure Relief Systems', *Plant/Operations Progress*, **8** (1) 40 (1989).

Grossel, S.S., 'Safe, Efficient Handling of Acids', *Chemical Engineering*, **105** (7) 88 (1998).

Jenett, E., 'Components of Pressure Relieving Systems', *Chemical Engineering Part I*, **70** (14) 125 (1963) and Part II, 70 (17) 151 (1963).

Keyser, D.S., 'Rupture Disc Selection', Loss Prevention, **6** (82), *CEP Technical Manual* (1972).

Kletz, T., *Learning from Accidents in Industry*, Butterworth, Heinemann Oxford (1994).

Kletz, T., *What Went Wrong*?, Gulf Publishing Company, Houston (1994).

Le Vine, R.Y., 'Electrical Safety in Process Plants', *Chemical Engineering*, **79** (9) 51 (1972) and **79** (9) 59 (1972).

Lees, F.P., *Loss Prevention in the Process Industries*, Vols. 1 & 2 Butterworth and Co., London (1980).

Louvar, J.F., and Crowl, D.A., 'Safety First', Chemtech, **20** (12) 752 (1990).

Ludwig, E.E., *Applied Process Design for Chemical and Petrochemical Plant*, Vol. 1 Second Edition, Gulf Publishing Company, Houston (1977).

McFarland, I., 'Safety in Pressure Vessel Design', *Ammonia Plant Safety*, 12 (43), CEP Technical Manual (1970).

Mead, W.J., *The Encyclopedia of Chemical Process Equipment*, Reinhold, New York (1964).

Mendoza, V.A., Smolensky, V.G. and Straitz III, J.F., 'Do Your Flame Arrestors Provide Adequate Protection?', *Hydrocarbon Processing*, **77** (10) 63 (1998).

Peterson, P. and Culter, H.R., 'Explosion Protection for Centrifuges', *Loss Prevention*, 7 (88) CEP Technical Manual (1973).

Proceedings of the Second Symposium on Chemical Process Hazards with Special Reference to Plant Design, The Institution of Chemical Engineers Symposium Series, 15 (1963).

Rearick, J.S., 'How to Design Pressure Relief System, Part I – Devices Codes and Causes of Overpressure', *Hydrocarbon Processing*, **48** (8) 104 (1969); 'Part II – Sizing Relief System and Disposal of Vented Material', *Hydrocarbon Processing*, **48** (9) 161 (1969).

Rego, A.J., 'Regulations pertaining to storage, usage and transportation of hazardous chemicals', *Chemical Business*, 559 (1991).

Reynolds, J.W., 'Checking up on Rupture Disk Holders', *Chemical Engineering*, **108** (11) 149 (2001).

Richter, S.H., and Turner, F., 'Properly Program the Sizing of Batch Reactor Relief Systems', *Chemical Engineering Progress*, **92** (4) 46 (1996).

Russell, W.W., 'Safety in Flange Joints', *Loss Prevention*, 8 (100), CEP Technical Manual (1974).

Saxena, S.K., 'Rupture Disk – Why It is the Preferred Safety Device', *Chemical Industry Digest*, **XVII** (7) 65 (2004).

Schmidt, M., 'Don't Get Burned – Know limits of flammable and combustible Fluids', *Chemical Engineering*, **109** (12) 58 (2002).

Simon, H., and Thomson S.I., 'Relief System Optimization', *Loss Prevention*, 6 (74), CEP Technical Manual (1972).

Smith, L.D., 'Easily Test Thermal Stability and Desirability', *Chemical Engineering Progress*, **90** (9) 67 (1994).

Spiegelman, A., 'Risk Evaluation of Chemical Plants', *Loss Prevention*, 3 (I) CEP Technical Manual (1969).

Swift, I., 'Designing Explosion Vents Easily and Accurately', *Chemical Engineering*, **95** (5) 65 (1988).

Thielsch, H., *Defects and Failures in Pressure Vessels and Piping*, Reinhold, New York (1965).

Venugopal, D., Hazardous chemicals – storage, handling and transportation, *Chemical Business*, **5** 39 (1991).

Walker, J.J., 'Sizing Relief Areas for Distillation Columns', *Loss Prevention*, 6 (104), CEP Technical Manual (1969).

Wilkie, F., 'Flame Arrestors not Simply a Matter of Choice', *Process Engineering*, **68** (4) 53 (1987).

Wong, W.Y., 'Safer Relief – Valve Sizing', *Chemical Engineering*, **96** (5) 137 (1989).

Yelland, A.E.J., 'Design Considerations when Assessing Safety in Chemical Plant', *The Chemical Engineer*, **201**, CE 214 (1966).

18

Chapter

Process Flow Diagrams

18.1 DEFINITION/MEANING

A pictorial representation of all unit operations and processes taking place in a chemical plant is shown with the help of following basic diagrams:

- Process Block Diagram (PBD)
- Process Flow Diagram (PFD)
- Piping & Instrumentation Diagram (P&ID) and is also known as an Engineering Diagram. Like the main manufacturing unit all the above diagrams are drawn for all utilities also. In case of electrical power as utility a single line diagram (SLD) is drawn.

18.2 PROCESS BLOCK DIAGRAM (PBD)

1. A Simplistic Representation of the Main Steps in the Form of Rectangular Block.
2. The Main Inputs (Raw Material and Utilities) and Outputs (Products, by Products) are shown around the Block.
3. Generally, such Diagrams are accompanied by a Brief Process Description.
4. Utility: These diagrams are useful for Quick Understanding of the Process. Figure 18.1 shows a Typical Block Diagram, for the manufacture of monochlorobenzene from benzene via chlorination.

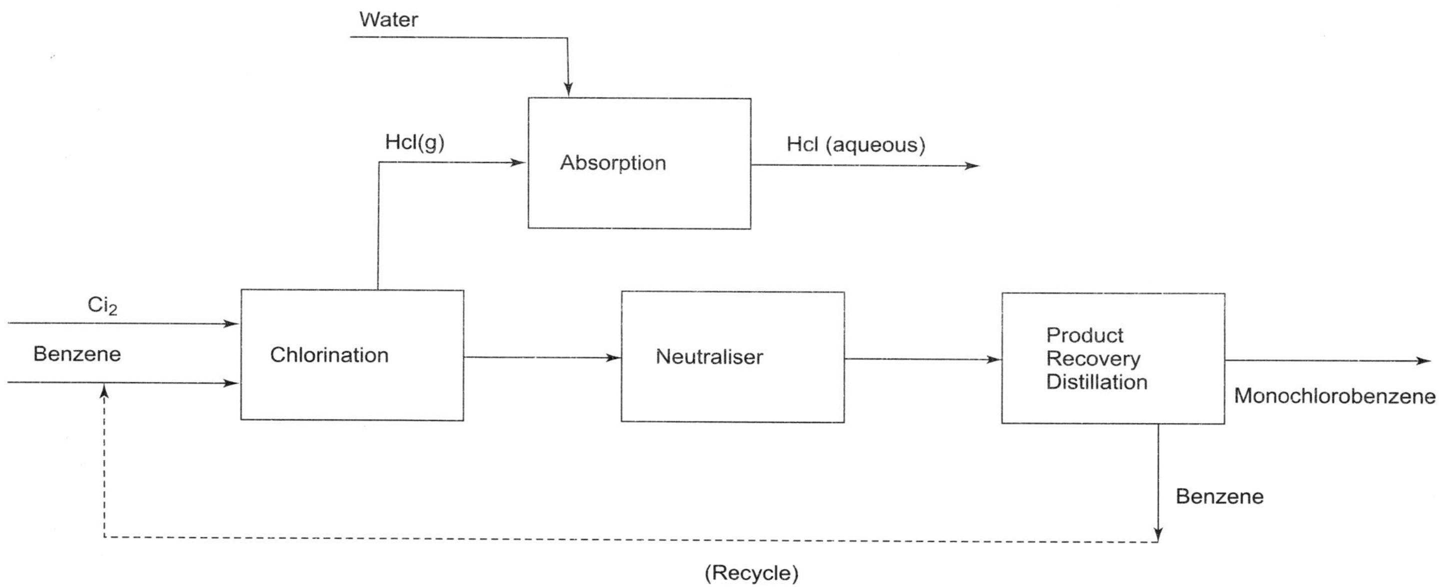

Fig. 18.1 Process Block Diagram

18.3 PROCESS FLOW DIAGRAM (PFD)

The entire process is represented sequentially by means of equipment symbols and lines indicating all equipment (inclusive of stand by) and flow directions.

18.3.1 The Main Objectives are

- to indicate pictorially the interrelationships among various unit operations and process units.
- to provide material balance at all major steps.
- to provide an energy balance in the form of specific heat, temperature and state of material (liquid, vapour or solid).

18.3.2 General Considerations

- It is desirable to show equipment in their relative positions and size, for instance, pumps, compressors are at ground level.
- Main process streams are always in bold, e.g., raw material to finished products.
- The streams are numbered within a diamond. Material and energy balance are represented in the same sheet.
- Equipments are numbered generally based on function or area or both for instance heat exchanger E 1001, reactor R 1001.
- PFDs are also shown drawn for utilities, effluent treatment process.
- While continuing PFD on other drawing sheet, one must indicate interconnection between various streams.

A typical PFD is exhibited in Fig. 18.2.

It is a good practice to exhibit below PFD flow rates and the state of material to present the material and energy balance. This is done by having diamonds on line having stream numbers within as indicated. Thus, we have below PFD:

Stream No.	1	2	3	....
Flow rate,	kg/hr			
Component				
A	a_1	a_2	a_3	and so on
B	b_1	b_2	b_3	
C	↓	↓	↓	
D				
↓				
Total	X	Y	Z	

Temperature, °C	30
Pressure, bar (abs)	1.01
State (L/V)	L

Some commonly used symbols in PFD are depicted in Annexure 18.1

18.4 PIPING AND INSTRUMENTATION DIAGRAM [P&ID]

18.4.1 Meaning and Definition

Mechanical aspects of each and every equipment are depicted in P&ID's to enable construction personnel to construct the plant and to enable maintenance personnel to maintain the same in order to sustain production as per the desired schedule. P&ID's are constructed with the help of PFDs.

P & ID exhibit all equipment along with the stand by equipment, all instruments, entire piping (piping includes pipes fittings, bends, valves, etc.).

18.4.2 Objectives are:

- to make pictorial presentation resembling a real plant as close as possible with all equipment, piping, instruments.
- to assist plant construction engineers.
- to assist commissioning engineers for mechanical commissioning.
- to identify all piping with or without insulation.

18.4.3 General Considerations

All valves fittings, instruments are presented with standard symbols. Lines are identified with line designation code. This includes, typically.

Line designation code: $X\ Y\ Z\ M\ Q$.

X: line diameter 25,100 mm.

Y: fluid: stream (s), boiler feed water (BFW); cooling water (CW); heat transfer fluid (HTF).

Z: area number in the plant, e.g. reaction zone 1, separation/product purification area 2, etc.

M: piping specification index.

Q: insulation type hot or cold.

- It is obvious that the line number would change due to any change in above parameters.
- Symbols, piping specification index, generally depend upon project engineering company

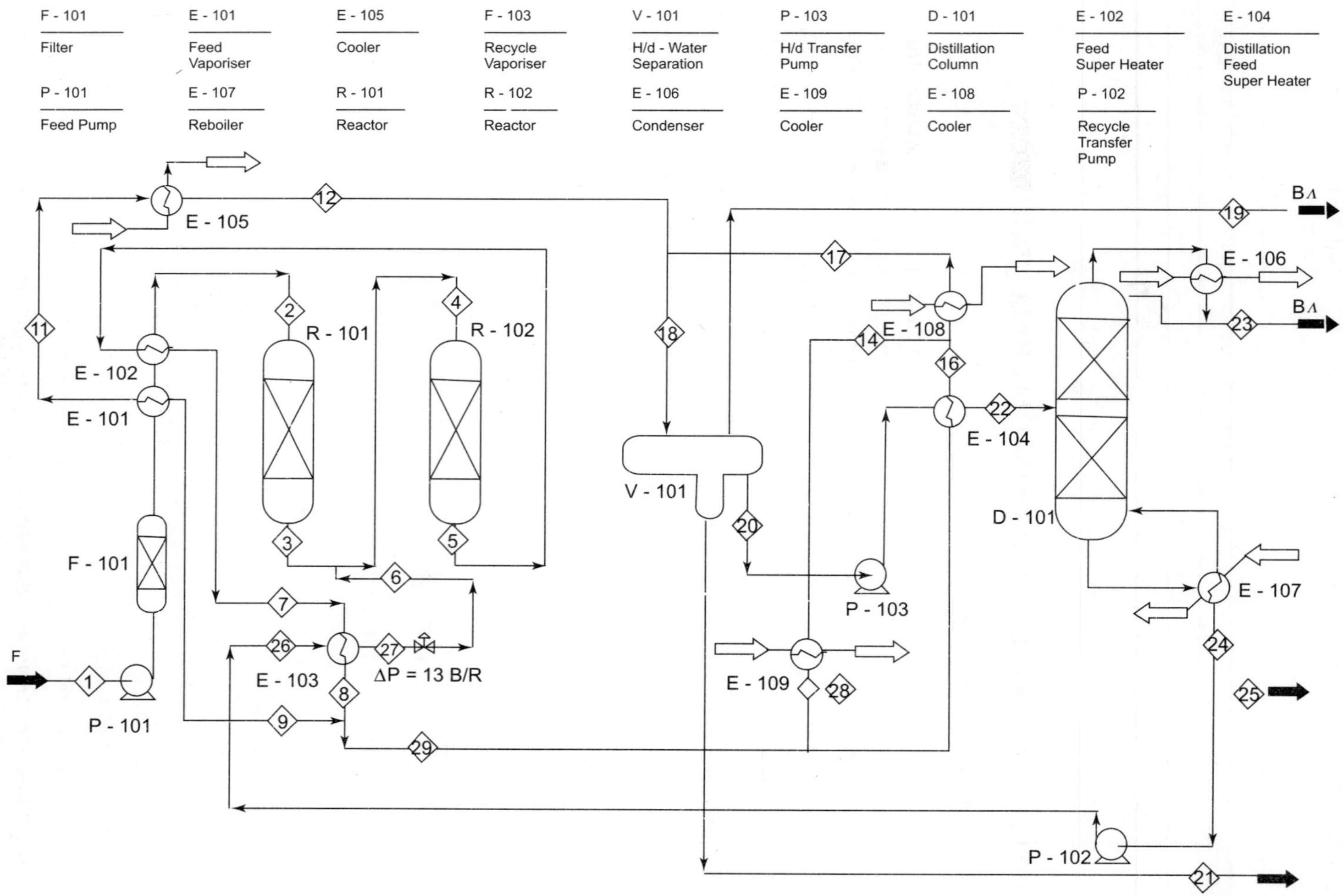

Fig. 18.2 Process Flow Diagram

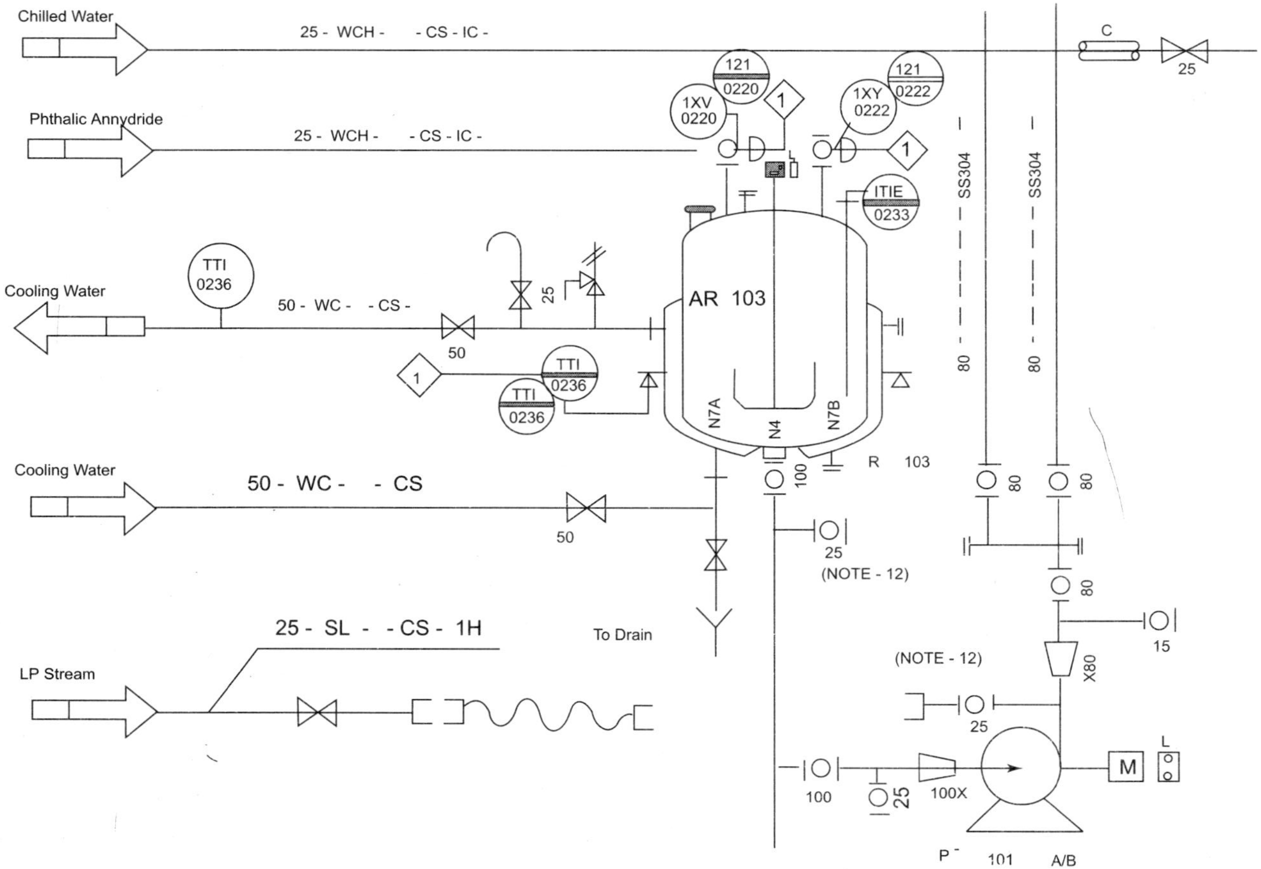

Fig. 18.3 Piping an Instrumentation Diagram

and hence all P & ID's need to be accompanied by legend sheet. A typical P & ID is shown in Fig. 18.3.

References and Recommended Readings

Austin, D.G., *Chemical Engineering Drawing Symbols*, George Godwin, London (1979).

Rase, R.E., and Barlow, M.H., *Project Engineering of Process Plant*, Wiley, New York (1957).

Walas, S.M., *Chemical Process Equipment Selection and Design*, Butterworth (1988).

Ludwig, E.E., *Applied Process Design for Chemical and Petrochemical Plants*, Vol. 1 Second Edition, Gulf Publishing Company, Houston (1977).

Sinnott, R.K., *Coulson Richardson's Chemical Engineering Series*, Chemical Engineering Design, Vol. 6, 4th Edition Butterworth Heinemann, Oxford (2005).

Annexure 18.1 Some Commonly used Symbols in PFD

Heat Exchanger	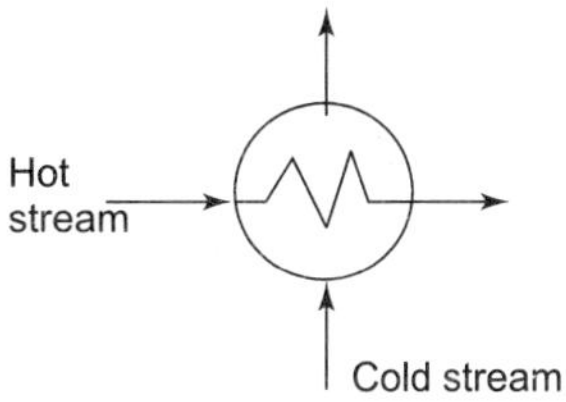
Heat Exchanger with cooling water	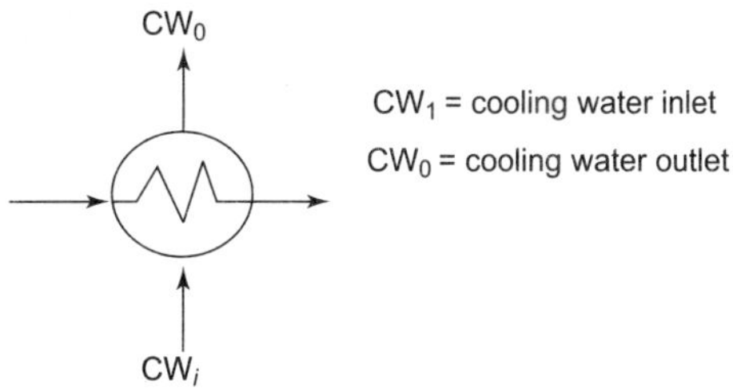
Heat Exchanger as reboiler	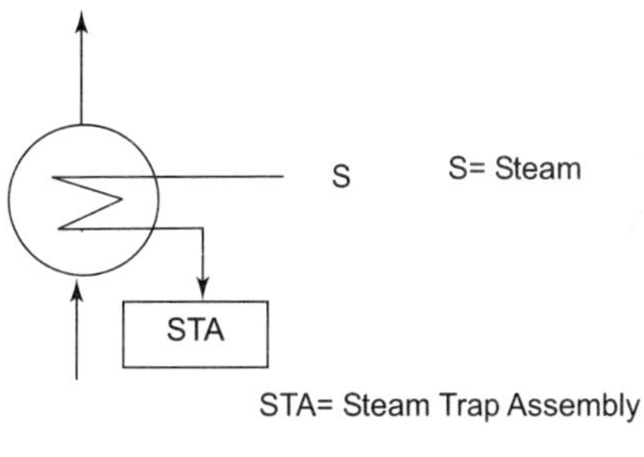
Fired heater/furnace	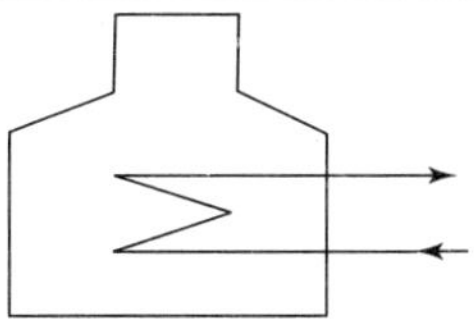
Kettle type reboiler	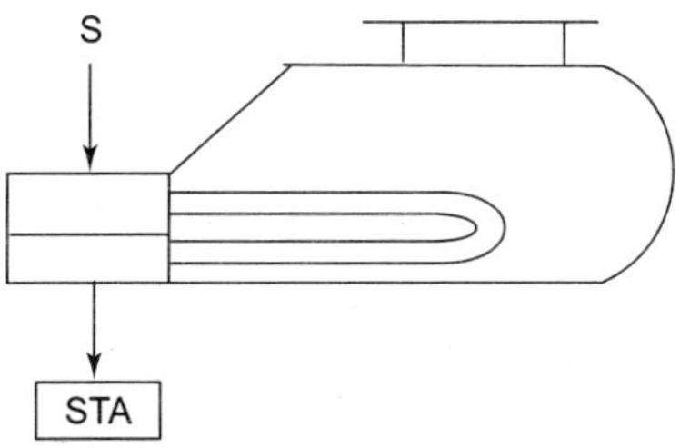

Packed column	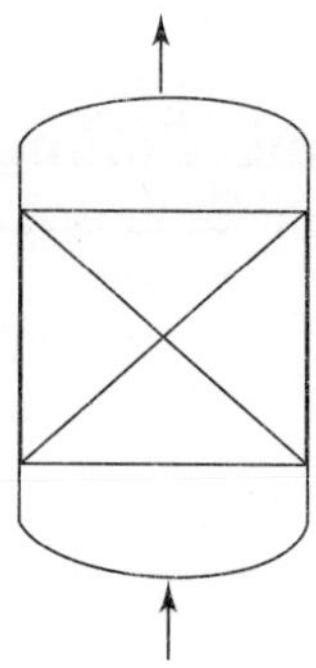
Packed column with feed in between	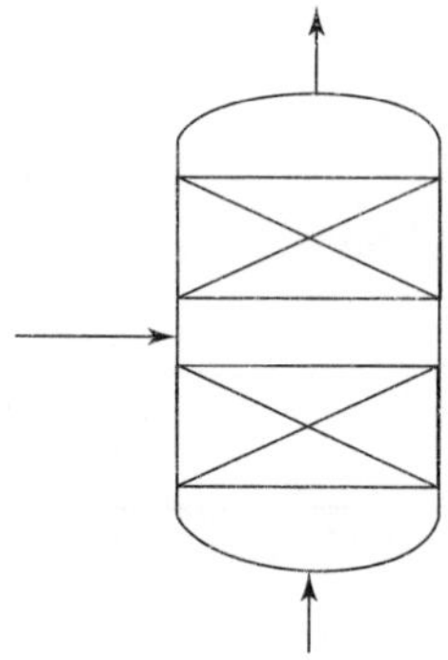
Plate column/Tray tower	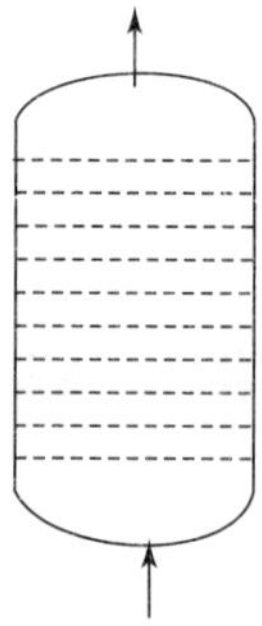
Packed bed reactor (adiabatic)	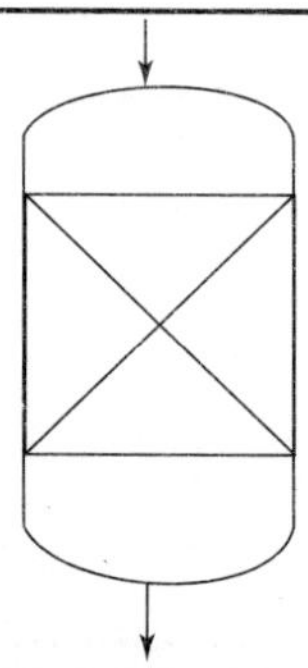

Packed bed Reactor with interstage cooler	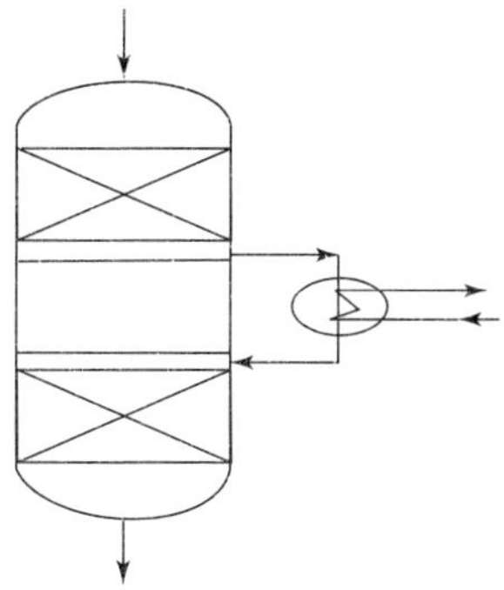
Isothermal tubular reactor	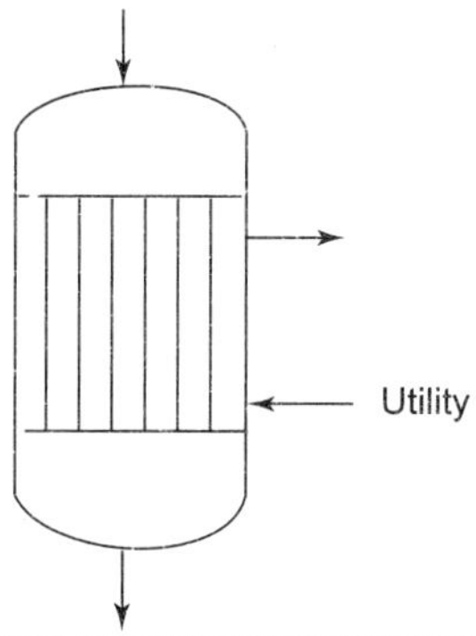
Pump	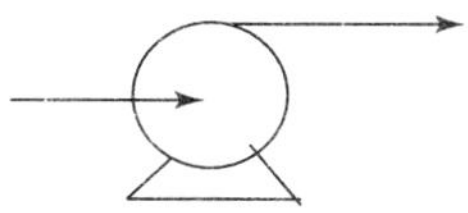
Compressor/Blower/Vacuum Pump	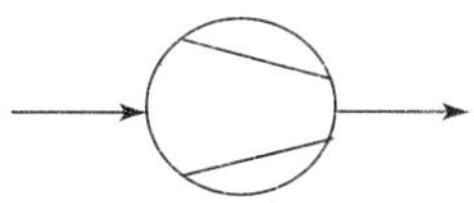
Expander	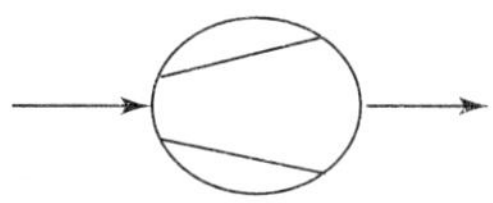
Ejector	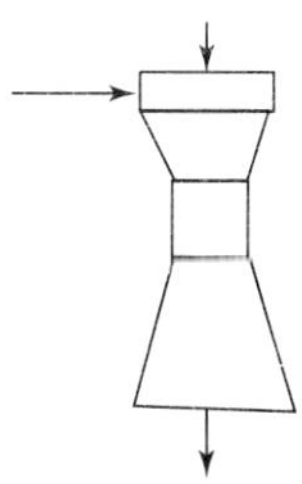

Turbine power generator	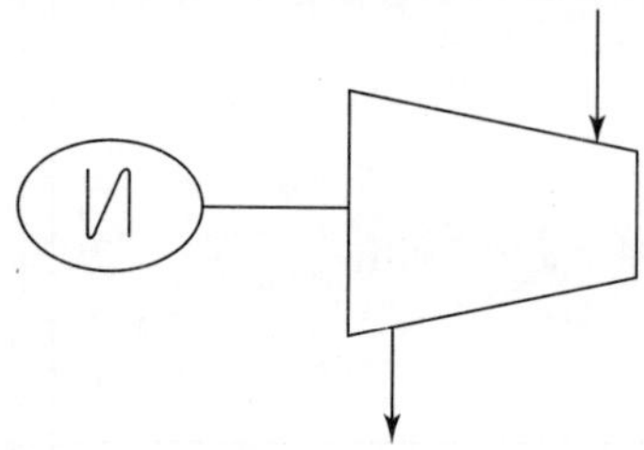
Storage day tank	
API Storage	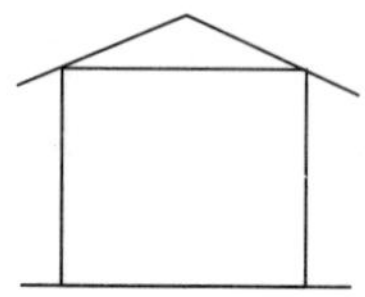
Mechanically agitated vessel/ dynamic mixer with heater/cooler	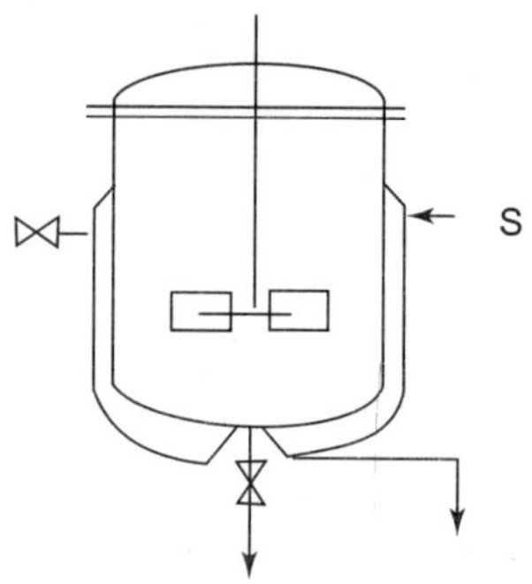
Filter	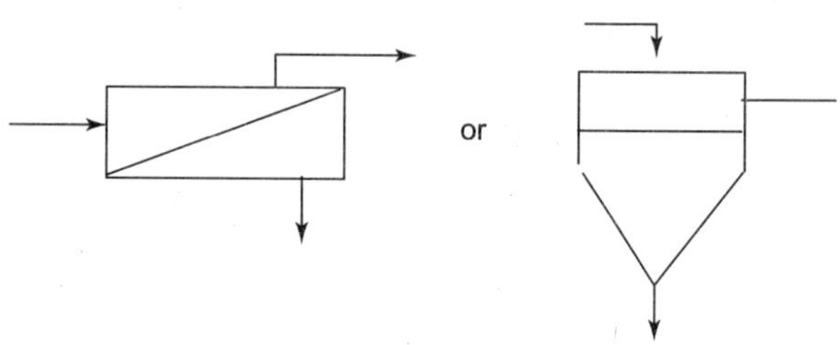

<table>
<tr><td>Centrifuge</td><td>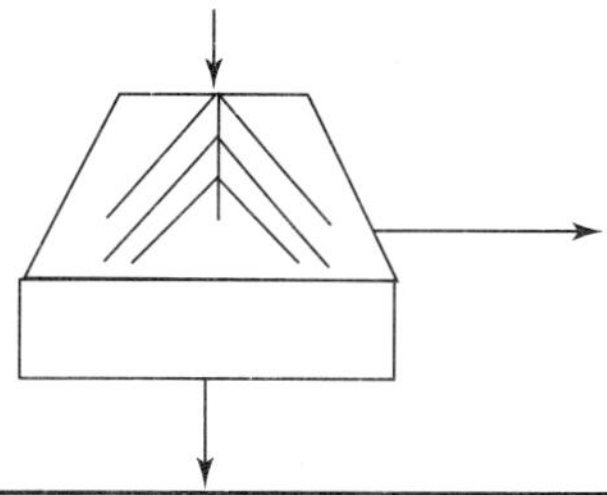</td></tr>
<tr><td>Drum</td><td>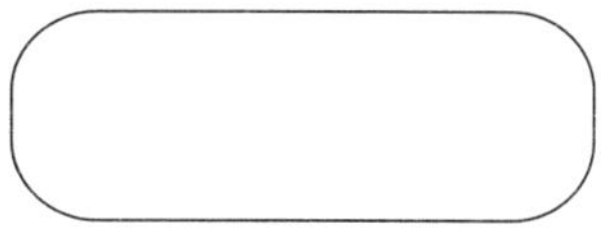</td></tr>
<tr><td>Drum with boot (separator)</td><td>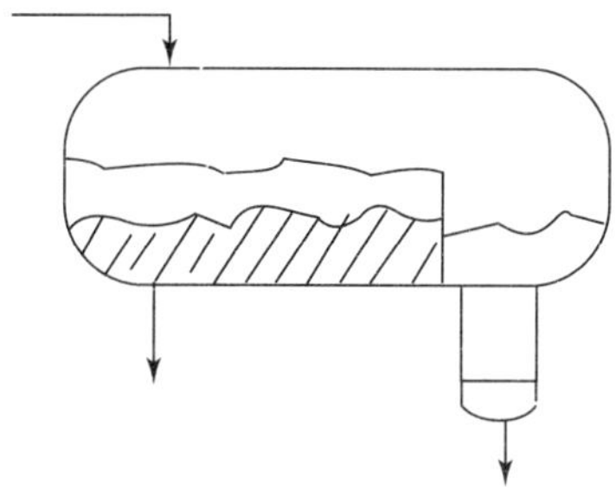</td></tr>
<tr><td>Static mixer</td><td>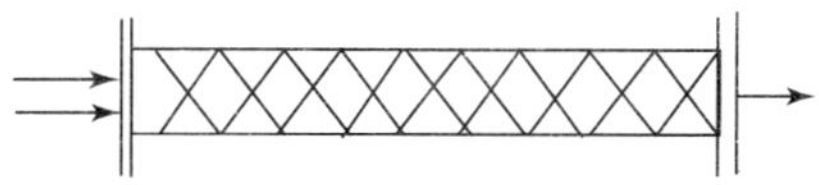</td></tr>
<tr><td>Screw conveyor</td><td>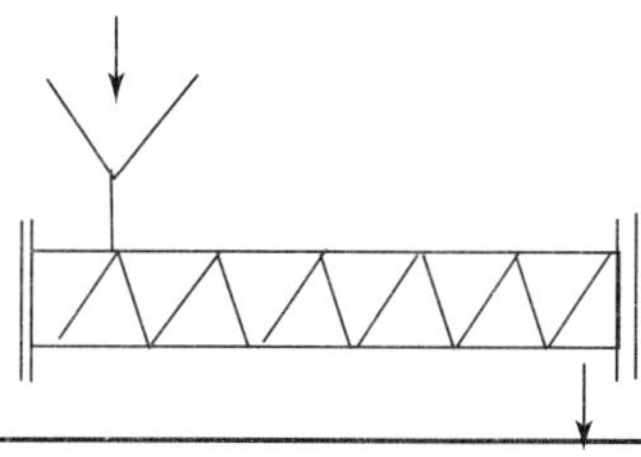</td></tr>
<tr><td>Cyclone</td><td>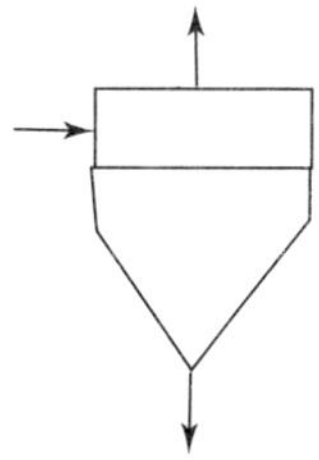</td></tr>
</table>

Annexure 18.2 Some Commonly used Piping Symbols in P&I Diagrams

Gate Valve	
Ball Valve	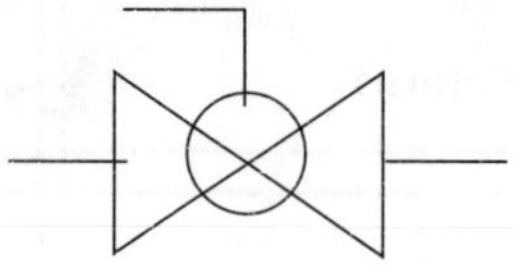
Globe Valve	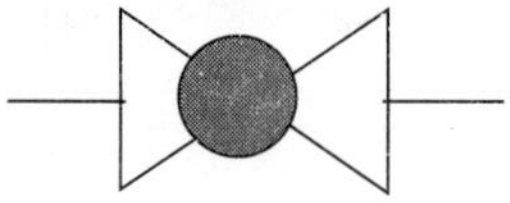
Needle Valve	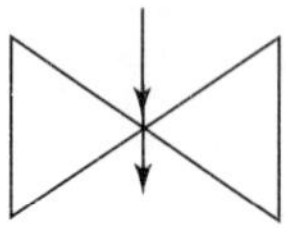
Three way Valve	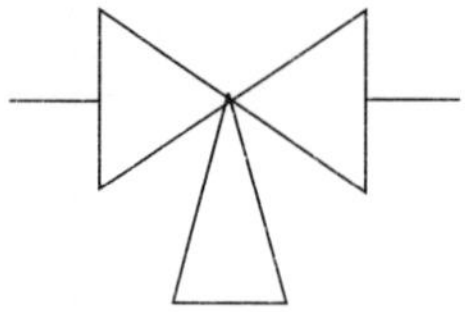
Plug Valve	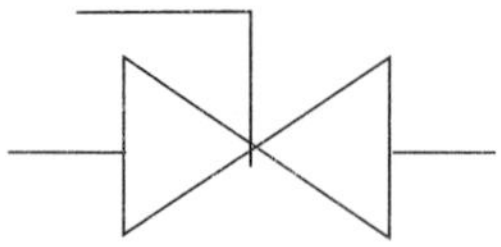
Steam Trap	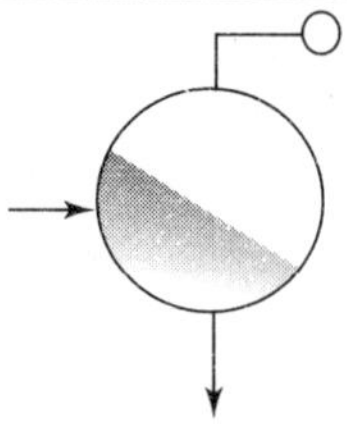
Strainer	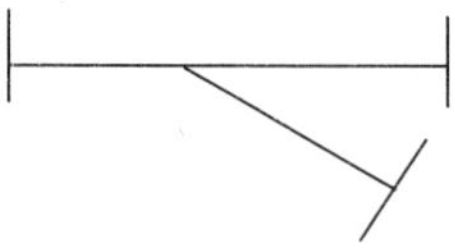

Sample Point	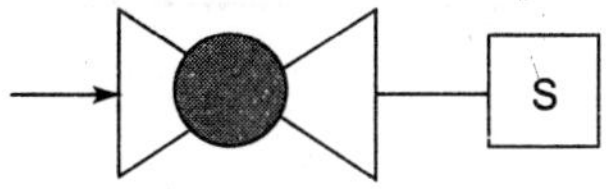
Check Valve (Non Return)	
Jacketed Line	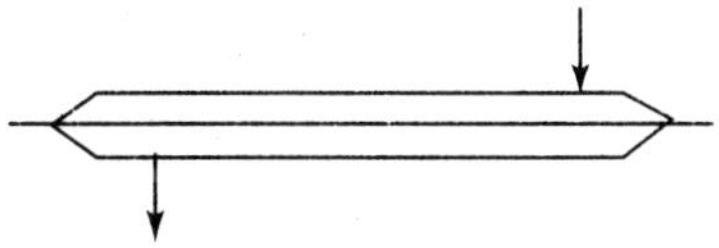
Pressure Safety Valve	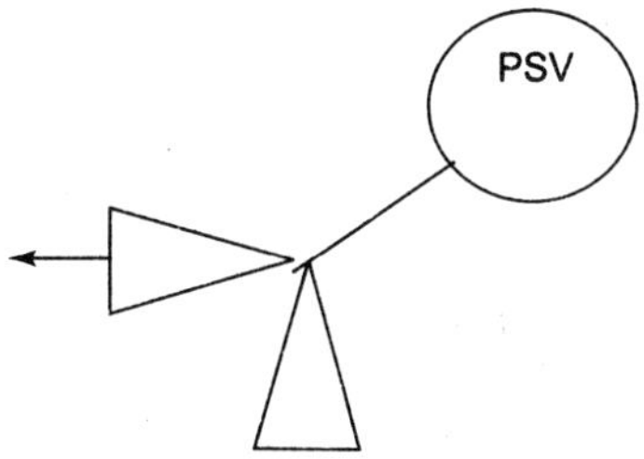
Rupture Disc	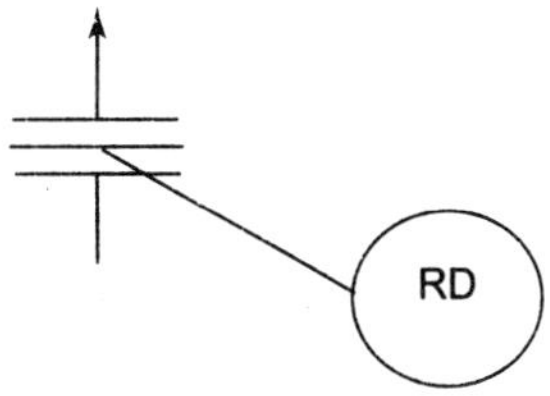
Solenoid Valve	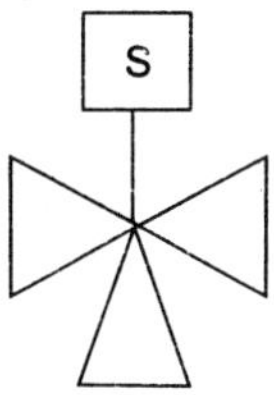
(a) Temperature Indicator Controller (b) Thermowell	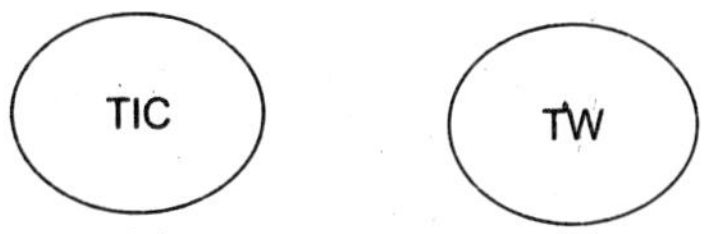 (a) (b)

(a) Pressure Indicator Controller (b) Pressure Indicator	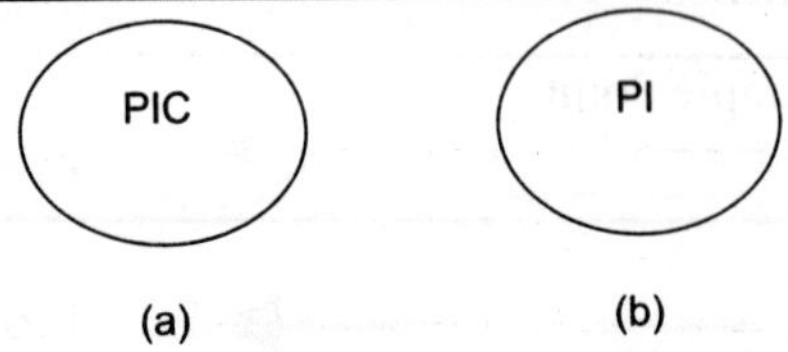
Vent with hood	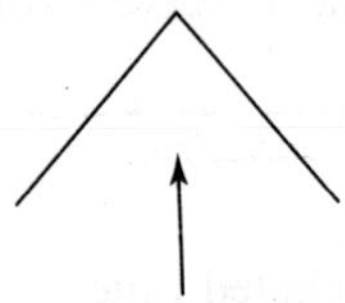
Orifice	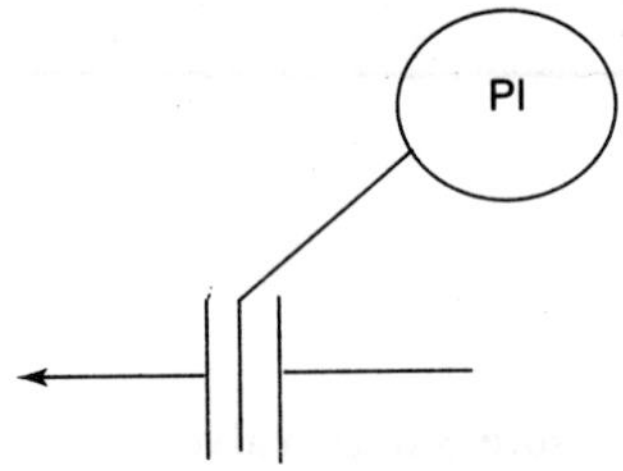
Rotameter	
Load Cell	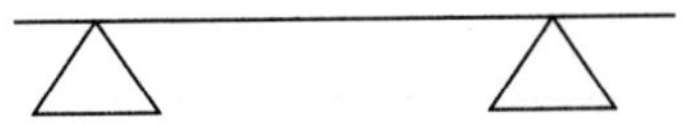
Flame arrester	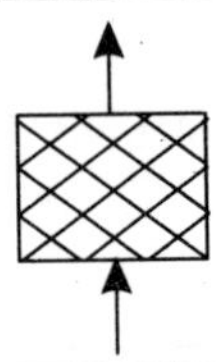

Lines

	Process Line
	Process Line with main material flows raw materials to product
	Pneumatic/air line
	Electrical supply/signal

Appendices

Appendix A

Table A1 Properties of Various Sections

Section	Type	Moment of Inertia I	Distance to Farthest Point y	Section Modulus $Z=\frac{I}{y}$	Radius of Gyration $k=\sqrt{\frac{I}{A}}$
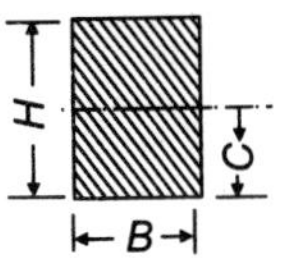	a	$\frac{bh^3}{12}$	$\frac{h}{2}$	$\frac{bh^2}{6}$	$0.289\,h$
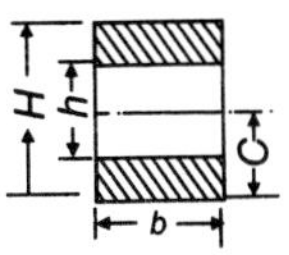	b	$\frac{b}{12}\left(H^3-h^3\right)$	$\frac{H}{2}$	$\frac{b(H^3-h^3)}{6H}$	$\sqrt{\frac{H^3-h^3}{12(H-h)}}$
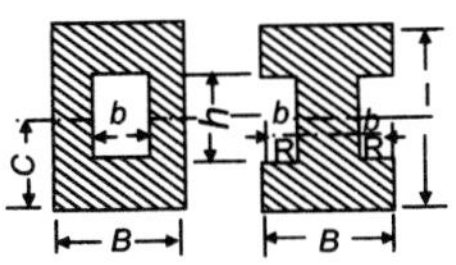	c	$\frac{BH^3-bh^3}{12}$	$\frac{H}{2}$	$\frac{BH^3-bh^3}{6H}$	$\sqrt{\frac{BH^3-bh^3}{12(BH-bh)}}$
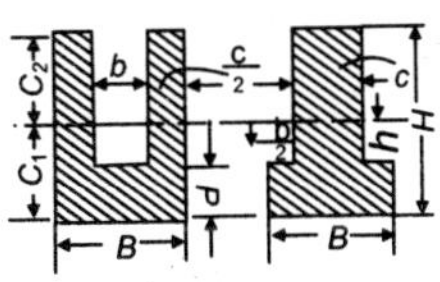	d	$\frac{Bc_1^3-bh^3+ac_2^3}{3}$ $h=c_1-d$	$c_1=\frac{aH^2+bd^2}{2(aH+bd)}$ $c_2=H-c_1$	$\frac{I}{c_1}$ and $\frac{I}{c_2}$	$\sqrt{\frac{1}{Bd+a(H-d)}}$

(Contd.)

Table A1 (*Contd.*)

Section	Type	Moment of Inertia I	Distance to Farthest Point y	Section Modulus $z = \frac{I}{y}$	Radius of Gyration $k = \sqrt{\frac{I}{A}}$
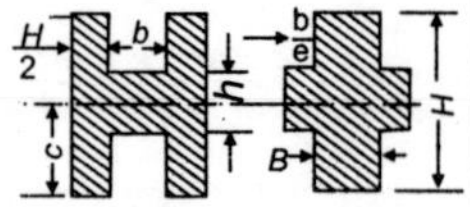	e	$\frac{BH^3 + bh^3}{12}$	$\frac{H}{2}$	$\frac{BH^3 - bh^3}{6H}$	$\sqrt{\frac{BH^3 - bh^3}{12(BH + bh)}}$
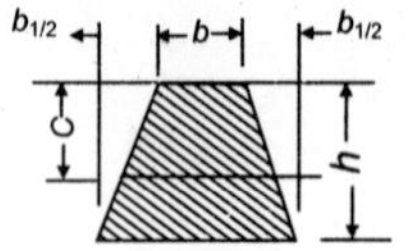	f	$\frac{(6b^2 + 6bb_0 + b_0^2)h^3}{36(2b + b_0)}$	$\frac{(3b^2 + 2b_0)h}{3(2b + b_0)}$	$\frac{(6b^2 + 6bb_0 + b_0^2)h^3}{12(3b + b_0)}$	$\sqrt{\frac{I}{A}}$
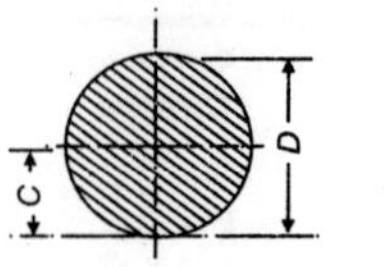	g	$\frac{\pi D^4}{64} = \frac{\pi R^4}{4}$	$\frac{D}{2} = R$	$\frac{\pi D^3}{32} = 0.0982D^3$	$\frac{D}{4} = \frac{R}{2}$
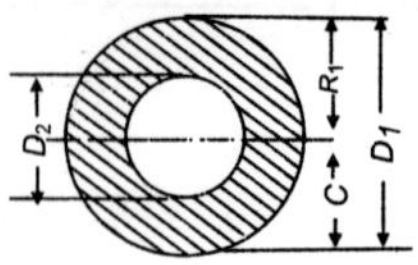	h	$\frac{\pi}{64}(D_1^4 - D_2^4)$ $= \frac{\pi}{4}(R_1^4 - R_2^4)$	$\frac{D_1}{2} = R_1$	$\frac{\pi(D_1^4 - F_2^4)}{32D_1}$	$\sqrt{\frac{D_1^2 + D_2^2}{4}}$ $= \sqrt{\frac{R_1^2 + R_2^2}{2}}$
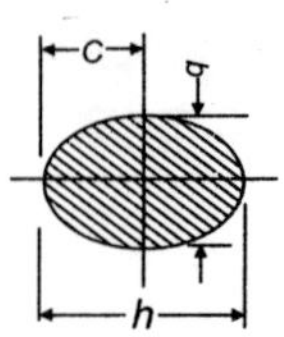	i	$\frac{\pi bh^3}{64}$	$\frac{h}{2}$	$\frac{\pi bh^3}{32}$	$\frac{h}{4}$

Appendix B

Table B1 Bending Moment and Deflection of Beams

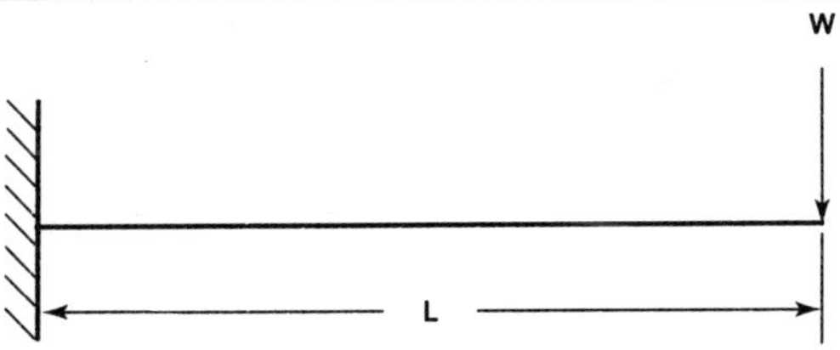

Cantilever—concentrated load at free end

M_{max} (at fixed end) = WL

δ_{max} (at free end) = $\dfrac{WL^3}{3EI}$

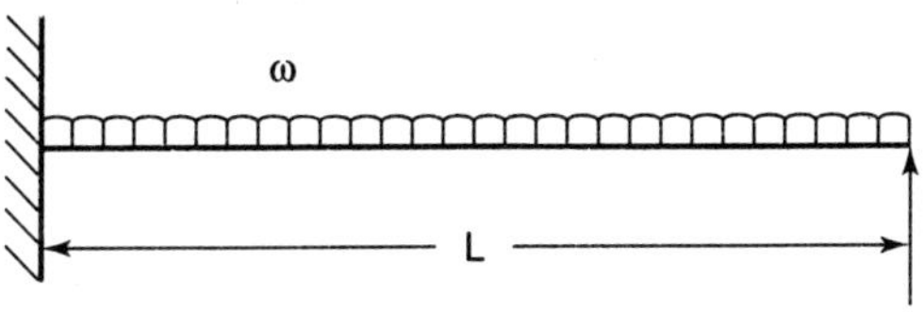

Cantilever—uniformly distributed load

M_{max} (at fixed end) = $\dfrac{\omega L^2}{2}$

δ_{max} (at free end) = $\dfrac{\omega L^4}{8EI}$

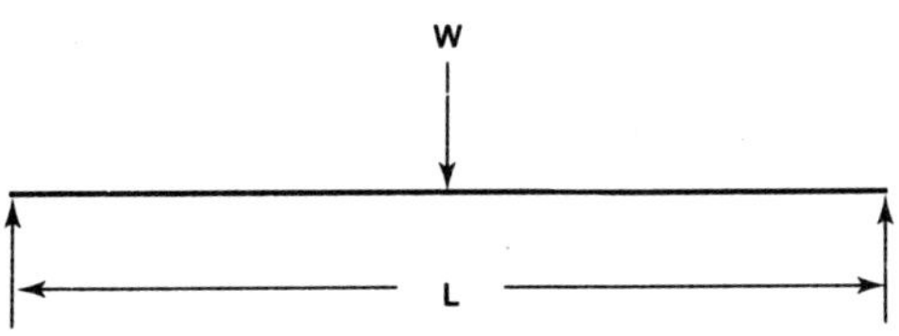

Simply supported—concentrated load at centre

M_{max} (at point at load) = $\dfrac{WL}{4}$

δ_{max} (at point of load) = $\dfrac{WL^3}{48\,EI}$

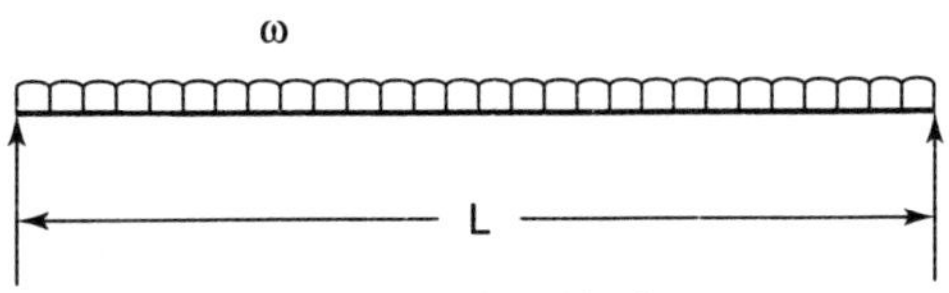

Simply supported—uniformly distributed load

M_{max} (at the centre) = $\dfrac{\omega L^2}{8}$

δ_{max} (at the centre) = $\dfrac{5\omega L^4}{384\,EI}$

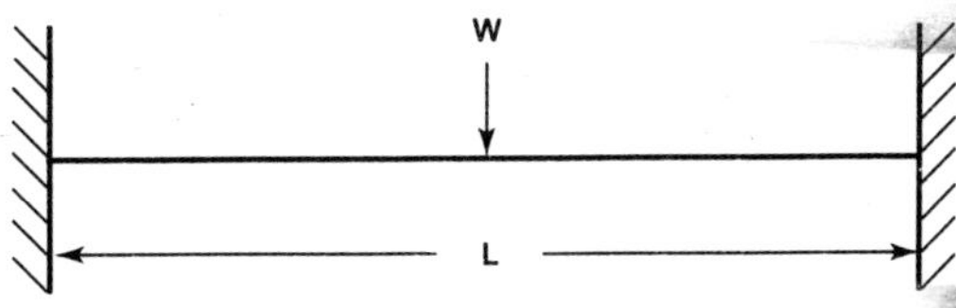

Fixed at both ends—concentrated load at centre

M_{max} (at the centre) = $\dfrac{WL}{8}$

δ_{max} (at the centre) = $\dfrac{WL^3}{192\,EI}$

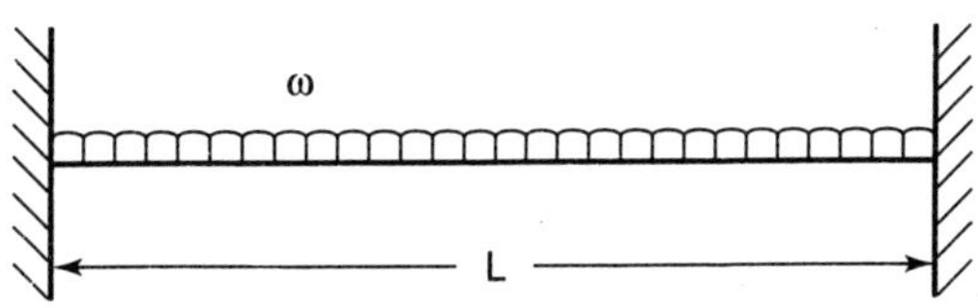

Fixed at both ends—uniformly distributed load

M_{max} (at the centre) = $\dfrac{WL^2}{12}$

δ_{max} (at centre) = $\dfrac{\omega L^4}{384\,EI}$

(Contd.)

Table B1 (*Contd.*)

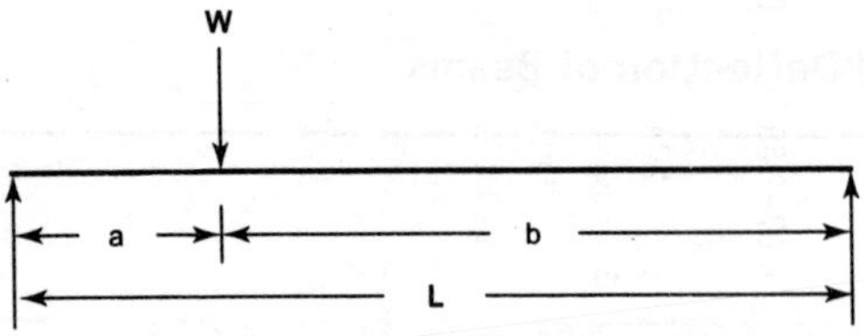

Simply supported—concentrated load at any point

$$M_{max} \text{ (at point of load)} = \frac{Wab}{L}$$

$$\delta_{max} \text{ (at point of load)} = \frac{Wa^2b^2}{3\,EIL}$$

Notation: W = concentrated load, ω = unit load, L = span, M = bending moment, δ = deflection, I = Moment of inertia.
$\omega = W/L$, E = Modules of elasticity

Appendix C

Table C1 Properties of Plastic Materials

Property	*ABS (Acryl-onitrile Butadiene Styrene*	*EP (Epoxies)*	*PC (Polycarbo-nate)*	*Polyesters*	*PE (Poly-ethylene)*	*PS, SAN, SBP, SRP (Poly-styrene)*	*SI (Silicones)*	*UF (Urea formalde-hyde)*	*Up (Urethanes)*	*PVAC, PVAl, PVB, PVC, PVCAC, PVFM*
Tensile Strength N/mm^2	28.1–56.3	28.1–210.9	56.3–140.6	5.6–351.5	7.0–38.7	10.6–140.6	5.6–246.1	38.7–91.4	1.2–70.3	3.5–63.3
Elongation (%)	2–300	0.5–70	0.9–130	0.5–310	15–1000	0.75–80	0–100	0.5–1.0	10–1000	2–450
Tensile Modulus 10^3 N/mm^2	1.6–7.2	0.007–21.4	2.5–13.0	2.1–14.0	0.098–1.30	1.05–9.80	0.006–21.0	7.0–10.5	0.07–7.0	3.5–4.8
Compressive Strength N/mm^2	49.2–154.7	7.0–281.2	87.9–133.3	84.4–351.5	0–38.7	28.1–154.6	0.7–126.5	175.8–316.4	140.6	7.0–154.7
Compressive Modulus 10^3 N/mm^2	1.2–2.7	–	2.1–3.20	–	0–1.10	0–3.70	–	–	0.028–0.7	0–4.8
Flexural yield Strength B/mm^2	35.2–189.8	7.0–421.8	94.9–210.9	56.3–562.4	0–4.9	35.2–182.8	0–246.1	70.3–126.5	0–63.3	0–119.5
Flexural Modulus 10^9 N/mm^2	0.14–9.10	–	2.4–8.4	0–14.10	0–2.5	0–12.60	–	9.10–11.20	0.07–2.50	0–2.80
Hardness Rockwell	275–M100	M80–M120	M70–R118	60 (bared) E89	D30 (shore) R15	R50–E60	– 40 (shore) M95	M–100 M–120	20A (shore) M28	10A (shore) M8J
Impact Strength Nm/mm	0.0108–0.544	0.044–0.544	0.0652–0.952	0.0108–0.87	0.0272 0.1088	0.0136–0.598	0–0.816	0.0136–0.217	0–0.272	0.0217–1.088

Please note: Values presented in Appendix C to F have been suitably rounded off while converting them to SI units.

Appendix D

Table D1 Modulus of Elasticity

Values of E for Ferrous Materials in 10^3 N/mm^2

Carbon	*Material*	*Design Temperature °C*						
		0	20	100	200	300	400	500
Steels	C < 0.30%	196	196	195	190	182	170	–
Carbon-Molybdenum	C < 0.30%	211	210	207	199	190	173	–
Steels & Chrom-molybdenum Steels (upto 3% Cr)		211	210	207	201	194	184	172
Intermediate Chrome molybdenum		193	193	190	186	180	173	160
Steels and Austenite Stainless Steels								

Note: Intermediate values may be obtained by interpolation.

Values of E for Aluminium and Its Alloys in 10^3 N/mm^2

Material Grade	*Design Temperature °C*								
	–200	–100	0	50	75	100	125	150	200
IB, N_3, N_4	78	74	71	70	70	69	68	57	64
H9	74	71	68	66	66	65	65	63	60
H15	83	79	75	74	74	73	71	71	68
A6	89	85	82	81	80	79	78	77	74

Notes: 1. Intermediate values may be obtained by interpolation.

2. Since aluminium and its alloys do not have a well-defined yield point, the above values of E are to be used with caution.

Values of E for Nickel and Nickel Alloys in 10^3 N/mm^2

Material	*Design Temperature °C*						
	20	*300*	*400*	*500*	*600*	*700*	*750*
Nickel	211	204	188	165	140	117	109
70% Nickel and 30% Copper Alloy	188	180	176	169	162	155	150
75% Nickel, 15% Chromium and 10% Ferrous Alloy	218	207	200	176	160	130	119

Values of E for Copper and Its Alloys in 10^3 N/mm^2

Material	*Composition*	*Design Temperature °C*								
		20	*50*	*100*	*150*	*200*	*250*	*300*	*350*	*400*
Copper	99.98% Cu	112	111	110	100	106	104	101	97	–
Commercial Brass	66% Cu, 34% Zn	98	97	96	95	91	89	86	85	–
Leaded in Bronze	88% Cu, 6% Sn, 1.5% Pb, 4.5% Zn	91	90	89	87	84	82	88	77	–
Phosphor Bronze	85% Cu, 12.5% Sn, 10% Zn	105	103	102	98	95	91	85	67	–
Muntz	59% Cu, 39% Zn	107	102	98	91	83	77	–	–	–
Cupio-Nickel	80% Cu, 20% Ni or 70% Cu, 30% Ni	133	131	129	126	124	121	118	115	142

Appendix E

Allowable Stress Values for Ferrous and Non-Ferrous Materials

Table E1 Allowable Stress Values for Carbon and Low Alloy Steel in Tension

Material Specification	Grade or Designation	Mechanical Properties			Allowable Stress Values in N/mm^2 at Design Temperature °C												
		Tensile Strength min N/mm^2 R	Yield Stress min N/mm^2 E	% Elongation min on Gauge Length = 5.65	Up to 250	Up to 300	Up to 350	Up to 375	Up to 400	Up to 425	Up to 450	Up to 475	Up to 500	Up to 525	Up to 550	Up to 575	Up to 600
1	2	3	4	5	6	7	8	9	10	11	12	13	14	15	16	17	18
							PLATES										
IS : 2002-1975	I	370	$0.55R_{20}$	26	95	87	78	75	72	59	43	36	–	–	–	–	–
IS : 2002-1975	2 A	420	$0.50R_{20}$	25	98	90	81	77	74	59	43	36	–	–	–	–	–
IS : 2002-1975	2 B	520	$0.50R_{20}$	20	121	111	100	95	83	59	43	36	–	–	–	–	–
IS : 2041-1986	20Mo55	480	280	20	143	132	123	119	115	112	108	77	56	37	–	–	–
IS : 2041-1986	$20Mn^2$	520	300	20	140	128	116	110	83	59	43	36	–	–	–	–	–
IS : 1570-1975	IS Cr 90 Mo 55	500	300	20	160	152	144	138	134	130	126	117	86	58	35	-	-
IS : 1570-1975	C15 Mn75	420	230	25	107	98	89	84	81	59	43	36	–	–	–	–	–

1	2	3	4	5	6	7	8	9	10	11	12	13	14	15	16	17	18
							FORGINGS										
IS : 2004-1975	Class 1	370	$0.50R_{20}$	-	86	79	71	68	65	59	43	36	–	–	–	–	–
IS : 2004-1975	Class 2	440	$0.50R_{20}$	15	102	93	85	80	77	59	43	36	–	–	–	–	–
IS : 2004-1975	Class 3	500	$0.50R_{20}$	21	117	107	96	91	83	59	43	36	–	–	–	–	–
IS : 2004-1975	Class 4	630	$0.50R_{20}$	15	147	134	122	115	83	59	43	36	–	–	–	–	–
IS : 1570-1975	20Mo55	480	280	20	143	132	123	119	115	112	108	77	56	37	–	–	–
IS : 2611-1986	15Cr 90 Mo55	500	300	20	160	152	144	138	134	130	126	117	86	58	35	–	–
IS : 1570-1975	10Cr 2 Mol	500	320	20	179	173	164	161	158	153	149	127	96	70	49	32	23
							TUBE PIPES										
IS : 3609–1984	1% Cr $\frac{1}{2}$ % Mo Tube normalised and tempered.	440	240	$950/R_{20}$	128	121	115	111	107	104	100	97	86	58	35	–	–
IS : 3609–1984	$2\frac{1}{4}$ % Cr 1% Mo tube normalized and tempered.	490	250	$950/R_{20}$	140	135	128	126	123	120	116	113	96	70	49	–	–
IS : 1570–1975	20 Mo55	460	250	$950/R_{20}$	128	118	110	106	103	100	96	77	56	87	–	–	–
IS : 1914–1975	320N/mm^2, min tensile Strength	320	$0.05R_{20}$	$950/R_{20}$	74	68	62	58	56	50	43	36	–	–	–	–	–
IS : 1914–1975	430N/mm^2, min tensile strength	430	$0.50R_{20}$	$950/R_{20}$	100	92	83	79	76	59	43	36	–	–	–	–	–
IS : 2416–1978	320N/mm^2, min tensile strength	320	$0.50R_{20}$	$950/R_{20}$	47	68	62	58	56	50	43	36	–	–	–	–	–
IS : 1978–1985	st. 18	316	176	-	82	75	67	64	62	59	43	36	–	–	–	–	–
	st. 20	337	197	-	92	94	76	72	69	59	43	36	–	–	–	–	–
	st. 21	337	211	-	98	90	81	77	74	59	43	36	–	–	–	–	–
	st. 25	422	246	-	115	105	95	90	83	59	43	36	–	–	–	–	–
IS : 1971–1985	st. 30	422	295	-	138	126	115	108	83	59	43	36	–	–	–	–	–
	st. 32	443	323	-	150	138	125	118	83	59	43	36	–	–	–	–	–
	st. 37	464	366	-	171	156	141	134	83	59	43	36	–	–	–	–	–

(Contd.)

Table E1 (*Contd.*)

1	2	3	4	5	6	7	8	9	10	11	12	13	14	15	16	17	18
							CASTINGS										
IS : 3038–1986	Grade 1	550	350	17	122	112	101	96	62	44	32	27	–	–	–	–	–
	Grade 2	470	250	17	96	88	82	80	77	75	72	58	42	28	–	–	–
	Grade 3	520	310	15	119	110	102	99	96	93	84	58	42	28	–	–	–
	Grade 4	490	280	17	112	106	101	97	93	91	88	85	65	44	26	–	–
	Grade 5	520	310	17	130	125	119	117	114	111	108	95	72	53	37	24	–
	Grade 6	630	430	15	172	163	155	149	143	140	135	67	49	35	26	17	9
IS : 2865–1975	CSW-C20	420	120	20	73	67	61	57	55	44	32	27	16	–	–	–	–
	CSW-C25	490	250	18	87	80	72	68	62	44	32	27	16	–	–	–	–
							RIVET AND STAY BAR										
IS : 1990–1975		370	055 R_{20}	26	86	79	71	68	65	59	43	36	–	–	–	–	–
		420	0.55 R_{20}	23	98	90	81	79	74	59	43	36	–	–	–	–	–
							SECTIONS, PLATES, BARS										
IS : 226–1975	st 42S	420	240	23	98	90	81	–	–	–	–	–	–	–	–	–	–
IS : 961–1975	st 55 HTW	500	240	20	117	107	96	–	–	–	–	–	–	–	–	–	–
IS : 2062–1975	st 42-W	420	230	23	98	90	81	–	–	–	–	–	–	–	–	–	–
IS : 3039–1986	Grade A	–	–	–	98	90	81	–	–	–	–	–	–	–	–	–	–
	Grade D	–	–	–	117	107	96	–	–	–	–	–	–	–	–	–	–
IS : 3503–1984	Grade 1	370	$0.55R_{20}$	26	86	79	71	68	65	59	43	36	–	–	–	–	–
	Grade 2	420	$0.55R_{20}$	25	98	90	81	77	74	59	43	36	–	–	–	–	–
	Grade 3	440	$0.55R_{20}$	23	102	93	85	80	77	59	43	36	–	–	–	–	–
	Grade 4	470	$0.55R_{20}$	22	117	107	96	91	83	59	43	36	–	–	–	–	–
	Grade 5	500	$0.55R_{20}$	21	121	111	100	95	83	59	43	36	–	–	–	–	–
IS : 3945–1984	Grade A-N	440	240	23	98	90	81	–	–	–	–	–	–	–	–	–	–
	Grade B-N	500	285	20	117	107	96	–	–	–	–	–	–	–	–	–	–

Table E2 Allowable Stress Values for High Alloy Steels in Tension

Material Specification	Designation	Product	Remarks	Mechanical Properties			Allowable Stress Values in N/mm^2 at Design Temperature °C							
				Tensile Strength N/mm^2 min R_{20}	Yield Stress N/mm^2 min E	Elongation percentage min. on gauge length 5.65	50	100	150	200	250	300	350	400
	04 Cr 19 Ni 19	Plates	Austen-	550	240	28	160	142.0	124.0	106.2	99.7	93.5	87.0	80.7
	04 Cr 19 Ni 9 Ti 20	sections	itic	550	240	28	160	142.8	125.6	108.3	106.4	106.0	106.0	108.0
IS : 1570–1961	04 Cr 19 Ni 9 Nb 40	bars	Stain-	550	240	28	160	142.8	125.6	108.3	106.4	106.0	106.0	106.0
	05 Cr 18 Ni 11 Mo 3	forging and seamless tubes	less Steel	550	240	28	160	145.0	130.0	115.0	112.4	112.3	112.3	111.2
	05 Cr 19 Ni 9 Mo 3 Ti 20			550	240	28	160	145.0	130.0	115.0	112.4	112.3	112.3	111.2
IS : 3444–1987	Grade 7, 8	Castings		470	210	21	140	129.4	118.8	108.3	106.4	106.0	106.0	106.0
	Grade 9, 11			470	210	13	140	134.7	123.4	115.0	112.4	112.3	112.3	111.2

Table E3 Allowable Stress Values for Aluminium and Aluminium Alloys in Tension

Material	Grade & Products	Condition	Mechanical Properties			Allowable Stress Values in N/mm^2 at Design Temp. °C						
			Tensile 0.2 Strength min N/mm^2	Proof Stress min N/mm^2	Elongation on $4/S_0$ min per cent	Up to 50	Up to 75	Up to 100	Up to 125	Up to 150	Up to 175	Up to 200
1	2	3	4	5	6	7	8	9	10	11	12	13
IS : 736–1986	PIB Plate	M	65	–	30							
IS : 737–1986	SIB Sheet, strip	0	65	–	30							
IS : 733–1983	EIB Bars, rod	M	65	–	25	13.6	12.7	11.6	10.5	9.5	8.4	7.4
IS : 1285–1975	VIB Extruded	M	63	–	95							
IS : 736–1986	PIB Plate	1/2H	100	–	8							
IS : 737–1986	SIB Sheet, strip	1/2H	100	–	8	21.1	20.7	20	19	16.5	14.1	11.2
IS : 737–1986	NS_3 Sheet, strip	0	100	–	30	23.0	22.2	20	18.9	16.9	14.8	12.6
IS : 373–1965	NS_3 Sheet, strip	1/2H	140	–	7	34.6	33.7	32.5	30.9	28.2	24.6	21.8
IS : 736–1986	NP_4 Plate	M	190	–	12							
IS : 737–1986	NS_4 Sheet, strip	0	175	–	18							
IS : 733–1983	NE_4 Bars, rod	M	175	–	18	43.5	43.5	43.2	42.0	38.0	32.7	24.6
IS : 738–1977	NT_4 Drawn, Tube	0	175	–	18							
IS : 1285–1975	NV_4 Extruded Tube	M	173	–	18							
IS : 737–1986	NS_4 Sheet, strip	1/4H	205	–	8	54.2	54.2	52.7	49.8	45.0	39.4	24.6
IS : 738–1977	NT_4 Drawn Tube	1/2H	235	–	5							
IS : 737–1988	NS_4 Sheet, strip	1/2H	235	–	5	59.7	59.5	57.8	54.0	48.5	42.8	24.6
IS : 737–1986	NS_5 Sheet, strip	0	220	–	18							
IS : 733–1983	NE_5 Bars, rods and sections	M	220	–	18	56.9	–	–	–	–	–	–
IS : 738–1977	NT_5 Drawn tubes	0	220	–	18							
IS : 1285–1975	NV_5 Hollow sections	N	220	–	18							
IS : 737–1986	NS_6 Sheet, strip	0	270	–	18							
IS : 738–1986	NT_6 Drawn tube	0	270	–	18	66.8	–	–	–	–	–	–
IS : 736–1986	NP_6 Plate	0	270	–	20							
IS : 733–1983	NE_6 Bras, rods	M	270	–	18							

Table E3 *(Contd.)*

1	2	3	4	5	6	7	8	9	10	11	12	13
IS : 1285–1975	NV_6 Extruded tubes	M	268	–	18							
IS : 733–1983	NE_8 Bars, rods and sections	O	270	–	16	70.2	–	–	–	–	–	–
IS : 1285–1975	HV_9 Extruded round tubes hollow section	M	110	–	15	29.7	29	27.8	26.7	25.3	19.3	13.3
IS : 1285–1975	HV_9 -do-	P	157	–	10	37.4	36.1	33.8	32.4	29.5	21.8	14.1
IS : 1285–1975	HV_9 -do-	WP	189	–	12	51.5	49.1	46.5	42.4	31.6	21.8	14.1
IS : 736–1986	HP_{30} Plate	W	205	110	15							
IS : 734–1975	HF_{30} Forging	W	190	110	18							
IS : 737–1986	HS_{30} Sheet, strip	W	205	110	15	51.8	50.1	48.6	47.1	45.2	39.4	28.1
IS : 733–1983	HE_{30} Bars, rods and sections	W	190	110	18							
IS : 738–1977	HT_{30} Drawn tube	W	220	110	16							
IS: 738–1977	HT_{30} Drawn tube	WP	315	250	7	–	–	–	–	–	–	–
IS : 736–1986	HP_{30} Plate	WP	300	235	8	–	–	–	–	–	–	–
IS : 734–1975	HF_{30} Forging	WP	300	250	10	72.9	71	68.5	66.0	55.6	43.6	31.0
IS : 737–1986	HS_{30} Sheet, strip	WP	300	255	8	–	–	–	–	–	–	–
IS : 733–1983	HE_{30} Bars, rods and sections	WP	300	250	10	–	–	–	–	–	–	–
IS : 1285–1975	HB_{15} Bolting alloy	WP	440	380	8	88.2	84.5	79.5	73.0	50.6	30.9	21.1
IS : 1285–1975	NB_6 Bolting alloy	1/2H	315	235	–	76.5	–	–	–	–	–	–
IS : 1284–1975	HB_{30} Bolting alloy	WP	300	250	–	58.5	57.0	54.8	52.7	44.3	34.4	23.2
IS : 617–1975	A3 Casting alloy	M (sand cast)	165	–	8	25.9	23.6	22.5	21.4	19.2	17.2	16.6
		M (chill cast)	189	–	12	29.6	26.8	25.6	24.6	21.8	19.7	17.6

Table E4 Allowable Stress Values for Copper Alloys

Material Specification	Grade & Products	Mechanical Properties			Allowable Stress Values in N/mm^2 at Design Temperature °C												
		Tensile Strength min N/mm^2	Yield Stress min N/mm^2	Elongation Per cent min N/mm^2	Up to 30	Up to 75	Up to 100	Up to 125	Up to 150	Up to 175	Up to 200	Up to 225	Up to 250	Up to 275	Up to 300	Up to 325	Up to 350
1	2	3	4	5	6	7	8	9	10	11	12	13	14	15	16	17	18
							PLATE, SHEET AND STRIP										
IS : 410-1967	CuZn 30	280	–	45	70.3	70.3	70.3	70.3	69.6	57.0	38.3	24.6	–	–	–	–	–
	CuZn 37	280	–	45	87.9	86.7	83.0	78.1	72.8	53.0	20.0	–	–	–	–	–	–
	CuZun 40	280	–	30	87.9	86.7	83.0	78.1	72.0	54.8	20.0	–	–	–	–	–	–
IS : 1972-1975	All grades	225	–	35	47.1	46.6	45.4	43.0	34.7	27.1	19.0	–	–	–	–	–	–
							BARS AND RODS										
IS : 288-1970	–	400	–	22	70.3	70.3	70.3	70.3	69.6	57.0	38.3	24.6	–	–	–	–	–
IS : 4171-1986	–	400	–	22	70.3	70.3	70.3	70.3	69.6	57.0	38.3	24.6	–	–	–	–	–
							BOLTING MATERIAL										
IS : 288-1970	–	400	–	22	17.6	17.6	17.6	16.7	15.4	14.8	14.1	–	–	–	–	–	–
IS : 4171-1986	–	400	–	22	17.6	17.6	17.6	16.7	15.4	14.8	14.1	–	–	–	–	–	–
							SECTIONS										
IS : 291-1970	Grade 1	350	–	22	87.9	86.7	83.0	78.1	72.8	53.8	20.0	–	–	–	–	–	–
	Grade 2	350	–	22	87.9	86.7	83.0	78.1	72.8	53.8	20.0	–	–	–	–	–	–
							TUBES										
IS : 407-1986	Alloy 1	290	–	20	70.3	70.3	70.3	70.0	69.6	57.0	38.3	–	–	–	–	–	–
	Alloy 2	290	–	20	87.9	86.7	83.0	78.1	72.8	53.8	20.0	–	–	–	–	–	–
IS : 1545-1984	ISBI 1	–	–	–	70.3	70.2	70.3	70.3	69.6	57.0	38.3	–	–	–	–	–	–
	ISBT2																
	ISABT	–	–	–	84.4	84.4	84.4	84.4	82.8	54.3	25.8	15.8	–	–	–	–	–
	ISABZT	–	–	–	87.6	86.7	85.3	83.4	80.9	70.9	46.4	31.6	18.8	–	–	–	–
IS : 2371-1984	CuNi 21	320	–	–	84.4	84.4	84.4	84.4	82.8	54.3	23.8	15.8	–	–	–	–	–
	Al2 As																

Tqble E4 (*Contd.*)

1	2	3	4	5	6	7	8	9	10	11	12	13	14	15	16	17	18
	CuNi 3I MN1 Fe	420	–	–	83.1	80.8	78.9	77.1	75.9	75.8	72.7	71.4	71.1	70.2	68.3	67.3	66.6
IS : 2501-1983		–	–	–	42.2	41.9	41.3	40.0	34.7	27.1	19.0	–	–	–	–	–	–
								CASTINGS									
IS : 318-1981	Grade 1	220	115	120	59.4	58.8	58.2	57.6	56.9	55.6	53.8	51.3	47.8	40.0	–	–	–
	Grade 2	190	110	75	52.1	50.9	49.6	48.4	47.7	46.5	45.8	–	–	–	–	–	–
	Grade 3	175	750	70	43.0	41.2	39.0	37.8	36.5	35.2	35.0	–	–	–	–	–	–

Appendix F

Table F1 Allowable Stresses for Flange Bolting Material N/mm²

Material	*Diameter (mm)*	*Specified Tensile Strength (N/mm^2)*	*Allowable Stress N/mm^2 for Design Metal Temperature not Exceeding (°C)*						
			50	*100*	*200*	*250*	*300*	*350*	*400*
Hot rolled carbon steel	Up to 150	440/520	58.70	56.20	54.50	48.50	–	–	–
1% Cr. Mo steel									
	Up to 63.5	860 Min	196.8	185.0	171.0	162.0	157.5	151.2	142.7
	over 63.5 to 102	790 Min	177.9	166.6	154.7	147.6	144.1	137.1	129.4
5% Cr. Mo steel	Up to 63.5,	710 Min	140.6	140.6	140.6	140.6	140.6	140.6	140.6
	over 63.5,	660 Min	–	–	–	–	–	–	–
1% Cr. Mo V steel	Up to 63.5	860 Min	196.8	190.5	184.9	179.3	172.9	168.0	160.3
	over 63.5	820 Min	179.9	172.3	166.6	162.4	155.4	152.6	145.5
13% Cr. Ni steel	Up to 102	710 Min	179.3	169.5	143.4	136.4	128.4	121.6	106.5
18/8 Cr. Ni steel	All (1) (2)		131.8	110.7	86.5	80.1	77.3	74.5	73.1
18/8 Cr. Ni Ti stabilized steel	All (1) (2)		131.8	115.3	101.9	94.9	91.4	87.9	85.8
18/9 Cr. Ni Nb stabilized steel	All (1) (2)		131.8	115.3	101.9	94.9	91.4	87.9	85.8
17/10 2 $\frac{1}{2}$ Cr. Ni Mo steel	All (1) (2)		131.8	111.8	95.6	88.6	85.1	80.8	79.4
18 Cr 2 Ni steel	Up to 102	860 Min	215.8	199.0	173.0	164.0	155.4	146.9	129.4

Appendix G

Table G1 Comparison of Design Features of Pressure Vessel Codes

Item	*IS-2825*	*ASME Code Sec. VIII Division I*	*BS-5500*	*AD-Merkblatter*
1	*2*	*3*	*4*	*5*
Scope	Unfired fusion welded pressure vessels, excluding a > Pressure exceeding 20 N/mm^2 b > Ratio of outside to inside dia. greater than 1.5 c > Internal operating pressure less than 0.1 N/mm^2 d > Internal diameter less than 150 mm e > Nominal water capacity of 500 litres or less	Pressure vessels welded, riveted, forged and brazed for containment of pressure either internal or external excluding: a > Fired process tubular heaters b > Parts or components of rotating or reciprocating mechanical devices. c > Structure for transportation of fluid. d > Piping components. e > Nominal water capacity of 120 gal. f > Internal/external operating pressure less than 15 p.s.i. g > Inside diameter less than 6 inch. h > Pressure exceeding 3000 psi.	Unfired fusion welded pressure vessels excluding a > Law pressure storage vessels b > Those in acids the stresses calculated are less than 10% of the design stress. c > Strip wound compound or other special designs of vessels appropriate for very high pressures. d > Transportable vessels.	a > AD-Merkblatter contains directives for material, manufacture, design and ancillary equipment of pressure vessels. It covers pressure vessels and pre-ssure parts which are wholly or predominantly under static load and for fluctuating loading (SI). b > Ratio of outside diameter to inside diameter less than 1.7.
Materials	Carbon and low alloy steels, high alloy steels, copper and copper alloys, aluminium and aluminium alloys, bolting alloys and casting alloys.	Carbon and low alloy steels, non ferrous metals, high alloy steels, cast irons, clad and lined material cast ductile iron and feritile steels.	Carbon, ferrite alloy (low and high alloys) and austenitic steels	Metallic material, e.g. carbon, low alloys, high alloy steel, cast iron, cast steel, aluminium and its alloys, clad steels, etc. and non-metallic materials, e.g. FRP, graphite, glass, safety factor Table 6.6.

(Contd.)

Table G1 (*Contd.*)

1	2	3	4	5
Design or allowable stress	a > Safety factor Table 5.2 b > Allowable stress values in tension for plates, forgings, tubes and pipes, casting, rivets and sections.	a > Safety factor Table 5.2 b > Allowable stress values in tension for plates, sheets, pipes and tubes, bars and forgings.	a > Safety factors Table 5.2 b > Allowable stress values in tension for plates, sections and bars, forging, pipes and tubes	
Design Pressure	Maximum working pressure including static head plus five per cent of maximum working pressure.	Maximum pressure at the most severe condition including the static head.	Pressure which will exist when the relaying device starts to relieve or set pressure of the relieving device whichever is higher or maximum pressure attained in service, including static head.	Design is based on permissible service pressure. The content at vessel during service for hydrctest or other loading need not be taken into account as long as stress in the wall is not increased by more than 5%.
Design Temperature	Mean metal temperature expected under operating conditions. For parts subjected to direct reaction and for the products at combustion, the design temperature shall not be less than the maximum surface temperature surface metal not to exceed maximum temperature.	Actual metal temperature expected under operation conditions.	Actual metal temperature with adequate margin to cover uncertainties of temperature prediction.	Highest wall temperature which can be expected during service at service pressure for unfired vessels. For fired vessels an allowance is taken considering the heating.
Thickness or cylindrical shell under external pressure	$t = \frac{PDi}{2fJ - P}$ **(not applicable unless D_0/D_1 ratio > 1.5)**	$t = \frac{PRi}{fJ - 0.6P}$ **(applicable when $t = R/2$).**	$t = \frac{PDi}{2f - P}$ **(applicable when $D_0/D_1 > 1.3$)**	$t = \frac{PD_0}{2kJ - P}$

(*Contd.*)

1	2	3	4	5
Thickness of cylindrical shell under external pressure.	a > for the given limits of L/D_0 the thickness. $t = D_0 \, [0.75P \, 2/D_0 \, K]^{2/5}$ where K is the ratio of modulus E of the material at design temperature to that at room temp. b > Assume value of t. Determine L/D_0 and D_0/t. From graph find the factor B, then $P_a = B/14.2 \, (D_0/t)$	Assume a value of t. Determine L/D_0 and D_0/t. From graph find the factor B, then $Pa = 4B/\{3 \, (D_0/t)\}$	Assume value of $P_{all} = \{(\Delta sft/R) \times (1/1 - r\alpha)\}$ values of various factors to be obtained from graphs.	Design the elastic buckling and plastic deformation as per formulae and design curves given.
Thickness of flat heads and covers. Thickness of dishead or domed ends concace to pressure	Choose a value of t that will make $Pa \geq P$ $t = CD\sqrt{P/F}$ $t = \frac{PD_0C}{2ft}$	$t = CD\sqrt{P/f}$ a > ellipsoidal head, $t = PD_0/2fJ = 0.2 \, P$ (Major to minor axis 2:1) b > Torispherical head $t - \frac{0.88 \, J \, PRC}{fJ - 0.1 \, P}$ Knuckle radius 6% at crown radius.	$t = CD\sqrt{P/f}$ Thickness determinant from P/f and h/D values with the helpot curves.	$t = CD\sqrt{P/K}$ Torispherical Head $t = \frac{D_0 \, pB}{4KS}$ Factor B is determined firm design curves.
Compensation for nozzles	a > For isolated nozzles no compensation up to maximum 250 mm diameter is required depending on the vessel diameter, thickness and factor $K = PD_0/18.2$ ft.	a > No compensation up to 50 mm nozzle. b > Compensation is determined by the area for are method. (Sec 6, 8.3, 2(a))	a > No compensation required if $P = d/D \sqrt{\frac{D}{2T_f}} \leq 0.1$ b > Compensation is determined from design curves and tables from p-value.	Compensation is based on weaking factors (UP) compensation is determined by graphs from V_p values. [Sec 6, 8.3, a(a)]

(*Contd.*)

Table G1 (*Contd.*)

1	2	3	4	5
	b > No compensation up to 50 mm mm isolated nozzle, when K has a value of unity or greater. c > Compensation is determined by the area for area method. (Sec 6, 8.3, 2(a)).			

Nomenclature

- T – thickness;
- D_i – inside diameter of shell
- J – joint factor
- R_i – inside radius of shell
- D_0 – outside diameter of shell or domed end.
- L – effective length of shell (maximum centre distance between two adjacent stiffening rings)
- c – edge fixity constant for flat heads or shape factor for domed ends.
- R_c – Crown radius of torispherical head
- d – diameter of nozzle
- T_r – Required thickness of shell or head
- p – design pressure;
- f – allowable stress at operating temperature
- J – joint factor
- K – allowable yield stress at operating temperature

Appendix H

Permissible Internal Pressure for Pipes

(a) Steel and Iron pipes – diameters between 0.6 and 12.5 cm

$$p = \frac{2f}{d_0}(t - 0.165) - 8.79$$

(b) Steel and Iron pipes—diameters over 12.5 cm

$$p = \frac{2f}{d_0}(t - 0.25)$$

(c) Non-ferrous seamless tubes and pipes

$$p = \frac{2f}{d_0}$$

where p – internal working pressure in N/mm^2

– permissible working stress in N/mm^2 (Tables H.1 & H.2)

t – pipe wall thickness in mm

d_0 – outside diameter of pipe in mm

Table H1 Values of 'f' for Non-Ferrous Materials (N/mm^2)

Material	*For Temperature in °C not to Exceed*						
	–20 to 65	*120*	*175*	*200*	*230*	*260*	*290*
Muntz Metal Tubing & High Brass Tubing	35	28.0	17.5	–	–	–	–
Muntz Metal Condenser Tubes	35	28.0	17.5	–	–	–	–
Red Brass Tubes	42	38.5	35.0	31.5	–	–	–
Copper Tubes	42	35.0	31.5	28.0	–	–	–
Copper Pipes	42	35.0	31.5	28.5	–	–	–
Admiralty Tubing	49	46.0	42.0	28.5	31.5	–	–
Admiralty Condenser Tubes	49	46.0	42.0	38.5	31.5	–	–
Steam Bronze	48	44.0	41.0	38.0	35.0	29.5	23.0

Table H2 Value of 'f' for Steel and Iron

Material	*For Temperature in °C not to Exceed*							
	–30 to 340	*370*	*400*	*425*	*455*	*480*	*510*	*500*
Seamless Medium Carbon Steel	840	840	730	858	445	310	185	–
Seamless Low Carbon Steel	660	630	605	500	410	310	185	–
Fusion Welded Steel	770	730	670	560	445	310	185	–
Fusion Welded Steel:								
Grade B	700	675	630	525	420	310	185	–
Grade A	630	620	590	485	400	310	185	–
Lap-Welded Steel	630	620	590	485	400	310	185	–
Butt-Welded Steel	630	620	590	–	–	–	–	–
Lap-Welded Wrought Iron	560	540	485	–	–	–	–	–
Butt-Welded Wrought Iron	560	540	485	–	–	–	–	–
Seamless Alloy Steel	770	770	770	755	740	700	560	350

Appendix I

Relevant Indian Standards

Material and Sections

IS-2712–1979	Specification for compressed asbestos jointing.
IS-800–1984	Code of practice for use of structural steel in general building construction.
IS-1972–1989	Specification for plate, sheet, strip for industrial purpose copper.
IS-3965–1981	Dimensions for wrought aluminium and aluminium alloys, bar, rod and section.
IS-2677–1979	Dimensions for plate, wrought aluminium and aluminium alloys.
IS-2676–1981	Dimensions for sheet and strip, wrought aluminium and aluminium alloys.
IS-1870–1965	Comparison of Indian and overseas standard for wrought steels for general engineering purpose.
IS-808–1989	Specification for rolled steel beam, channel and angle sections (revised).
IS-1730–1989	Dimensions for steel plate, strip and sheet for structural and general engineering purpose.
IS-4687–1980	Specification for gland packing asbestos.
IS-4688–1968	Dimensions for proofed cotton duck gland packing.

Machine Elements

IS-2535–1978	Specifications for basic rack and modules of cylindrical gears for general engineering and heave engineering.
IS-4460–1967	Method for rating of machine cut spur gears and helical gears.
IS-2636–1972	Specification for wing nuts.
IS-3824–1983	Part I – Rolling bearings, methods of evaluating dynamic load ratings of: 1st-radial ball bearings.
	Part II – Radial roller bearings.
	Part III – Thrust ball bearings.
	Part IV – Thrust roller bearings.
IS-3823–1988	Rolling bearings, methods of evaluating static load rating of Part, 1st – radial ball bearings.
	Part II – Radial roller bearings.
	Part III – Thrust ball bearings.
	Part IV – Thrust roller bearings.

IS-3132–1965	Recommendations for shafts diameters for chemical equipments.
IS-1363–1992	I Specification for black hexagonal bolts M5 – M64
	II Specification for black hexagonal screws M5 – M65
	III Specification for black hexagonal nuts M5 – M64
IS-2585–1968	Specification for black square bolts and buts (dia. range 6 to 39 mm) and black square screws (dia. range 6 to 24 mm).
IS-3138–1966	Specification for hexagonal bolts and nuts (M42 to M150).
IS-1691–1980	Specification for cast-iron and mild steel flat pulleys.
IS-2693–1989	Dimensions for cast iron flexible couplings.
IS-3653–1966	Dimensions for forged and type rigid couplings.
IS-2293–1974	Dimensions of Gib-head keysand keyways.
IS-2403–1991	Specification for transmission steel roller chains and wheels.
IS-2494–1974	Specification for V-belts for industrial purposes.
IS-3142–1982	Specification for V-grooved pulleys for V-belts groove sections A, B, C, D and E.
IS-1367–1979	Part I to XVIII technical supply conditions for threaded fastners.

Tubes, Pipes and Valves

IS-3382–1965	Specification for stainless tell milk pipes and fittings.
IS-3589–1991	Specification for electrically welded steel pipes (200 mm to 2,000 mm nominal dia).
IS-3516–1996	Specification for cast iron pipe flanges and flanged fittings, class 9, for petroleum industry.
IS-6011–1970	Specification for carbon steel tubes for use on board-ships for pressure services.
IS-2501–1985	Specification for copper tubes for general engineering purposes.
IS-2416–1986	Specification for boiler and superheater tubes for marine and naval purpose.
IS-3233–1965	Glossary of terms for safety and relief valves and their parts.
IS-778–1984	Specification for gunmetal gate, globe and check valves for general engineering purposes.
IS-4854–1969	Glossary of terms for valves and their parts.
	Part I – Screw down stop, check and globe valves and their parts.
	Part II – Valves and cocks and their parts.
	Part III – Butterfly valves.
IS-407–1981	Specification for brass tubes for general purposes (second revision).
IS-1545–1982	Specification for tubes, solid drawn copper alloy.
IS-2678–1987	Dimensions for wrought aluminium and aluminium alloys, drawn tubes.

IS-2673–1979	Dimensions for wrought aluminium and aluminium alloys, extruded tubes (round).
IS-1239–1990	Part I – Specification for mild steel tubes.
	Part II – Specification for mild steel tubulors.
IS-6418–1971	Specification for cast iron and malleable cast iron flanges for general engineering purposes.
IS-6392–1971	Specification for steel pipe flanges.

Pressure Vessels and Components

IS-3503–1966	Steel plates for pressure vessels.
IS-4682–1968	Code of practice for lining of vessels and equipments for chemical process.
	Part I – Rubber lining.
	Part II – Glass enamel lining.
	Part III – Lead lining.
	Part IV – Plasticized PVC lining.
IS-4682–1968	Part V – Epoxide resin lining.
	Part VI – Phenolic resin lining.
	Part VII – Corrosion & heat resistant metals.
IS-4049–1988	Specification for formed ends for tanks and pressure vessels.
	Part I – Based on outside diameter.
	Part II – Based on inside diameter.
IS-3133–1989	Manholes and inspection openings for chemical equipment.
IS-5428–1985	Gauges for pressure vessels and boilers.
	Part I – Tubular glasses for level gauges.
	Part II – Protector glasses for level gauges.
	Part III – Reflex glasses for level gauges.
	Part IV – Circular sight glasses for level gauges.
	Part V – Port gauge glasses for level gauges.
IS-6202–1971	Specification for flat glass oil level gauges for oil storage tanks.
IS-4864–1968	Specification for shell flanges for vessels and equipments.
IS-4865–1968	Specification for shell flanges for vessels and equipments welded shell flanges for no-pressure service.
IS-4866–1968	Specification for shell flanges for vessels and equipments welded shell flanges for carbon steel pressure vessels and equipments.
IS-4867–1968	Specification for shell flanges for vessels and equipments welded neck shell flanges for carbon steel for pressure vessels and equipments.

IS-4868–1968	Specification for shell flanges for vessels and equipments welded shell flanges for stainless steel pressure vessels and equipments.
IS-4869–1968	Specification for shell flanges for vessels and equipments welded shell flanges with hub for stainless steel pressure vessels and equipments.
IS-4870–1968	Specification for shell flanges for vessels and equipments Flat gaskets for shell flanges.
IS-6076–1990	Dimensions of vacuum flanges.
IS-6838–1973	Dimensions of 'O' rings and grooves for vacuum flanges.
IS-2825–1969	Code for unfired pressure vessels.
IS-804–1967	Specification for rectangular pressed steel tanks.
IS-803–1976	Code of practice for design, fabrication and creation of vertical mild steel cylindrical welded oil storage tanks (first revision).

Chemical Equipment

IS-4503–1967	Specifications for shell and tube type heat exchangers.
IS-5036–1969	Filter presses, recessed plate type, and plate and frame type.
IS-6034–1989	Edge type vacuum filters.
IS-5675–1987	Rotary disc vacuum filters.
IS-4072–1992	Centrifuges, basket type.

Appendix J

Dished End Dimensions

(A)Torispherical Dished End

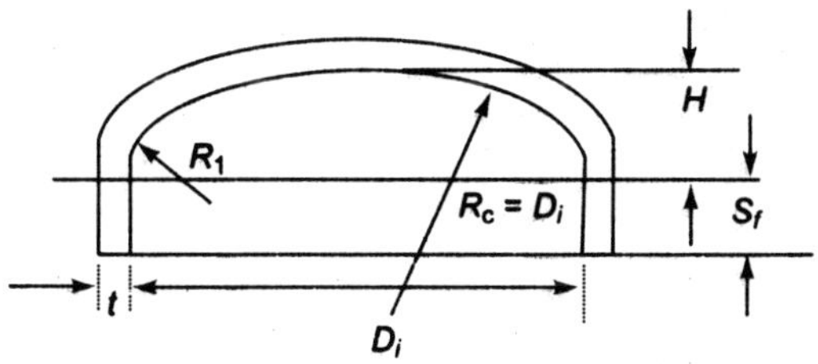

$$H = D_i - \{[(D_i - R_1)^2 - (D_i/2 - R_1)^2]\}^{0.5}$$

$$S_f = 40 \text{ to } 100 \text{ mm}$$

Approximate Volume of the dished end having $R_1 = 0.06\ R_c$ is given by:

$V = 0.081\ D_i^3$; m^3 and D_i is expressed in m.

The volume of the straight flanged portion = $(\pi/4)D_i^2 \times S_f$

For estimation of material take off we need to know the weight of the head. This can be estimated as follows:

The weight, W = Area × t × density

$= (\pi/4)B^2 \times t \times \rho$

Where B is the blank diameter

$$B = 1.024\ D_0 + 0.67\ R_1 + 2\ S_f$$

And $D_0 = D_i + 2t$

(B) Ellipsoidal Dished End (2:1)

$$H = D/4$$

The volume = $0.13\ D_i^3$

The blank diameter, $B = 1.18\ D_0 + 2\ S_f$

The material take off can now be estimated as in above (A).

Volume of a horizontal vessel

The vessel consists of a cylindrical shell and dished ends.

The volume of a cylinder when full is given by

$V = (\pi/4) D^2 \cdot L;$ D = inside diameter of the shell, m

L = length of a straight portion, m

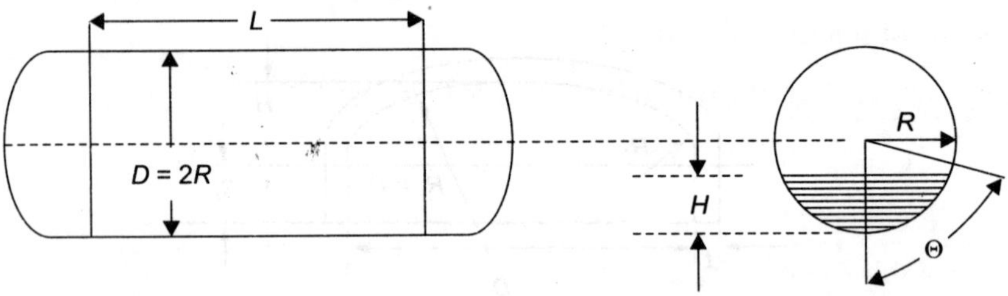

Volume, V_C at any height, H, from the bottom line of a tank is given by:

$$V_C = L(D/2)^2 \{(0.01745)\Theta - \sin\Theta\cos\Theta\}$$

Wherein $\cos\Theta = (1-2\beta)$

where Θ is expressed in degrees and $\beta = H/D$

The dished ends could be torispherical or ellipsoidal. The volume held in both dished ends when partially filled $H/D \le 1$ we can use following expressions.

Ellipsoidal (2:1) dished ends: (having head depth = $D/4$)

$$V_E = 0.26\beta^2(3 - 2\beta)D^3;\ m^3$$

Torispherical ends

$$V_T = 0.16\beta^2(3 - 2\beta)D^3;\ m^3$$

Thus the total volume at height H from the bottom will be

Total volume, V is sum of volume in a cylindrical shell (V_C) and liquid volume (V_H) in both heads (V_H could be equal to either V_E or V_T depending upon type of the head)

$$V = V_C + V_E \quad \text{or} \quad V = V_C + V_T$$

Illustration

- Calculate liquid held in a horizontal tank having a diameter 1.5 m and length 3 m at 0.4 m height (H) from the bottom. The tank has standard ellipsoidal dished ends.

We have, $D = 1.5$ m, $L = 3$ m and $H = 0.4$ m

Therefore $\beta = H/D = 0.4/1.5 = 0.267$ and $\cos\Theta = (1 - 2 \times 0.267)$

We, therefore, have $\Theta = \cos^{-1}(1 - 2 \times 0.267) = 62.225$ degrees

The volume of liquid in cylindrical portion is :

$$V_C = 3\ (1.5/2)^2\{(0.01745) \times 62.225 - \sin(62.225).\cos(62.225)\}$$

$$= 1.137\ \text{m}^3$$

The volume of liquid in dished ends is:

$$V_E = 0.26(0.267)^2\ (3 - 2 \times 0.267)(1.5)^3$$

$$= 0.154\ \text{m}^3$$

The total liquid held is = 1.137 + 0.154 = 1.291 m^3

It is always desirable to calibrate the tank by measuring dry dip with respect to any nozzle flange located on top of the tank.

Index